T0327378

Thermal Management of Electric Vehicle Battery Systems

Automotive Series

Series Editor: Thomas Kurfess

Title	Author(s)	Date
Automotive Aerodynamics	Katz	April 2016
The Global Automotive Industry	Nieuwenhuis and Wells	September 2015
Vehicle Dynamics	Meywerk	May 2015
Vehicle Gearbox Noise and Vibration: Measurement, Signal Analysis, Signal Processing and Noise Reduction Measures	Tůma	April 2014
Modeling and Control of Engines and Drivelines	Eriksson and Nielsen	April 2014
Modelling, Simulation and Control of Two-Wheeled Vehicles	Tanelli, Corno and Savaresi	March 2014
Advanced Composite Materials for Automotive Applications: Structural Integrity and Crashworthiness	Elmarakbi	December 2013
Guide to Load Analysis for Durability in Vehicle Engineering	Johannesson and Speckert	November 2013

Thermal Management of Electric Vehicle Battery Systems

İbrahim Dinçer
University of Ontario Institute of Technology, Canada

Halil S. Hamut
The Scientific and Technological Research Council of Turkey–Marmara Research Center, Turkey

Nader Javani
Yildiz Technical University, Turkey

This edition first published 2017

Registered Office
John Wiley & Sons Ltd, The Atrium, Southern Gate, Chichester, West Sussex, PO19 8SQ, United Kingdom

For details of our global editorial offices, for customer services and for information about how to apply for permission to reuse the copyright material in this book please see our website at www.wiley.com.

Library of Congress Cataloging-in-Publication Data

Names: Dinçer, İbrahim, 1964- author. | Hamut, Halil S., author. | Javani, Nader.
Title: Thermal management of electric vehicle battery systems / İbrahim Dinçer, Halil S. Hamut, Nader Javani.
Description: Chichester, West Sussex, United Kingdom : John Wiley & Sons, Inc., [2017] | Includes bibliographical references and index.
Identifiers: LCCN 2016041896| ISBN 9781118900246 (cloth) | ISBN 9781118900222 (epub) | ISBN 9781118900215 (Adobe PDF)
Subjects: LCSH: Electric vehicles–Batteries–Cooling.
Classification: LCC TL220 .D56 2017 | DDC 629.25/024–dc23 LC record available at https://lccn.loc.gov/2016041896

A catalogue record for this book is available from the British Library.

Cover Design: Wiley
Cover Images: Courtesy of the authors

Set in 10/12pt, Warnock by SPi Global, Chennai, India

Printed in the UK

Contents

Preface

Over the last few decades, concerns over the dependence and price instability of limited fossil fuels as well as environmental pollution and global warming have encouraged researchers, scientists and engineers to conduct more proactive research on vehicles with alternative energy sources. Today, electric vehicles (EVs) are starting to replace their conventional counterparts, due to the recent improvements in battery technologies, as they offer diversification of energy resources, load equalization of power, improved sustainability as well as lower emissions and operating costs.

Through this transition towards EVs, the vehicle related problems are mainly composed of the battery and its performance. In order to achieve the most ideal performance, the discrepancy between the optimum and operating conditions of the batteries need to be reduced significantly, which requires the effective use of thermal management systems (TMSs). Since EVs have a wide range of battery characteristics, size and weight limitations, and variable loads, achieving the most optimal battery thermal management system design, configuration and operation play a crucial role in the success and wide adoption of this technology.

In this book, electric vehicles, their architectures, along with the utilized battery chemistries, are initially introduced to the readers to provide the necessary background information followed by a thorough examination of various conventional and state-of-the-art EV battery TMSs (including phase change materials) that are currently used or potentially proposed to be used in the industry. Through the latter chapters, the readers are provided with the tools, methodology and procedures to select/develop the right thermal management designs, configurations and parameters for their battery applications under various operating conditions, and are guided to set up, instrument and operate their TMSs in the most efficient, cost effective and environmentally benign manners using exergy, exergoeconomic and exergoenvironmental analyses. Moreover, a further step is taken over the current technical issues and limitations, and a wider perspective is adopted by examining more subtle factors that will ultimately determine the success and wide adoption of these technologies and elaborate what we can expect to see in the near future in terms of EV technologies and trends as well as the compatible TMSs. Finally, various case studies in real-life applications are presented that employ the tools, methodology and procedures presented throughout the book to further illustrate their efficacy on the design, development and optimization of electric vehicle battery thermal management systems.

The book includes step-by-step instructions along with practical codes, models and economic and environmental databases for readers for the design, analysis, multi-criteria assessment and improvement of thermodynamic systems which are often not included in other solely academic textbooks. It also incorporates a large number of numerical examples and case studies, both at the end of each chapter and at the end of the book, which provide the reader with a substantial learning experience in assessment and design of practical applications. The book is designed to be an invaluable handbook for practicing engineers, researchers and graduate students in mainstream engineering fields of mechanical and chemical engineering. It consists of eight chapters with topics that range from broad definitions of alternative vehicle technologies to the detailed thermodynamic modelling of specific applications, by considering energy and exergy efficiency, economic and environmental considerations and sustainability aspects.

Chapter 1 introduces the current alternative vehicles in terms of their configurations and architectures, hybridization rates, energy storage systems as well as the emerging grid connections. It also examines the individual thermal management systems of the vehicle and elaborates on the sustainability issues.

Chapter 2 provides in-depth information in the existing and near future battery chemistries and conducts evaluations with respect to their performance, cost and technological readiness. Moreover, different battery management methodologies and techniques are provided with the focus on battery state estimation and charge equalization to increase the performance and longevity of the cells. The steps that are necessary to develop, manufacture and validate the batteries from cell to pack levels are also provided at the end of the chapter.

Chapter 3 introduces and classifies the basic properties/types of phase change materials, their advantages/drawbacks as well as methods to measure and improve their heat transfer capabilities. Moreover, novel methods to replace liquid battery TMSs with lighter, cheaper and more effective PCM alternatives are also presented for various applications.

In Chapter 4, a walkthrough of the necessary steps is provided to develop representative models of the battery from a cell to a pack level, achieve reliable simulations, form correct set ups of the data acquisition hardware and software, as well as to use the right procedures for the instrumentation of the battery and the vehicle in the experimental set up. The main focus is given to the battery cell and submodule simulations to provide the fundamental concepts behind the heat dissipation in the cell and heat propagation throughout the battery pack. The chapter is designed to provide an in-depth understanding of the cell electrochemistry as well as the thermodynamic properties of the thermal management systems before more detailed analysis are conducted in the next chapters.

In Chapter 5, various types of state-of-the-art thermal management systems are examined and assessed for electric vehicle battery systems. Subsequently, step-by-step thermodynamic modelling of a real TMS is conducted and the major system components are evaluated under various parameters and real life constraints with respect to energy and exergy criteria to provide the readers with the methods as well as the corresponding results of the analyses. The procedures are explained in each step and generic and widely used parameters are utilized as much as possible to enable the readers to incorporate these analyses to the systems they might be working on.

A walkthrough of developing exergoeconomic and exergoenvironmental analyses are presented in Chapter 6, in order to give the readers the necessary tools to analyze the investment costs associated with their system components and assess the economic feasibility of the suggested improvements as well as the environmental impact (using LCA). Procedures to determine the associated exergy streams are shown and a databank of investment/operating cost and environmental impact correlations are provided for the readers to provide assistance in modelling their system components with an extensive accuracy without the need of cumbersome experimental relationships. Finally, the vital steps for conducting a multi-objective optimization study for BTMS is carried on, where the results from exergy, exergoeconomic and exergoenvironmental analyses are used according to the developed objective functions and system constraints in order to illustrate the methods to optimize the system parameters under different operating conditions with respect to various criteria using Pareto Optimal optimization techniques.

Furthermore, various case studies are provided in Chapter 7 that employ the tools, methodology and procedures presented throughout the book and conduct analyses on real-life applications to further illustrate their efficacy on the design, development and optimization of electric vehicle battery thermal management systems.

Finally, Chapter 8 presents a wider perspective on electric vehicle technologies and thermal management systems by examining the remaining outstanding challenges and emerging technologies that might provide the necessary solutions for the success and wide adoption of these technologies. Furthermore, various TMS technologies that are currently under development for an extensive range of applications are introduced to provide an indication of what the BTMS might incorporate in the future.

We hope that this book brings a new dimension to EV battery thermal management systems and enables the readers to develop novel designs and products that offer better solutions to existing challenges and contribute to achieving a more sustainable future.

İbrahim Dinçer
Halil S. Hamut
Nader Javani
May 2016

Acknowledgements

The illustrations and supporting materials provided by several past/current graduate students and postdoctoral research fellows of Professor Dincer for various case studies in the book are gratefully acknowledged, including Satyam Panchal, Sayem Zafar, Masoud Yousef Ramandi, David MacPhee and Mikhail Granovskii.

We also acknowledge the support provided by the Natural Sciences and Engineering Research Council of Canada, General Motors Canada in Oshawa, and Marmara Research Center of The Scientific and Technological Research Council of Turkey (TUBITAK Marmara Research Center).

Last but not least, we warmly thank our families for their support, motivation, patience and understanding.

İbrahim Dinçer
Halil S. Hamut
Nader Javani
May 2016

1 Introductory Aspects of Electric Vehicles

1.1 Introduction

Energy is used in all aspects of life, and it is considered an essential part of the existence of the ecosystem and human civilization. Thus, energy-related issues are one of the most important problems that we face in the twenty-first century. With the onset of industrialization and globalization, the demand for energy has increased exponentially over the past decades. Especially with a population growth of faster than 2% in most countries, along with improvements on lifestyles that are linked to energy demand, the need for energy is ever-increasing. Based on the current global energy consumption pattern, it is predicted that the world energy consumption will increase by over 50% before 2030. Thus, based on this pervasive use of global energy resources, energy sustainability is becoming a global necessity and affects most of the civilization (Dincer, 2010).

Currently, the world relies heavily on fossil fuels such as oil, natural gas and coal, which provide almost 80% of the global energy demands, to meet its energy requirements. It is estimated that most of large-scale energy production and consumption of energy causes degradation of the environment as they are generated from these sources. It is believed that climatic changes driven by human activities (especially greenhouse gas emissions) have significant direct negative effects on the environment and contribute to over 160,000 deaths per year from side effects associated with climate change, which is estimated to double by 2020. Moreover, the nominal prices of retail gasoline have increased approximately five times more between the years of 1949 and 2005 (Asif and Muneer, 2007; Shafiee and Topal, 2006). These aforementioned reasons have motivated researchers, scientists, engineers and technologists to look for more efficient, cheaper and ecofriendly options for energy usage. As the transportation sector is a major contributor to this problem, several alternatives to conventional vehicles are developed which can be competitive in many aspects, all while being significantly more efficient and environmentally benign. Among these alternatives are electric and hybrid electric vehicles, which are two of the leading candidates to replace conventional vehicles in the future.

Over the last few decades, concerns over the dependence and ever-increasing prices of imported oil, as well as environmental pollution and global warming, have led scientists to conduct more proactive research on vehicles with alternative energy sources. Today, approximately 15 million barrels of crude oil per day are used in the United States alone. About 50% of this crude oil is used in the transportation sector, a sector where 95% of the energy supply comes from liquid fossil fuels (Kristoffersen *et al.*, 2011). Moreover, the increasing demand and relatively static supply for petroleum

Thermal Management of Electric Vehicle Battery Systems, First Edition.
İbrahim Dinçer, Halil S. Hamut and Nader Javani.

and stricter pollutant regulations have caused an increase and instability in crude oil prices. Furthermore, since the majority of the crude oil reserves are located in a few countries, some of which have highly volatile political and social situations, it presents a problem for diversified energy supply and potential cause for political conflict. In addition, the conventional vehicles using these fossil fuels cause excessive atmospheric concentrations of greenhouse gasses (GHG), where the transportation sector is the largest contributor in the United States with over a quarter of the total GHG emissions.

It is important to note that electric vehicle (EV) and hybrid electric vehicle (HEV) technologies have been improved significantly, due to recent enhancements in battery technology, and they now compete with conventional vehicles in many areas. They offer solutions to key issues related to today's conventional vehicles by diversification of energy resources, load equalization of power, improved sustainability, quiet operation as well as lower operating costs and considerably lower emissions during operation without significant extra cost. Especially, with plug-in hybrid electric vehicles (PHEVs), it has become possible to achieve further energy consumption and emission reductions as well as potential applications for performing ancillary services by being able to draw and store energy from the electric grid and utilizing it in the most efficient operational modes for both the engine and the motor. Thus, hybrid and electric vehicles are currently considered some of the best alternatives for conventional vehicles.

1.2 Technology Development and Commercialization

It would be agreed by many experts in the industry that the history of EV and/or HEV is composed of three main periods. At the dawn of mechanic traction, until the beginning of twentieth century: steam, internal combustion and electric motors (EMs) had very similar market penetration. At the time, EVs had various advantages compared to the alternatives since steam vehicles were highly dangerous, dirty and expensive, and internal combustion vehicles were newly developed and still had certain technical issues. Moreover, since the cities were considerably smaller with a very small percentage of paved roads, electric range was not a significant limitation to the users. However, with the extension of the modern road networks and large distribution of petrol stations along with mass production; internal combustion technology become significantly cheaper and the predominant technology in the vehicle market.

First HEVs were developed as early as 1899 by Porsche due to the higher efficiencies that can be achieved when internal combustion motors are operated with combination of electric traction motors. Moreover, the second resurge is triggered with the development of power electronics. The research of motor control for EVs was founded in the 1960s. With the Arab oil embargo of 1970s, which increased the oil prices significantly, U.S. interest in federal policy to decrease fossil fuel consumption in the transportation sector began, which also led to average fuel economy standards to mandate an increase in efficiency standards in passenger cars. Among these, the Clean Air Act of 1965 also triggered numerous research institutes and firms to conduct research on electric vehicles. Thus, the interest in EVs and HEVs increased and various prototypes were built to reduce the fuel consumption, which established the foundation of today's modern hybrid and electric vehicles. However, they have not attained significant developments and were not able to penetrate into the vehicle market mainly due to the low energy density and high prices of the batteries at the time, which made them inferior to conventional vehicles in many aspects. At the end of 1970s, fewer than 4,000 battery electric vehicles were sold worldwide and it was not until the late 1980s and early 1990s that the research accelerated again due to oil prices and environmental concerns, which resulted in a significant comeback for EVs in the vehicle market, both in commercial and passenger vehicles (de Santiago *et al.*, 2012).

Even during the years 1990–2005, European automakers were still highly concentrating on further developments of ICEs on various topics (especially on variable-valve-timing and direct fuel injection systems) since over 80% of the patents were awarded on this technology against only 20% for the technologies associated with EVs and HEVs (Dijk *et al.*, 2013). Meanwhile, Japan had a considerable rise in EV and HEV patent applications in the early 1990s, which plummeted significantly after 1995, showing that the majority of the researchers and most of auto makers did not find electric propulsion technology profitable during this period compared to ICE vehicles. The main reasons behind the failure of this technology to become widespread can be listed as using lead-acid batteries at the time (which have very low energy densities and limited lifetime), unsatisfied customers (mainly with respect to price and range) and lobbying efforts from the auto industry (especially on loosening up the emission regulations). Thus, between the years of 1995 and 2000, only a few thousands of EVs and HEVs were sold worldwide.

During this time, the biggest successes of EV and HEV technologies were Toyota and Honda, which realized a business opportunity in this market and moved towards the mass commercialization of low emission vehicles utilizing alternative powertrains regardless of the relaxed emission regulatory measures. This included launching the Toyota Prius in Japan (in 1997), Prius II in California (in 2000) and Prius III worldwide (2004). Toyota subsequently sold over 1 million Prius between the years 1997 and 2007. In 1996, General Motors introduced EV1, a pure battery electric vehicle and leased it to a limited number of customers. However, the vehicle was not very successful due to various negative customer feedback, such as "range anxiety" and the fear of becoming stranded with a discharged battery. Meanwhile, most other car manufacturers started allocating significant R&D resources towards this technology after 2005, based on the heightened climate change concerns and peak oil prices during that time.

In 2012, around 113,000 EVs were sold in the world, more than twice of the previous year, mainly in the United States, Japan and China, and 20 million EVs are projected to be on the roads by 2020. Currently, Chevrolet Volt, Nissan Leaf and Toyota's Plug-in Prius are the most widely sold electric vehicles in the world. With the government incentives, significant increase in R&D and infrastructure for electric vehicle technologies and reduction in battery costs, the market penetration of these vehicles is expected to become more prominent in the near future.

In addition, there were significant national and local government involvements in the market preparation and the provision of infrastructure along with the allocation of R&D funds in this area in order to increase the market penetration of EVs and HEVs. Until 2005, the U.S. federal government provided a flat $2,000 tax deduction for all qualifying hybrids, which then replaced with a tax credit–based system on an individual model's emission profile and fuel efficiency from a few hundred to several thousand dollars. In addition, many states also offered additional incentives on top of the federal tax credit. Today, as compiled from various sources on the internet, many countries provide tax incentives for EVs and HEVs; Finland (€5M), France (€450M), Italy (€1.5M), Holland (12% of vehicle cost), India (20%), China (60,000 RMB), Spain (€6,000+), Sweden (€4,500) and United States ($7,500) being the leading countries in this regard.

It should be noted that during the past two decades fuel cell technology has started finding applications in many sectors, including transportation sector. Even though the inverse process of the one occurring in hydrogen fuel cells, which is the decomposition of water into hydrogen and oxygen using electricity was discovered in as early as 1800, the actual phenomenon of fuel cell was not discovered until 1838. However, it was still not until 1933 that the technology reached its adolescence, where the first practical use of fuel cells was established by converting air and hydrogen directly into electricity. This technology was later used in submarines of the British Navy (1958) and the Apollo Spacecraft. In 1960s, fuel cells that could be used directly with air as opposed to pure oxygen were developed (Andújar and Segura, 2009).

Fuel cells developed since 1970 have offered several advantages, such as less expensive catalysts, increased performance and longer lifetime. Thus, after a century of its invention, fuel cells became an important candidate for a paradigm shift in the field of electric power generation due to achieving high efficiencies and low emissions. In the last two decades, the specific powers of fuel cells have increased as much as two orders of magnitude and are started to be considered for various applications, especially the automotive sector.

Currently, a large majority of the vehicles using fuel cells are utilized for research and development and testing. The first commercially available fuel cell vehicle model, FCX Clarity, was developed by Honda in 2007 and was manufactured in series. Since then, various models of vehicles have been developed by different manufacturers including Fiat Panda, Ford HySeries Edge, GM provoq, Hundai I-Blue, Peugeot H2Origin and Toyota FCHV-adv. Moreover, due to their relatively high levels of emissions per liter of fuel consumed, this technology was also adopted in motorcycles and ships.

As the densities of the cities increased considerable the advantage of ICs reduced due to the health issues associated with the negative environmental impact of this technology. It is expected by many that the European Commission will eliminate the conventional fueled vehicles in cities by the year 2050 which will enable all electric and fuel cell operated vehicles to dominate the market in close future.

1.3 Vehicle Configurations

In order to be able to elaborate further on electric vehicles and their subsystems; first the definition and characteristics of different vehicle configurations is necessary to be clearly understood. Thus, in the next sub-sections, a brief description of various commonly used vehicle configurations is provided to convey the readers with the fundamentals of the basic vehicle configurations.

1.3.1 Internal Combustion Engine Vehicles (ICEV)

Internal combustion engine vehicles (ICEV), which are generally referred to "conventional vehicles" from now on, have a combustion chamber that converts chemical energy (of the fuel) to heat and kinetic energy in order to provide rotation to the wheels and propel the vehicle. ICEVs have relatively long driving range and short refueling times

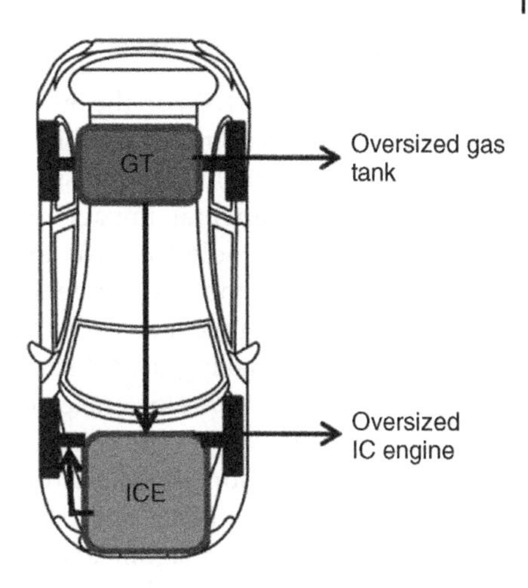

Figure 1.1 Illustration of internal combustion engine vehicle configuration.

but face significant challenges with respect to oil consumption and associated cost and environmental impacts. The vehicle configuration for ICEVs is illustrated in Figure 1.1.

The main advantages of ICEVs are listed as follows:

- The vehicle can store high volume of liquid fuel (typically gasoline or diesel) onboard in a fuel tank.
- The utilized fuel has high energy density sufficient to travel several hundred miles without refueling.
- It has short refueling times.

There are some drawbacks of these vehicles as follows:

- The vehicle is not satisfactorily efficient with less than 20% energy of the gasoline used as propelling power.
- The remainder of the energy is lost to the engine and to the driveline inefficiencies as well as idling.
- It is a significant contributor to environmental pollution and global warming, mainly due to hydrocarbon fuels utilized.

ICEVs have a plethora of moving parts, which makes the system complicated and hard to maintain (from regular oil changes, periodic tune-ups, to the relatively less frequent component replacement, such as the water/fuel pumps as well as the alternator) and reduces the system efficiently considerably. Moreover, it needs a fueling system to introduce the optimal fuel-air mix and an ignition system to have a timely combustion, a cooling system to operate safely, a lubricating system to reduce wear, an exhaust system to remove the heated exhaust products. Even though significant advancements have been made on ICEs in the past decades, they require fossil fuels which have unstable and ever-increasing prices, have political and social implications and causes environmental pollution and global warming.

In the past decades, substantial advancements have been made in using alternative fuels, including alcohol fuel derived from biological sources, such as food crop which mitigates the negative environmental effects; however these resources are also used very inefficiently due to the nature of the combustion process and the mechanical linkages (Electrification Roadmap, 2009).

1.3.2 All Electric Vehicles (AEVs)

All electric vehicles (AEVs) on the other hand, use the electric power as their only source to propel the vehicle. Since the vehicle is only powered by batteries or other electrical energy sources, virtually zero emissions can be achieved during operation. However, the overall environmental impact depends significantly on the method of energy production, thus a cradle-to-grave analysis is usually needed in order to get a much realistic measures of the environmental impact. Since they do not incorporate an ICE and its corresponding mechanical or automatic gearbox, the mechanical transmissions can be eliminated, making the vehicle much simpler, reliable and more efficient. Thus, EVs can attain over 90% efficiencies (in the battery) compared to 30% efficiencies of ICEs. Moreover, they can utilize regenerative breaking which increases their efficiency even further. In addition, they have the advantages of having quite operation and using electricity that can be generated from diverse resources. As the energy portfolio in many countries become significantly more diverse with various forms of renewable energy (especially solar and wind), the benefits of AEVs will become more much apparent in the future. The vehicle configuration for AEVs is provided in Figure 1.2.

The main advantages of AEVs are listed as follows:

- The vehicle is propelled using an efficient electric motor(s) that receive power from an onboard battery.
- Regenerative breaking is used to feed the energy back to the battery when the brakes are used.

There are some drawbacks of these vehicles as follows:

- It has the largest size batteries compared to HEVs or PHEVs since batteries are the only source of energy.
- The vehicle has limited range compared to conventional (ICE) vehicles.
- Full charging can take up to 7 hours in Level 2.

However, the specific energy of gasoline is incredibly high compared to that of electric batteries. Thus, in order to provide the same energy levels, the battery pack becomes significantly large, which adds considerable weight and cost to the vehicle. Thus, AEVs have very limited driving ranges and higher costs compared to ICEVs, which are the

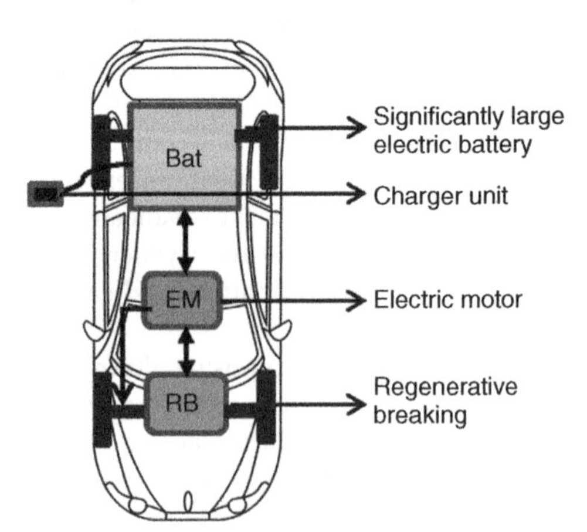

Figure 1.2 Illustration of the electric vehicle configuration.

main barriers of this technology to widely enter the vehicle market. However significant research is being conducted to increase the capacities associated with the batteries, supercapacitors and reduced-power fuel cells to overcome these issues.

1.3.3 Hybrid Electric Vehicles (HEVs)

Hybrid electric vehicles (HEV) on the other hand combines a conventional propulsion system with an energy storage system, using both ICE and electric motor as power sources to move the vehicle and therefore represent an important bridge between ICEVs and EVs. Hybrids are closer to conventional cars since they depend solely on fossil fuels for propulsion. The EM and the battery are generally used for maintaining engine efficiency by avoiding idling and providing extra power, therefore reducing its size. Thus, HEVs can achieve improved fuel-economy (compared to ICEVs) and longer driving range (than pure EVs). The vehicle configurations for HEVs and PHEVs are provided in Figure 1.3.

The main advantages of HEVs are listed as follows:

- The vehicle has both a battery/EM and an ICE/fuel tank.
- Either EM or both ICE and EM provide torque to the wheels depending on the vehicle architecture.
- A/C and other systems are powered during idling.
- Efficiency gains of 15–40% can be attained.

There are some drawbacks of these vehicles as follows:

- The vehicle still relies heavily on the ICE.
- All electric range is usually limited to 40–100 km.
- It costs more than its conventional counterparts.

Plug-in HEVs (PHEVs) are closer to AEVs based on the large size of the battery pack but can even have longer driving range since they be recharged simply by plugging into an electric grid. The success of Toyota Prius on the market shows that PHEVs are a real alternative to conventional vehicles. By having the appropriate energy generation mix of electricity and the suitable driving applications, both HEVs and PHEVs can

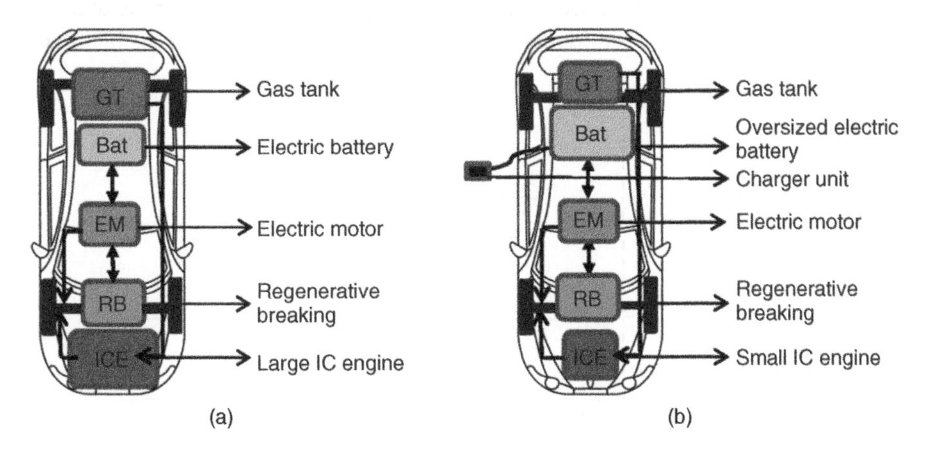

Figure 1.3 Illustrations of (a) hybrid and (b) plug-in hybrid electric vehicle configurations.

use significantly less gasoline and produce fewer tailpipe emissions than conventional vehicles.

The main advantages of PHEVs are listed as follows:

- Batteries can be charged/recharged by plugging into the electric grid.
- The vehicle is ideal for commuting and doing errands within short distances.
- There is no gasoline consumption or emissions during all electric mode.

There are some drawbacks of these vehicles as follows:

- Batteries used in the vehicle are larger and more expensive than HEV batteries.
- Charging may take up to 4 hours in Level 2.

Moreover, unlike EVs that can have their full capacity withdrawn at each cycle, an PHEV battery has a capacity draw that ranges around 10% of the nominal operating level (which is 50% state of charge) in order to deal with charge/discharge current surges without going into overcharge above 75% and deep discharge below 25% state of charge (SOC). Thus, only around half of the battery capacity is being used in PHEVs. The energy management modes for these vehicles are listed as follows.

Charge Depleting Mode (CD-mode): In this mode, the battery SOC is controlled in a reducing fashion when the vehicle is being operated. After charging PHEVs through conventional electrical outlets, they operate in charge-depleting mode (CD-mode) as they drive until the battery is depleted to the target state of charge, which is generally around SOC of 35%. In this mode, the engine may be on or off, however a portion of the energy for propelling the vehicle is provided by the energy storage system (ESS).

Charge Sustaining Mode (CS-mode): In this mode, the battery SOC is controlled to remain within a narrow operating band. After the previous operation (where the battery is depleted to the targeted SOC), the vehicle shifts to charge-sustaining mode (CS-mode) by utilizing the internal combustion engine to maintain the current SOC. PHEVs can be further categorized based on their functions in CS-mode. The conceptual illustrations of CD and CS modes are provided in Figure 1.4.

Electric Vehicle (EV) Mode: In this mode, the operation of the IC engine is prohibited and therefore the ESS is the only source of energy to propel the vehicle. Range-extended PHEVs act as a pure EV in CD-mode using only the electric motor, whereas blended

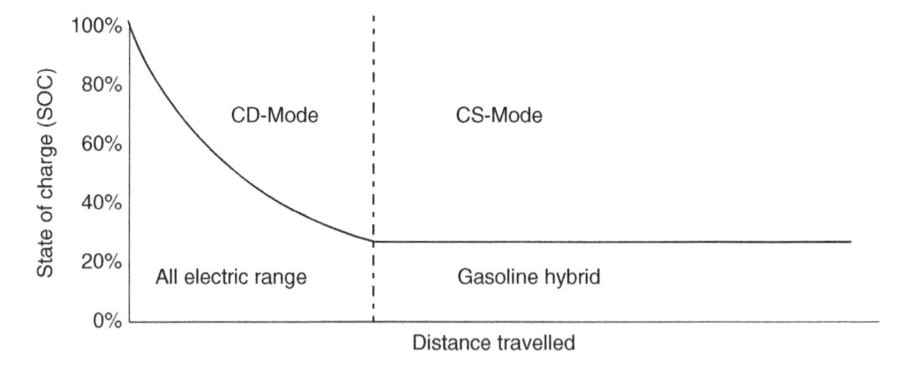

Figure 1.4 Conceptual illustration of battery discharge.

PHEVs use the electric motor primarily with the occasional help of the engine to provide additional power.

Engine Only Mode: Finally, after CS-mode, if the vehicle is still driving, it enters the engine-only mode where the operation of the electric traction system does not provide tractive power to the vehicle.

Finally, the factors effecting the use and market penetration of the aforementioned technologies are shown in Figure 1.5. The green and red colors of the arrows indicate some of the enabling and disabling factors in the development or integration of the different powertrain technologies.

Moreover, the electric configurations can also be mapped into a fit-stretch scheme of technical form and design of innovation in the x-axis and user context and functionality on the y-axis. The more innovation is similar to the established practice, the higher the fit and the smaller stretch. Combining these two dimensions makes it easier to compare different technologies with each other on a multi-dimensional facet.

Figure 1.6 shows two pathways showing that the alternative fuel vehicles may be used an additional vehicle which is more sustainable or can be used in combination

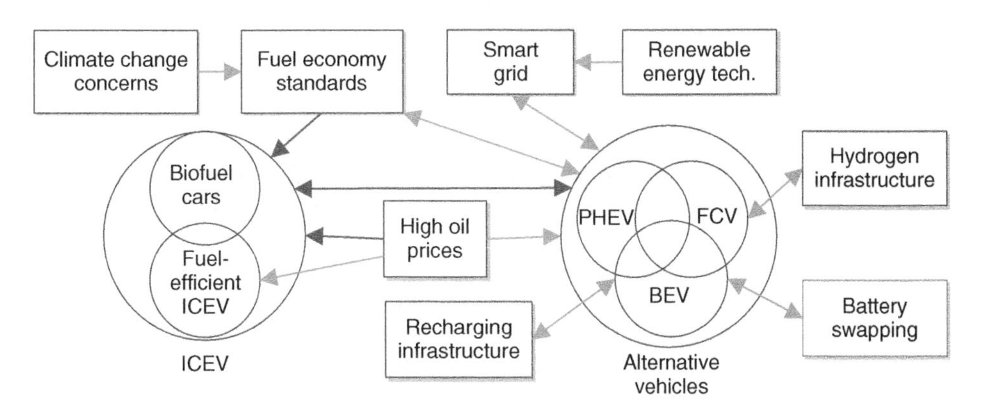

Figure 1.5 The factors influencing the market penetration of various technologies (adapted from Dijk *et al.*, 2013).

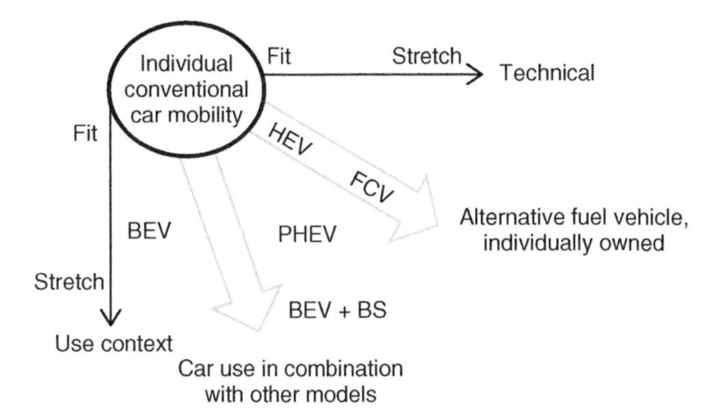

Figure 1.6 Fit-stretch pattern for different powertrain technologies (adapted from Hoogma, 2000).

Figure 1.7 Sodium borohyride fuel cell vehicle (courtesy of TUBITAK Marmara Research Center).

with other transport modes, the difference being the degree in mobility patterns and travel behavior. Thus, in the upper pathway, they remain mostly unchanged, where even though the vehicles have better efficiencies and lower emissions, the users do not change their travel behavior accordingly. The second pathway considers more active planning, wide range of transport modes and reduced sense of ownership of the vehicle as well as technological, infrastructural and regulatory reinforcements.

1.3.4 Fuel Cell Vehicles (FCVs)

Fuel cell vehicles can be considered as a type of series hybrid vehicle where the fuel cell acts as an electrical generator using hydrogen. The electricity produced by the fuel cell can either (or both) used to power the EM or stored in the energy storage system (such as battery, ultracapacitor, flywheel) (Chan *et al.*, 2010). The National Research Council (NRC) of Canada report on alternative transportation technologies showed that unlike biofuels or advanced ICE vehicles, FCVs can set the GHG emissions and oil consumptions at a steady downwards trajectory. An example of a fuel cell vehicle using sodium borohydride is shown in Figure 1.7.

1.4 Hybridization Rate

Electric and hybrid electric vehicles have considerable advantages over conventional vehicles in terms of energy efficiency, energy source options and associated environmental impact. Electric vehicles can be powered either directly from an external power station, or through stored electricity, and by an on-board electrical generator, such as an engine in HEVs. Pure electric vehicles have the advantage of having full capacity withdrawn at each cycle, but they have a limited range (Hamut *et al.*, 2013). HEVs on the other hand, have significantly higher ranges, as well as the option of operating in electric only mode, and therefore they will be the main focus of the analysis.

Hybrid electric vehicles take advantage of having two discrete power sources; usually primary being the heat engine (such as diesel or turbine, or a small scale ICE) and the auxiliary power source is usually a battery. Their drivetrains are generally more fuel efficient than conventional vehicles since the auxiliary source either shares the

Table 1.1 Characteristics of vehicles with different hybridization rates (adapted from Center for Advanced Automotive Technology, 2015).

Hybridization Characteristics	Micro Hybrid	Mild Hybrid	Full Hybrid	Plug-in Hybrid
Vehicle Examples	Mercedes Benz A-class, Smart car, Fiat 500, Peugeot Citroen C3, BMW 1 and 3 series, Ford Focus and Transit	BMW 7 series ActiveHybrid, Honda Civic and Insight, Mercedes Benz S400 BlueHybrid	Toyota Prius and Camry Hybrid, Honda CR-Z Chevrolet Tahoe Hybrid, Ford C-Max	Chevrolet Volt, Toyota Prius Plug-in, Porsche Panamera S E-Hybrid
Engine	Conventional	Downsized	Downsized	Downsized
Electric Motor	Belt Drive/Crankshaft	Belt Drive/ Crankshaft	Crankshaft	Crankshaft
Electric Power	2–5kW	10–20kW	15–100kW	70kW+
Operating Voltage	12V	60–200V	200+	200+
Fuel Savings	2–10%	10–20%	20–40%	20%+

power output allowing the engine to operate mostly under efficient conditions such as high power for acceleration and battery recharging (dual mode), or the auxiliary sources furnish and absorb high and short bursts of current on demand (power assist). Moreover, in both architectures, the current is drawn from the power source for acceleration and hill-climbing, and the energy from braking is charged back into the HEV battery for reuse which increases the overall efficiency of the HEVs. Currently, a wide range of configurations exist for HEVs based on the role and capability of their battery and electric motor as shown in Table 1.1.

These hybridization rates can provide various functionalities in different extends to the HEV such as engine stop/start operation, adjustments of engine operating points, regenerative braking and various levels of hybrid electric propulsion assist as shown in Figure 1.8. More information regarding different hybridization rates are provided in the next subsections.

1.4.1 Micro HEVs

Micro-HEVs have a starter-generator system coupled to conventional engine, where limited-power electric motor helps the ICE to achieve better operations during startup which is used as a starter alternator and combine automatic engine stop/start operation with regenerative breaking. They have typical generator capacities up to 5 kW and conventional 12 V batteries to reduce the fuel consumption of the vehicle, usually between 2% to 10% in urban driving cycles (depending on the vehicle, drivetrain and driving conditions), and are currently only found in light-duty vehicles. Moreover, the electric motor does not provide additional torque to the engine when the vehicle is in motion.

1.4.2 Mild HEVs

Mild HEVs provide electrically-assisted launch from stop and charge recuperation during regenerative breaking, but have a more slightly larger electric motor (than Micro

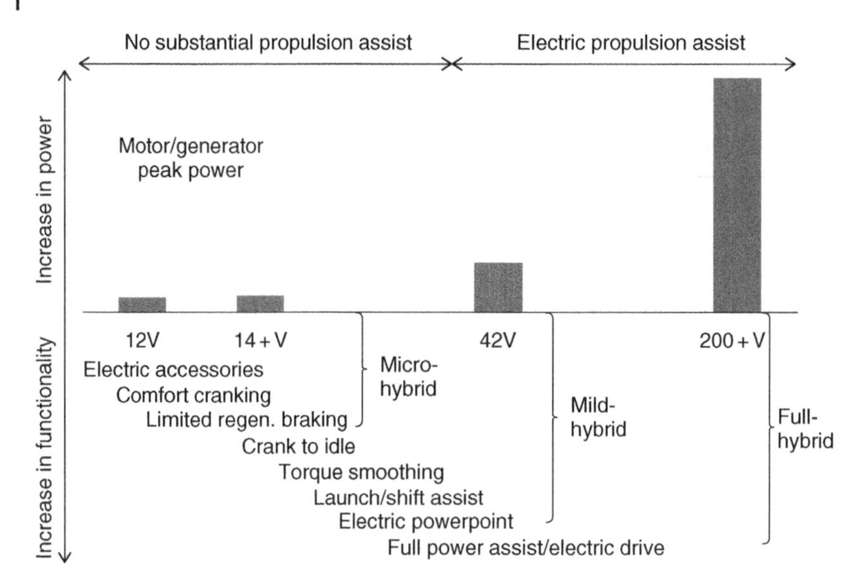

Figure 1.8 Hybrid classification based on powertrain functionality (adapted from Karden *et al.*, 2007).

HEVs) with 6–12 kW power and around 140 V operating voltage which assists the ICE. They still do not provide a sole source of driving power use the electric motor to boost the ICE during acceleration and breaking by providing supplementary torque, since it cannot run without the ICE (the primary power source) as they share the same shaft. With this configuration, fuel efficiencies of up to 30% (usually between 10 to 20%) can be acquired and can reduce the size of the ICE. Among the vehicles available in the market, GMC Sierra pick up, Honda Civic and Accord and Saturn Vue are known as some examples for Mild HEVs.

1.4.3 Full or Power-Assist HEVs

In full (or power-assist) hybrids, the electric motor can be utilized as the sole sources of propulsion since they have a fully electric traction system and provide power for engine staring, idle loads, full-electric launch, torque assistance, regenerative breaking energy capture and limited range and unlike Mild HEVs, they can split power path by either running the ICE or the electric motor or both. When used in full electric mode, the vehicle achieves virtually zero emissions during operation.

Full HEVs usually have a high capacity energy storage system with used power around 60 kW and operating voltage above 200V, this configuration with a wide range of architectures (series, parallel or combinations). As a result, this configuration can reduce the fuel efficiency up to 40% without any significant loss in driving performance (usually between 20 to 50%). However, they usually require significantly larger batteries, electric motors and improved axillary system (such as thermal management system) than the aforementioned configurations (Tie and Tan, 2013).

1.4.4 Plug-In HEVs (or Range-Extended Hybrids)

Plug-in HEVs are very similar to full HEVs (can use both fuel and electricity for propulsion) with the additional feature of the electrochemical energy storage being able to be charged by being plugged into an off-board source (such as the electrical grid) instead of using fossil fuels alone. They can either be used as a BEV with limited-power ICE or to extend the driving range by having ICE act as a generator that charges the batteries, which is also called "range extended EV".

In PHEVs, since the vehicle has an alternative energy unit and a battery that can be charged from the grid, the mass of the battery is significantly smaller than EVs (and typically have batteries larger capacity than HEVs), thus enabling the PHEVs to operate more efficiently in electric-only mode (due to the reduction in power required to propel the vehicle) than similar EVs. PHEV chargers must be light-weight, compact and highly efficient in order to maximize the effectiveness of the electric energy from the grid. By utilizing the stored multi-source electrical energy from the grid and stored chemical energy in the fuel tank together or separately, PHEVs can achieve even better driving performance, higher energy efficiencies, lower environmental impact and lower cost than conventional HEVs, mainly depending on the driving behavior and energy mix of the electricity generation.

The electrical power requirement depends on various factors (especially vehicle weight) and is above 70 kW. Since the power is drawn from the grid (instead of the ICE), the efficiency and vehicle performance could be improved significant in short distances and urban drive as the vehicle can be driven in electric motor mode. Thus, plug-in HEVs become very desirable for both in city driving and highway patterns.

1.5 Vehicle Architecture

In all hybrid electric vehicles, the arrangement between the primary and secondary power sources can be categorized as parallel, series, split parallel/series (and even complex) configurations. The hybrid vehicles configurations can be seen in Figure 1.9. There are complex trade-offs among these configurations in terms of efficiency,

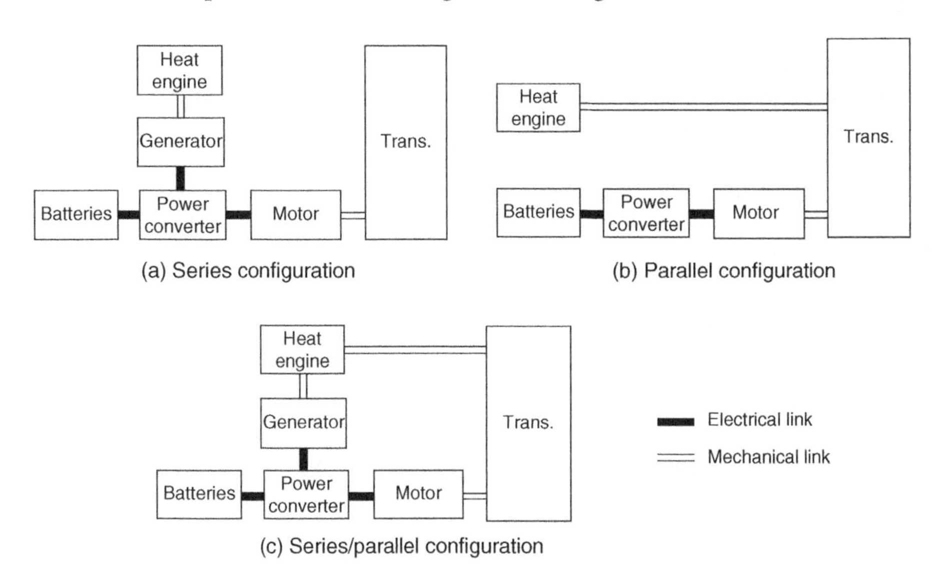

Figure 1.9 Hybrid vehicles configurations in (a) series, (b) parallel and (c) series/parallel.

drive-ability, cost, manufacturability, commercial viability, reliability, safety and environmental impact, and therefore the best architecture should generally be selected based on the required application, especially driving conditions and drive cycles.

1.5.1 Series HEVs

In a series configuration the engine generally provides the electrical power through a generator to charge the battery and power the motor. Conceptually, it is an engine-assisted EV which extends the driving range in order for it to be comparable with conventional vehicles. In this configuration, the output of the heat engine is converted to electrical energy that, along with the battery, powers the drivetrain. The main advantage of this configuration is the ability to size the engine for average rather than peak energy needs and therefore having it operate in its most efficient zone. Moreover, due to a relatively simplistic structure and the absence of clutches, it has the flexibility of locating the engine-generator set. In addition, it can reserve and store a portion of its energy through regenerative breaking. On the other hand, relatively larger batteries and motors are needed to satisfy the peak power requirements and significant energy losses occur due to energy conversion from mechanical to electrical and back to mechanical again. This configuration is usually more suitable for city driving pattern with frequent stop and run conditions. In general, this configuration has worse fuel economy (due to power conversion) as well as cost (due to extra generator) compared to the parallel configuration but has a flexible component selection and lower emissions (due to the engine working more efficiently).

1.5.2 Parallel HEVs

In a parallel configuration (such as Honda Civic and Accord hybrids), both the engine and motor provide torque to the wheels, hence much more power and torque can be delivered to the vehicle's transmission. Conceptually, it is an electric assisted conventional vehicle for attaining lower emissions and fuel consumption. In this configuration, the engine shaft provides power directly to the drivetrain and the battery is parallel to the engine, providing additional power when there is an excess demand beyond the engine's capability. Since the engine provides torque to the wheels, the battery and motors can be sized smaller (hence, the lower battery capacity) but the engine is not free to operate in its most efficient zone. Thus, a reduction of over 40% can be achieved in the fuel efficiency. This configuration is usually desirable for both city driving and highway conditions.

1.5.3 Parallel/Series HEVs

Finally, in a split parallel/series powertrain (such as Toyota Prius, Toyota Auris, Lexus LS 600h, Lexus CT 200h and Nissan Tino), a planetary gear system power split device (shown in Figure 1.9c) is used as well as a separate motor and generator in order to allow the engine to provide torque to the wheels and and/or charge the battery through the generator. This configuration has the benefits of both the parallel and series configurations in the expense of utilizing additional components. However, the advantages of each configuration are solely based on the ambient conditions, drive style and length, electricity production mix as well as the overall cost.

1.5.4 Complex HEVs

Lastly, complex configuration is very similar to the parallel/series configuration with the main difference of having a power converter as well as the motor/generator and motor which improves the vehicle's controllability and reliability compared to the previous system. The main disadvantage of this configuration is the need for a more precise control strategy.

1.6 Energy Storage System

Once the various types of vehicle configurations and architectures are examined, the use of the most appropriate energy storage system for the intended application becomes one of the main selection criteria in HEVs. Therefore, descriptions along with the advantages and drawback of these ESSs are briefly described below.

1.6.1 Batteries

Battery is a portable storage device which usually incorporates multiple electrochemical cells that are capable of converting the stored chemical energy into electrical energy with high efficiencies and without any gaseous emission during operation stage. In batteries, the chemical reactions take place throughout the bulk of the solid, thus the material should be designed in order to allow the ingress and removal of the reaction species throughout the material over hundreds/thousands of cycles to deliver a practical rechargeable battery (Whittingham, 2012). All types of batteries contain two electrodes, an anode and a cathode as shown in Figure 1.10.

Several battery chemistries have been developed in the past decades. However, among the ones available, Li-ion chemistry currently dominates the market in a wide range of applications. These batteries are available in four different geometries, namely small and large cylindrical, prismatic and pouch. Note that cylindrical cells are produced in high volumes and with high quality and can retain their shape, while other formats require

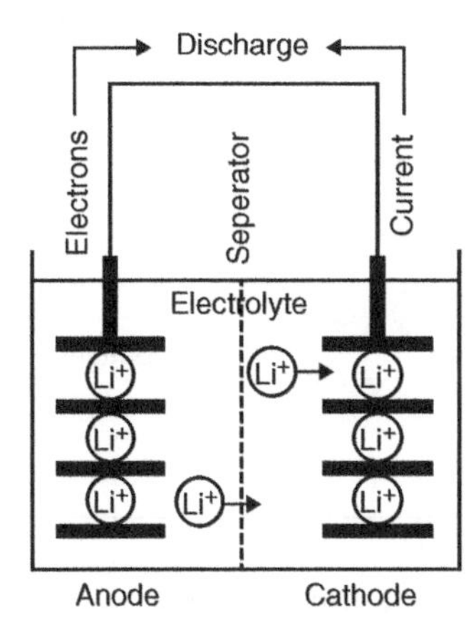

Figure 1.10 Images and schematics of a common battery (courtesy of TUBITAK Marmara Research Center).

and overall battery enclosure to retain their expansion. Moreover, cylinder volumes have the advantage of being robust and structurally durable (against shock and vibration); however their heat transfer rates reduce with increasing size. Moreover, it is very hard for the large ones and is almost impossible for the small ones to be replaced. Prismatic cells on the other hand are encased in semi-hard plastic cover and have better volume efficiencies. They are usually connected with threaded hole for bolt and have easy field replacement but require retaining plates at the ends of the battery. The soft pouch packaging has high energy/power densities (without the extra packaging) and usually has tabs that are clamped, welded or soldered. Like prismatic cells, they also require retaining plates and have poor durability unless additional precautions are taken, which would in turn increase the volume and the weight of the cells (Pesaran *et al.*, 2009).

Currently, they offer the most promising option to power HEVs and EVs in a relatively efficient manner. The most important characteristics of batteries are the battery capacity (which is proportional to the maximum discharge current) measured in Ah, the energy stored in the battery (capacity x average voltage during discharge) measured in kWh as well as the power (voltage x current) measured in kW. The maximum discharge current (typically represented by the index of C) indicates how fast the battery can be depleted and is affected by the batteries chemical reactions and the heat generated. Another important parameter in batteries is the state of charge (SOC) which displays the percentage of the charge available in the battery (Tie and Tan, 2013).

Batteries are currently the most commonly used technology for EVs and HEVs due to being able to deliver peak and average power at excellent efficiencies, but have inherently low specific energy, energy density and refueling/charging rates (compared to fossil fuels), which limits their range, increases their size and cost which in turn prevents their wide-spread adoption. Their power and energy characteristics with respect to the alternative ESSs are provided in Figure 1.11.

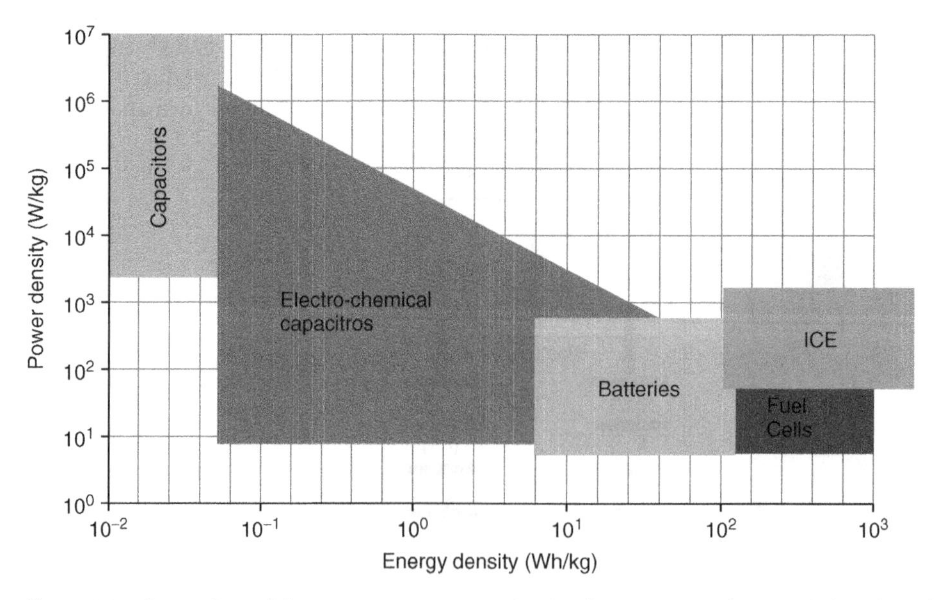

Figure 1.11 Comparison of the power versus energy density characteristics of various ESSs (adapted from Guerrero *et al.*, 2010).

Typical battery electric vehicles achieve between 3–6 mile kWh^{-1} depending on various factors including the vehicle design and driver behavior. As an example, current technology requires roughly 150 kg of Li-ion cell (or over 500 kg lead acid cells) in order to travel a range of 200 km for an average passenger car under a non-demanding drive cycles. In order to double this range, the power, weight and the corresponding cost must also be almost doubled, which presents important limitations. Currently, some of the common technical demands on the batteries are to have high discharge power, high battery capacity and cycling capability, good recharging capability and high power capacity for electric vehicle applications. Further information regarding different battery chemistries, their performance, cost and environmental impact will be described in Chapter 2.

1.6.2 Ultracapacitors (UCs)

(H)EV requirements are becoming more and more demanding as the technology improves and as they become widely available in the market. Even though significant improvement have been made on the battery technologies in terms of charge rate, capacity and battery life, there are still significant barriers to fuller use of electric vehicles. In this regard, ultracapacitors (also known as supercapacitors) can provide potential benefits in many areas where battery technologies currently face challenges.

Unlike batteries where the electric energy is stored as chemical energy, in capacitors it is stored in terms of surface charge and therefore a large surface area is required to attain high storage capacities. Since the capacitor material's structural integrity is not damages through charging/discharging process, pure capacitors can be virtually charged/discharged millions of times without any significant degradation of the materials.

Ultracapacitors (shown in Figure 1.12) have similar structures with normal capacitors but with much higher capacitance (with a factor of 20 times) than capacitors and much shorter charging times than electric batteries. They are hybrids between batteries and capacitors, involving both surface charge and some Faradaic reaction in the bulk of the material. Their characteristics include virtually maintenance-free operation, longer

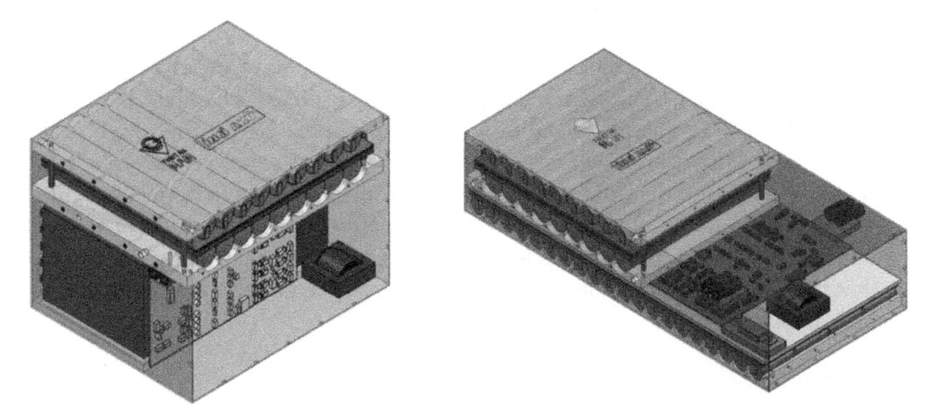

Figure 1.12 Illustrations of powerpacks using ultracapacitors with VRLA and li-ion batteries, respectively (courtesy of TUBITAK Marmara Research Center).

cycle-life and insensitivity to changes in the environmental temperatures, shocks and vibrations and can be used as standalone or in conjunction with an onboard battery.

There are currently three types of ultracapacitors that are mainly used in electric vehicles, namely electric double layer capacitors (EDLC)-carbon/carbon (more power density but lowest energy density with 5–7 Wh/kg), pseudo-capacitors and hybrid capacitors (both with 10–15 Wh/kg energy density) based on the energy storage mechanism and their electrode materials. The lifetime of ultracapacitors can reach up to 40 years which is the longest in all energy storage systems.

Electric double layer capacitors have different ways of storing energy than conventional electrochemical energy storages, where in EDLC the energy is stored directly in the electric field. The main advantage of this technology is its surface phenomenon without faradic reactions which implies very fast kinetics, ensuring a high power performance as well as a considerable cycle life.

On the other hand the energy density is significantly lower (25 times less) than a similar sized Li-ion battery. However, with the current advancements in nanotechnology (especially in carbon nanotubes), the ion-collection surface area of the ultracapacitor can be increased considerably, increasing the associated energy storage capacity up to a quarter of the energy storage capacity of a conventional Li-ion battery. Moreover, electric battery/ultracapacitor hybrid technologies are also being developed in the past couple of years which combines an ultracapacitor and a lead-acid battery in a single unit cells in order to improve the power and cycle life of the lead-acid battery. Although there are numerous vehicles driven only by ultracapacitors which exist today, they are still at the developmental stage (with prototypes mostly) and are used in relatively limited applications.

1.6.3 Flywheels

The flywheel is an energy conversion and storage device (also known as electromechanical battery) which stores energy in a rotatory mass. This principle has been used for a while to stabilize the output voltage of synchronous generators, but the recent developments have made this technology compatible to be also used in the transportation sector. Flywheels usually consists of a high strength carbon fiber wheel, magnet floating bear supporting device, motor/generator (for electric/kinetic energy conversion) and power electronics control device.

Flywheels can attain cycle efficiencies around 90%. They have around 40 Wh/kg specific energy, but much higher specific power than ordinary chemical batteries, which enables them to be charges much faster (Vazquez *et al.*, 2010). Their high peak power is only limited by the power converters. No chemical reactions take place in flywheels (which prevents the possibility of gas emissions or waste materials harmful to the environment) and therefore they are virtually maintenance free and have almost infinite number of charge/discharge cycles. This makes them very attractive for applications that use high number of charge/discharge cycles such as regenerative breaking in EVs and HEVs.

1.6.4 Fuel Cells

In this ESSs, chemical energy (hydrogen gas) is converted into mechanical through either by burning hydrogen (in internal combustion engines) or by reacting with oxygen

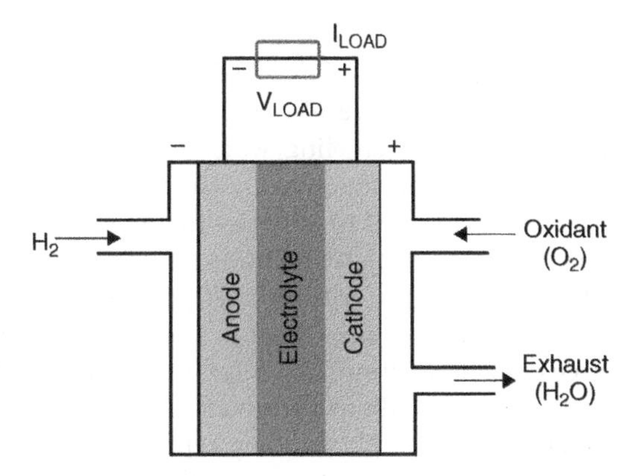

Figure 1.13 A basic configuration of a proton exchange membrane fuel cell.

(in fuel cells) to produce electricity (without the need to go through inefficient thermodynamic cycles) that can be used for the vehicle propulsion and powering accessories. As briefly mentioned in Section 1.3.4, FC vehicles emit only water vapor and can be highly efficient. The proton exchange membrane fuel cell (PEMFC) can achieve much higher energy density than any of the aforementioned technologies due to much lower atomic mass of the hydrogen and has the favorable characteristics including high energy efficiency, low operation noise and environmental compatibility. The configuration of hydrogen fuel cell and its electrochemical reaction are shown in Figure 1.13.

The European Commission co-funded project HyWays which explored a range of hydrogen scenarios for the European Union and concluded that, by the year 2050, if 80% of road vehicles were hydrogen-fuelled this would result in 50% less CO_2 emissions, compared to the extension of current scenarios (Offer *et al.*, 2010).

Even though FC technology seems like a newly emerging energy sources, it dates back to 1839, where the first FC is assembled by Sir William Grove. After decades of research, it was started to be used mainly in aerospace applications in 1960s and become more widely explored in the 1980s.

Today, if a comparison was to be made between current ICE, electric battery and fuel cell vehicle technologies; in order for a diesel vehicle to travel 500 km, a tank that weighs over 40 kg would be needed (with a volume of less than 50 L), whereas this becomes 830 kg for a Li-ion electric battery (for a potential usable energy density of 120 $Whkg^{-1}$ and 100kW of electrical energy) and 125 kg for hydrogen (based on 700 bar compressed gaseous hydrogen vessel and 200 kWh chemical energy) to achieve the same range. Moreover, refueling the vehicle would take somewhere between 30 minutes (with up to 80 kW DC fast charging) to several hours for the electric battery as opposed to 3 to 5 minutes for the hydrogen vehicle (Eberle and von Helmolt, 2010).

However, the technology is not still not mature enough for practical EV applications due to operational problems related to electro-catalysis in direct FCs as well as issues associated with hydrogen generation, storage and distribution as well as system complexity and manufacturing cost. The manufacturing cost is mainly associated with expensive membrane and raw materials (used as catalysts) and fabrication processes (especially for collector plates). In addition, when starved from fuel or oxygen, significant performance degradation and the cell voltage drop leading to cell reversal which accelerates the corrosion of carbon components which is harmful to the FC stack and components. Moreover, the fact that hydrogen tanks are characterized by high specific energy but low volumetric energy density requires a bulkier hydrogen tank (than equivalent gasoline tanks), which creates a drawback to for it to be utilized in vehicles. Furthermore, FCs are also produced in low quantities and may require additional infrastructure for refueling, which increases their cost significantly. However, hydrogen tanks can provide significantly larger range (50 – 100 mile $kg^{-1}H_2$) and be refueled in minutes (as opposed to batteries that may take hours) which make it more comparable to conventional vehicles in terms of refueling.

FCs normally perform the best performance on pure hydrogen or at least hydrogen rich gas, which requires them to store on board hydrogen. Hydrogen can be stored via three possible solutions, namely compression (at pressures 700 times the atmospheric), cryogenic system (liquefaction at −253°C) and hydrogen absorbing materials (through metal, charcoal and by holding captive in solid matrix). Moreover, FC hybrid vehicles that uses hydrogen fueled internal combustion engines are also receiving significant attention since they can operate at a very lean stoichiometry which enables them to achieve top brake efficiencies (over 45%) while permitting Euro 6 emissions without any after-treatment.

As a result, FC hybrid vehicles can improve vehicle performance and fuel economy, and hydrogen, like electricity, can be produced from any primary energy source including many of the renewables and therefore can also assist in breaking the link between oil and transport.

1.7 Grid Connection

Charging capabilities, strategies and power flow play a significant role in gaining wide acceptance of plug-in electric and hybrid electric vehicles in the market since most important barriers related with the cost and cycle-life of the batteries, obstacles related to the use of chargers and the lack of charging infrastructure. Charging systems can be divided into off-board (with unidirectional) and on-board (with bidirectional) power flow. On-board charger system can be conductive or inductive and off-board charger system can be designed for high charging rates with Level 1 (convenience), Level 2 (primary) and Level 3 (fast) power levels (Yilmaz and Krein, 2013). Inductive chargers have preexistent infrastructure and are inherently safer. On the other hand, conductive chargers are lighter, more compact and allow bidirectional power, thus can achieve higher efficiencies. The impact of charging the vehicles on the grid can significantly affected by picking the optimum times to charge the vehicles (smart charging) and the ability to feed back the charge when the grid has the peak load (V2G).

1.7.1 Charger Power Levels and Infrastructure

Adequate and well-structured charging and its associated infrastructure is imperative for electric vehicles to be able to create a solution for customers' "range anxiety" and a successfully penetration into the vehicle market as they are the main contributor of the "chicken-and-egg" problem of the EV development. The problem describes the reluctance of vehicle manufacturers to introduce alternative vehicles in the absence of supporting infrastructure and similarly the reluctance of fuel producers to invest in infrastructure when no alternative vehicles are available. Moreover, selecting the appropriate location and the type of charging infrastructure has an important impact on the vehicle owner and the grid. In addition, charger power level also plays an important role for the user since it has a significant impact on acquiring the necessary power at a given time span, the associated cost as well as the impact on the grid. There are 3 levels of charging equipment currently available. Level 1 charging is the slowest method that would require no additional infrastructure for home and business sites (expected to be integrated to the vehicle) and uses a standard 120 V/15 A single-phase outlet. Level 2 charging, which is currently the primary method for dedicated private and public facilities, offers a faster charging from 208 V or 240 V and the associated infrastructure can also be onboard to avoid redundant power electronics. Otherwise, it may require dedicated equipment and installation for home and public units which costs between $1000 and $3000. Lastly, unlike the first two levels which are typically used for overnight charging (which utilizes low off-peak rates) Level 3 can provide fast charging in less than 1 hour with 480 V or higher three-phase circuit and is usually allocated in refueling stations. The characteristics of charging power levels are shown in Table 1.2.

The infrastructure cost for this level is reported between $30,000 and $160,000 and they can overload the distribution equipment. Although the number of charging stations are very limited today, with the further enhancement of the electric vehicle technology (especially in terms of increasing all electric range and reducing total cost) and the associated penetration of them in the market, the number of charging equipment/stations will be increased significantly throughout the world.

1.7.2 Conductive Charging

Currently, chargers for EVs are mainly plug-in connections where the user needs use insert a plug into the car's receptacle to charge the batteries. These systems use direct contact and a cable, that is either fed from the outlet (in Levels 1 and 2) or from a charging stations (in Levels 2 and 3), between the EV connector and charge inlet. A concept Level 3 charging station is shown in Figure 1.14.

This technology has several disadvantages such as the cable and connector delivering 2–3 times more power than standard plugs in the houses, which poses a risk for electrocution, especially under wet environments. Moreover, during cold climates, the plug

Table 1.2 Charging power levels.

Power Level Types	Charger Location	Typical Charging Location	Expected Power Level	Charging Time	Vehicle Technology
Level 1 120 VAC (US) 230 VAC-EU)	On-board 1-phase	home/office	1.4 kW (12A) 1.9 kW (20A)	4–11 hours 11–36 hours	PHEVs (5–15 kWh) EVs (16–50 kWh)
Level 2 240 VAC (US) 400 VAC (EU)	On-board 1- or 3-phase	Private or public outlets	4kW (17A) 8 kW (32A) 19.2 kW (80A)	1–4 hours 2–6 hours 2–3 hours	PHEVs (5–15 kWh) EVs (16–30 kWh) EVs (3–5 kWh)
Level 3 (208–600 VAC/VDC)	Off-board 3-phase	Commercial	50 kW 100 kW	0.4–1 hour 0.2–0.5 hour	EVs (20–50kWh)

Source: Yılmaz and Krein (2013).

Figure 1.14 Level 3 charging station concept.

Table 1.3 Charging infrastructure costs.

		Low ($)	Base Case ($)	High ($)
Home	1.4 kW	25	75	550
	7.7 kW	500	1,125	4,000
Away	1.4 kW	1,050	3,000	9,000
	7.7 kW	2,500	5,000	15,000
	38.4 kW	11,000	20,000	50,000

Source: Peterson and Michalek (2013).

in charge point may become frozen onto the vehicle. In addition, the long cables can have tripping hazards and may look aesthetically unappealing. The cost of the charging infrastructure, including the installation and equipment costs, are provided in Table 1.3. It should be noted that this cost can vary significantly based on several factor such as the availability of existing outlet, maintenance and even potential vandalism (for public charging points).

1.7.3 Inductive Charging

An inductive charger on the other hand, transfers power magnetically and is explored mainly for Level 1 and 2 devices. This technology requires large air gaps, high efficiency and a large amount of power and eliminates the aforementioned disadvantages by not using any cables. In inductive charging, a power supply produces high frequency alternating currents in the transmitter pad or coil that transfer power to the receiving coil inductively, where the receiver electronics converts it to direct current (DC) to charge the battery. The main operating parameters of inductive charging systems are power level, maximum charging distance, efficiency, charging tolerances and size and weight.

The main advantage of this technology is that instead of deep charging/discharging the battery, the vehicle can be often topped-off while being parked at home/work (static

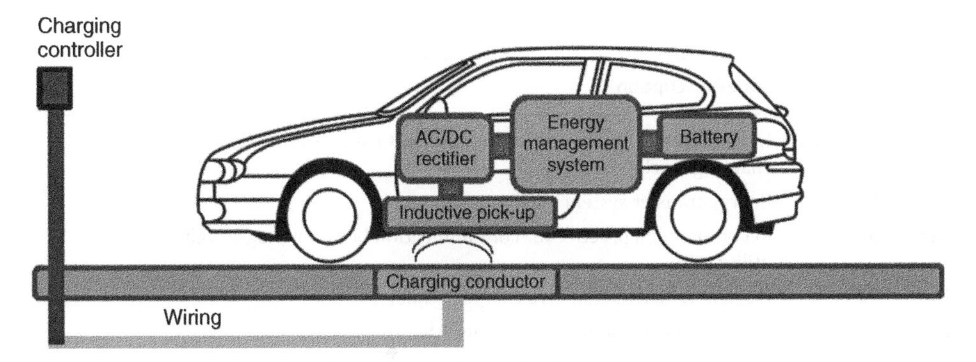

Figure 1.15 Illustration of the inductive charging technology.

inductive charging). Moreover, the technology also leads way into semi-dynamic and dynamic charging, where the vehicle can be charged wirelessly as it is traveling at low and "regular" speeds respectively. This would provide solutions to many key issues associated with availability of charging stations and electric range. Thus, methods for charging/discharging of electric vehicles with the emphasis on simplicity/convenience, cost effectiveness, high efficiency and flexibility have gained even a wider importance among industrial and academic communities. An illustration of the concept is shown in Figure 1.15.

Inductive power transfer (IPT) has acquired global recognition as a method for applications with no physical contact, through the weak or loose magnetic coupling. This method can offer high efficiency (up to 85–90%), robustness and high reliability without being significantly affected by dust or chemicals. Currently many IPT systems with a wide range of topologies and levels of complexity in control have been research and tested. Even though some of these technologies are focused on improving the contactless power flow in unidirectional applications, there are also some bidirectional systems under development for EV application that can enable regenerative breaking and V2G applications. Such systems include a coupled magnetic circuit to facilitate bidirectional power transfer while operating as a voltage source. However, it is currently not profitable to sell the electricity back into the grid using IPT. The main disadvantages of this technology however include lower charging efficiency and power density as well as manufacturing complexity and size, and cost.

1.7.4 Smart Grid and V2G/V2H/V2X Systems

Even though EVs and PHEVs can provide significant benefits in terms of reduction of fossil fuel consumption and related emissions, they still needed to be plugged into the grid to get the energy to charge up the battery, which can increase the electricity demand especially when they are in growing numbers. Most conventional charging method for PHEVs is plugging it to the household outlet (so called V0G) to be charged when needed, which can add significant load to the grid, especially as the number of PHEVs increase in the future. In order to provide common grounds and methods/procedures for EV charging, various standards on energy transfer, connection interface and communications have been developed over the years (and still continue to do so) which are summarized in Table 1.4.

Table 1.4 Vehicle charging standards.

Standard	Code/Description
NEC Article 625	EV charging system (wires and equipment used to supply electricity for charging an electric vehicle)
SAE J2293	Energy transfer system for EVs
SAE J2836	Recommended practice for communication between plug-in vehicles and utility grid
SAE J1772	Electric vehicle conductive charge coupler
SAE J1773	Electric vehicle inductively coupled charging
IEC 62196	Plugs, socket outlets, vehicle couplers and vehicle inlets, conductive charging of electric outlets
IEEE 1547.3	Interconnecting distributed resource with electric power system

Source: Young *et al.* (2013).

Moreover, their load on the system can be reduced significantly and even feed electricity back to the grid with the implementation of indirect charging and bi-directional power transfer systems. This concept of integrating the battery powered vehicles into the grid and charge when the electricity demand is at its lowest, when there is excess capacity (and/or related other metric) is commonly called smart charging (so called V1G).

In this regard, integration of distributed resources load and generation/storage device between the EVs and the grid is commonly called vehicle-to-grid (V2G) system (although smart charging and V2G are used widely starting to be used interchangeably). The vehicle can also communicate with the building, as opposed to the grid, as home generators during periods of electrical service outage (or even for the purpose of self-generated renewable energy use) which is commonly called V2B. Finally, there are systems that include both along with additional features such as storage of power to a remote site or to other PHEVs, which are commonly called V2X. These interactions are represented in Figure 1.16.

The aforementioned charging schemes are also provided in Table 1.5. As shown in this table, communication between the grid and the vehicle exists in smart charging, usually

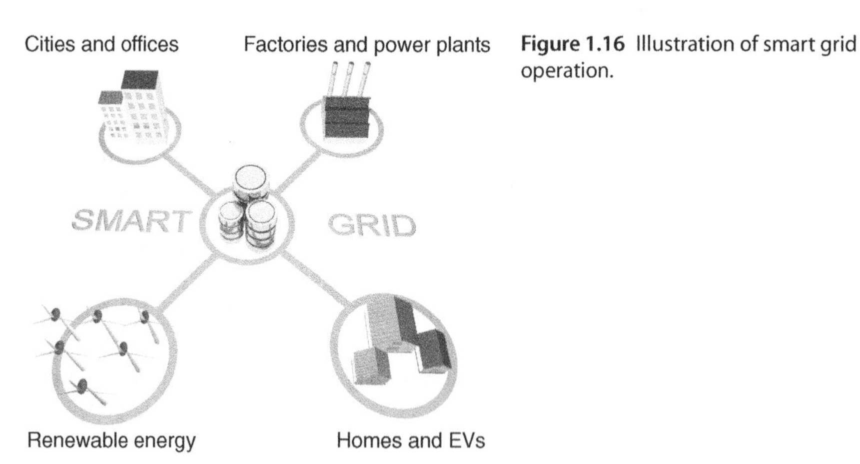

Figure 1.16 Illustration of smart grid operation.

Table 1.5 Charging schemes for electric vehicles.

Features	V0G	V1G	V2G	V2B	V2X
Real-time communication		√	√	√	√
Communication with the grid		√	√		√
Communication with the building/Home generator feature				√	√
Provide power to a remote site					√
Transfer energy to other PHEVs					√
Timed charging		√	√	√	√
Backup source			√	√	√
Controllable load		√	√	√	√
Bidirectional grid ancillary service			√		√
Load shifting for renewables			√	√	√

Source: Young *et al.* (2013).

through advanced metering infrastructure. In addition, energy stored in the batteries can also be transferred back to the grid and building in V2G and V2B schemes respectively. Finally, V2X has all the previous with the addition of providing power to remote sites and/or other PHEVs.

Thus, these concepts can provide solutions and link two critically important problems; the petroleum dependency of the transportation sector and the imbalance between electricity supply and demand, in ways that may address significant problems in both issues. Moreover, smart grid/V2G systems can enable PHEVs to have even more impact on enhancing the reliability, technical performance, economics and environmental impact of the grid operations by provision of capacity and energy based ancillary services and the reduction of the need for peaks and load levelization and can even generate revenues to the owners of these vehicles. In turn, this can help reducing the petroleum use, strengthening the economy, enhancing natural security and reducing the carbon footprint.

In order to have such a system where the electricity resources could be utilized better, vehicles must incorporate a power connection to the electricity grid, a control or logical connection for communication with the grid operators and high accuracy metering on the vehicle to tract energy transfers. The control of the grid operator is essential must be overridden in order to prolong the battery life and have the vehicle ready for operation. Since most of these vehicles stay idle in parking lots or garages over 90% of the time in the United States, the size of these resources can be quite large. However, in order for V2G (or V2X) systems to be successful, the requirements of the grid system operator and the vehicle owner must be satisfied. The grid system operator demand industry standard availability and reliability from these systems, whereas the vehicle owner desires a quick returns on the additional hardware cost associated with the system.

The literature studies show that in 2020, with a quarter of people in 13 regions of the United States having EVs/HEVs, 160 new power plants would be required if all the EV/HEV owner plugs their vehicle to the grid around 5 p.m. On the other hand, smart-grid technology can utilize these vehicles to provide valuable generation capacity at peak times (along with ancillary services) and enable the demand for electricity to be supplied within the existing capacity by better utilizing the daily load. Moreover, since the electricity price is lower during the charged off-peak hours than the generated peak hours, the owners of these vehicles would be able to make revenues from this process. However, currently the estimated profit from this technology ranges significantly, from -$300 to $4600 profit per vehicle per year with most estimates ranging of $100 - $300. Since this may not be economically adequate to gain significant participation by individuals or aggregator organizations, governments may need to support these technologies with policies in order to reinforce customer and business participation.

Even though V2G/V2B/V2X systems can have the aforementioned positive impact on the efficiency, cost and environmental impact of the energy used from the grid or the house, it can also reduce the capacity of the battery as a result of cycling based on the number cycles, depth of discharge (DOD) and the actual chemistry of the battery. Even though currently not enough data are available to demonstrate the exact impact of these systems on the battery degradation, some studies have determined that using the battery for V2G/V2B/V2X energy incurs approximately half the capacity loss per unit energy processed compared to that associated with the more paid cycling encountered while driving. Moreover, new standards and certifications as well as updates/modifications to the building codes and electrical regulations will be necessary to be able to utilize these technologies.

In addition, this technology would also help with the integration of renewable energies (especially solar and wind) and the transformation of the electricity system used today. In wind energy, the power generated from the wind turbines fluctuate significantly due to wind gusts, cloud cover, thermal cycles, the movement of weather fronts and seasonal changes, whereas in solar energy, this changes are mainly based on the time, season and the associated solar irradiation. A concept of renewable energy integration into smart grid systems is shown in Figure 1.17.

V2G systems could help level the daily fluctuations of these renewable energy sources and help with the integration of these intermittent resources into the grid. Studies

Figure 1.17 A conceptual representation of integrating renewable energy into a smart grid system.

show that in the next century, the installed renewable energy capacity could increase by up to 75% with V2G capable EVs, however, this depends on the electric vehicle storage capacity, through more vehicles and/or larger batteries. In addition, charging directly with solar energy would avoid both the DC/AC conversion and transmission losses. When installed in parking lots, a significant portion of personal vehicle and city passenger transportation energy demand could be provided through solar PVs, especially during summer time.

1.8 Sustainability, Environmental Impact and Cost Aspects

EVs, conventional HEVs and PHEVs provide significant reduction in emissions compared to conventional vehicles (CVs) with ICEs, while having competitive pricing due to government incentives, increasing oil prices, and high carbon taxes combined with low-carbon electricity generation. The emissions of CVs increase significantly for short distance travels due to the inefficiencies of the current emissions control systems during cold starting of the gasoline vehicles. It is estimated that vehicles travelling fewer than 50 km per day are responsible for more than 60% of daily passenger vehicle kilometers travelled in the United States. Powering this distance with electricity would reduce gasoline use significantly and yield a considerable reduction of emissions. Even when traveling with the use of gasoline in HEVs and PHEVs, the efficiency of the ICE is significantly higher than the ICE of CVs. However, the reduction in fuel and emissions depends primarily on the energy generation mix used to produce the electricity. The balance of the 2006 US electricity mix is composed of coal (49%), nuclear (20%), natural gas (20%), hydroelectric (7%), renewable (3%) and other (1%). Therefore, for the U.S. average GHG intensity of electricity, PHEVs can reduce the GHG emissions by 7–12% compared to HEVs. This reduction is negligible under high-carbon scenarios of electricity production and 30–47% under the low-carbon scenarios. When PHEVs are compared against CVs, the reduction in GHG emissions is about 40% for the average scenarios, 32% for high cases and between 51–63% for low-carbon based scenarios. The detailed life cycle GHG emissions (g CO_2–eq/km) for CVs, HEVs and PHEVs under various scenarios are shown in Figure 1.18. The number after PHEV (PHEV30 or PHEV90) represents the all-electric range of the vehicle in km.

When the emissions for PHEVs are examined, the majority of emissions come from the operational stage. A large portion is due to the gasoline used for traveling, followed by electricity used for traveling based on the carbon-intensity of the electricity generation source. When the emissions from the electric power increase significantly under a high-carbon scenario (coal-based generation capacity), the reduction in volatile organic compounds (VOCs) and CO are offset by a dramatic increase in SO_x and slight increase particulate emissions (PM10). However, the total GHG emissions are still lower compared to CVs since the increase in upstream emissions has a lower magnitude than the decrease in tailpipe emissions (Bradley and Frank, 2009). The GHGs associated with most battery materials and production generates a relatively small portion of the emissions and accounts for 2–5% of the life cycle emission from PHEVs. Moreover, the GHG emissions from the vehicle end-of-life are not shown since they are relatively negligible. The reduced fuel use and GHG emissions for PHEVs depend significantly on vehicle and battery characteristics, as well as the recharging frequency. Using PHEVs

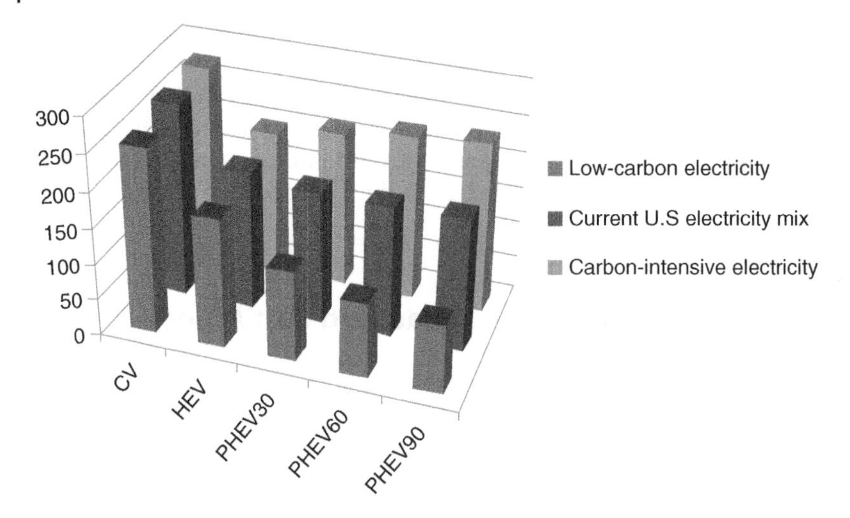

Figure 1.18 Life-cycle GHG emissions sensitivity of CVs, HEVs, PHEV30 and PHEV90 under different carbon intensity scenarios (data from Samaras and Meisterling, 2008).

also has a significant impact on the operating costs of the vehicle. PHEVs in all-electric mode can reduce the gasoline consumption by half, by shifting 45–77% of the miles from gasoline to electricity, which would reduce the operating costs assuming the electricity cost per mile is significantly less than the gasoline cost. Battery life also has a significant role on the cost associated with PHEVs since replacing the battery would increase the life cycle cost of a PHEV by between 33% and 84%. However, the overall cost savings would be based on the overall cost of the vehicle, range and driving behavior, as well as economic incentives such as taxes on carbon emissions and gasoline.

Even though EVs and HEVs compete with conventional vehicles in terms of performance and cost with much less environmental impact, their benefits depend mainly on the battery technology utilized in these vehicles. Although many battery technologies are currently being analyzed for EVs and HEVs, the main focus has been mainly on lead-acid, NiCd, NiMH and Li-ion battery technologies. Thus, in order to understand the effects of EVs and HEVs, further analysis is needed for these battery technologies based on various criteria.

1.9 Vehicle Thermal Management

Thermal issues associated with EV and HEV battery packs and under hood electronics can significantly affect the performance and life cycle of the battery and the associated system. In order to keep the battery operating at the ideal parameter ranges, the discrepancy between the optimum and operating conditions of the batteries need to be reduced significantly by implementing thermal management systems (TMS) in EVs and HEVs. These systems are utilized to improve the battery efficiency, by keeping the battery temperature within desired ranges. Thus, freezing and overheating of the electrochemical systems in the battery can be averted which can prevent any reduction in power capability, charge/discharge capacity and premature aging of the battery. Most electric and hybrid electric vehicle thermal management systems consist of four different cycles to keep the associated components in their ideal temperate range in order to operate safely and efficiently. Even though the components and structure of these loops may vary from vehicle to vehicle, their purposes are usually the same; creating an efficient and robust system that is not adversely affected by internal and ambient temperature variations. Generally, the overall vehicle thermal TMS is composed of the radiator coolant loop, power electronics coolant loop, drive unit coolant loop, and air-conditioning (A/C) and battery loop. A brief description of these loops is provided as follows.

1.9.1 Radiator Circuit

In the radiator loop, the engine is kept cool by the mixture of water and anti-freeze pumped into the engine block to absorb the excess heat and draw it away from the crucial areas. When this superheated engine coolant leaves the engine block, it returns to the radiator. The radiator has a very large surface area through the internal chambers where the excess heat of the coolant is drawn out through the walls of the radiator.

As the vehicle moves, the front of the radiator is also cooled by the ambient air flowing through the car's grill. The loop also includes a surge tank, which acts as a storage reservoir for providing extra coolant during brief drops in pressure, as well as to absorb sudden rises of pressure as shown in Figure 1.19. Next, a coolant pump is used for moving the coolant back and forth to the radiator. When the ICE is off, the coolant heating control module is used to provide heat to the coolant. A portion of the heat in this loop is also transferred to the passenger cabin with help of the heater core.

1.9.2 Power Electronics Circuit

The power electronics coolant loop is mainly dedicated to cooling the battery charger and the power inverter module to ensure the main under-hood electronics do not

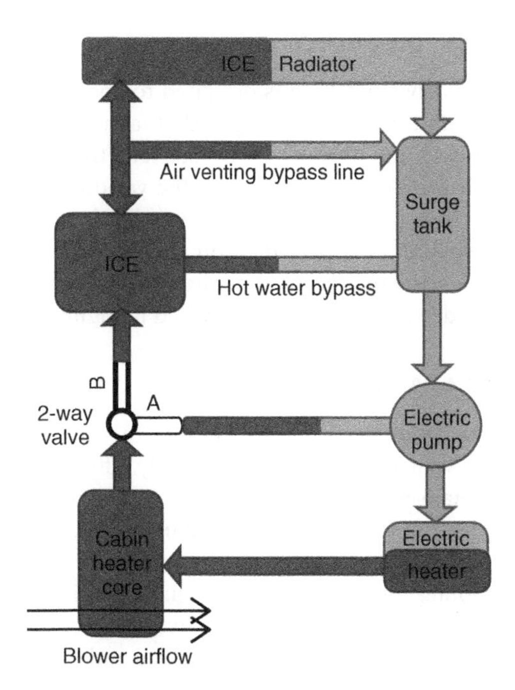

Figure 1.19 Simplified radiator circuit of a HEV (adapted from WopOnTour, 2010).

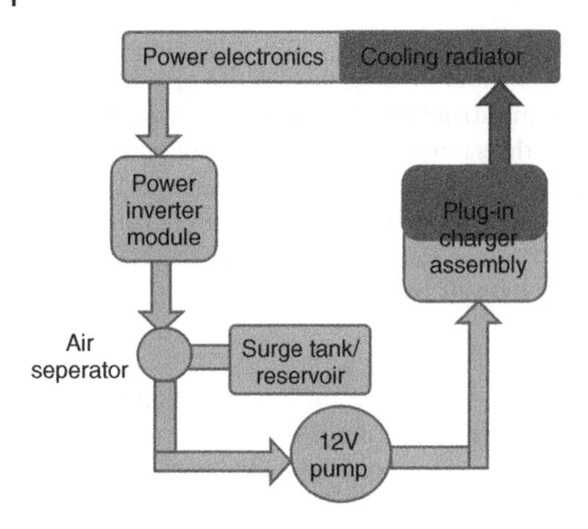

Figure 1.20 Simplified power electronics cooling circuit of a HEV (adapted from WopOnTour, 2010).

overheat during usage. The power inverter module converts direct current (DC) from the high-voltage battery into 3-phase alternating current (AC) motor drive signals for the motor generator units. The module is also responsible for converting AC to DC for charging operations during regenerative braking.

In these operations, a large amount of heat is generated in the system. In order to prevent overheating, the loop incorporates a high flow electric pump to produce and control the coolant flow which passes through the plug-in battery charger assembly, the radiator, and the power inverter module before it flows back to the pump as shown in Figure 1.20. This loop also includes a coolant pump for the circulation of the coolant and an air separator to ensure that the coolant does not have any air bubbles that would affect the cooling performance before traveling through the major electronic parts.

1.9.3 Drive Unit Circuit

The drive unit loop is designed to cool the two motor generator units and electronics within the drive unit transaxle that are used to propel the vehicle using electric power (in addition to generating electricity to maintain high voltage battery state of charge).

It provides lubrication for the various associated parts. Significant heat is generated in these parts due to high power levels during normal operation. The drive unit uses a system of pressurized automatic transmission fluid to cool the electronics in the loop, especially the motor generator units to prevent overheating. The simplified diagram of the drive unit circuit is provided in Figure 1.21.

1.9.4 A/C Circuit

Even though all of the circuits mentioned above have significant roles in enabling the vehicle to operate as robustly, efficiently and safely as possible, in EVs and HEVs, a majority of the focus is given to A/C and battery cooling loops due to its direct effect on the battery performance, which has significant impact on the overall vehicle performance, safety and cost. For this reason, various studies are conducted in this cooling loop to optimize their operating conditions of the associated components, the cabin and the battery. Thus, different cooling systems and configurations will be analyzed based on various criteria and operating conditions.

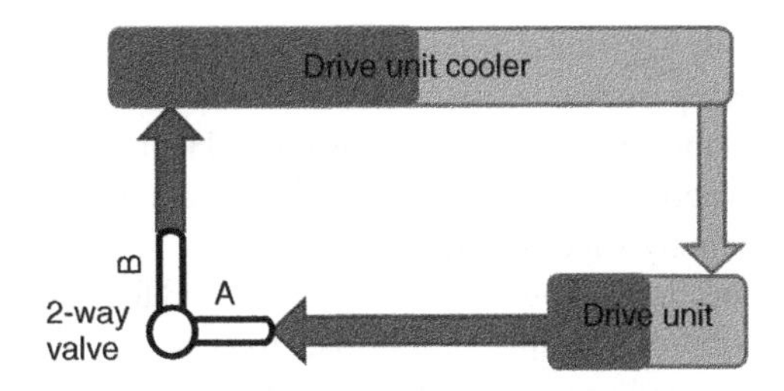

Figure 1.21 Simplified drive unit circuit of a HEV.

The main goal of the A/C cycle is to keep the battery pack at an optimum temperature range, based on the cycle life and performance trade-off, in a wide spectrum of climates and operating conditions as well as keeping even temperature distributions with minimal variations within cells, while keeping the vehicle cabin at desired temperatures. Meanwhile, the system should also consider trade-offs between functionality, mass, volume, cost, maintenance and safety.

Since the main focus will be the A/C and battery loops, they will be called the thermal management systems (TMSs) for the rest of the analysis. They will be categorized based on their objective (providing only cooling vs. cooling and heating), method (passive where only the ambient environment is used vs. active cooling where a built-in source is utilized for heating/cooling), and heat transfer medium (air distributed in series/parallel or liquid via direct/indirect contact).

A passive cabin air cooling system utilizes the conditioned air to cool the battery in warm ambient conditions. It was used on early EV and HEV battery packs (Honda Insight, Toyota Prius and Nissan Leaf) mainly due to cost, mass and space considerations. This is a very effective cooling method for the battery at mild temperatures (10°C to 30°C) without the use of any active components designated for battery cooling. It is highly efficient since it utilizes the heat from the vehicle air conditioning. The ideal battery operating temperature (for Li-ion) is approximately 20° *C* on the low end, which is highly compatible with the cabin temperature. However, air conditioning systems are limited by the cabin comfort levels and noise consideration, as well as dust and other contaminants that might get into the battery, especially when air is taken from outside. Certain precautions should be taken in this system to prevent toxic gases from entering the vehicle cabin at all situations. In independent air cooling, the cool air is drawn from a separate micro air conditioning unit (instead of the vehicle cabin) with the use of the available refrigerant. Even though this may provide more adequate cooling to the battery, the energy consumption as well as cost and space requirements associated with installation of the blower and the micro air conditioning unit increases significantly. The rate of heat transfer between the fluid and the battery module depends on various factors such as the thermal conductivity, viscosity, density and velocity of the fluid. Cooling rates can be increased by optimizing the design of air channels; however it is limited by the packaging efficiency due to larger spacing between the cells. Air can flow through the channel in both serial and parallel fashions, depending on whether the air flow rate splits during the cooling process. In series cooling, the same air is exposed to the modules since the air enters from one end of the pack and leaves from the other. In parallel cooling however, the same air flow rate is split into equal portions where each portion flows over a single module. In general, parallel airflow provides a more uniform temperature distribution than series.

Refrigerant cooling is a compact way of cooling the battery, with more flexibility compared to a fan with ducts, by connecting the battery evaporator parallel to the evaporator in the cooling loop. Heat generated by the battery is transferred to the evaporating refrigerant. This system only requires two additional refrigerant lines, namely suction and pressure lines. The battery evaporator uses some portion of the compressor output that was reserved for the air conditioning, and thus this might cause conflict in some conditions. However, the compressor work needed to cool the battery is usually considerably lower than the air conditioning evaporator need.

Liquid cooling utilizes the previous cooling method with the incorporation of an additional liquid cooling loop specifically for the battery that connects to the refrigerant. This additional cooling loop usually has water or a 50/50 water-glycol mixture and it is kept cool via different procedures depending on the cooling load and ambient conditions. The coolant can be cooled either by ambient air through the battery cooler (if the ambient temperature is low enough) or by transferring the heat to the refrigerant through the chiller. Both methods increase the efficiency of the system since the additional compressor work (that is used in refrigerant cooling) is no longer needed. A simplified diagram of an A/C circuit with liquid battery cooling is shown in Figure 1.22.

In addition, battery cooling can also be done with phase change materials (PCM) integrated cooling systems. PCMs have significant advantages over the aforementioned TMSs, due to their simple design, light weight and compact size, safety and relatively low cost, especially when the integration is considered from the outset and it is improved with the addition of aluminum foam and fins. PCMs are capable of keeping the magnitude and uniformity of the cell temperatures under stressful operating conditions without the need of a complicated system or fan power. Moreover, the heat transfer

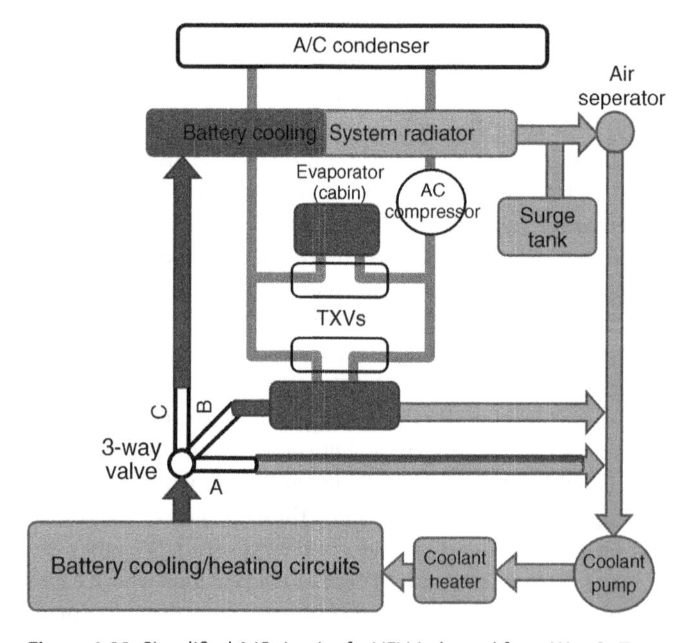

Figure 1.22 Simplified A/C circuit of a HEV (adapted from WopOnTour, 2010).

associated with adding PCMs to a cell can prevent the propagation of thermal runaway, when the cell temperature reaches critical levels. Furthermore, PCMs can be used to have both an active and passive role (complementary/secondary) in thermal management of the battery packs which can reduce the complexity and cost of the system.

1.10 Vehicle Drive Patterns and Cycles

A driving cycle is a time series compilation of vehicle speed (speed versus time curve) established in order to represent typical driving patterns (usually to represent in a specific location and mode of driving) to estimate fuel consumption, emissions and the impact of traffic. The experienced speeds, acceleration, start conditions, gear changes, temperature and loading are some of the important recorded conditions. The first drive cycles were developed in 1950s by Los Angeles County Air Pollution Control District for emissions measurement of typical Los Angeles driving and more realistic drive cycles were achieved in 1969, mostly representing “typical” home to work driving in Los Angeles which formed the basis of Urban Dynamometer Driving Schedule (UDDS) and Federal Test Procedure (FTP). In the later years, similar data collection procedure was established for highways which resulted in Highway Fuel Economy Test (HWFEET). These cycles were used by EPA to publish city and highway driving fuel economy numbers.

The drive patterns and cycles play an important role in electric vehicles since they determine the power and energy requirements and have major implications for the real life battery performance and electricity infrastructure. Information regarding how much of the time the vehicle is at motion and at rest during the day and how long the trips are play an important role on determining when and how often the drivers would need to charge their vehicles. In addition, the drive cycle information together with the equations of motion for the vehicle can provide the power profile requirements for the battery. A drive cycle testing of a vehicle on a dual axle chassis dynamometer and its drive characteristics are shown in Figure 1.23. Drive cycles represent the velocity against time relationships for a given vehicle in a certain type used in a specified manner. There are usually two types of tests, namely highway and urban driving where the initial is characterized by relatively constant velocities over 60 mph and the latter has frequent stops and starts with velocities averaging in 20–30 mph. Generally regional driving

Figure 1.23 Vehicle performance testing using a preloaded drive cycle (courtesy of TUBITAK Marmara Research Center).

cycles have to be developed in order to imitate real-world condition in specific parts of the world to conduct representative analyses. However, this still presents significant challenges to cope with extremes that lie outside of the test capabilities as traditional evaluations generally achieve limited success.

1.11 Case Study

In this section, a case study is provided to show the readers how the aforementioned information is taken into consideration in real life to make decisions on selecting the most appropriate vehicle technologies to be utilized in a country. In this regard, the market penetration of Turkey is selected, since the country has interesting and unique features in terms of the ratio of the country's energy import/exports, its technological capability and governmental regulations. Thus, a Strengths, Weaknesses, Opportunities, Threats (SWOT) analysis of H&EV Turkish market penetration is conducted in the light of the recent domestic and global developments in this area to provide information on factors that played key roles on the decision making process along with their underlying considerations and reasons.

1.11.1 Introduction

During the past decade there has been increasing interest in the deployment of HEVs and EVs. For example, in 2012, around 113,000 EVs were sold in the world, more than twice of the previous year, mainly in United States, Japan and China and 20 million EVs are projected to be on the roads by 2020. With the significant increase in R&D and infrastructure for electric vehicle technologies and reduction in battery costs, the market penetration of these vehicles is expected to become more prominent in the near future. However, in Turkey – the country that has one of the fastest growing economies in the world with the largest increase in the energy demand among the OECD countries – the hybrid and electric vehicle market is still in its initial stages. Therefore, transportation sector is among the highest contributors to this energy demand with respect to the liquid fuels used by conventional vehicles. Since the country has limited reserves of oil, this increase in energy demand is not able to be met through domestic energy production alone and increases the energy imports which possess a threat on the country's economic growth, national security and industrial well-being. Moreover, in the last decade, the effects of rising fossil fuel prices and environmental awareness became more prominent within Turkish industries, research and development organizations, and Turkish society as a whole. Thus, significant work has been done by the government in the past years in terms of policies and legislation to encourage the entrance of H&EVs in the market and various efforts are made to develop these vehicles domestically in the country. However, significant technical knowledge, tremendous investments and abundant infrastructure will be required to achieve these goals.

1.11.2 Research Programs

The Turkish automotive industry's awareness and interest on electric vehicles have been increasing in the past decades, and research projects on electric vehicle technologies and system components are being carried out by several research institutions, programs and platforms.

One of the most important platforms in this area, the National Automotive Technology Platform (OTEP), was formed in 2008 in order to determine a vision for the Turkish automotive industry and identify strategic research areas to be addressed and increase the countries international competitiveness. Moreover, in 2010, "Turkish Automotive Industry Vision and Strategic Research Program for 2023" document was initiated by the associated working groups in the platform. Moreover, an "Electrical Vehicles" group was also established in order to generate new ideas and provide advancements in a wide range of electric vehicle technologies, subcomponents and infrastructure (IA-HEV, 2010–2012). In 2011, with the progress of these groups and the introduction of EVs in the vehicle market, various joint projects and ventures between universities/institutes and private companies have been initiated.

Since 2012, the government launched major research programs by supporting research and development projects at universities, research institutes regarding EVs and subcomponent technologies. Among these, the main focus has been the electric motors and battery technologies in H&EVs along with energy management systems and dynamics and control of these vehicles along with attention on internal combustion engine performance and emission control in HEVs. The duration for the selected projects were determined to be 2–3 years, with an approximate budget of $6.5–7.5M. Moreover, a support programme for industry has been established for the development of electric motors/generator and driver systems (up to $10.5M with 5 projects), energy management, control system, hardware and algorithm (up to $5.9M with 3 projects), vehicle electronics and electromechanical system components (up to $32.5M with 8 projects) and innovative vehicle components and systems (up to $43.6M with 16 projects) for H&EVs (IA-HEV, 2010–2012).

In parallel with the support programs described above, various government incentives such as purchasing guaranty, tax benefits, infrastructure development (such as increasing the number of charging stations and improving their access) and new legislation and implementation programs and are also started to be employed.

1.11.3 Government Incentives

1.11.3.1 Tax Benefits

Turkey implements two different taxation measures for vehicles in the market. The first is a tax on an initial new vehicle purchase (special consumption tax), whereas the second one (motor vehicle tax) is an annually paid based on the engine cylinder volume and the age of the vehicle. The special consumption tax (SCT) for conventional vehicles is increased in the beginning of 2014 with respect to the previous years. Depending on the engine volume, the tax has increased from 40% to 45% for under 1600cc, from 80% to 90% for between 1600cc and 2000cc and from 130% to 145% for over 2000cc as shown in Table 1.6. With the new regulations, the current prices of conventional vehicles are expected to increase approximately by 10%. Since the SCTs on EVs have not changed (with maximum of 15% on passenger vehicles), EVs gathered a wider economic advantage based on the new tax system compared to conventional vehicles when a new vehicle is purchased, especially since electric vehicles are also exempt from the motor vehicle tax (MTV).

Table 1.6 Special consumption tax classification categories for new vehicle sales[a].

	Conventional		Electric Only	
Vehicle Type	**Engine Cylinder Volume (cc)**	**Special Consumption Tax (%)**	**Electric Motor Power (kW)**	**Special Consumption Tax (%)**
Passenger Vehicle	<1,600	45	<85	3
	1,600–2,000	90	85–120	7
	>2,000	145	>120	15
Motorbike	<250	8	<20	3
	>250	37	>20	37

a) The vehicle sales tax reduction includes only battery electric vehicles and battery electric motorbikes and excludes HEVs and plug-in electric vehicles (PHEVs).
Source: ODD (2013).

1.11.3.2 EV Supply Equipment and Charging Infrastructure

In addition, various installation efforts are being made in Turkey to install EV supply equipment across Turkey (especially in Istanbul), however they are a very small in quantities and are mostly done by a few private companies. Aside from these, In January 2013, Sabanci University became the first university in Turkey with a charging station. In June, 2013 the first domestically developed charging station producer Gersan Electric Incorporated Company started building charging stations in pilot areas of Istanbul with a target number of 60 to 65 units. In September 2013, legal ground was established in fuel station areas to build electric charging, CNG, LPG and hydrogen filling stations. In November 2013, Izmir metropolitan municipality installed charging stations to a number of parking stations where electric vehicle owners can charge their vehicles for free in order to increase the number of electric vehicles in the city. It is estimated that over 100 charging stations currently exist in Turkey.

In addition to the aforementioned incentives, implementation programs and new legislation are also are announced. A number of electric vehicles per year for a 5-year period are stated to be purchased by the national ministry. More incentives are projected to be announced for other public institutions to purchase EVs. Based on the targets set by the Turkish government, the electricity grid infrastructure will be strengthened, and electricity tariff deregulation will be completed in the following years. Moreover, access to charging stations near residences, car parks, and shopping centers will be increased and awareness projects will be executed concerning EV technology and EV usage. Furthermore, legislation regarding the recycling of EV batteries will be revised and the capacity of test centers will be improved in the future. Furthermore, several legal and policy instruments were also established to encourage the use of H&EVs in Turkey including a strategy document regarding energy efficiency in transportation by the Ministry of Energy and Natural Sources and automotive industry strategy document and action plan by the Ministry of Science, Industry and Technology.

1.11.3.3 EV Developments in the Turkish Market

The funding support in electric vehicle R&D projects, reduction in their consumption and vehicle taxes along with increased efforts on building associated infrastructure encouraged many private companies to start conducting research on electric battery and motor, emission reduction methods and vehicle system integration domestically.

Among these, in 2009, the Turkish bus manufacturing company TEMSA introduced the Avenue Hybrid, which had a series hybrid powertrain that enabled 25% fuel reduction and lower CO_2 emissions compared to its standard conventional buses. The vehicle was also quitter due to not having a gearbox and claimed to have a better riding experience due to lack of vibrations that exists in the conventional version. The same year Otokar announced a concept hybrid urban bus Doruk 160LE Hibra with electric battery and diesel engine which claimed to have a 20% reduction in fuel consumption.

In November 2009, TOFAŞ started developing the all-electric version of the vehicle, Doblo EV which became "the first electric vehicle designed and developed for mass production in Turkey" and introduced its prototype in 2010. The vehicle has 105 kW maximum power output and can travel 150 km on a single charge and has regenerative breaking. The vehicle can be charged approximately in 7 hours, but this time can be reduced to 1 hour with fast charging.

In addition, Fluence Z.E. production has begun solely at the Oyak Renault Bursa Plant in 2010. The vehicle uses 22 kWh Li-ion battery, 70 kW electric motor that provides an all-electric range of 185 km and can speed up to 135 km/h. It can be charged in 10–12 hours using a household outlet, but is also compatible with fast charging stations for much quicker charging times (Renault, 2014). The batteries are also designed for "quick drop" technology which can be switched in a battery exchange facility. However, the production was stopped in 2013 due to the low number of sales in Turkey and Europe.

At the beginning of 2013, Derindere Motor Vehicles (DMA) launched its pure electric vehicle "DMA All Electric", the first Turkish vehicle with type approval certificate for electric vehicles in Turkey, and made the first test drives. The vehicle will be able to be charged in 8 hours on 220V, would be compatible with European Standard type 2 charging stations and travel approximately 280 km in one charge. It will come both with buying (US $53,500) and renting (US $1200 per month) options for operational fleets. The company is targeting to produce 100 vehicles per month. The motor and the battery for the vehicles are currently being imported; however the ECU is developed in Turkey. The vehicle has a 40 kWh Li-ion battery and 62 kW electric motor (with 225–325 Nm torque) and will have 3 year 100,000 km warranty (DMAOTO, 2013).

In addition, after a year of research, In mid-2013 Malkoçlar Automotive has developed a pure electric vehicle with all the R&D and manufacturing done in Turkey (with the exception of the electric motor) that can travel 100 km with an energy cost of under a dollar. The vehicle has 2 versions, one with a 2-seater and another with 4 seats and is mainly designed for inner city traveling and commercial use. It is mostly made out of plastic and had aluminum construction space frame for collusion safety. It weighs approximately 800 kg and can travel up to 130 km/hr with maximum range of 150 km. Charging from the regular outlet will take approximately 6–7 hours with a fast charging option of 45 minutes. The vehicle is currently under testing and waiting for type approval certificate with a plan to be sold for US $13,500 in 2014 after the all tests are successfully completed (Elektriklioto, 2013).

Finally, initiated by the Ministry of Science, Industry and Technology, TUBITAK MRC has taken imperative steps in early 2015, towards establishing a Turkish National Car

Brand and the Industrialization of its first products, in line with the the Supreme Council for Science and Technology agenda and 2023 National Technology Targets. In this regard, TUBITAK MRC has been working on vehicle design, engineering, testing and certification as well as widening its corresponding workforce and infrastructure to introduce the first vehicles to the Turkish market before 2020.

1.11.3.4 HEVs on the Road

The number of vehicles on the Turkish roads is increasing rapidly. Even though the total fleet of vehicles on the road reached up to 18 million at the end of 2013, only a negligible percentage of them are currently electric vehicles. The number of vehicles between 2005–2013 can be found in Figure 1.24.

Meanwhile, the passenger car sales have also increase significantly in the past years reaching to 664,655 in the year 2013 (19.48% increase compared to 2012) as provided in Table 1.7. The light-commercial market on the other hand, shrank and therefore the combined total passenger car and light-commercial market had a 9.72% increase from 777,761 units in 2012 to 853,378 units in 2013. When the passenger car market is examined according to the engine volumes, the passenger cars under 1600 cc received the highest share of sales every year due to the lower tax rates (compared to larger engine sizes). In 2013, only 31 EV passenger cars were sold in Turkey compared to 184 the year before.

When the passenger car market is examined according to average emission values, even though the total emissions for the passenger cars increased in 2013 due to high number of sales, cars that have emission values under 140 gCO_2/km limit has accounted for more than 75% of the vehicle sales (Table 1.8). This is primarily a result of the lower tax values for the engine volumes $\leq$1,600 cc, which also helps in bringing down the increase in total fleet emissions average of the vehicles in Turkey.

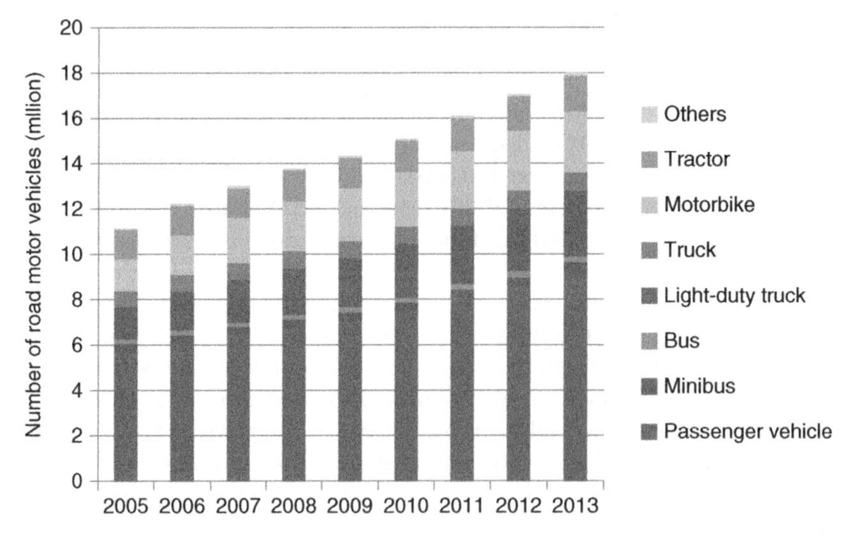

Figure 1.24 Total vehicle fleet according to the vehicle types between 2005 and 2013 (data from road motor vehicle statistics report, 2013).

Table 1.7 Passenger car market according to the engine/electric motor size for 2009–2013.

Engine Size	Engine Type	2009	2010	2011	2012	2013	SCT Tax Rates[a]	VAT Tax Rates
≤1600 cc	Gas/diesel	304,755	412,162	530,069	514,861	625,621	45%	18%
1601 cc to ≤2000 cc	Gas/diesel	56,766	87,246	52,396	35,850	33,035	90%	18%
≥2001 cc	Gas/diesel	8268	10,376	11,054	5385	5968	145%	18%
≤85 kW	Electric	0	0	0	184	31	3%	18%
86 kW to ≤120 kW	Electric	0	0	0	0	0	7%	18%
≥121 kW	Electric	0	0	0	0	0	15%	18%
Total		**369,819**	**509,784**	**530,069**	**556,280**	**664,655**		

a) 2014 SCT tax rates.
Source: ODD Press Summary (2013).

Table 1.8 Passenger car market according to average emission values for 2001–2013.

Average Emission Values of CO_2 (g/km)	2011 Cumulative		2012 Cumulative		2013 Cumulative		2013/2012
	Units	%	Units	%	Units	%	%
<100 g/km	3820	0.60	18,635	3.30	56,570	8.51	203.57
≥100 to <120 g/km	172,652	29.10	173,218	31.10	238,816	35.93	36.19
≥120 to <140 g/km	223,020	37.60	202,118	36.30	216,016	32.50	7.83
≥140 to <160 g/km	109,013	18.40	118,107	21.20	116,245	17.49	1.29
≥160 g/km	85,014	14.30	44,202	7.90	37,008	5.57	16.28
Total	**593,519**	**100.00**	**556,280**	**100.00**	**664,655**	**100.00**	**19.48**

Source: ODD Press Summary (2013).

1.11.3.5 Turkey's Standing in the World

When the countries that have high share of H&EVs are examined, it can be seen from Figures 1.25 and 1.26 that out of the 18 countries with the highest electrified market share in 2013, 12 of them have the highest gasoline prices in the world. When the remaining 6 countries are analyzed, 4 out of the 6 (Japan, United States and Spain and Estonia) have among the highest incentives for H&EVs. Among those, Japan pays one-half of the price gap between EV and corresponding ICE vehicles up to 1 million yen (around 10,000 USD). Spain provides incentives up to 25% of vehicle purchase price before taxes, up to 6,000 Euros (around 8,200 USD) along with possible additional incentives up to EUR 2000 Euro (around 2,500 USD) per EV/PHEV. The United States has incentives with up to 7,500 USD tax credit for vehicles along with additional incentives depending on the state. Finally, Estonia has grants for purchasing electric cars that are 50% of the purchase price (as a part of the ELMO program) up to 18,000 Euros (25,000 USD). Among the remaining 2 countries, even though Austria does not have significant incentives for electric cars, the country has high taxes for purchase of a vehicle along with taxes on fuel consumption and CO_2 emissions which the electric vehicles are exempt from (up

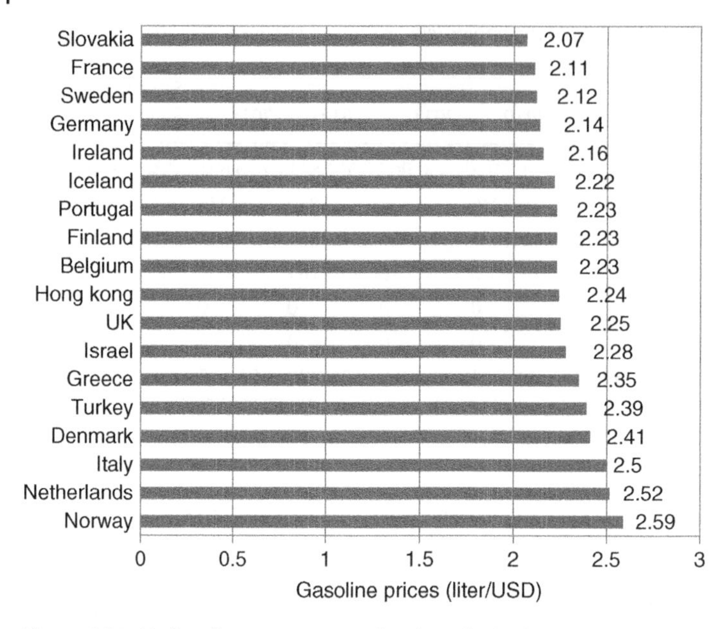

Figure 1.25 Market share percentages for electrified vehicle compared to all vehicles in 2013 (data from ABB, 2014).

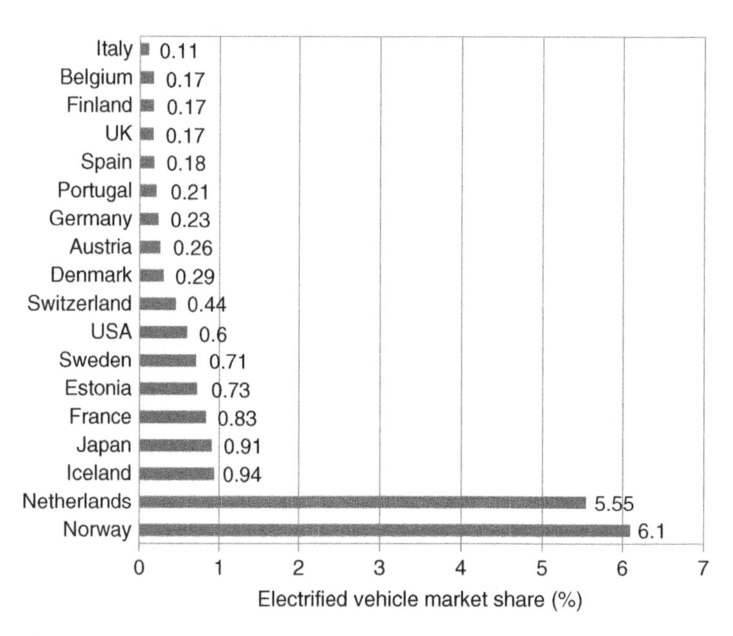

Figure 1.26 Gasoline prices as of 07, july 2014 (data from GlobalPetrolPrices, 2014).

to 16%, can be over 1,000 USD per month). Finally, Switzerland also has reduction or even exemption from vehicle taxes as well as taxes associated with CO_2 emissions.

Thus, it can be seen that, gasoline price and tax incentives play a major role in the penetration of H&EVs in the market. Turkey, however, has a unique situation in this regard

Table 1.9 SWOT analysis for domestic and global market penetration of H&EVs.

Strengths

Global

- The reduction in the fossil fuel consumption of the transportation sector, being one of the largest sectors that depend on the limited, globally unevenly distributed and increasingly priced fossil fuels[a)]
- The reduction of foreign oil dependency of many countries that do not possess fossil fuel resources and mitigation of corresponding economic and political implications
- Various advantages such as the diversification and more efficient use of energy sources, load equalization of power, quite operation and lower operating costs
- High sustainability and low environmental impact with respect to conventional vehicles

Domestic

- The success of Turkish market in the light commercial vehicles[b)]
- The existence of large capable motor and battery manufacturers and sub-industries that are currently being utilized solely for the conventional vehicles[c)]
- The advantage of the geographical location and the potential of being an international hub.

Weaknesses

Global

- The inability of EV and HEV costs to compete with conventional vehicles without the incentives provided by majority of the governments[n)]
- Long charging times and the need for charging infrastructure[o)]
- The limited all electric range[p)]
- The lack of proper standards for the H&EV technology.

Domestic

- The lack of domestic R&D and manufacturing capabilities in value added aspects of technology.
- The incompatibility of apartment layouts that mainly prevents home charging
- The lack of serious investments for the charging infrastructures in the country[q)]
- Fossil fuels being the main contributor of the electricity generation mix of the country[r)]
- The lack of social knowledge and public relations on the EV and HEV technologies

Opportunities

Global

- The rapid advancement in the battery technology which increases the vehicle range and reduces the associated cost and emissions of the vehicle[d)]
- The relatively seamless integration of electric vehicle technology with renewable energy sources[e)]
- The efficient, low cost and safe use of the electric grid with the introduction of EVs[f)]

Domestic

- The large tax incentives for electric vehicles by the Turkish government[g)]
- Turkey having considerably high fossil fuel prices[h)]
- Turkey having high energy imports and correspondingly high current deficits[i)]
- New electric vehicle related funding support for R&D and investment subsidies by the government[j)]
- The recent established targets of emissions and energy efficiency of the country[k)]
- The current plan on establishing a carbon tax for all road vehicles[l)]
- The pre-existing research on the regulations and infrastructure for EVs in Europe and the United States
- The integrability of EVs with the county's abundant renewable energy sources[m)]

Threats

Global

- The possibility of rapid improvements on alternative vehicles technologies, especially on hydrogen fuel cells[s)]
- The possibility of developing cheap, efficient and environmentally benign conventional petroleum alternatives[t)]
- The possible alterations of people's positive perspectives on EVs due to witnessing unsuccessful attempts with the technology[u)]

(*Continued*)

Table 1.9 (Continued)

Domestic

- The years of experience acquired by U.S., Europe and Japan on the EV technology[v)]
- Large R&D budget allocations on (H)EVs, their subcomponents and infrastructure in the United States, China and various European countries[w)]
  The possible undercutting of domestic sales in the Turkish Market by EVs and HEVs designed and manufactured by foreign countries

a) Over 71% of the U.S. crude oil consumption is used in the transportation sector, a sector where 93% of the energy supply comes from liquid fossil fuels in the United States.
b) Turkey had the sixth place for automotive sales among the OECD countries in 2013.
c) In terms of motor, Arçelik, Femsan, Tepaş and Gems; In terms of electric battery, İnci Akü, Mutlu Akü and Yiğit Akü have valuable infrastructure for EV and HEV components.
d) Major breakthroughs have been made on the LCO, NMC and NCA battery technologies and a wide range of research in currently being conducted on the Zinc-air, Lithium-sulfur and Lithium-air batteries.
e) Batteries are compatible with the intermittent characteristics various renewable energy resources (especially solar and wind) and can reduce the DC/AC conversions and transmission losses when charged by them.
f) Since electric grids work under capacity for a large portion of the day, the demand and supply in the grid can be optimized in the near future with the integration of smart grid and vehicle to grid (V2G) technologies.
g) Taxes on electric vehicles range between 3% to 15% in Turkey.
h) Turkey has the fifth highest price of gasoline in the world with $2.39 per liter.
i) By the end of 2013, Turkey is ranked the second among the 57 countries in terms of having the economy with highest ratio of current deficit and over 50% of the trade deficit is caused by oil and oil derived products. Turkey imports over 90% of the country's energy consumption and exports 8 billion USD worth of energy with respect to 60 billion USD worth of imports.
j) TUBITAK provided support for over 30 research programs related to EV technologies, its subcomponents and infrastructure in the next 3–4 years with 65–70M Euros.
k) Turkey is targeting 130 g/km CO_2 emissions established by the OECD countries for the year 2015.
l) New regulations are expected to be effective sometime in 2014 to increase the special consumption and motorized vehicle taxes based on the vehicle emissions to reduce the climate change and air pollution.
m) Turkey has 88 Btoe potential in solar energy potential and 166 TW technical wind energy potential, which is significantly higher than European renewable energy averages.
n) Some of the major electric vehicle incentives include Finland with €5M, France with €450M and Italy with €1.5M incentive pools. Moreover, tax cuts are being applied to 12% of vehicle cost in Holland, 20% in India, 60,000 RMB in China, €6,000+ in Spain, €4,500 in Sweden and $7,500 in the United States. Furthermore, Japan provides government tax cuts to the half of the cost difference between electric and comparable ICE vehicle and Denmark and Germany provides exemption from road taxes for electric vehicles.
o) Electric vehicles can be charged in an hour to travel on average 3–8 km in Level 1 (120V AC), 15–30 km in Level 2 (240 or 208 V AC). In Level 3 (DC fast charging), the vehicle can be charged in 20 minutes to travel approximately 100–130 km.
p) Electric vehicles can have an all-electric range in between 60 km (Scion iQ EV) and 425 km (Tesla Model S - 85 kWh).
q) In 2014, it is estimated to have between 100–200 charging stations in Turkey
r) Over 70% of Turkey's electricity is produced from fossil fuels.
s) Toyota has stated to have a hydrogen vehicle in 2015 that can travel 480 kms in one tank and can be fueled in under 3 minutes.
t) Important developments have been made in the researched alternative fuels such as are biofuels, propane, compressed and liquefied natural gas (CNG and LNG), ammonia and compressed air (CAV).
u) In June 2, 2011, a Chevrolet Volt test vehicle caught fire in the test parking lot after crash testing. In late 2013, 3 Tesla Model S vehicles caught fire within a 6-week time frame due to battery penetration with metal objects (2) and fast collusion (1). Both incidents largely occupied the headlines in the press.
v) As of September 2012, 2.2 million electric vehicles were sold in the United States, 1.5 million in Japan and 0.45 million in Europe and a total of 4.5 million word wide.
w) $2.5B has been spent on R&D on electric vehicles in 2012, its subcomponents and infrastructure by countries who are leading this technology.

due to the discrepancy between gasoline price, taxes/incentives and the electric vehicle sales in the country, especially when compared to countries in similar conditions. Even though it is a country with one of the highest fossil fuel prices and conventional vehicle taxes in the world, the number of electric vehicles sold is negligible (with a few hundred) in the market, especially when it is compared to similar countries in this regard. In Norway, for example (one of the most similar country in terms of the taxes and incentives), is placed on the other end of the spectrum where electric vehicles play a significant role in the countries market with the largest fleet of plug-in electric vehicles per capita in the world (over 3% of vehicles on the road are plug-in electric vehicles). Thus, the reasons behind the current situation of Turkish market should be analyzed thoroughly in order to understand the conditions that make the H&EV sales the way they are in the country.

1.11.3.6 SWOT Analysis

In order to sum up to issues and provide a better understanding of the current state of electric vehicles, and potential strategies to increase their market penetration, a SWOT analysis, (a strategic planning tool used to evaluate the strengths, weaknesses, opportunities and threats) is performed. Within SWOT analysis, the strengths and weaknesses are seen as internal factors which are controllable, and can be acted upon. The opportunities and threats are external, uncontrollable factors. These form the external environment within which the organization operates and may include various key parameters. Conditions associated with both Turkey and the rest of the world are analyzed since the penetration of the Turkish market is highly correlated to the developments that are both domestic and global. The summarized results of the SWOT analysis are provided in Table 1.9. More information and associated data regarding some of the elements are provided in the footnotes.

1.12 Concluding Remarks

It becomes more obvious each passing day that the concerns over the dependence and ever-increasing prices of limited oil, the associated energy security problems, as well as environmental pollution and global warming with respect to transportation sector are becoming more pressing. Thus, there is a general agreement from the policy makers and the scientific and technological community that electric vehicles are currently one of the best alternatives in this regard due to the certain advantageous characteristics, such as diversification of energy resources, load equalization of power, improved sustainability, as well as lower operating costs and considerably lower emissions during operation, of this technology.

Currently, the transportation sector is undergoing a transformation from conventional internal combustion engine vehicles to electric and (plug-in) hybrid electric vehicles, and even hydrogen fuel cell vehicles to a certain extend. Moreover, electric vehicle technologies are starting to compete with conventional vehicles since they can attain superior efficiency, endurance, durability, acceleration capacity and simplicity and are more environmentally benign. However, they still have low energy storage capability, high price and long charging time which are the main barrier to penetrate into the vehicle market.

In the next decade, it is expected that the specific energies of the batteries will be significantly increased since only a small portion of the theoretical limit is currently being used. It is expected that most of these improvements will be achieved with modifications in the chemistry and reduction in the battery dead weight and volume. These improvements will also have a significant impact on reducing the associated cost of the (hybrid) electric vehicles. In addition, in areas where the electric batteries fail, flywheels (mostly in niche applications) and supercapacitors will be the utilized. Moreover, with the introduction of semi-dynamic inductive charging and V2X/smart grid applications, electric vehicles would be able to better intertwine with the energy structure and improve its utilization. As the oil availability peak across the globe, renewables will play a larger role in electricity generation, which will increase the impact of electric vehicles even further.

Nomenclature

Acronyms

BEV Battery electric vehicle
DOD Depth of discharge
EV Electric vehicle
FC Fuel cell
FCV Fuel cell vehicle
GHG Global greenhouse gas
HEV Hybrid electric vehicle
ICE Internal combustion engine
ICEV Internal combustion engine vehicle
IPT Inductive power transfer
PHEV Plug-in hybrid electric vehicle
SOC State of charge
TMS Thermal management system
UC Ultracapacitors
V2G Vehicle-to-grid

Study Questions/Problems

1.1 What would be some tangible strategies to mitigate the chicken-and-egg problem of developing electric vehicles and associated infrastructure?

1.2 What are the benefits and difficulties of integrating EVs and HEVs with renewable energy sources?

1.3 Are EVs and HEVs always better for the environment compared to conventional ICE vehicles? What are the key variables in this consideration?

1.4 Consider your daily driving pattern. Which HEV architecture would be most compatible with it and why?

1.5 How balanced/efficient is the grid power use in your city/country and would it make sense to use (H)EVs in the near future in terms of cost and energy savings there?

1.6 What are the main competitive advantages and barriers for market penetration of EVs and HEVs in your city/country?

1.7 Which vehicle hybridization rates are most commonly available in the industry and what would be the underlying reasons?

1.8 Which energy storage system would be dominating the market in the next decades? Explain your reasons.

References

ABB. (2014). Electric vehicle market share in 19 countries. Available at: http://www.abb-conversations.com/2014/03 [Accessed August 2016].

Andújar JM, Segura F. (2009). Fuel cells: History and updating. *A walk along two centuries, Renewable and Sustainable Energy Reviews* **9**:2309–2322.

Asif M, Muneer, T. (2007). Energy supply, its demand and security issues for developed and emerging economies. *Renewable & Sustainable Energy Reviews* **11**:1388–1413.

Bradley TH, Frank AA. (2009). Design, demonstrations and sustainability impact assessments for plug-in hybrid electric vehicles. *Renewable and Sustainable Energy Reviews* **13**(1):115–128.

Chan C C, Bouscayrol A, Chen K. (2010). Electric, Hybrid, and Fuel-Cell Vehicles: Architectures and Modeling. *IEEE Transactions on Vehicular Technology* **59**:589–598.

Center for Advanced Automotive Technology. (2015). Hybrid and Battery Electric Vehicles, *HEV Levels.* Available at: http://autocaat.org [Accessed August 2016].

de Santiago, J, Bernhoff H, Ekergård B, Eriksson S, Ferhatovic S, Waters R, Leijon M: . (2012). Electrical Motor Drivelines in Commercial All-Electric Vehicles: A Reviev. *Vehicular Technology IEEE Transactions 2012,* **61**(2): 475–484.

Dincer I: . (2010). Renewable energy and sustainable development: a crucial review. *Renewable and Sustainable Energy Reviews* **2010**, 4: 157–175.

DMAOTO. (2013). Technical Specification Document. *Changes Changing the World.* Available at: http://www.dmaoto.com [Accessed August 2016].

Dijk M, Renato J, Orsato RK. (2013). The emergence of an electric mobility trajectory. *Energy Policy* **52**:135–145.

Eberle U, Helmolt R. (2010). Sustainable transportation based on electric vehicle concepts: a brief overview. *Energy & Environmental Science* **3**:689–699.

Electrification Roadmap, Revolutionizing Transportation and Achieving Energy Security. (2009). *Electrification Coalition.* Available at http://www.electrificationcoalition.org/ [Accessed June 2015].

Elektriklioto. (2013). Turkey's first national electric car to start its engine soon. Available at: www.elektriklioto.com/2013/01 In Turkish [Accessed August 2016].

GlobalPetrolPrices, Gasoline Prices as of 07 July 2014. (2014). Available at: http://www.globalpetrolprices.com [Accessed July 2015].

Guerrero CPA, Li J, Biller S, Ziao G. (2010). Hybrid/Electric Vehicle Battery Manufacturing: State-of-the-Art. *Proceedings of the 6th annual IEEE Conference on Automation Science and Engineering*. Canada.

Hamut HS, Dincer I, Naterer GF. (2013). Exergy Analysis and Environmental Impact Assessment of Using Various Refrigerants for Hybrid Electric Vehicle Thermal Management Systems, Chapter 46. In: *Causes Impacts and Solutions to Global Warming*, I Dincer, CO Colpan, F Kadioglu (eds.). Springer-Verlag New York.

Hoogma R (2000). *Exploiting Technological Niches.* Twente University, Enschede (PhD Thesis).

IA-HEV Annual Reports. (2013). Hybrid and Electric Vehicle Technologies 2010–2013. Available at: http://www.iahev.org [Accessed August 2016].

Karden E, Ploumen S, Fricke B, Miller T, Snyder K (2007). Energy Storage Devices for Future Hybrid Electric Vehicles. *Journal of Power Sources* **168**:2–11.

Kristoffersen TK, Capion K, Meibom P. (2011). Optimal charging of electric drive vehicles in a market environment. *Applied Energy* **88**:1940–1948.

ODD Annual Reports. (2013). Automotive Distributors' Association Press Summary. Available at: http://www.odd.org.tr/web_2837_1/neuralnetwork.aspx?type=35 [Accessed August 2014].

Offer GJ, Howey D, Contestabile M, Clague R, Brandon NP (2010). Comparative analysis of battery electric, hydrogen fuel cell and hybrid vehicles in a future sustainable road transport system. *Energy Policy* **38**(1):24–29.

Pesaran AA, Kim G, Keyser M (2009). Integration Issues of Cells into Battery Packs for Plug-in and Hybrid Electric Vehicles. *EVS 24 Stavanger*, Norway.

Peterson SB, Michalek JJ. (2013). Cost-effectiveness of plug-in hybrid electric vehicle battery capacity and charging infrastructure investment for reducing US gasoline consumption. *Energy Policy* **52**:429–438.

Renault. (2014). ZOE Price and Specifications. Available at: http://www.renault.co.uk/ [Accessed August 2016].

RenaultRoad Motor Vehicle Statistics Report. (2013). *General Directorate of Public Security*. Turkish Statistical Institute.

Samaras C, Meisterling K. (2008). Life Cycle Assessment of Greenhouse Gas Emissions from Plug-in Hybrid Vehicles. *Implications for Policy Environmental Science and Technology* **42**:3170–3176.

Shafiee S, Topal E (2008). An econometrics view of worldwide fossil fuel consumption and the role of US, *Energy Policy* **36**:775–786.

Tie SF, Tan CW (2013). A review of energy sources and energy management system in electric vehicles, *Renewable and Sustainable Energy Reviews* **20**:82–102.

Vazquez S, Lukic SM, Galvan E, Franquelo LG, Carrasco JM (2010). Energy storage systems for transport and grid applications, *IEEE Transactions on Industrial Electronics* **57**(12):3881–3895.

Whittingham MS. (2012). History, Evolution, and Future Status of Energy Storage. *Proceedings of the IEEE Special Centennial Issue* **100**:1518–1534.

WopOnTour. (2009). The Chevrolet Volt Cooling/Heating Systems Explained. Available at: http://gm-volt.com/2010/12/09 [Accessed August 2016].

Yilmaz M, Krein PT. (2013). Review of Battery Charger Topologies, Charging Power Levels, and Infrastructure for Plug-In Electric and Hybrid Vehicles. *Power Electronics* **28**:2151–2169.

Young K, Caisheng W, Li Y, Strunz K. (2013). Chapter 2: Electric Vehicle Battery Technologies. In: *Electric Vehicle Integration into Modern Power Networks*, R Garcia-Valle, JAP Lopes (eds). Springer, New York.

2

Electric Vehicle Battery Technologies

2.1 Introduction

Batteries are the most commonly utilized electrical energy storage devices in EVs; and the performance, cost, safety and reliability of the vehicle is closely tied to the characteristics and usage of its battery. It is estimated that approximately one-third of the EV's cost lies on the battery pack, and various studies concluded that over 45% of the EVs cost is the battery related costs (Petersen, 2011). The performance of the battery itself depends mainly on the battery chemistry, thus a considerable amount of research is being conducted to improve the current battery chemistries and to develop chemistries of the future that allows for higher performance, longer range, more reliable and safer batteries. However, these chemicals degrade over time which impacts the overall performance of the vehicle and therefore this battery degradation process should be mitigated by conditioning the battery and controlling the charge/discharge profiles under different loads and environmental conditions. For this reason, battery management systems that constantly obtain data from the battery, estimate the state, equalize the charge and thermally manage the battery cells are needed to enhance the safety, cycle-life and performance of the battery while reducing the associated cost. Thus, in this chapter, state of the art technologies associated with the current and future battery chemistries, battery management systems and associated applications are described.

2.2 Current Battery Technologies

Automotive sector continuously strives to achieve lower fuel consumption, emissions and manufacturing costs while improving reliability, safety and comfort. If EVs would want to compete against conventional vehicles, they would need to fulfill and even exceed the aforementioned criteria. Since the batteries are at the heart of these vehicles and provide the necessary energy demands under various configurations, conditions and drive cycles; selection of the appropriate battery technology for the right application is crucial in electric vehicles. Even though no battery technology would meet all needs of the vehicle, trade-offs need to be made to optimize the battery utilization.

Battery technologies are evaluated based on their capacity to provide sufficient energy and power for acceleration under various operating and ambient conditions while being compact, long lasting, low cost and environmentally friendly. The battery technologies that are currently available on the market as well as the ones that are under development

Thermal Management of Electric Vehicle Battery Systems, First Edition.
İbrahim Dinçer, Halil S. Hamut and Nader Javani.

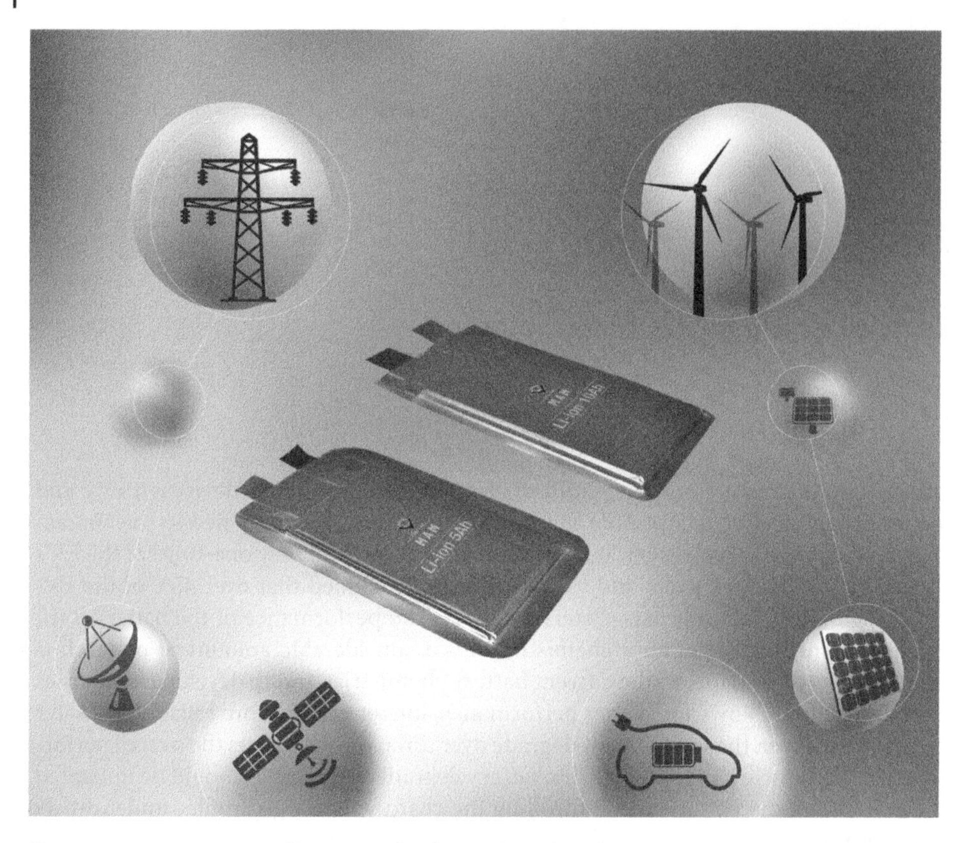

Figure 2.1 Representation of battery technology to be utilized in various applications (courtesy of TUBITAK Marmara Research Center).

to be potentially used in the near future are provided in the following sections. A representation of using battery technology for various applications is provided in Figure 2.1.

As mentioned briefly in the previous chapter, batteries are made up of electrochemical cells with negative and positive electrodes along with an electronically insulating (but ionically conductive) separators and electrolytes that are present both throughout the electrodes and in the separator. Each electrode is composed of an active material as well as various additives (to enhance the electronic conductivity) and a polymer binder. Today, only a limited number of battery technologies are commonly used for electric and hybrid electric vehicles, such as lead-acid (used in original versions of EV1 and RAVEV) nickel cadmium (used in Peugeot 106, Citroen AX, Renault Clio and Ford Think), nickel metal hydride (used in Toyota Prius and Highlander, Ford Escape, Honda Insight and Saturn Vue) and Li-ion (used in Tesla Roadstar, Chevrolet Volt, BMW i3, i8 and majority of the vehicles developed after 2013). Thus, comparisons among these batteries are provided with respect to various criteria as listed in Table 2.1. More detailed information regarding the specifications, performance characteristics and application areas for these battery chemistries are provided in the next sections.

Table 2.1 Battery characteristics for today's most common battery chemistries

Cell Chem.	Specific Energy (Wh/kg)	Specific Power (W/kg)	Open-Circuit Voltage (V)	Operating Temp. Range (°C)	Cycle Life[a)]	Cost[b)] ($/kWh)	Environmental Impact[c)] (mPts)
Pb-acid	35 – 50	80 – 300	2.1	−30 – 60	500 – 1000+	100 – 150	503
NiCd	50 – 80	200 – 500	1.3	−20 – 50	800 – 2000	200 – 300	544
NiMH	75 – 120	200 – 2000	1.25-1.35	−20 – 50	1500+	300+	491
Li-ion	100 – 200+	800 – 2000+	2.5	−20 – 55	1000 – 3000+	300 – 600	278

a) **Defined as battery capacity falling below 80% of its initial rated capacity.**
b) **Production cost are highly dependent on the volumes**
c) **Based on eco-indicator 99.**

2.2.1 Lead Acid Batteries

Lead acid (Pb-acid) battery is the oldest commercially available secondary battery that dates back to mid-1800 and considerable progression has been made on these batteries in terms of obtaining higher performance levels. Today, they are commonly used in ICEVs due to being cheaper and simple to use (shown in Figure 2.2), and therefore, it is widely known and utilized in the industry. These batteries utilize metallic lead and lead oxide as anode and cathode respectively, and during discharge both electrodes are converted into lead sulfate. The half reactions at the negative and positive plates along with the full cell discharge reaction of the battery are shown in equation 2.1 respectively:

$$Pb + HSO_4^- \rightarrow PbSO_4 + H^+ + 2e^+ \quad (2.1a)$$

$$PbO_2 + HSO_4^- + 3H^+ + 2e^- \rightarrow PbSO_4 + 2H_2O \quad (2.1b)$$

$$Pb + PbO_2 + 2H_2SO_4 \leftrightarrow 2PbSO_4 + 2H_2O \quad (2.1c)$$

Discharging a battery changes the valence charge of the lead and results in release of electrons and formation of lead sulfate crystals. Even though this battery chemistry provides the best cost option, it has very low energy content and therefore valve regulated lead acid (VRLA) is usually used more often due to their higher ampere-hour (Ah) turnover and virtually maintenance free operation. Even though VRLA has relatively higher cycle life, it still has issues under partial state of charge.

Figure 2.2 Typical lead-acid battery.

Currently, their applications are limited to industrial and other low speed vehicles (as well as applications including automotive starting, lighting and ignition) since the weight (due to low specific energies, under 40 $Whkg^{-1}$) and lifetime (3–5 years) is still the biggest limiting factor. Moreover, the battery chemistry has low potential for charging with high currents which shortens its cycle-life even more. Since lead acid battery chemistry is also reaching its limits in specific energy, significant breakthroughs are needed in this technology in order for these batteries to be implemented in a wider range of applications. This can be done through bipolar configuration or by altering the lead electrodes with the help of novel lighter materials such as carbon where lead is no longer the supporting material (Conte, 2006).

2.2.2 Nickel Cadmium Batteries

Nickel Cadmium (NiCd) batteries dates back to as early as 1899 and use nickel oxyhydroxide for anode and metallic cadmium for cathode. However, they were only introduced in large volumes in early 1960s and were the preferred chemistry for a wide range of high performance applications (such as phones, toys, power tools, medical instrumentation, etc.) in the last quarter of the twentieth century due to having significantly higher energy density and longer cycle life than lead acid chemistry. The reactions at the negative and positive plates and the full cell discharge reaction of this battery chemistry are shown in equation 2.2 respectively:

$$Cd + 2OH^- \rightarrow CdOH_2 + 2e^- \tag{2.2a}$$

$$NiO_2 + 2H_2O + 2e^- \rightarrow NiOH_2 + 2OH^- \tag{2.2b}$$

$$Cd + NiO_2 + 2H_2O \leftrightarrow CdOH_2 + NiOH_2 \tag{2.2c}$$

This chemistry has the advantage of having a wide temperature range, relatively low internal resistance (lower than NiMH) along with sustaining high discharge rates without negatively impacting the battery capacity as well as being virtually maintenance-free. However, they are replaced by NiMH and Li-ion batteries due to having lower energy density, shorter life cycle, more pronounced "memory effect" (due to the size increase in crystalline formation through charging before it is fully discharging) and environmental impact (due to the poisonous cadmium). The memory effect can cause higher impedance and prevents the battery from discharging beyond this last point, and can lead to rapid self-discharging of the battery. This effect is less prevalent in modern NiCd batteries and can be reduced by reconditioning of the battery through numerous complete discharge and charge cycles of the battery. The rated voltage for the alkaline batteries is 1.2 V and commonly has energy densities and life time (1500+ at deep discharge levels) than those of lead-acid batteries. However, NiCd batteries cost significantly higher (3–4 times more expensive) than lead acid batteries and have very high self-discharge rates which can reach up to 10% of the rated capacity per month in some applications.

2.2.3 Nickel Metal Hydride Batteries

Nickel metal hydride (NiMH) batteries, as shown in Figure 2.3, were treated as a good candidate for EV applications between 1990s and 2000s since they have relatively high specific energy and cycle-life and are composed of non-toxic recyclable materials. Moreover, when overcharged, NiMH batteries use the excess energy to split and recombine

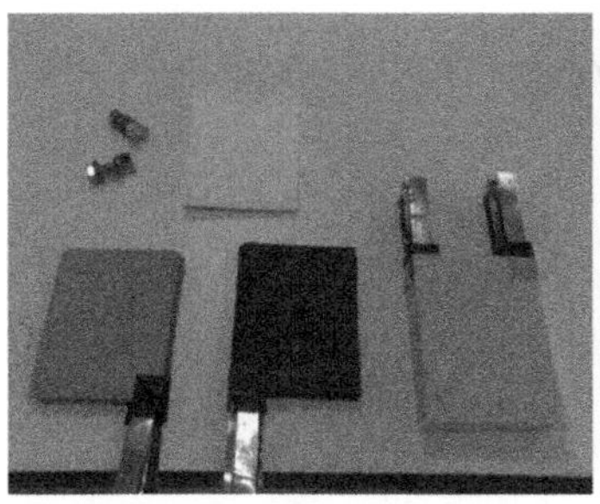

Figure 2.3 A NiMH battery cell and pack (courtesy of TUBITAK Marmara Research Center).

water which makes it maintenance free. This battery chemistry is based on the release and absorption of hydrogen (OH^-) by a nickel oxide anode and a metal-hydride cathode (Gerssen-Gondelach and Faaij 2012). The half reactions at the negative and positive plates along with the full cell discharge reaction of the battery chemistry are shown in equation 2.3 respectively:

$$MH + OH^- \rightarrow M^+ + H_2O + e^- \tag{2.3a}$$

$$NiOOH + H_2O + e^- \rightarrow Ni(OH)_2 + OH^- \tag{2.3b}$$

$$NiOOH + MH \leftrightarrow Ni(OH)_2 + ME = 1.35V \tag{2.3c}$$

These batteries have been used for many decades (since 1960s) in combination with hydrogen, iron, cadmium, zinc and most popularly metal hydride. It was initially introduced as a substitute for nickel cadmium chemistry and has similar cell structure with the same positive electrode and electrolyte; however cadmium electrode is replaced with metal hydride alloy for the anode. When charged, this alloy receives hydrogen ions and stores it in a solid state, which increases its rate of self-discharge, making NiMH unfavorable in this aspect. On the other hand, this chemistry has very low voltage depression (so called memory effect) under partial discharge utilization, especially compared to NiCd chemistry.

NiMH chemistry first became commercially available for electric vehicles in the EV1 vehicle. Currently, it is mostly used Toyota Prius' battery pack. In this chemistry, the material intercalates protons and has a theoretical specific capacity of 289 mAh g^{-1}. One of the biggest advantages of this chemistry is the presence of side reactions that provide a measure of overcharge potential and cell balancing. Moreover, the oxygen can diffuse across the cell from the potentials present at the end of a positive charge to the negative electrode where it is recombined. This oxygen that is transferred to the cell also assists in capacity balancing and therefore major problems can arise in other chemistries (such as Li-ion) that do not have this mechanism. However, the oxygen reactions produce considerable heat, and when combined with the heat of the exothermic charging process, increase the temperature during charging, as opposed to NiCd where the charging process is endothermic (Cairns and Albertus, 2010).

In addition, this chemistry is relatively safe in terms of ignition due to using an aqueous electrolyte rather than an organic solvent and a cell misuse usually only effects the cycle life of the battery. However, considerable abuse of the cells can still produce dangerous gasses. This issue is usually solved with a unidirectional gas vent to that helps in

Figure 2.4 Sample pouch type lithium-ion battery (courtesy of TUBITAK Marmara Research Center).

removing the hydrogen, oxygen or water vapor in NiMH cells. Moreover, it has a comparatively high lifetime; however it is mainly reduced due to the degradation of the metal hydride electrode and the loss of water from the electrolyte. On the other hand, the main dominant failure mechanisms of Li-ion cells are not present in this chemistry. Thus, it is usually sufficient to adopt module level monitoring system, which avoids complex and costly monitoring systems at cell level.

Finally, even though NiMH does not really excel compared to any other specific chemistry (since Pb-acid is cheaper, NiCd has higher cycle life, Li-ion has higher energy density), it provides a good compromise and has significant tradeoffs between battery maximum power and energy capacity. Furthermore, even though it has significantly higher specific energies than lead-acid batteries, it is still not sufficient to satisfy the high demand of current battery powered vehicles. In addition, it has low potential for cost reduction due to a high share of relatively expensive nickel in the battery.

2.2.4 Lithium-Ion Batteries

Current EVs and HEVs have significant demands from the battery technologies as they incorporate additional demanding functionalities (such as power-assistance, engine load-point shifting, regenerative breaking and electric auxiliaries) which cannot be easily satisfied with the previous chemistries due to their relatively lower energy and power densities. Lithium on the other hand, is the lightest metallic element and has a considerably low redox potential which provides cells with relatively high voltage and energy density. In addition, Li^+ has a small ionic radius which is beneficial for diffusion in solids (Ellis and Nazar, 2012). These properties make Li-ion chemistry a very good candidate for electric vehicle applications.

Primary lithium batteries (an example is shown in Figure 2.4) become commercially available during 1970s. In 1990s rechargeable lithium metal batteries were tried to be developed but faced significant issues in terms of safety and long charging times. The popularity of lithium batteries were increased and conquered the market by introducing graphite (as opposed to the lithium metal) and lithiated transition metal oxide as anode and cathode materials. As a result, $LiCoO_2$, $Li[NiCoAl]O_2$, $Li[MnNiCi]O_2$and $LiMn_2O_4$ became the most widely used lithium chemistries in the market.

The current predominant lithium-ion chemistry encompasses the intercalation of lithium in each electrode, where at the positive electrode the reaction (for $LiCoO_2$) is written as

$$Li^+ + e^- + CoO_2 \leftrightarrow LiCoO_2 \quad (2.4a)$$

whereas, at the negative electrode (for a carbon electrode), the reaction is

$$Li^+ + e^- + C_6 \leftrightarrow LiC_6 \quad (2.4b)$$

Today, a wide range of Li-ion chemistries are available and are identified with respect to the composition of their cathode. Some of the widely utilized ones are lithium-cobalt-oxide ($LiCoO_2$), lithium-manganese-oxide ($LiMn_2O_4$), lithium-iron-phosphate ($LiFePO_4$ or LFP) as well as nickel-cobalt-aluminum-oxide ($LiNiCoAlO_2$ or NCA) and lithium- nickel-manganese-cobalt ($LiNiMnCoO_2$ or NMC). The most common ones among them are shown in Table 2.2.

Among these, $LiFePO_4$ has the lowest specific energy and short calendar life; however it is the safest chemistry with good life expectancy and low material costs. $LiMn_2O_4$ is commonly used in cell phones and hybrid and electric vehicle applications due to having high specific energy and power and relatively low cost. However, the chemistry is highly affected by high temperatures and therefore significant thermal management is required during operating stage. $LiCoO_2$ chemistries can reach even higher specific energies, but they are considerably expensive due to their high Co content. Thus, they are mostly used in consumer products instead, especially in notebook computers. Finally, considerable research is being conducted on high energy density chemistries such as $LiNiMnCoO_2$ and $LiNiCoAlO_2$; however, more time would be needed before they can be utilized safely and at a low cost for EV applications.

Among the anode materials, most common ones are graphite, $Li_4Ti_5O_{12}$ (or LTO) and silicon, with graphite being the most commonly used due to having comparatively high specific energy and low cost. However, it has unstable solid electrolyte interface (SEI) layer at high state of charges and temperatures, which can reduce its output power considerably. LTO has higher cycle and calendar life, but half of the specific capacity. Finally, silicon is also a good candidate since it has low voltage for an anode and high theoretical specific capacity (up to 10 times of the conventional carbon anodes). Thus, the use of silicon anodes that utilize silicon nanotubes or a comparable process would provide considerably higher energy storage and longer battery life. However, silicon expands and contracts significantly during lithium insertion and extraction, which degrades the mechanical integrity of the electrodes and negatively affects the stability of SEI. Further information on the characteristics of these chemistries is provided in Table 2.3.

Table 2.2 Energy densities of some common Li-ion chemistries with respect to composition of their cathode.

Chemistry	Size	Wh/L theoretical	Wh/L actual	%	Wh/kg theoretical	Wh/kg actual	%
$LiFePO_4$	54208	1980	292	14.8	587	156	26.6
$LiFePO_4$	16650	1980	223	11.3	587	113	19.3
$LiMn_2O_4$	26700	2060	296	14.4	500	109	21.8
$LiCoO_4$	18650	2950	570	19.3	1000	250	25.0

Source: Whittingham (2012).

Li-ion batteries have limited cycle life (especially at high temperatures) and have safety issues (especially for large, multi-cell modules). At extreme operational conditions, such as reaching very high temperature and/or the upper limit of the charging process, oxygen may be released from the cathode and may react with a flammable electrolyte causing thermal runaway and even explosions. Thus, $LiFePo_4$ chemistry has also being widely used despite of its relatively lower nominal voltage, since it has stronger covalent bonds than the aforementioned chemistries which make it intrinsically safer. In addition, improved separators (such as ceramic parts and high-boiling electrolyte) are also used to improve the safety on a cell level. Moreover, some commercial lithium-ion chemistries also incorporate C, Co and Sn as the anode material in order to attain further stability where the graphite has a fragile structure (Etacheri, 2011; Dixon, 2010).

In order for hybrid and electric vehicles to be able to genuinely compete with conventional vehicles and provide an easy transition from these vehicles without changing people's habits, they would be required to offer at least the same performance and range as conventional vehicles. Currently, the energy densities of the Li-ion batteries are constraint by the weight of the active materials for the anode and the metal oxide for the cathode (typically graphite with 170 mAh g^{-1} and $LiCoO_2$ with 130 mAh g^{-1} respectively) (Padbury and Zhang, 2011). Aforementioned research on the electrodes will likely to improve the energy density, and many sources expect yearly energy density improvements of up to 6%, but even at the best case scenario of doubling it, it would still not be enough to compete the conventional vehicle that are in the market today. During charging, lithium deposits in the form of lithium dendrites which raises internal safety issues such as thermal instability and formation of internal short circuits. Thus, today mostly Li-ion chemistries that do not contain metallic lithium are on the market. Finally, even though these battery chemistries are mostly evaluated with respect to their specific energy and power density, cycle life, cost and environmental impact; various engineering aspects such as the volume change upon full intercalation with lithium or the ability

Table 2.3 Comparison of Li-ion battery cathode and anode materials.

Material	Specific capacity mAh/g	Voltage vs. L+/Li, V	Characteristics
$LiCoO_2$	160	3.7	Good capacity and cycle life / Expensive, unsafe during fast charging
$LiMo_2O_4$	130	4.0	Acceptable rate capability, low cost / Poor cycle and calendar life
$LiFePO_4$	140	3.3	Good cycle life and power capability low cost, improved abuse tolerance / Low capacity and calendar life
NMC	180	4.2	High capacity, lowest cost / Life cycle (less than NCA)
NCA	185	4.2	Highest capacity, low cost / Safety concerns
Graphite	372	<0.1	Low cost, flat and low potential profile / Low volumetric density, high sensitivity to electrolytes, easy exfoliation
LTO	168	1.0–2.0	Highest cycle life / High cost and low energy density
Silicon	3,700	0.5–1.0	Very high energy / In early experimental stage, large volume expansion

Source: Wu *et al.* (2012).

to indicate the SOC of the battery can also have a major impact in the decision making process to utilize them in the specified applications.

It should be noted that, even though most of these aforementioned battery chemistries can provide considerably energy (and power) for EV propulsion; the energy density of gasoline is 13,000 Wh/kg and the average tank-to-wheel efficiency of the U.S. fleet is around 12.6%, which makes the usable energy density of gasoline for automotive application to be approximately 1,700 Wh/kg (Girishkumar *et al.*, 2010). On the other hand, on average, electric propulsion systems have battery-to-wheel efficiencies of around 90%. Even then, in order for battery electric propulsion systems to be compatible with conventional vehicles, energy densities of an order of magnitude higher than the ones currently available with Li-ion systems would be needed. Thus, radically different chemistries and approaches would be required to fulfill the very demanding requests of energy storage for today and the future. The practical specific energies of some of the most commonly used and researched EV batteries are shown in Figure 2.5.

2.3 Battery Technologies under Development

Since their introduction in 1991, Li-ion batteries have transformed portable electronic devices and have been playing a major role in the electrification of transport. Especially, over the past decade and a half, Li-ion battery technology has become widely accepted in the market and reached to a high level of maturity and reliability which enables it to support more demanding applications. On the other hand, even after considerable developments, the highest energy storage of the chemistry by itself is insufficient to meet the demands and long term needs of the transportation sector. Thus, various efforts on the exploration of new configurations, electrochemistries and materials are currently being conducted.

Among these, there are also novel approaches employed with respect to the fabrication of separators to improve the wetting by the electrolyte solutions. Moreover,

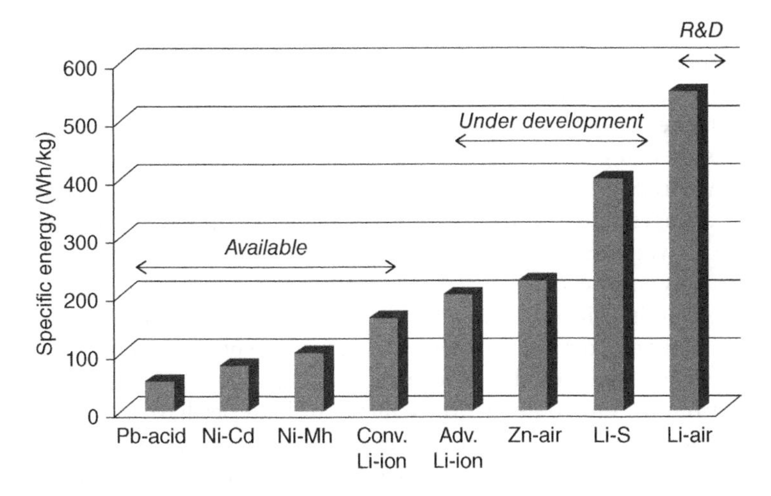

Figure 2.5 Practical specific energies of some of the most commonly used and researched rechargeable batteries for EV applications (adapted from Bruce *et al.*, 2012).

developments of Li-ion batteries which use gel, polymetric or glassy matrices serve as the electrolyte system, multiple types of Li-insertion cathodes, new solvents (namely ionic liquids) and salts, and large variety of additives are also studied together with lithium-ion batteries (Etacheri, 2011). Ionic salts, for example, have favorable features such as being composed of ionic ions, having high conductivity, environmental compatibility and even more importantly high thermal stability. Therefore, ionic liquid based solutions are not flammable (stable up to 400°C) and can go under virtually unlimited structural variations. However, they are still under research and have the disadvantages of not being stable at low, reducing voltages and therefore more testing and analysis would be needed before they become widely available. Today, there are three types of anodic reactions that are being studied for rechargeable lithium-ion batteries, namely intercalation, conversion and alloying. Intercalation-type electrodes may accept a maximum of one lithium ion equivalent per mole of the host compound, which limits the specific capacity and reduce the energy density of the Li-ion battery chemistry (Scrosati *et al.*, 2011).

For the conversion approach, the most important materials currently being utilized are graphite and $Li_4Ti_5O_{12}$. When passing to the conversion chemistry, significantly more lithium can be exchanged, leading to a much higher specific capacities. Currently, nanoparticles of transition metal oxides are used for this approach where the capacity can reach up to three times larger than graphite. Lithium-sulfur (Li-S) chemistry also plays an important part in this with a theoretical energy density of 3730 $Whkg^{-1}$, which is almost an order of magnitude higher than the intercalation based chemistry. The main problem with the Li-S cells is the shuttle mechanism that impedes full capacity extraction of the sulfur cathode, which makes it virtually impossible to fully recharge a Li-S battery. In addition, there are also issues associated with the high solubility of the cell reaction which causes severe corrosion, leading to loss of active materials, low usage of the sulfur cathode, low overall Coulombic efficiency as well as considerable reduction in capacity upon cycling (Scrosati *et al.*, 2011). Several advancements have been made such using lithium sulfide-carbon composite cathode and optimization its fabrication, however, more technological breakthrough would be required before this chemistry becomes commercially available.

Finally, for the alloying approach, various studies are also currently being conducted to incorporate tin (Sn) and silicon (Si) to the conventional lithium batteries which can reach even higher capacities than conversion process. Furthermore, they offer the characteristics of being cheap, environmentally benign and able to alloy with lithium ($Li_{4,4}Sn$ and $Li_{4,4}Si$). However, their main drawback is the large volume expansion-contraction changes upon lithiation (volume change of up to 3 times for $Li_{4,4}Si$ and 2.5 for $Li_{4,4}Sn$), which may lead capacity fading through mechanical stress and limits it cycle life to only a few cycles.

It is worth mentioning that, in principle, oxygen is also a cathode material and can be used with metal anode materials to produce energy. Usually, this configuration yields the highest energy density since the cathode active material (oxygen) is not stored in the battery but is accessed from the environment. Thus, metal-air batteries have attracted considerable interest because of their extremely high energy densities compared to the conventional rechargeable batteries. However, significant problems such as high polarization (internal resistance) at low current densities (due to air electrode) and a considerable low cycle-life also exist with these types of batteries. Moreover, since metal-air cells are not closed systems, both gases and contaminants may enter the electrodes and electrolyte which can cause significant problems.

In the next section, more detailed information regarding the most promising battery chemistries are provided. Even though most these batteries are still under research and very limited practical applications currently exist, these technologies are predicted to be able to satisfy the heavy demands of electric vehicles in the near future.

2.3.1 Zinc-Air Batteries

Zinc is a relatively inexpensive, naturally benign and an abundant element with low equilibrium potential, flat discharge voltage and a long shelf life. Thus, Zinc-air batteries have played an important role in negative electrode materials in primary alkaline batteries for the past couple of decades and have been combined with a range of positives electrodes (including nickel, hydrogen and air). It is one of the chemistries that is currently used in the market (mostly in hearing aids) due to having higher energy densities than conventional batteries. However, it is not high enough to satisfy the electric vehicle demands. In the Zn-air chemistry, oxygen is reduced in the alkaline aqueous electrolyte at the air electrode upon discharge as follows:

$$O_2 + 2H_2O + 4e \rightarrow 4OH^- \tag{2.5}$$

With this chemistry, specific energies over 300 Wh kg^{-1} (with theoretical maximum of 1084 Wh kg^{-1}) can be achieved in practical batteries (Lee *et al.*, 2011). Moreover, all components of Zn-air batteries are stable towards moisture, which enables the assembly of the cell to be carried out under ambient air conditions, making the manufacturing process easier (especially compared to Li-air). The main drawback however is associated with the long term cycling of the system due to the ZnO precipitation which occurs when the zincate concentration significantly exceeds the solubility limit. Thus, having a reversible electrochemical system with a soluble product requires intricate changes. Currently, one method to get closer to achieve this goal is to include additives that reduce the zinc solubility in order for the product to remain close to the reaction site. This chemistry can also be charged through series of mechanical and electrochemical steps with the help of a special infrastructure using removable zinc-anode cassettes. Even though this method adds significantly to the cost of the battery, it can reduce the charging times to less than 5 minutes (to change the zinc anodes) and the battery can exceedingly prolong the battery life due to always having new anodes (Dixon, 2010).

The chemistry also has some drawbacks in terms of the formation of dendrites (which can lead to short circuits), absorption of carbon dioxide in the electrode structures as well as the drying out or flooding (based on the humidity level of the environment) of the cell. Electrode problems can be mitigated by mechanical recharging of the cell and regenerating Zn externally. In addition, the bi-functional air electrode related issues can be resolved by using separate electrodes for charge and discharge processes, which may not be appropriate for EV applications. Thus, the issues related to reversible cycling (for Zn electrode) and capacity (for air electrodes) would be necessary in the future to be compatible with the EV technology. That being said, the stage of development is still considerably ahead of many the comparable metal air battery chemistries and has the potential to reach the required cost levels to be competitive with the current technology. Therefore, this chemistry can be a strong candidate to be utilized in the next generation of EVs.

2.3.2 Sodium-Air Batteries

Sodium is significantly cheaper than lithium; and sodium-air batteries have theoretical average specific energy of 1980 $Whkg^{-1}$ (1,690 $Whkg^{-1}$ with oxygen and 2,271 $Whkg^{-1}$ without), which is approximately 4 times the state-of-the-art Li-ion batteries. In addition, they have very suitable redox potential ($E^{\circ}_{Na^+/Na}$=−2.71 V versus standard hydrogen electrode) (Ellis and Nazar, 2012). Furthermore, even though they have lower capacity than lithium air batteries, their open-circuit potential is still relatively high and they are significantly easier to extract and process. The primary limiting agent and mechanism for capacity/cycleability is often considered to be associated with the discharge product (Li_2O_2) and the blocking of oxygen transport respectively.

The usually lower voltage of sodium compounds provides significant benefits for negative electrode development by increasing the overall voltage and therefore the associated energy density. Moreover, there are abundant sodium resources (sodium being the sixth most abundant element) in both the earth's crust (2.3%) and the oceans (1.1%) for any potential wide adoption of this chemistry (Sun *et al.*, 2012). The open-circuit voltage of sodium-oxygen cell is 2.3V and 1.95V for Na_2O_2 and Na_2O as the product respectively, but is still lower than that of Li-air cells (3.1V). The potentials for reaction of sodium with oxygen are given in equation 2.6;

$$Na^+ + O_2 + e^- \rightarrow NaO_2 \qquad E = 2.263\ V \tag{2.6a}$$

$$Na^+ + O_2 + e^- \rightarrow NaO_2 \qquad E = 2.263\ V \tag{2.6b}$$

$$4Na^+ + O_2 + 4e^- \rightarrow Na_2O \qquad E = 1.946\ V \tag{2.6c}$$

Finally, the charge-discharge curves of this chemistry exhibits promising and stable plateaus with a considerable round trip efficiency. It can sustain high current rates and deliver high power densities, especially with the appropriate selection of air cathodes. However, the cycle life still is the major barrier for its commercialization. Therefore, in order to improve the cycle life of these batteries, air-breathing membranes that selectively permeate oxygen from ambient air will be needed since sodium metal is highly reactive to moisture.

2.3.3 Lithium-Sulfur Batteries

Lithium-sulfur (Li-S) batteries have a sulfur cathode in which sulfur is typically paired with carbon. Usually metallic lithium is used for the anode even though various other materials are also utilized. This chemistry has been investigated since the 1940s and after extensive efforts made on this chemistry, it is also an important candidate for EV applications, especially since sulfur is currently inexpensive, can be produced in high amounts and is environmentally friendly. The cell operates by reduction of sulfur at the cathode on discharge in order to form various polysulphides that combine with lithium to produce Li_2S as can be seen in equation 2.7.

$$2Li + S \rightarrow Li_2S \tag{2.7}$$

This reaction has considerably high theoretical specific energy of 2,600 Wh kg^{-1} which is more than 4 times the specific energies of Li-ion cells that are currently commercially sold. This chemistry was first explored in 1960s, however, the sulfur containment was found to be difficult and significant issues with cycle life was present. Later research using cathode materials that incorporates with FeS_2 and FeS yielded the development of Li/FeS_2 cells for various vehicles used in defense applications (Cairns and Albertus, 2010).

On the other hand, there are several issues inherent to the cell chemistry that needs to be resolved, such as having very low electronic conductivity, limited capability, fast capacity fading and poorly controlled Li/electrolyte interface, which prevent achieving high electrochemical utilization. Moreover, lithium polysulfides can be formed as Li-S cells containing organic solvent-based electrolytes are discharged, which in turn reduces the capacity and cycle life of the cells . Thus, significant research and development efforts are currently being conducted to increase the life cycle of this chemistry and it is expected that over 1000 cycles may be commercially available by the year 2020. Even if specific energies of 800 Wh kg^{-1} is achieved with the advancements in Li-S cell technology, it would provide significant advantages over the current battery chemistries with respect to performance, cost and environmental impact and reach all-electric ranges up to 300 miles per charge when used in midsize EVs.

2.3.4 Aluminum-Air Batteries

Aluminum-air cell is a primary metal-air battery with an aluminum anode and an air-breathing cathode in contact with an aqueous electrolyte, which usually tends to be sodium hydroxide, potassium hydroxide or sodium chloride (Egan *et al.*, 2013). It is a technology that has been studied for several decades due to various advantages such as geological abundance of the aluminum (being the most abundant metal element on Earth), rapid mechanical rechargeability, environmental benignity and strong recyclability. Aluminum has a high energy density of 8.1 kWh kg^{-1} and a theoretical potential of 2.35 V in alkali electrolyte which makes it a promising anode candidate. However, the use of it has been mainly limited to niche markets and the interest in this chemistry is declined as the expectations of the competing battery chemistries increased. The ideal overall discharge reaction of the aluminum-air cell is:

$$4AL + 3O_2 + 6H_2O \rightarrow 4AL(OH)_3 \qquad E_{cell} = 2.7\ V \tag{2.8}$$

The main limiting factor of this technology is the electrochemical and corrosion properties of the aluminum alloy electrode. This impedes the storage of wet aluminum-air cells by forming a thin compact oxide layer which reduces the battery performance significantly via constraining the ability to achieve reversible potential and causing delayed activation of the anode (Smoljko *et al.*, 2012). The anodic self-corrosion problem has been the major challenge faced by this chemistry since it leads to considerably lower utilization efficiencies and increases the likelihood explosions due to hydrogen buildup. Thus, significant attempts have been made to inhibit the hydrogen generation reaction by alloying aluminum with various elements, adding inhibitors to the electrolyte and/or using high purity aluminum. However, they have not been very successful and caused other issues like higher material costs and complexity.

2.3.5 Lithium-Air Batteries

In Li-air (LiO_2) batteries, lithium is applied as anode material and oxygen from the ambient air acts as cathode. Note that the battery chemistry has received much less attention from the scientific community until recently. Li-air has the highest theoretical density among the metal-air chemistries and is an order of magnitude higher than the conventional Li-ion batteries on the market today, based on the utilization of pure lithium metal anode and the absence of stored oxygen in the system. In addition, the chemistry has the low atomic mass at the anode (due to lithium metal) and has low electronegativity which enables electrons to be donated more readily producing positive ions. Thus, when lithium is used as the anode material for a metal/air battery to gain an even better performance than the previously mentioned metal-air chemistries (3842 $mAhg^{-1}$ for Li, compared to 815 $mAhg^{-1}$ for Zn, 2965 $mAhg^{-1}$ for Al and for 1673 $mAhg^{-1}$ for S).

Li-air batteries were initially proposed around 1970s after being discovered by Littauer and Tsai in Lockheed in 1976, but the project was shut down due to unwanted reaction of lithium with water. In the 1990s, the studies were continued with the incorporation of organic electrolytes, but the chemistry has not gained worldwide attention until 2009 as a possible chemistry for EV applications.

A typical Li-air cell would be composed of a lithium metal anode, a porous carbon cathode and an electrolyte with the typical reactions provided in equation 2.9. The main difference between this process and most of the typical chemistries is that the reactants are no longer have to be carried on-board (oxygen can be received from the ambient air) and that the supply of O_2 is virtually infinite.

$$Li \rightarrow Li^{+} + e^{-} \tag{2.9a}$$

$$2Li^{+} + O_2 + 2e^{-} \rightarrow Li_2O_2 \tag{2.9b}$$

This described oxidation of 1 kg of lithium metal releases 11,680 Wh/kg, which is highly compatible with the energy density of gasoline (13,000 Wh/kg). Since the well-to-wheel efficiencies of average conventional cars are very low, an energy density of 14.5% would be sufficient for Li-air batteries to be in par with the conventional vehicles in terms of total usable energy density. It should be noted that this reported theoretical specific energy is based on the mass of lithium alone and in reality, it should be less since all metal-air cells gain mass (O_2) as they discharge. Currently, existing metal-air batteries have energy densities up to 50% of their theoretical density. However, it would not be possible to expect the same for Li-air, since the main factor limiting the practical energy density of this cell is the need for excess lithium in the anode. This excess lithium compromises the volumetric energy density of the cell since lithium is significantly lighter and therefore the overhead of the battery structure and the electrolytes creates a significantly larger impact. In addition, Li-air batteries would be able to go through 50 cycles with moderate loss in capacity, which translates to over 25,000 miles in its lifetime, significantly lower than the average distance covered during the life of an ICE. Moreover, the power density of Li-air technology is significantly low (about 0.1–1 mA cm^{-2}). Thus, in order to supply enough power for full vehicle propulsion, the current densities need to be increased by an order (preferably by two orders) of magnitude or the cell DC resistance needs to be reduced to below 10 Ωcm^2. Finally, the cycle electrical efficiency of this technology is around 60–70%, which is much lower than the 90% mark of the practical propulsion batteries. In addition, the oxygen discharge products are very reactive toward the environment (especially to electrolyte solvents), which might require replacing oxygen-discharge product from peroxide to oxide.

Even though, currently, lithium-air batteries have greater energy densities than any other known chemistry, there are still various issues associated with this technology. These range from the reactivity of the lithium anode to the poor reversibility and efficiency of the oxygen electrode, including poor kinematics of Li_2O_2 oxidation, formation of side products and electronic isolation of Li_2O_2. Unlike the Zinc-air chemistry (where the oxygen reacts to form OH^- and moves across the cell to the negative electrode), in Li-air chemistry Li travels across the cell and is deposited in the positive electrode. Thus, while the Li electrode shrinks during discharge and the air electrode swells (Cairns and Albertus, 2010). Moreover, the chemical heterogeneity of solid electrolyte interface (SEI) can result in having a brittle and morphologically heterogeneous structure, leading to uneven current distributions at the metal-electrolyte interface during cycling. This in turn can cause lithium dendrite/moss formation, which may eventually lead to shorts between the anode and the cathode, making this technology difficult to be used for vehicle propulsion. In order to prevent this, several approaches are being currently studied including developing highly Li-ion conductive artificial protective layers.

There are currently two categories with respect to the type of the electrolyte being liquid (fully aprotic, aqueous and mixed) or all solid. The first category faces the problems of mechanical instability of the protecting film, high interfacial resistance and issues associated with the solubility of the reaction products. The second category on the other hand, has sensitivity to moisture and issues related to electrochemical reaction species which in turn can have significant impact on the reliability, safety and the cycle-life of the battery. In order to reduce the moisture content along with other unwanted contents of air (such as carbon dioxide, particulates, NO_x and SO_x), an oxygen diffusion membrane that is permeable to oxygen but not to water must be placed. In addition, while the oxygen is needed in the air electrode, it should not contact the surface of Li electrode as it would be rapidly reduced. Thus, significant difficulties still exist with both approaches.

In order to reduce the size and cost of this technology and make it safer; the prevention of the dendritic lithium deposit formation would be necessary. Moreover, the current density needs to be increased to 100 mA cm^{-2} levels and change the oxygen-discharge product from peroxide to oxide (Peled *et al.*, 2011). More studies need to be conducted in future on to have more stable anodes that are protected from moisture, high oxygen diffusivity and electrical conductivity in cathodes, oxygen solubility and diffusivity in electrolytes and to have catalysts capable of reducing over potentials (Whittingham, 2012).

2.4 Battery Characteristics

2.4.1 Battery Cost

Aside from the battery characteristics, production costs have a significant role in selecting the most appropriate battery technology for the application. However, due to relatively recent commercialization and widespread usage of these battery technologies in EV and HEV applications (and thus currently having low production volumes), it is difficult to compare the costs associated with these battery technologies. The cost corresponding to each kWh of the current battery technologies are estimated in Table 2.1. The battery costs are mainly due to manufacturing costs (based on production volume and manufacturing technology) and material expenses (mostly negative and positive electrodes, separator, cell hardware and others). Even though the total costs of these batteries are significantly higher than the cost of the ICEs, they are predicted to be reduced significantly when these vehicles are produced commercially in large quantities. The cost breakdown associated with a typical plug-in hybrid electric vehicle (PHEV) cell is shown in Figure 2.6.

Among the compared battery chemistries, Pb-acid batteries are cheapest to produce and they are being extensively recycled which reduces the overall cost of the technology even further. The cost associated with NiCd batteries is significantly higher than lead-acid batteries mainly due to the recycling cost of the materials, especially cadmium, which is an environmental hazardous substance that is highly toxic to all higher forms of life. However, it has a remarkable cycle-life performance which reduces the total cost over time. NiMH batteries cost relatively lower than NiCd based on a higher capacity and lower amounts of toxic materials. Since they do not require cell monitoring, have a good endurance against misuse and relatively simple production; they are inexpensive compared to most other compared chemistries. However, they may require additional maintenance which may increase the operating costs. In future large-scale applications, on the other hand, the overall costs can be reduced significantly.

The costs of Li-ion batteries are the highest among the aforementioned batteries due to the high cost of raw materials and material processing. Moreover, the cost of necessary incorporation of battery management system, thermal management system and associated packaging can also be as high as the cell cost. Furthermore, additional costs associated with power electronics to integrate the battery package to the vehicle and the grid also needs to be considered. The typical battery package and associated power electronics for an EV is provided in Figure 2.7.

It should be noted that these battery chemistries and its corresponding auxiliary components are expected to be funded substantially for EV and HEV development and therefore the costs are predicted to be reduced considerably in the near future.

2.4.2 Battery Environmental Impact

Electric vehicles play a significant role in decarbonization of the transportation sector in the face of climate change since over 10% of global greenhouse gas (GHG) emissions are

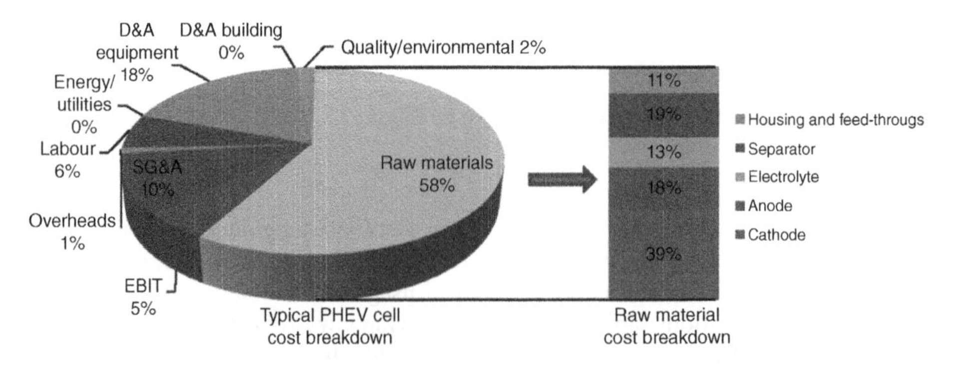

Figure 2.6 Typical PHEV cell cost structure (adapted from Roland Berger, 2012).

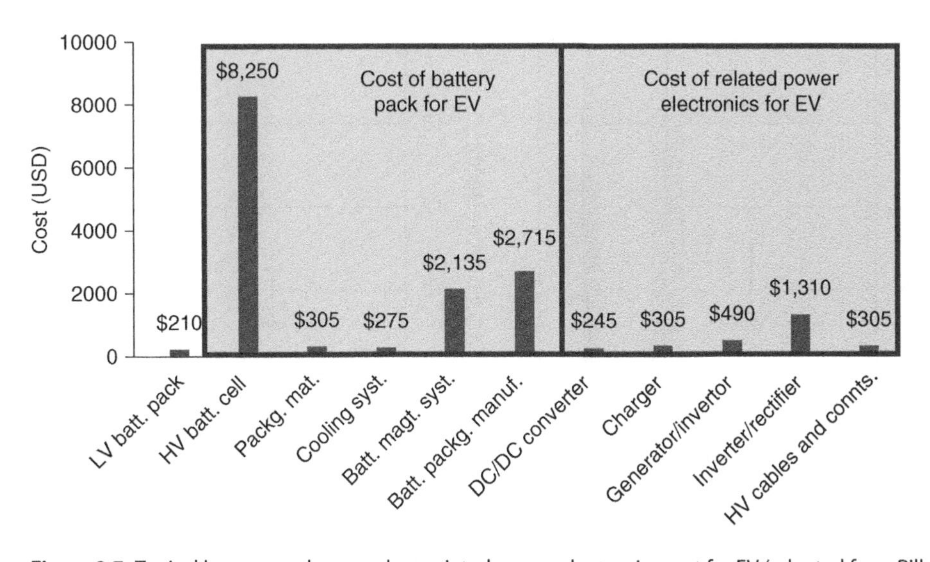

Figure 2.7 Typical battery package and associated power electronics cost for EV (adapted from Pillot, 2013).

resulted from road transport (OECD, 2010). Since the electric battery is the main component that differentiates these vehicles from the conventional ones and plays a key role in reducing the associated emissions; environmental concerns has a significant role in advancing the battery technologies based on customer behavior, regulatory limitations and cost (such as carbon tax and government incentives). Even though the substitution of battery technologies with conventional energy sources in the transportation industry reduces the associated environmental impact, the content and magnitude of this impact depends heavily on the electricity mix, battery technologies and the operating conditions. Several approaches have been taken in order to reduce the environmental impact associated with the battery technologies that includes attempts to eliminate the solvent and using water instead. Moreover recycling, which includes discharging the cells, venting the electrolyte solvent, shredding the entire package and recovering the flashed off solvent is also considered (Guerrero *et al.*, 2010).

Generally, life cycle assessment (LCA) is performed in order to assess the overall environmental impact of the different battery technologies in various stages of their life. It is an established but still evolving method and a very useful tool in the product development stage to identify the steps/procedures with the highest environmental concerns caused by products, processes, or activities and find more environmentally benign alternatives. "ISO 14044 Environmental management – Life cycle assessment – Requirements and guidelines" (ISO, 2006) and the general rules of International Environmental Product Declaration comprises the procedure, general rules as well as the basic foundation for conducting life cycle assessments. Since scope, system boundary and the depth of detail is varied significantly based on the main purpose of the conducted research, the important aspect of LCA is to set the right parameters for the associated study to obtain the most useful results andbreak conclusions.

LCA is based on a perspective called the functional unit, which specifies the capacity, cycle life and overall range and the corresponding weight. Then the system boundary

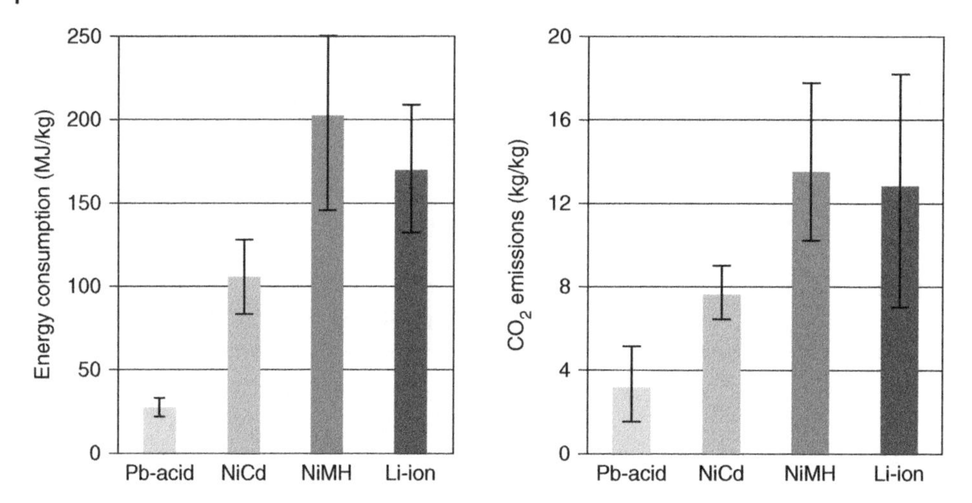

Figure 2.8 Energy consumption (cradle-to-grave) and average CO_2 emissions associated with typical EV battery chemistries (adapted from Argonne National Laboratory, 2010).

is selected, energy and material flows are quantified and cut-off assumptions are made (if necessary) in order to determine how far the materials will be tracked back (usually to the point of extraction). More detailed information regarding LCA will be provided in Chapter 6. The environmental impact assessments for the analysis are usually chosen with respect to five largely accepted impact categories:

- global warming
- acidification
- ozone depletion
- photochemical smog
- eutrophication

In electric vehicles, batteries are usually the main focus of the LCA analysis since they incorporate a considerable portion of the total impact and still have significant room for improvement. The main components of the batteries that are usually studied are the cathode, anode, separator, cell packaging, electrolyte, cell electronics, module assembly, transports as well as the phases associated with utilization and recycling. In most batteries, the global warming impacts are dominated by energy use in manufacturing, electronics and the cathode which can make up as much as 90% of the total impact. The energy consumption and average CO_2 emissions (per kg of battery technology) for current battery technologies are illustrated in Figure 2.8.

However, this does not provide a fair comparison among currently used battery chemistries for EVs since various additional characteristics (including battery efficiency and mass) also need to be considered in practical vehicle applications. Thus, the total environmental impact of battery package with different chemistries providing the same all-electric range is compared (with respect to eco-indicator 99 under a European electricity mix) in Figure 2.9. It should be noted that the energy losses due to efficiency

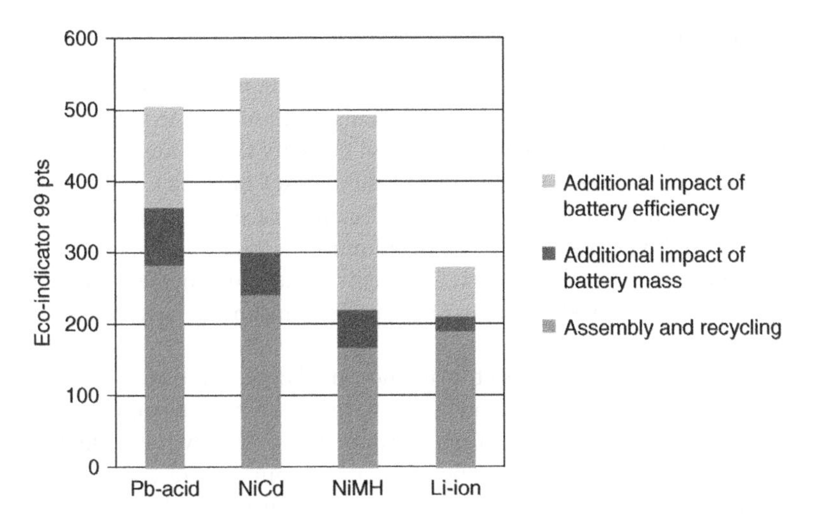

Figure 2.9 Environmental impact of the evaluated technologies based on Eco-indicator 99 (adapted from Bossche *et al.*, 2006).

as well as the utilized mass of the battery have a significant effects on the environmental impact.

Among the analyzed battery technologies, NiCd has the most environmental impact mainly due to the presence of the environmentally hazardous material cadmium. This is followed by lead-acid based on its energy storage capacity, rather than the chemical content of the battery. The high environmental impact during the production stage is reduced by the excess recyclability since this battery technology has been utilized the longest. However, since lead-acid has the lowest specific energy density among the analyzed batteries, it may require additional mass and/or multiple charging to cover the same range with other batteries, producing more environmental impact. NiMH has relatively lower impact than the previous batteries since it has a significantly higher energy density than lead-acid and better recyclability than NiCd technologies. This environmental impact is only reduced further with Li-ion since it can store 2–3 times more energy than NiMH in its lifetime and has an order of magnitude less nickel and an insignificant amount of rare earth metals. However, these predictions are very sensitive to the battery characteristics and even a small change in battery efficiency can lead to large changes in global warming potential (GWP) and reduction of life-time can increase the impact in all categories considerably.

In the environmental assessments, the impact associated with the operating phase is can contribute up to 40% of the global warming potential and up to 45% of eutrophication for the currently used battery technologies (based on the European electricity mix) (Majeau-Bettez *et al.*, 2011). However, this impact is very sensitive to the content of the electricity mix and would increase significantly in countries with carbon intensive electricity generation mix such as China. The remainder is associated with assembly and recycling, which is based on the emissions with respect to the energy and materials used to assemble the batteries as well as the likelihood of recycling.

2.4.3 Battery Material Resources

The large scale substitution of conventional vehicles with EVs will increase the demand for corresponding raw materials considerably; especially lithium due to being the chemistry with the highest practical specific energy and capacity that is used in commercial vehicles today. As this chemistry dominated the market in electric batteries, significant research has been conducted to estimate the amount of lithium available on earth and it corresponding price. The price of lithium was highly steady during 1970–1990, before the wide use of lithium in electric vehicle batteries. As these batteries started to be more commonly used, its prices fluctuated significantly, especially after 1997.

Lithium appears naturally in mineral or salt form and can also be embedded in hard clays and mineralized lithium and underground brines have been explored comprehensively. The resources of lithium are also available in seawater; however they are in a lower content (170 ppb) than sodium and it would be industrially complicated and expensive to separate lithium from the other seawater mineral salts. Currently, there are just a few laboratories that are working on this subject and could only achieve production costs around $80/kg, which are significantly larger than salt lake brines (<$3/kg) or spodumene ($6/kg-$8/kg) (Grosjean *et al.*, 2012). On the other hand, production of lithium from salt lake brines have significant drawbacks including low grades, high dispersions of composition, uncertainty rate and long durations associated with building new facilities. Furthermore, lithium mining from hard-rock minerals can provide good lithium grades, high recovery times and quick process duration, but should be weighed against the cost and environmental impact associated with mining damages and thermochemical processes.

In addition, lithium can also be obtained from other ores enclosed in pegmatite rocks that contains 1%–6% lithium by weight as well as some of the "soft"-rock evaporates. Moreover, it can also be extracted from geothermal and oilfield brines since it would require virtually no additional cost as the main goal would be to produce heat and electricity from oil and natural gas.

Based on the latest estimations of lithium reserves, a lithium amount between 37.1 Mt and 43.6 Mt are predicted (62% brines and 38% rock minerals), mostly located in South America (43.6%), North America (25%) and Australasia (25%). Today, the Salar de Atacama in Chile alone holds over 20% of the world's known reserves and supplies approximately 50% of the global demand. In addition, The U.S. Geological Survey identifies extensive unclassified lithium deposits in various parts of the world such as Austria, Afghanistan, India, Spain, Sweden, Ireland, and Zaire (Electrification Roadmap, 2009). A picture of the mine that is the one of the world's largest hard-rock producer of spodumene is shown in Figure 2.10.

On the other hand, in Europe, where the world's largest potential electric vehicle related lithium end-users are established (in countries such as Germany, UK and France), only less than 3% of the lithium resources are currently located (Scrosati *et al.*, 2011). Thus, a considerable discrepancy between supply and demand of lithium can be seen, which is likely to cause significant trade imbalance as EVs penetration to vehicle market increases in the near future. In addition, even though these locations have significant lithium reserves, there are still barriers to meet a rapid short time rise

Figure 2.10 Processing plant at galaxy lithium mine in Ravensthorpe, Western Australia.

in lithium demand due to geostrategic and geoeconomics bottlenecks as well as the process duration for extraction and treatment of lithium. Thus, in order to mitigate the imbalance between the short-term supply and demand and prevent any associated price increase, exploration efforts should be conducted from diverse range of lithium resources in nature.

When these numbers are compared to the needs of EVs with lithium ion battery, it can account for 12.3–14.5 billion new electric vehicles, which is 10 times the current world number of automobiles. Thus, based on the current technology and supply of lithium, it would be safe to assume there are enough resources of lithium for EVs in the future.

Lithium, cobalt and nickel are the main constituents of electric vehicle technologies. The global lithium production (as Li_2CO_3) was between 25,900 and 28,200 metric tons in 2012 and the reserves are assessed to be 13×10^6 tons (USGS, 2013), which is roughly enough for the production of 5×10^6 vehicles/yr (based on 40 kWh batteries). In addition, if all the existing lithium reserves are used in the manufacturing of EV batteries, it would correspond to approximately 2.4×10^9 vehicles.

Furthermore, the EVs are projected to have a lifespan of approximately 10 years. Based on this low lifespans and significant increase in production, accumulation of battery waste would be unavoidable without and efficient and wide scale recycling. Recycling will play an important role in both increasing the available resources and reducing the energy during material extraction (shown in Figure 2.11).

Lead acid batteries are highly recycled (up to 95% in some cases) and 60–80% of the battery has recycle content. On the other hand, the present levels of recycling for lithium is low and only accounts for less than 3% of the total battery production cost. However, due to valuable metals such as cobalt and nickel in the Li-ion battery, it still makes economic sense to recycle these batteries. Finally, battery chemistries such as phosphate or manganese have virtually no valuable metal in them and thus create a negative economic value for recycling. Thus, recycling these materials in the long term would be mainly due to ecological benefits and compliance with environmental regulations.

Figure 2.11 Lead-acid batteries waiting to be recycled.

2.4.4 Impact of Various Loads and Environmental Conditions

Even though different battery chemistries have varying characteristics, their battery performance depends heavily on the applied load (and therefore charge/discharge rate) and the operating conditions (especially temperature). Batteries generally work efficiently over a narrow range of discharge rates (typically C/8 – 2C), operating temperatures (typically 20°C to 45°C) and uniformity (typically under 5°C) which is usually difficult to maintain due to different conditions and ambient temperatures. In Ontario, Canada for example, the historical average maximum and minimum temperatures are 24.9°C during summer days and –13.4°C during winter nights respectfully, with local temperature extremes of 41.7°C and –50.4°C in 2014 (Climatemps, 2014). The historical average temperatures can be seen in Figure 2.12.

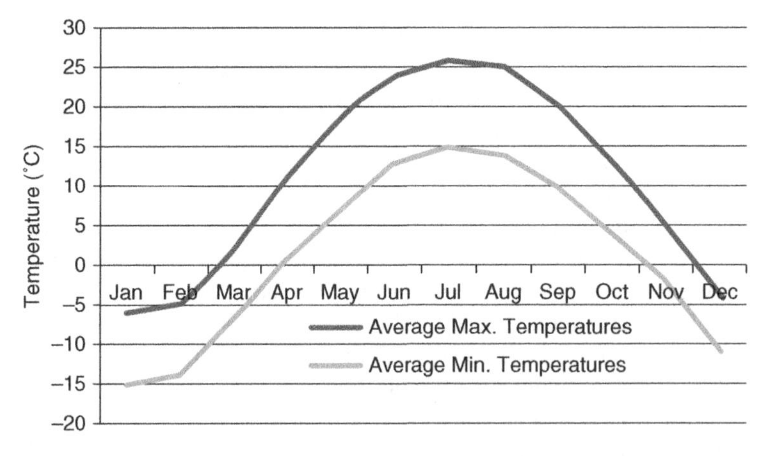

Figure 2.12 Average historical minimum and maximum temperatures in Ontario, Canada (adapted from Climatemps, 2014).

Thus, functioning outside of these specified ranges affect the round trip efficiency, charge acceptance and power and energy capability of the battery. Since the battery performance and efficiency directly affect the vehicle performance, such as range, power for acceleration and fuel economy, as well as reliability, safety and life cycle cost; considerable focus has been given to mitigate the negative impacts of excessive loading and harsh environmental conditions. In order to achieve this objective, several types of battery management (BMS) and thermal management systems (TMSs) are currently used in EVs and HEVs.

In this regard, the applied loads on the vehicle and the corresponding charging/discharging rates have significant effect on the delivered capacity of the battery. This relationship can be defined by Peukert's law, as given in equation 2.10a, which shows how much the battery capacity can be diminished (mostly temporarily), due to the rise in the internal resistance of the battery and decrease in recovery rates, with respect to the rate it is discharged.

$$C_p = I^k t \tag{2.10a}$$

where C_p is the capacity at a one-ampere discharge rate (in Ah), I is the discharge current (in A), t is the associated time (in h) and k is the Peukert constant of the battery chemistry (usually between 1.1 – 1.4) which demonstrates how well the battery holds up under high rates of discharge. The Peukert constant close to 1 indicates the battery performs well over high discharge rates whereas larger numbers indicate higher capacity lost.

$$t = H\left(\frac{C}{IH}\right)^k \tag{2.10b}$$

In order to determine the how long the battery will last at a particular discharge rate, the equation can be adjusted as shown in equation 2.10b, where H is the discharge time (in h) at the based on the specified ampere-hour and C is the theoretical battery capacity (in Ah) based on the specified discharge time. The real life results can deviate from the equation to a certain extent under dynamic loading and varying temperature conditions. However, it is a useful mathematical derivation due to being easy to use (requires only few empirical test parameters) and being accurate over a wide of discharge rate (C/20 to 10 C) and environmental conditions (from 18°C to 60°C) (Cugnet *et al.*, 2010).

Even though the charge/discharge rate is one of the most important indicator of the performance and the lifetime of the battery, in order to have a more thorough understanding of the battery behavior, the environmental conditions are also needed to be taken into consideration. The performance and the lifetime characteristics of the current battery technologies are presented in Table 2.1. However, the changes in the operating temperatures can have a varying impact on the characteristics of each battery chemistry. For example, lead-acid has very good high temperature operation, but the electrolytes would freeze when operated at low temperatures at 50% state of charge level. The battery can be brought to a full state of charge at the expense of reduced battery life. For NiCd and NiMH, discharge performance at the lower limit would be poor. At the upper limit, the charge acceptance is minimal and can suffer permanent capacity loss when fully discharged at high temperatures. Li-ion cannot operate at low temperatures (below −20°C) due to significant reduction in cell conductivity as a result of freezing of the electrolytes. It can operate up to 45°C with significant efficiency, after which electrolytes may become unstable, resulting in an exothermic electrolyte oxidation which can lead to thermal runaway.

As seen in these battery chemistries, even though high ambient temperatures can increase the performance of the battery to a certain degree by increasing the rate of chemical reactions; it increases the unwanted side reactions which reduce the battery operation and shelf life through increasing the self-discharge rate. In addition, these adverse chemical reactions also lead to passivation of the electrodes, corrosion and gassing which also has significant impact on cell health degradation. These in turn can cause faster rate of aging for the battery pack, especially when the cells have non-uniform temperatures (and therefore different health levels) in the same unit since total cycle-life of the battery unit (e.g., submodule or module) is determined by the cell with the highest degradation. Even the batteries designed specifically for high temperature reactions are effected from this heat induced failures caused by the increased rate of side chemical reactions. This relationship can be defined by the following Arrhenius equation, which is a concise, but relatively accurate characterization of temperature-dependent degradation rates of electrochemical cells.

$$k = Ae^{-Ea/RT} \tag{2.11a}$$

where k is the rate constant, A is the pre-exponential factor (also called frequency factor) related to the frequency of collisions between molecules (which is usually assumed to be constant over small temperature ranges), and R is the universal gas constant (8.314 J/(mol.K)). E_a is the activation energy (in J/mol), which is the minimum energy molecules must possess to form a product, and T is the absolute temperature of the reaction (in K), respectively. The equation can be rearranged in the natural logarithm form to determine the slope and y-intercept from the Arrhenius plot easier, as follows:

$$\ln k = \frac{-E_a}{RT} + \ln A \tag{2.11b}$$

Moreover, the pre-exponential factor can be eliminated if two temperature and/or rate constants are known. Thus, the previous equation can be rearranged to demonstrate the effect of temperature on multiple rate constants as shown in equation 2.11c. A convenient rule of thumb (for many common reactions at room temperature) is that the reaction rate doubles for every 10°C increase in temperature, provided that A is sufficiently larger than RT.

$$\ln \frac{k_2}{k_1} = \frac{E_a}{R}\left(\frac{1}{T_1} - \frac{1}{T_2}\right) \tag{2.11c}$$

2.5 Battery Management Systems

As seen in previous sections, electric and thermal management of EV cells, modules and package have significant impact on the efficiency, cost and safety of the battery packs since discrepancies in cell and module temperature and voltages can lead to faster degradation of the battery performance and can lead to safety problems. Generally, the capacity and the voltage of the battery cells utilized in EV battery packs are comparatively small; thus many cells and modules are connected in series in order to satisfy the high voltage requirements of today's EVs and achieve a long range. Due to manufacturing inconsistencies as well as operational variations (such as SOC, self-discharge current, resistance and capacity differences), these cells have different performance characteristics, which can result in reduced lifetime and performance as well cause hazardous situations during extreme charge/discharge cycles and temperatures. Moreover, since most of the cells are connected in series, the least charged cell determines the end of discharge even if there is still energy stored in other cells of the battery, thus reducing the usable capacity of the pack. Furthermore, this charge unbalancing exacerbates over time when not properly managed. For these reasons, monitoring equipment that can measure the individual cell and modules and provide balancing is needed since virtually all battery types (especially Li-ion) must be operated within the safe and reliable temperature and voltage windows.

When the instructions on the most of the battery manufacturers are observed, the preferred operating temperature range for current EV Li-ion batteries is generally set to −20–55°C for discharging and 0–45°C for charging (no less than −30°C and more than 90°C), whereas the usual operating voltages vary based on the specific chemistry, but are generally within 1.5–4.2V. When the temperature of the cell is over 120°C, combustible gases would be produced and after 130°C, the separator would start melting and the positive material will start decomposition if even higher temperatures are reached, depending on the battery chemistry. Temperatures over 200°C can result in production of oxygen with the decomposition of the positive electrode and can cause thermal runaway. When Li-ion batteries are charged below 0°C, it would cause metallic lithium deposit on the carbon negative electrode surface, which reduces the life-time of the battery. At even lower temperatures, the cathode would break down and lead to short circuits. At the same time, over-discharging the batteries will lead the lattice to collapse (due to phase change), resulting in low battery performance. In addition, extreme low voltage or overdischarge can also reduce the electrolyte, produce combustible gases and create safety risks in the battery. Meanwhile, exceedingly high voltage (or overcharge) can cause the positive electrode to decompose and produce considerable heat in the battery, accelerating the capacity fade and causing internal short circuits (Languang *et al.*, 2012). Thus, a battery management system (BMS) is needed in order to keep the batteries operating within their optimal parameters to prevent the aforementioned issues in the pack.

Even though there is no definite description of the tasks that need to be performed by BMS, it usually serves to the three main tasks:

- Battery cell/pack monitoring and damage protection
- Safe operation and prolonged life of the batteries by enabling them to operate within the proper voltage and temperature interval
- Performance maximization and signaling users and/or external devices

However, unlike classical industrial cycling, battery monitoring becomes significantly more cumbersome in automotive batteries since these batteries are not always completely charged. Moreover, discharging of the battery virtually never starts from full SOC and is performed with a wide range of different current rates with the possibility of overdischarging (Meissner and Richter, 2003). Thus, having a capable BMS plays a key role in improving the performance and safety of the vehicle and prolonging the life of the battery EV applications.

In battery management systems, battery current, voltage and temperature are the main inputs and are used for state determination (such as state of charge, stage of heath and state of function) as well as safe operation of the battery. BMSs can also have additional inputs which can be both analog (such as accelerate and brake pedal sensors) and digital (such as the start key ON/OFF signals). Its outputs are mainly thermal management module (fan and electric heater), balancing module (capacitor and/or dissipation resistance), voltage safety management (main circuit contactor), communication module and general digital inputs (charging indicator, failure alarm). In addition BMS operates auxiliaries such as cooling fans in air TMSs and pumps in liquid TMSs and manages all the electrical and electromechanical circuitries for executing the pre-charge process or for coordinating the insulators like the shutdown in case of hazardous conditions. The configuration of a typical BMS is shown in Figure 2.13.

Depending on the application, different management topologies can be used in BMSs. These can range from utilizing separately and directly on each cell, altogether in a single device or in some intermediate form. This topology has significant impact on the cost, reliability, ease of installation/maintenance and even the measurement accuracy of the system. The BMSs are categorized as centralized, master-slave, modular and distributed based on their functionality. Centralized BMSs are located in an assembly where a single controller is connected to the battery cells through a multitude of wires, making it compact, cheap and easy to maintain. This configuration is very similar to the centralized system except that it is divided into multiple identical modules, which gives it the additional advantages of proximity to the battery and ease of expansion (by adding more modules). However, it is slightly more expensive and can occupy more space. In master-slave configuration, the initial (slave) modules measure the voltage of a number of cells and the later (master) modules handle the computation and communication. The main advantage of this configuration compared to the previous one is having a slightly lower cost due to having higher number of slaves that are less expensive (since they only perform the same simple measurements). Finally, in distributed BMSs, the

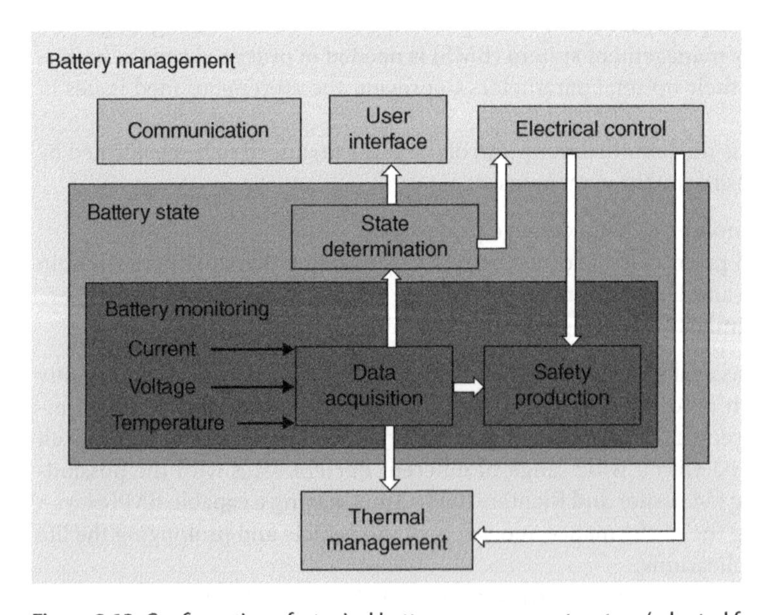

Figure 2.13 Configuration of a typical battery management system (adapted from Xing *et al.*, 2011).

electronics are placed directly on top of the measured cells with a single communication cable between the battery and a controller which reduces the number of communication wires significantly. They are simple to install and maintain, but are more expensive than the aforementioned topologies (Andrea, 2010).

The need for BMS also depends on the battery chemistry used in the packs. In lead acid chemistry, the final charging voltage plays the most important role and the voltage monitoring can be performed at a module level. Even though the estimation of SOH and SOC is complicated in this chemistry, there are a plethora of reliable models already created since this is one of the oldest and most worked on battery chemistries. In NiMH, the temperature dependency of this chemistry plays an important role in the BMS. Moreover, due to relative flatness of the charge/discharge curves, SOC estimation becomes cumbersome. In addition, BMS would also need to take the so called "memory effect" into consideration, thus at least a current sensor along with a voltage sensor would be needed in the BMS for accurate SOC estimation of this chemistry. In Li-ion chemistry, a cell level monitoring system is unavoidable to avoid overcharging, temperature non-uniformities and prevent accelerated cell aging. Thus, the BMS associated with this chemistry tend to be more complex and costly. Besides, the need of BMS for Li-ion batteries also show difference based on the specific chemistry used in these batteries. For instance, $LiFePo_4$ is one of the popular choices for battery chemistry in EV application due to its good compromise of energy density, cost, safety and cycle life. However, in this chemistry, having a fully capable BMS becomes imperative since it has a very flat charging/discharging curve in the 20% - 80% region, which makes the accurate SOC estimation very difficult in this region.

Finally, the main BMS functions can be listed as battery parameters detection, battery state estimation, on-board diagnosis (OBD), battery safety regulation and notification, charge control, battery equalization, thermal management, networking and data storage and more. The key issues associated with BMS are provided in the following sections.

2.5.1 Data Acquisition

One of the key tasks of EV BMS is to capture voltages, current and temperature at different points in the battery and convert them into digital values to estimate the battery status in later stages. Among these, voltage measurement is a complex process as the EV pack is composed of hundreds of cells in various (series/parallel) configurations with numerous channels to measure the voltages. Since there is an accumulated potential (which is different in each cell) when the cell voltage is measured, it makes it virtually impossible to have unified compensation or elimination methods. This issue becomes considerably more prominent in certain battery chemistries such as $LiFePo_4$ since their slope for open circuit voltage is comparatively gentle and the utmost corresponding SOC rate of change per mV voltage reaches 4% for the most part, which makes the accuracy of the cell voltage data acquisition an important necessity.

As for temperature and current measurements, it is important to monitor the temperature of the battery in real time to ensure that the battery is within it optimal operating window. Thus, heat-variable resistors are often placed in the seam of the batteries. Moreover, the measuring off the current usually plays a significant role in accurately estimating SOC and therefore needs to be measure accurately, generally using a Hall sensor. However, a resistor is required to convert the current signal into voltage before sending further downstream.

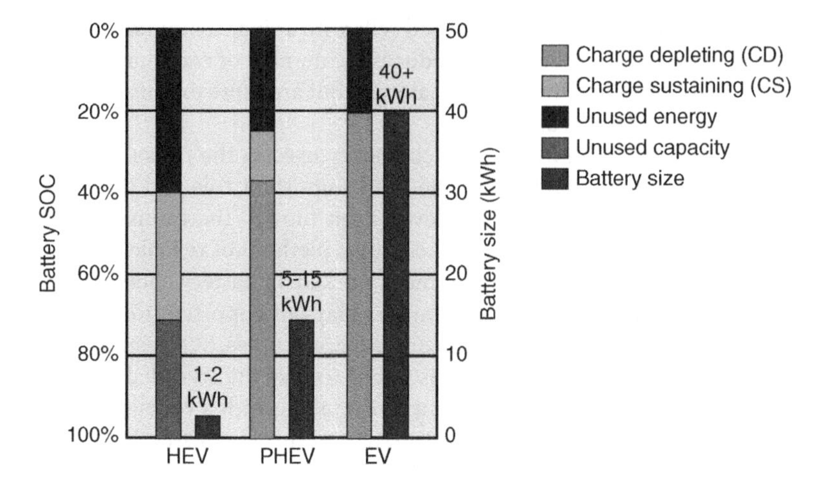

Figure 2.14 Battery size and operating characteristic of various electric vehicle configurations (adapted from Chrysler Group LLC, 2012).

2.5.2 Battery States Estimation

Having an accurate estimation of the battery states enables evaluating the safety and reliability of the operating environment and yields data regarding the charge/discharge status, which plays a special role in cell balancing. Usually the battery states include state of charge (SOC), state of health (SOH) and state of function (SOF). State of charge is analogous to fuel usage indication in gasoline cars (compares to a fully charged battery) but can be very hard to estimate due to not being able to measure aging, varying environmental conditions, charge/discharge cycles directly. State of health compares the utilized battery to a new one. It is not a physical quantity, but is a very useful measure that deals with the battery aging and degradation and provides information on the remaining life of the battery. Finally, state of function measures the performance of the battery in meeting the real demands (such as delivering power to the vehicle) and is determined both by SOC and SOH along with the operating temperature.

2.5.2.1 SOC Estimation Algorithm

Estimating SOC is crucial in EV applications in order to avoid unpredicted system interruption and permanently damaging the internal structure of the batteries, but it cannot be measured directly since it involves the measurement of various characteristics such as the battery voltage, current, temperature and other information related to the analyzed battery. The precision of the SOC is essential since it indirectly indicates its residual capacity and plays an important role in safely charging and discharging the battery, maximizing its driving distance and cycle life, and managing the power distribution strategy for all types of electric vehicles. Different vehicle configurations utilize different SOC ranges with respect to their application, use and battery size. The SOC ranges used in each type of electric vehicle along with their typical battery size are shown in Figure 2.14.

Even though there is no generally accepted definition of SOC, it is usually considered to be the ratio between the remaining and total charge of the battery (when it is full) at the same specific standard condition (or present capacity expressed in terms of its rated capacity). The estimation of SOC for a battery pack requires the determination of SOC for each module and the consisting cells and the whole package could be treated as a single cell due to the self-balancing characteristics of parallel connection. Current SOC estimation algorithms are as follows:

Discharge Test Method: This is one of the most dependable methods to calculate the battery SOC by precisely finding the remaining charge of the battery; however, it requires significantly long periods of time and completely drains the battery, making it nearly impossible to use for real vehicle applications.

Coulomb Counting Method: Coulomb counting is a relatively straightforward and popular method for low power applications which characterizes the energy in a battery in Coulombs simply by accumulating the charge transferred in or out of the battery. It can provide adequate precision if the initial SOC is relatively precise, however, this method is very sensitive to measurement errors caused by the offset temperature drifting of the current sensor (error accumulation glitch) and may lead to large SOC error over time. Thus, the reference point must be compensated and the SOC estimation should be updated under different measure voltages (Xing, 2011).

Open Circuit Voltage Method: Another method is the relationship between SOC and open-circuit voltage (OCV), but are not suitable for Li-ion batteries since the dependence of OCV to SOC is very small in this chemistry, requiring very accurate measurement of OCV which must be done in steady-state. Thus, this method is not realistic for real time SOC estimation in EVs and can only be used in special cases such as when the vehicle is in parking rather than driving.

Artificial Neural Network Model Method: This universal method does not require any significant expertise in the modelling complex systems due to adopting a "black box" approach to various sources of data and therefore can be used for all battery chemistries as long as sufficient sample data are provided to train the network and estimates SOC through non-linear mapping characteristics of the neural network. However, the method is considerably affected by estimation errors and needs extensive computation which in turn requires powerful processing chips.

Fuzzy Logic Method: In this method, a certain level of uncertainty/ambiguity is allowed in processing incomplete and noisy data as opposed to targeting to achieve precise information. This method simulates the fuzzy thinking of human beings with respect to a plethora of test curves, experience and reliable fuzzy logic theories in order to determine the SOC. However, it requires a complete understanding of the battery chemistry and requires extensive hardware for calculations (Zhang and Lee, 2011).

Traditional and Extended Kalman Filter Method: Kalman Filter Method filters measurements of system input and output to produce an intelligent estimation of a dynamic system's state using minimum mean squared error estimate of the true state. The traditional method is used for linear problems, while the extended version linearizes the prediction by using partial derivatives and Taylor series expansion where after linearization the remaining process becomes highly similar to traditional Kalman filter method (Xing, 2011).

Electrochemical Impedance Spectroscopy (EIS) Method: EIS is used mainly to provide information with respect to electrochemical reactions associated with the batteries that cannot be obtained by usual measuring devices. This method requires an electrochemical model of the battery and triggering AC signals at various frequencies to calculate the values for the modelled components associated with the battery. However, this method requires the system to stay at steady state levels throughout testing and therefore is not very compatible with EV applications.

2.5.2.2 SOH Estimation Algorithms

SOH describes the physical condition of the battery with respect to its internal (e.g., loss of rated capacity) and external (temperature extremes) behavior. Similar to SOC, there is no industry definition of SOH and how it should be determined, however, it reflects the battery's ability to deliver specified performance compared to a fresh battery and is usually taken as the ratio of the present condition of the battery cell compared to its ideal conditions. It can be derived by several characteristics such as battery capacity, internal resistance, AC impendence, self-discharge rate and power density and is usually used to characterize the ability to drive a specific distance or range. The battery SOH is usually reduced due to battery aging and degradation due to durability problems (including damages caused by collusion, battery short circuit etc.) and when the battery SOH levels reach to 80%, the BMS warns the driver to replace the batteries in the vehicle. Usually, the most dominant inputs that have the highest impact on the cycle-life and the safety of the batteries are temperature extremes, high potential or overcharge/overdischarge rates. Cell aging can happen under various scenarios including loss of active material and/or lithium inventory, kinetic degradation or increase in polarization resistance, formation of parasitic phases and lithium plating.

Two mainly used SOH estimation methods are "Durability Model-based Open-loop SOH Estimation Method" which uses battery durability model to directly determine the capacity loss and variations in the impedance, and "Battery Model-based Parameter Identification Closed-Loop SOH Estimation Method" which utilizes optimal state estimation technologies and identifies main modelling variables in order to obtain SOH of the batteries (Languang *et al.*, 2012).

2.5.2.3 SOF Estimation Algorithms

SOC and SOH are important indicators that provide useful information on the how much the utilized battery varies from its fully charged version and its brand new version, respectively. However, they do not provide much help on the battery's current capability in terms of performing the required tasks. Thus, SOF is usually utilized to illustrate how the performance of the battery meets the real demand while it's employed. Even though calculating SOF of a cell would be relatively straightforward if SOC and SOH of the cells are calculated; determining the SOF of the battery module/pack is usually more tangible and useful, but much harder to calculate. The relationship between SOF, SOC and SOH are shown in Figure 2.15. Performance profile, voltage limit and operating temperature would also be needed to develop a better evaluation.

2.5.3 Charge Equalization

It should be initially noted that the battery can be charged by different methods in EV applications. Constant voltage charging is the simplest charging method and is suitable for all kinds of batteries. As the voltage remains the same, the current usually starts large and decreases gradually to zero when the battery is fully charged. This method

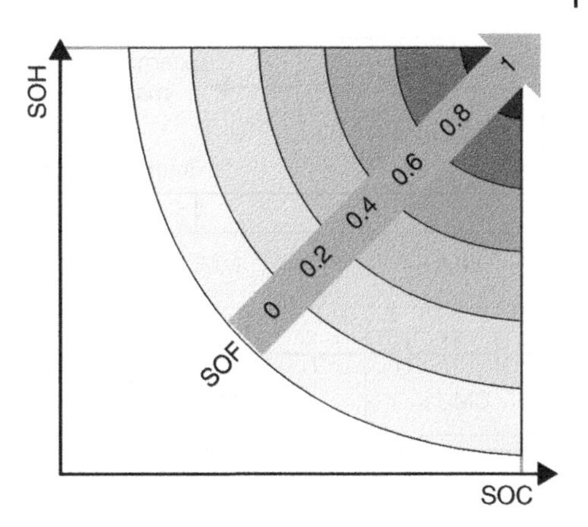

Figure 2.15 Qualitative dependency of SOF on SOC and SOH at a given temperature (adapted from Meissner and Richter, 2003).

has certain advantages since the current through the battery is self-limiting. However, it has higher costs due to requiring the charger to provide fairly large currents during the initial stages of the charging process and may not be provided by most residential and parking structures. On the other hand, in constant current method, the current is fixed by controlling the voltage applied to the battery, which provides a linear increase of SOC in the battery. However, unlike the previous method, it requires the accurate estimation to determine when the battery is fully charged in order to stop the charging process. This is usually done through using various parameters including the temperature, voltage and charging time. In real life applications, the combination of these two methods is usually used. At the initial stage, battery is charged at constant current (initially in low currents followed by high currents). After the battery reaches to a certain state of charge threshold, the charging switches into constant voltage and can be used to maintain the battery voltage as long as the DC charging supply is still available. The constant voltage and current charging process is illustrated in Figure 2.16.

Another concern is that, even though the battery pack is composed of cells with same type and specification, there would still be some dissimilarity in terms of among the cells (in term of voltage, SOC, capacity loss, internal resistance cycle-life, etc.) over time due to manufacturing and chemical offset of the cells. These differences would increase significantly at temperature extremes and non-uniformity. When the battery consists of a large number of cells in series, its failure rate will be higher than any single cell due to

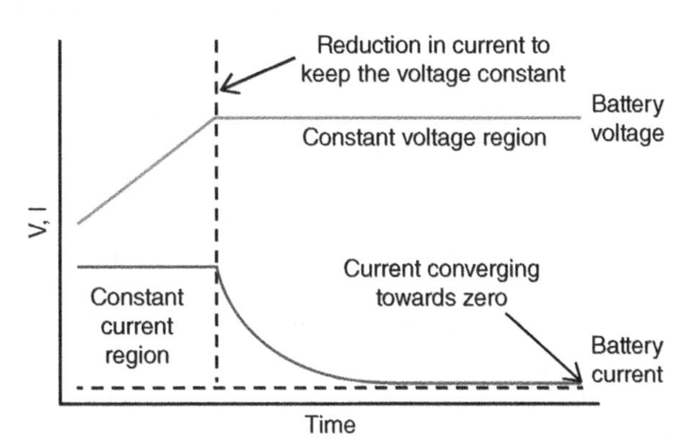

Figure 2.16 Battery voltage and current during recharge (adapted from A123, 2013).

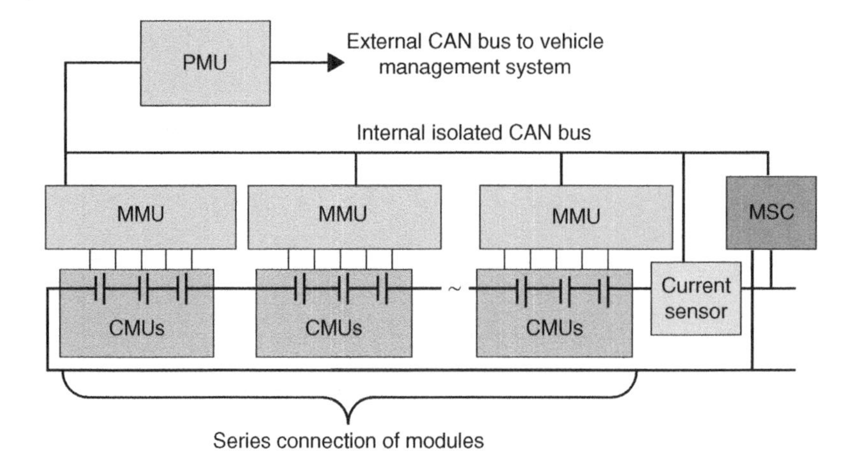

Figure 2.17 Hierarchical structure of a typical BMS (adapted from Brandl *et al.*, 2012).

the series network. Thus, in order to reduce this effect and prolong the battery life, an effective cell balancing architecture as well as method would be needed by the BMS.

2.5.3.1 Hierarchical Architecture Platform/Communication

In electric vehicle applications, various parallel and series connections of a plethora of cells are established to satisfy the high energy and power requirements of the vehicle. In these configurations, the battery is divided into nested levels called cell, module and pack, which lead to a corresponding hierarchical architecture of the BMS systems in order to have a flexible and redundant battery management system. These layers are usually called the Cell Monitoring Unit (CMU) and Module Management Unit (MMU), where the initial monitors each individual cell in the module and incorporates the hardware implementation of the charge equalizer circuit whereas the latter provides higher services to the Pack Management Unit (PMU), the unit that supervises all the battery string and controls the amount of energy stored in each cell by estimating the associated SOC. Usually, the connection between each CMU and MMU is done via a dedicated bus whereas a shared Controller Area Network (CAN) is used between MMU and PMU. Finally, PMU is linked to Vehicle Management System (VMS). Finally, a battery protection unit also exists to prevent the battery by disconnecting the current flow through the Main Switch Unit (MSU) contactor. This configuration is shown in Figure 2.17.

2.5.3.2 Cell Equalization

Due to current manufacturing processes, the battery cells show certain differences. Cell capacity variation from a couple percentage points to 15% is common. Moreover, other variations (including internal resistance and charge/discharge characteristics) are usually inevitable. In addition, the discrepancies in cell capacities become larger over time with charge/discharge cycling of the batteries. In EVs where many cells are connected in series, the cells with the lowest remaining charge suppress the total system capacity and therefore reduce the performance and/or overall range of the vehicle. Thus, cell equalization is highly important to fully utilize the potential of the battery chemistry (Cheng *et al.*, 2011) as shown in Figure 2.18. It should be noted that not all layers of these platforms have to be utilized in BMS and most commonly, the first layer is omitted due to

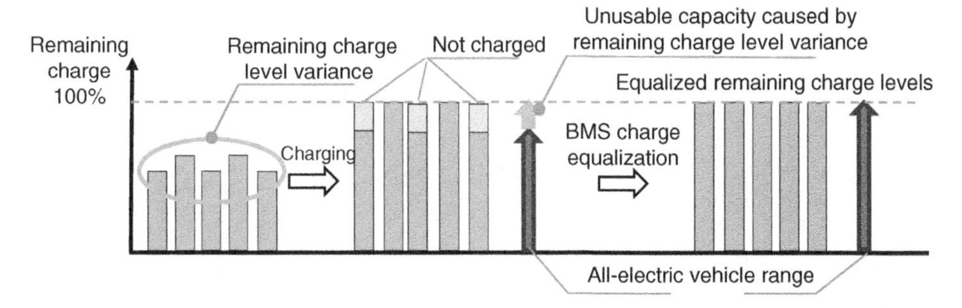

Figure 2.18 The effects of charge imbalance on all-electric vehicle range (adapted from Honda, 2015).

impact on the cost and self-discharge rates associated with providing each cell with a dedicated CMU.

Equalization is usually achieved through dissipative and non-dissipative methods. Both of these methods are devoted to reducing the cell voltage imbalance. In dissipative cell balancing, the module compares the maximum difference in cell voltages and if this value is higher than the preset threshold, charging stops and the cell with the highest voltage dissipates the excess energy or current in order to level the outstanding capacity among cells within the battery unit. This is usually done through heat by using the resistors or other means until the difference diminishes. This is a popular method in EVs due to its simplicity and ability to achieve equalization without consuming substantial levels of energy; however, it is a very inefficient process.

On the other hand, non-dissipative equalizers are generally implemented by transformer, inductor or capacitor to transfer the charge from the highest voltage cell to the lowest voltage cell. Even though the efficiency of this method is higher, the exchange of charge and energy among the cells makes this more complicated than the previous method and therefore has not started to be widely used in many applications. Since self-discharge has the largest impact on the battery uniformity, equalization is primarily utilized to compensate for the non-uniformity associated with this phenomenon, which is typically around 3–5% in Li-ion chemistry. The most commonly used cell equalization and SOC determination methods and summary of their advantages/drawbacks are listed in Table 2.4.

2.5.4 Safety Management/Fault Diagnosis

Finally, in order to ensure the battery safety, various safety measured would be necessary in EVs since the electrical risks have more significant consequences in these vehicles compared to their conventional alternatives due to their requirement to deliver high power and high energy from the traction battery. After determining the SOC and SOH of the battery, BMS calculates the maximum charge/discharge current at any given moment based on the algorithm and transfers this data to the ECU. This plays an important role in keeping the battery operating in its safe zone and inhibits the battery to violate its operating specifications. The main tasks of BMS for safety include overcharge, deep discharge and over temperature protection, power line cut-off in case of electrical short-circuits and loss of electrical isolation. Moreover, the BMS also stores the battery data for better maintenance and improved serviceability. The diagnostic system usually captures and stores the atypical operating status of the

Table 2.4 The main SOC determination and cell equalization methods used in EV BMSs.

	Method	Input	Advantage	Disadvantage
SOC Determination	Discharge Test	Charge remaining and capacity	Simple and accurate	Offline, time intensive, energy loss, modifies the battery state
	Coulomb Counting	Discharging current, time	Online and easy to operate	Hard to determine initial SOC, accuracy is highly effected by temperature, battery history, discharge current, and cycle life, Doesn't take self-discharge of the battery into account,
	Open Circuit Voltage	Rest time and voltage	Easy to implement and accurate	Requires long rest times, sensitive to different discharge rates and temperatures, only suitable in very high and low SOCs for some Li-ion chemistries
	Artificial Neural Network Model	Current, voltage, cumulative charge, initial SOC	Online, can be utilized without knowledge on cell internal structure	Requires high amount of training data
	Fuzzy Logic	AC impedance, voltage, current	Allows complex systems to be modelled easily	Requires extensive hardware for calculations, may result in low accuracy
	Kalman Filter	Current, temperature, internal resistance, coulomb efficiency, self-discharge rate	Online, high accuracy, dynamic, insensitive to noise and initial SOC value error	Computationally intensive, high complexity, volatility if the gain is undesirable
Equalization	Passive (Dissipative)	Cell-to-heat	Very simple and cheap	Low efficiency and slow
	Active (Non-dissipative)	Module-to-cell	Relatively simple, moderate efficiency and high speed	Switch network, high isolation voltage of the DC/DC
		Cell-to-cell (distributed)	Moderate efficiency and speed	Bulky and complex control
		Cell-to-cell (shared)	High efficiency and speed	Switch network
		Cell/module bypass	High efficiency, speed and reliability	High current switches, implementation complexity, reduced battery efficiency during normal operation

battery and provides information to the driver. Thus, in International Electrotechnical Commission (IEC) requires that all BMS for EVs must have battery fault diagnosis functions.

2.5.5 Thermal Management

BMS is also in charge of keeping the battery operation within preset limits in order to prevent the reduction in performance and/or fast aging of the battery. Accurately monitoring the cell temperature is crucial since cell equalization problems are often related to poor thermal management issues (Qiang *et al.*, 2006), since equalization requires the modules to have almost uniform electrical characteristics, which considerably depend on their given temperature. Thus, BMS acquires data on the ambient and battery temperatures, controls cooling/heating operations and transmits this information to ECU during undesired operating conditions (Cheng *et al.*, 2011).

2.6 Battery Manufacturing and Testing Processes

Battery chemistries play a paramount role in the performance, cost and safety of EVs and therefore considerable research and development are being made in producing better cell chemistries and management systems to reduce the cost and increase the capacity, efficiency and cycle life of the batteries. The snapshots from the production of lithium-ion pouch cells are shown in Figure 2.19.

Even though the battery research conducted on cell chemistry is highly important; manufacturing these units and integrating them into modules and packs is an essential part of EV design and production process to ensure a safe and long operation. Thus, the manufacturing processes from individual cells to the entire battery package and corresponding testing procedures need to be examined to understand the steps taken before the battery pack is placed into the vehicle.

2.6.1 Manufacturing Processes

With the rapid developments in electric battery chemistries and the associated increase of EV market penetration in the developed countries, the need for producing cheap, reliable and long lasting battery cells becomes more important than ever. In addition to the battery chemistry, the shapes of the battery cells also play an important role on the performance and the cost of the battery. Since Li-ion cells are the fastest growing chemistry in the market, most of the manufacturing processes will be explained with respect to this chemistry. The cell manufacturing methodology for a typical lithium-ion cell, to be used in EV applications, is shown in Figure 2.20.

Furthermore, even though most of the information provided in this chapter was regarding batteries in a cell level, they have to be placed in modules that encompass numerous cells that are electrically and mechanically joined to scale-up the battery capabilities and to simplify the battery control functions. This requires compatibility with various cell types, integration of series-parallel electrical connections, installation of various auxiliary members and successful connection of the battery cells. Thus, significant research is also currently being conducted on developing and manufacturing battery modules that can be used for various applications.

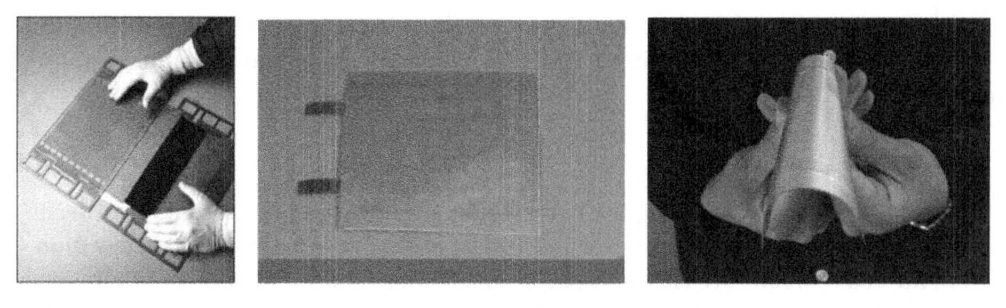

Figure 2.19 Production of typical pouch Lithium-ion cells (courtesy of TUBITAK Marmara Research Center).

Although the main cell chemistry and development are virtually identical for all types of Li-ion automotive cells, cell size/geometry and the ratio of the included materials differentiate the cell production in terms of being used in different applications. After the cells are produced, they are, along with their necessary supplementary components (such as frames, foams additional cooling plates etc.), placed into an assembly line and are inspected for their properties (size, tab shape, position), conditions (physical damage, defects etc.) and electrochemical characteristics (capacity, voltage, etc.) using various optical, electrical, ultrasonic or x-ray devices. Then these components are assembled and aligned into a stack and clamped/strapped with end plates in order to form a battery module (Li *et al.*, 2010). The cell joining process in these modules also plays an important

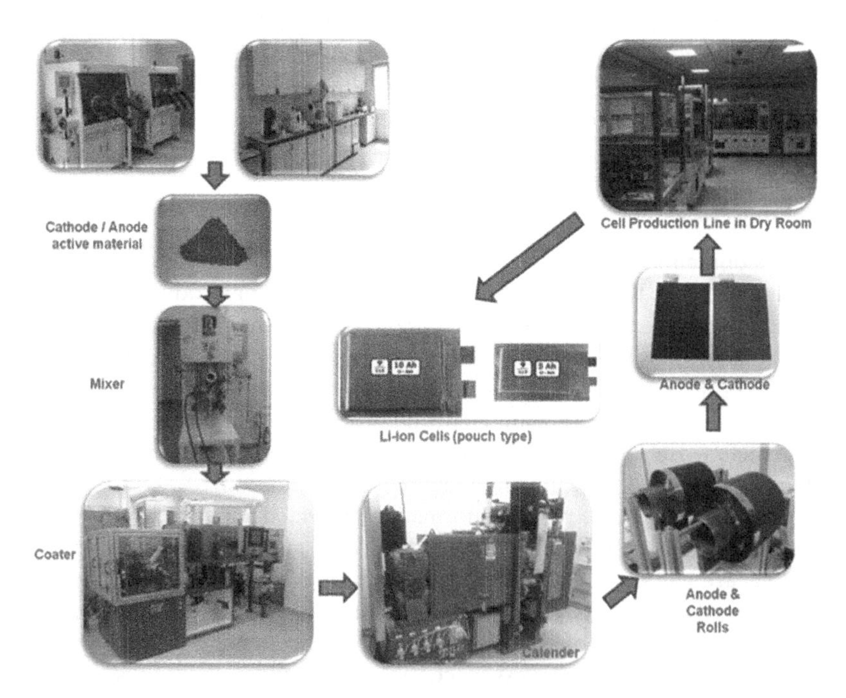

Figure 2.20 Conventional cell production methodology for Li-ion pouch cells (courtesy of TUBITAK Marmara Research Center).

role due to the difficulties associated with welding multiple layers of thin, considerably conductive and different materials with utter reliability.

Among those, resistance welding relies on the electrical resistance at the metal interface to produce fusion of the materials and is a fast and automatable process. However, the process has certain drawbacks since generally used aluminum and copper electrodes are not very compatible with this process as they have high thermal and electrical conductivities and dissimilar melting temperatures. Laser welding overcomes these drawbacks and can heat the materials using intense laser beam (non-contact) in milliseconds, but requires precise joint fit-up to have successful welds. Moreover, the high reflectivity and thermal conductivity of these battery materials cause additional difficulties. In addition, ultrasonic welding can also be used, especially for dissimilar materials and can have good performance at low temperatures in battery materials. This process is also environmentally benign due to the absence of any filler metals or gases. However, it cannot be used for joints under a certain thickness and has sensitivity to surface conditions. But they are considered to be superior with respect to the other welding techniques, particularly for pouch prismatic cells. Finally, the cells can also be joined together mechanically by fasteners (nuts, bolts, screws, etc.) or integral joints. Even though welding can have various advantages in joining the battery cells, they may not be suitable for connecting modules because of concerns associated with maintenance and serviceability. Thus, mechanical joining can improve disassembly for maintenance/repair in these cases. However, improper use of these joints can cause excessive heating and short circuit of the battery (Lee *et al.*, 2010).

Finally, the manufactured battery package should be produced with modular, standardized and common parts, supplementary equipment (such as cell electronics and high and low voltage circuits), minimum interconnection cables, capable cooling devices and efficient and effective welding methods as illustrated in Figure 2.21. A service plug (also called service disconnect) is usually placed in the electrical path of the battery stack to split it into two electrically isolated halves when needed. This prevents the technicians to be exposed to high potential electrical danger in the main terminals. The battery pack also contains a number of relays, or contactors, which govern the distribution of the battery pack's electrical power to the output terminals.

2.6.2 Testing Processes

After the EV battery pack is produced, it goes through rigorous testing procedures to validate the safety and performance of the battery package and its components and to ensure it can successfully fulfill its intended purpose. EV battery packs are large and complex systems composed of a large number of cells and modules (due to the relatively low energy densities of the battery cells) used as rechargeable electric storage to supply the necessary energy and power requirements of the vehicle. Even though controlled release of the battery's energy can provide useful electric power in forms of current and voltage, significant dangerous conditions, such as release of toxic materials, fire, high-pressure event and their combinations, can occur when this energy is released in an uncontrolled fashion. This energy release can be caused by both severe physical (such as crushing, puncturing and burning) and electrical abuse (shortened cells, extreme discharge rates, overcharging and internal heating) of the battery. Thus, the battery is tested to confirm that it can survive any static and dynamic mechanical stresses and the effects of environmental conditions it may be subjected to in the

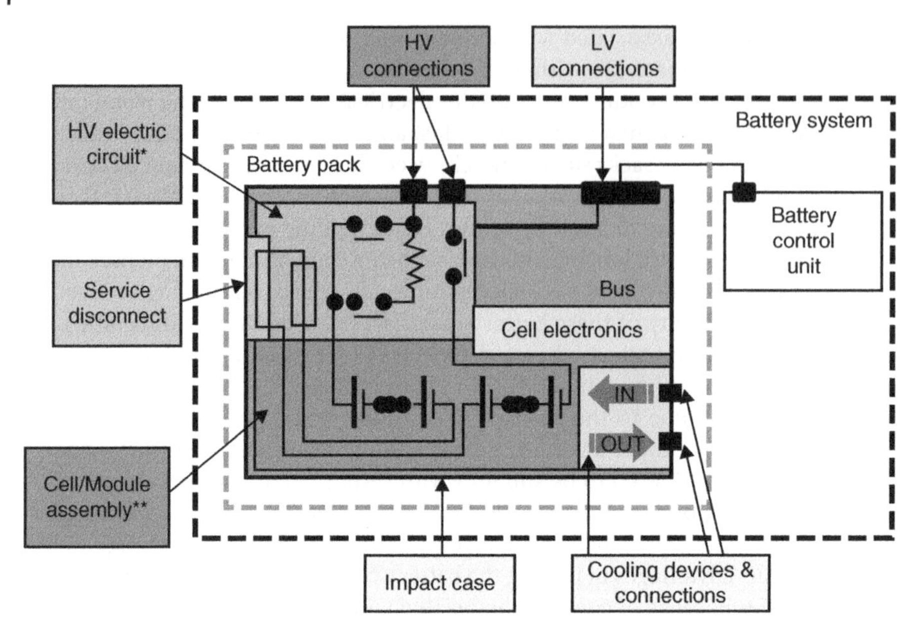

Figure 2.21 Typical battery pack diagram for EV applications (adapted from ISO 12405-2, 2012).

vehicle. Even though various management and safety systems are put in place in the battery pack to identify the conditions, prevent the occurrence and the reduce severity of undesired events, significant testing has to be conducted before the batteries can be commercially sold. Some of the typical testing equipment for EV batteries are shown in Figure 2.22.

Since conducting all tests for a large and powerful battery pack at once can be a long, complex and misleading process, dividing the battery into manageable assembly levels and testing with respect to the category and severity of the abuse associated with that level can improve the time and effectiveness of this process. Initial tests are usually conducted on a cell level in order to test the charging ability, mechanical integrity and thermal stability of the cell. Once multiple cells are put together to develop a module, simple charging and discharging test are conducted to ensure that the connections between the cells are robust and can handle the intended loads without failing or shedding excessive heat. Finally, with the development of the battery package, tests are conducted to confirm the proper operation and functionality of the communication and safety devices. The typical tests conducted on most EV batteries are provided in Table 2.5.

It should also be noted that these tests and procedures are generally not only conducted for verification purposes in terms of satisfying the manufacturer's specifications, but also utilized to check their compatibility with the arbitrary limits set by the engineers to observe the components when faced with undesirable operating conditions and excessive loads, to determine the failure modes and safety factors.

Figure 2.22 Typical cell/module testing equipment (courtesy of TUBITAK Marmara Research Center).

Table 2.5 Basic tests typically conducted on the EV batteries.

Name	Category	Abuse level[a]	Min. assembly level	Name	Category	Abuse level	Min. assembly level
Shock	Mechanical	1	Module	Overheat/Thermal Runaway	Thermal	2	Module
Drop	Mechanical	3	Pack	Thermal Shock	Thermal	2	Cell
Penetration	Mechanical	3	Cell	Elevated Temperature Storage	Thermal	2	Cell
Roll-over	Mechanical	1	Module	Extreme-cold Temperature	Electrical	2	Cell
Immersion	Mechanical	3	Module	Short Circuit	Electrical	3	Module
Crush	Mechanical	3	Cell	Partial Shot Circuit	Electrical	3	Module
Simulated Fuel Fire	Thermal	3	Cell	Overcharge	Electrical	2	Module
Thermal Stability	Thermal	3	Cell	Overdischarge	Electrical	2	Module
Compromise of Thermal Insulation	Thermal	2	Module	Vibration	Mechanical	1	Cell

a) Abuse Level 1: Passive protection activated, 2: Defect/damage, 3: Leakage with Δmass<50%
Source: Doughty and Crafts (2006).

2.7 Concluding Remarks

EV batteries are required to have high energy densities for long driving ranges, and high power density for rapid acceleration, as well as to be durable and inexpensive for making good business and economic sense. A wide range of battery technologies are currently available for electric vehicle applications, with their corresponding benefits and drawbacks, and alternative chemistries are also currently investigated to have a comparable technology and ease the transition between the conventional and electric vehicles by improving the performance, reducing the cost and environmental impact. The fast technological advancements for better chemistries are expected to result in production of cost effective and durable batteries. However, the reliability of these batteries still creates significant concerns due to issues associated with thermal stability and capacity loss in high current applications. Thus, finally capable BMS are described to utilize the battery within the safe and reliable operating temperature and voltage windows.

Nomenclature

Acronyms

BMS	Battery management system
CAN	Controller area network
CMU	Cell monitoring unit
EV	Electric vehicle
GHG	Global greenhouse gas
HEV	Hybrid electric vehicle
ICE	Internal combustion engine
IEC	International electrotechnical commission
MMU	Module management unit
MSU	Main switch unit
OCV	Open-circuit voltage
PHEV	Plug-in hybrid electric vehicle
PMU	Pack management unit
SEI	Solid electrolyte interface
SOC	State of charge
SOF	State of Function
SOH	State of health
TMS	Thermal management system
VMS	Vehicle management system
VRLA	Valve regulated lead acid

Study Questions/Problems

2.1 In your opinion, do the battery chemistries that are currently utilized in commercially available electric vehicles have sufficient performance characteristics to fully replace conventional vehicles? Please explain your answer.

2.2 What are the most common problems that prevent/limit the batteries that are still under research to be used in practical vehicle applications? What are some of the approaches/methods that are taken to resolve these issues?

2.3 Which components of the EV battery pack have the highest cost and environmental impact? How are these factors change with respect to different battery chemistries?

2.4 If a lead acid battery with a 10 hour discharge rating of 100 Ah gets discharged at 5 A, approximately how long would the battery last? Please explain.

2.5 Please derive equation 2.11c by completing the steps of 2.11b for two different temperature values.

2.6 The rate of constant for a reaction is 9.16×10^{-3} s^{-1} at 15°C. If the temperature is raised by 5°C, what would the new rate constant be? (E_a is 88 kJ/mol).

2.7 Please list the main tasks of the EV battery management system and the key functions performed to successfully complete these tasks.

2.8 What are the main inputs used in the EV battery package and the accessories interacted by the system to safely operate the battery and the vehicle.

2.9 Please briefly describe the concepts of state of charge (SOC), state of health (SOH) and state of function (SOF) and their relationship with each other.

2.10 Which state of charge estimation and charge equalization method do you think is the most appropriate for EVs or HEVs in your city/country? Please explain your reasoning.

References

A123 System User Manual. (2013). Cylindrical Battery Pack Design, *Validation, and Assembly Guide.*

Andrea D. (2010). Battery Management Systems for large Lithium-Ion Battery Packs. *Artech House, Norwood, MA.* **02062.**

Argonnne National Laboratory. (2010). *A Review of Battery Life-Cycle Analysis: State of Knowledge and Critical Needs,* Center of Transporation Research.

Brandl M, Gall H, Wenger M, Lorenthz V, Giergerich M, Baronti F, Fantechi G, Fanucci L, Roncella R, Saletti R, Saponara S, Thaler A, Cifrain M, and Prochazka W. (2012). Batteries and Battery Management Systems for Electric Vehicles . *Proceedings of the Conference on Design,* Automation and Test in Europe.

Bossche PV, Vergels F, Mierlo JV, Matheys J, Autenboer WV. (2006). SUBAT: An assessment of sustainable battery technology. *Journal of Power Sources* **162**:913–919.

Bruce PG, Freunberger SA, Hardwick LJ, Tarascon JM. (2012). Li-O2 and Li-S batteries with high energy storage. *Nature Materials* **11**:19–29.

Cairns EJ, Albertus P. (2010). Batteries for Electric and Hybrid-Electric Vehicles. *Annual Review Chemical Biomolecular Engineering* **1**:299–320.

Cheng KWE, Hongje BPD, Ding K, Ho HF. (2011). Battery-Management System (BMS) and SOC Development for Electrical Vehicles. *IEEE Transactions on Vehicular Technology* **60**:76–88.

Chrysler Group LLC. (2012). *Overview of Battery Management System – A Controls Engineering Perspective*, EV Battery Conference.

Climatemps. (2014). *Ottawa, Ontario Climate & Temperature*. Available at: http://www.ottawa.climatemps.com/ [Accessed March 2015].

Conte FV. (2006). Battert and battery management for hybrid electric vehicles: a review. *Elektrotechnik & Informationstechnik* **123**/10:424–431.

Cugnet M, Dubbary M, Liaw BY. (2010). Peukert's Law of a Lead-Acid Battery Simulated by a Mathematical Model. *ECS Transactions* **25**(35):223–233.

Dixon J. (2010). Energy Storage for Electric Vehicles. *IEEE International Conference on Industrial Technology*, pp. 20–26, March 14–17, 2010. DOI: 10.1109/ICIT.2010.5472647 ISBN: 978-1-4244-5695-6

Doughty HD, Crafts CC. (2006). *FreedomCAR Electrical Energy Storage System Abuse Test Manual for Electric and Hybrid Electric Vehicle Applications. Sandia National Laboratories*. Available at: http://prod.sandia.gov/ [Accessed March 2015].

Egan DR, Leon CP, Wood RJK, Jones RL, Stokes KR, Walsh FC. (2013). Developments in electrode materials and electrolytes for aluminum-air batteries. *Journal of Power Sources* **236**:293–310.

Electrification Roadmap, Revolutionizing Transportation and Achieving Energy Security. (2009). *Electrification Coalition*. Available at http://www.electrificationcoalition.org/ [Accessed June 2015].

Ellis BL, Nazar LF. (2012). Sodium and sodium-ion energy storage batteries. *Current Opinion in Solid State and Materials Science* **16**:168–177.

Etachari V, Marom R, Elazari R, Salitra G, Aurbach D. (2011). Challenges in the development of advanced Li-ion batteries: a review. *Energy & Environmental Science* **4**:3243–3262.

Gerssen-Gondelach SJ, Faaij APC. (2012). Performance of batteries for electric vehicles on short and longer term, *Journal of Power Sources* **212**(15):111–129.

Girishkumar G, McCloskey B, Luntz AC, Swanson S, Wilcke W. (2010). Lithium-Air Battery: Promise and Challenges. *The Journal of Physical Chemistry Letters* **1**(14):2193–2203. DOI: 10.1021/jz1005384

Grosjean C, Miranda PH, Perrin M, Poggi P. (2012). Assessment of world lithium resources and consequences of their geographic distribution on the expected development of the electric vehicle industry. *Renewable and Sustainable Energy Review* **16**:1735–1744.

Guerrero CPA, Li J, Biller S, Xiao G. (2010). Hybrid/Electric Vehicle Battery Manufacturing: The State-of-the-Art. 6th Annual IEEE Conference on Automation Science and Engineering. Toronto, ON. Canada, August 21–24, 2010.

Honda. (2015). Available at: http://world.honda.com/EV-neo/battery/ [Accessed January 2015].

ISO 12405 – 2. (2012). *Electrically propelled road vehicles—Test specification for lithium-ion traction battery packs and systems—Part 2: High energy applications.* Available at: http://www.iso.org/iso/catalogue_detail?csnumber=55854 [Accessed January 2015].

ISO14040. (2006). *"Environmental Management – Life cycle assessment – Principals and Framework"* International Organization for Standardization Geneva, Switzerland.

Lee SS, Kim TH, Hu SJ, Cai WW, Abell JA. (2010). Joining Technologies for Automotive Lithium-ion Battery Manufacturing – A review. Proceedings of the ASME 2010 *International Manufacturing Science and Engineering Conference*, Pennsylvania, USA

Li S, Wang H, Lin Y, Abell J, Hu S.J. (2010). Benchmarking of high capacity battery module/pack design for automatic assembly system. Proceedings of the ASME 2010 *International Manufacturing Science and Engineering Conference*, Pennsylvania, USA.

Languang L, Xuebing H, Jianqiu L, Jianfeng H, Minggao O. (2012). A review on the key issues for lithium-ion battery management in electric vehicles. *Journal of Power Sources* **226**:272 – 288.

Majeau-Bettez G, Hawkins TR, Strømman AH. (2011). Life Cycle Environmental Assessment of Lithium-Ion and Nickel Metal Hydride Batteries for Plug-in Hybrid and Battery Electric Vehicles. *Environmental Science and Technology* **45**:4548 – 4554.

Meissner E, Richter G. (2003). Battery Monitoring and Electrical Energy Management Prediction for Future Vehicle Power Systems. *Journal of Power Sources* **116**:79 – 98.

OECD/TF. (2010). *Reducing Transport GHG Emissions – Trends and Data.* Available at http://www.itf-oecd.org/ [Accessed June 2015].

Padbury R, Zhang X. (2011). Lithium-oxygen batteries- Limiting factors that affect performance. *Journal of Power Sources* **196**:4436 – 4444.

Peled E, Golodnitsky D, Mazor H, Goor M, Avshalomov S. (2011). Parameter analysis of a practical lithium- and sodium-air electric vehicle battery. *Journal of Power Sources* **196**:6835 – 6840.

Petersen J. (2011). *Global autos: Don't believe the hype—analyzing the costs & potential of fuel-efficient technology.* S. Shao (Ed.). Bernstein Global Wealth Management.

Pillot, C. (2013). *Micro hybrid, HEV, P-HEV and EV market, 2012 – 2015.* Impact on Battery Business, Avicenne Energy. Available at http://www.avicenne.com/

Qiang J, Ao G, Zhong H. (2006). Battery Management System for Electric Vehicle Applications. IEEE International Conference on Vehicular Electronics and Safety :134,138.

Roland Berger Strategy Consultant. (2012). Technology & Market Drivers for Stationary and Automotive Battery Systems, *Batteries, Nice France, October* 24 – 26, 2012.

Scrosati B, Hassoun J, Sun Y. (2011). Lithium-ion batteries. *A look into the future.* Energy & Environmental Science **4**:3287 – 3295.

Smoljko I, Gudic S, Kuzmanic N, Kliskic M. (2012). Electrochemical properties of aluminum anodes for Al/air batteries with aqueous sodium chloride electrolyte. *Journal of Applied Electrochemistry* **42**:969 – 977.

Sun Q, Yang Y, Fue Z W. (2012). Electrochemical properties of room temperature sodium-air batteries with non-aquous electrolyte. *Electrochemistry Communications* **16**:22 – 25.

USGS Mineral Commodity Summaries. (DATE?) Available at: http://minerals.usgs.gov/minerals/pubs/commodity/lithium/mcs-2013-lithi.pdf [Accessed August 2016].

Whittingham MS. (2012). History, Evolution, and Future Status of Energy Storage. *Proceedings of the IEEE Special Centennial Issue* **100**:1518 – 1534.

Wu H, Chan G, Choi J W, Ryu I, Yao Y, McDowell MT, Lee SW, Jackson A, Yang Y, Hu LB, Cui Y. (2012). Stable cycling of double-walled silicon nanotube battery anodes through solid-electrolyte interphase control. *Nature Nanotechnology* 7:309–314.

Xing Y, Ma EWM, Tsui KL, Pecht M. (2011). Battery Management Systems in Electric and Hybrid Vehicles. *Energies* **4**:1840–1857. DOI: 10.3390/en4111840

Zhang J, Lee J. (2011). A review on prognostics and health monitoring of Li-ion battery. *Journal of Power Sources* **196**:6007–6014.

3

Phase Change Materials for Passive TMSs

3.1 Introduction

There are two basic types of heat storage methods, such as sensible (based on temperature difference) and latent (based on phase change). The phase change materials (PCMs) are considered the main players of latent heat storage materials. The term phase change material (PCM) commonly refers to materials with high latent heat of fusion (LHF) during a predictable, stable and virtually constant temperature phase change (melting/solidifying) process. These materials release/gain isothermal energy during the phase transitions and can have heat storage capacities more than an order of magnitude (per unit volume) higher than most sensible storage materials, especially within small temperature differences. Thus, when the PCMs with suitable phase change temperatures are used for the intended application, they can accumulate and release substantial amount of heat energy when faced with slight changes in temperature. For these reasons, PCMs have attracted considerable attention in various thermal management applications, especially in the past decade. Thus, in this chapter, PCMs will be introduced and the advantages/drawbacks of different type of phase change materials and methods of improving their heat transfer capabilities will be evaluated. Moreover, various applications that are currently being used and are intended for the future are examined. Finally, case studies are presented to familiarize the readers with PCMs and PCM based battery thermal management systems. More detailed thermodynamic approach for utilizing, evaluating and improving PCMs in battery thermal management applications will be explained in the following chapters.

3.2 Basic Properties and Types of PCMs

Before providing in-depth information regarding different types of PCMs and their methods of utilization/improvement for desired applications, it is important that the readers first fully understand what is commonly referred as phase change materials in academia and industry, since this term is often used improperly in various texts. Thus, at the risk of boring some of the readers who are familiar with this concept, this is done by simply breaking down the phrase and providing the associated basic thermodynamic relations. A "phase" is commonly defined as "a set of states of a macroscopic physical system that have relatively uniform chemical composition and physical properties" (such as density, crystal structure, etc.) (Ivancevic and Ivancevic, 2008). Thus, the term

Thermal Management of Electric Vehicle Battery Systems, First Edition.
İbrahim Dinçer, Halil S. Hamut and Nader Javani.

"phase change" refers to the transformation of this thermodynamic system between phases (solid, liquid, gas), with the distinctive characteristics of an abrupt change in single or multiple physical properties, especially the heat capacity, with a slight change in thermodynamic variable (most commonly temperature). As the phase of a material is changed, its molecules reposition themselves, which changes the entropy of that system. The thermodynamic principles necessitate that the material has a transfer of thermal energy (or heat) due to this change in entropy, which is referred as the latent heat of the material (when considered as the unit mass). Thus, even though virtually any substance that goes through this process can be considered as a PCM, when "phase change materials" are being mentioned, what is commonly referred is the substances with high heat of fusion that enables the storage/release of large amount of energy along with certain predictable melting/solidifying temperatures over long cycles. In order to illustrate the magnitude of their thermal capabilities, the heat store capacity of PCMs are compared to sensible heat of various materials (over a 15°C temperature difference) as shown in Figure 3.1.

Thermodynamically speaking, the heat stored in a PCM from a given temperature below the melting point (T_i) to a temperature above (T_f) are provided in Equation 3.1, where the first and last terms on the right hand side refer to the sensible heat storage and the middle term ($ma_m\Delta h_m$) to the latent heat storage, which increases the energy storage considerably.

$$Q = \int_{T_i}^{T_m} mC_p dT + ma_m \Delta h_m + \int_{T_m}^{T_f} mC_p dT \tag{3.1a}$$

$$Q = m\left[C_{sp}(T_m - T_i) + a_m \Delta h_m + C_{lp}(T_f - T_m)\right] \tag{3.1b}$$

where m is the mass of heat storage medium, C is the average specific heat, h is the heat transfer coefficient and a_m is the fraction melted. In these equations, the thermochemical systems rely on the energy gained/released in breaking/reforming molecular bonds to be in a virtually reversible reaction (this assumption will be further scrutinized in Chapter 5), which makes the effectiveness of thermal management system depend mainly on the quantity of storage material, the endothermic heat reaction and the extent of conversion. It should be noted that the specific heats of solid and liquid PCMs only

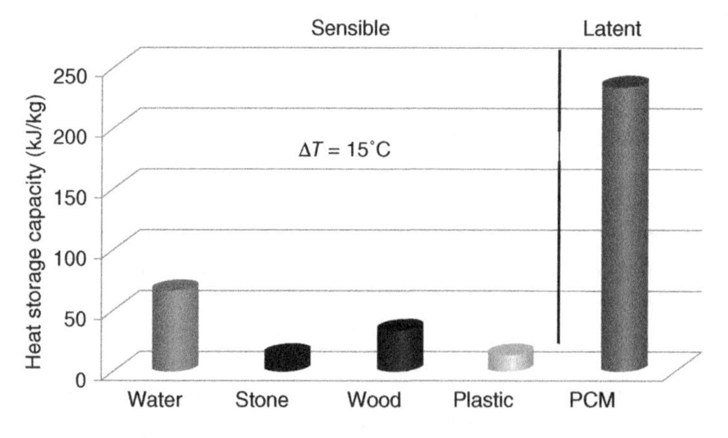

Figure 3.1 The heat storage capacities (sensible versus latent) of various materials.

become the key contributors in the effectiveness of a thermal management system if the operating span of the storage unit highly exceeds the transition zone of the PCM (Dincer and Rosen, 2010). In that case, the system operates more as sensible heat storage unit (as the relative contribution of latent heat is reduced), which reduces the overall performance of the PCM.

In order to provide a wider perspective to the readers, it is worth mentioning that the energy storage (or thermal management) processes can be categorized into physical storage, via sensible heat and latent heat, and chemical (also called thermo-chemical) storage. Sensible heat is generally utilized in applications where there is a significant mass flow of the cooling/heating medium (where water being the cheapest and most commonly used option), the system is not very limited in terms of space and where an inexpensive solution is needed. They are mainly used in domestic systems, district heating and industrials needs. These systems have the advantage of requiring smaller heat exchange surface area between the storage and the heat transfer fluid due to better heat exchanger surface contact. However, they have low energy density and variable discharging temperatures which may require effective designs and additional considerations. Chemical storage systems incorporate thermal storage solutions that utilize a reversible chemical reaction in the medium for heat transfer where chemical reactions store and release thermal energy by making use of the enthalpy of reaction (ΔH) and can offer higher heat storage/transfer capabilities than sensible and latent systems. However, currently they are mostly used for niche areas and are still under development for many other practical applications. Latent heat systems (PCMs) on the other hand have a good combination of both high energy storage/transfer capability and commercial availability. They enable target-oriented discharging temperature and go through narrower temperature fluctuations during charge/discharge which improves the effectiveness of the utilized thermal management system. The main characteristics of the aforementioned systems are compared in Table 3.1.

When PCMs are further analyzed, it can be seen that PCMs can go through three types of phase transitions under latent heat systems, namely solid-solid, solid-liquid and liquid-gas and can be found in various formats as shown in Figure 3.2. Among these, solid-solid and liquid-gas PCMs do not have much real-world use since they have high transition temperatures and large volume changes, respectively (Hyun *et al.*, 2014). Thus, when PCMs are being discussed in this chapter, it will refer exclusively to solid-liquid PCMs as they have considerable utilization in research and applications due to their high latent heat capacities and high thermal conductivity, which enables them to be a great candidate for a wide range of thermal management systems and products that is related to the content of this book.

Table 3.1 Typical parameters of thermal energy storage (TES) systems.

TES System	Capacity (kWh/t)	Power (kW)	Efficiency (%)	Storage Period (h,d,m)	Cost (€/kWh)
Sensible Heat	10–50	1–10,000	50–90	d/m	0.1–10
Latent Heat	50–150	1–1000	75–90	h/m	10–50
Chemical Reactions	120–250	10–1000	75–99	h/d	8–100

Source: (Hauer, 2011).

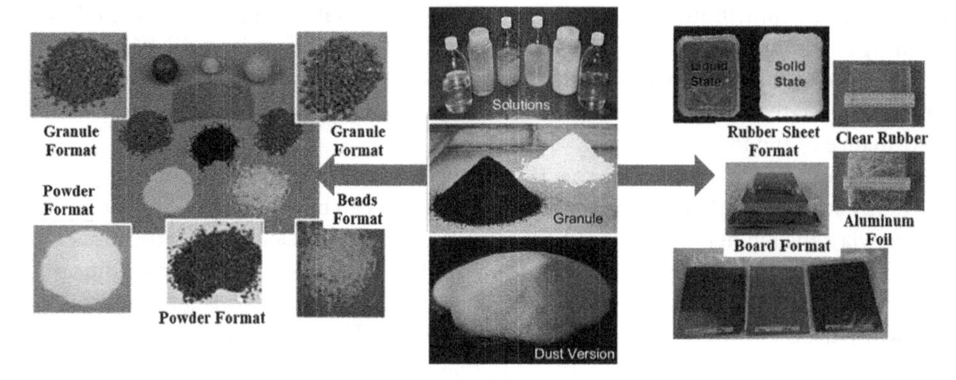

Figure 3.2 Different types of phase change materials (courtesy of PCM Products).

Before the characteristics of these different types of PCMs and their specific advantages and drawbacks are discussed, it is imperative to understand the main aspects that are used in the evaluation and selection of PCMs for various applications. Thus, these aspects are briefly described below (Dincer and Rosen, 2010).

- **Thermal behavior:** The incorporation of PCMs into a system that leads to a considerable improvement in the thermal performance is a challenging process as it usually leads to increasing the thickness of the surrounding heat source/sink which also increases the overall weight of the system. Thus, the thermal behavior of the material plays a significant role in order to increase the associated thermal inertia and the resulting capacity to store energy. The thermal behavior of a PCM depends on various characteristics which will be described in the following sections. However, it should be noted that the heat transfer to/from a storage unit depends mostly on the thermal conductivities of the solid and liquid PCM. Nevertheless, the heat transfer phenomena during phase change of the material is a highly complex process due to various factors including the moving solid-liquid interface, the density and conductivity dissimilarities among the phases, and the induced movements in the liquid phase.
- **Freezing or solidification:** Even though a material may have considerable latent heat of fusion, thermal conductivity and change phase within the desired temperature range; it may not be a suitable candidate for PCM applications if it has incongruent freezing behavior, crystallize exceedingly slow, form viscous mixtures, or are unstable over the utilized temperature ranges. Thus, the freezing behavior of the material plays a key role in selecting the PCM for the desired applications. However, with the help of the current advancements in this field, materials with near-congruent freezing behavior could be acceptable if the solidification problem can be overcome by increasing the nucleation rates.
- **Supercooling:** Another important issue with many PCMs (especially salt hydrates) is supercooling of the material where the temperature falls significantly below the melting point before the freezing begins, usually due to the slow rate of nucleation and/or growth of these nuclei. As a result, the melt does not solidify at the thermodynamic melting point when being cooled, reducing the effectiveness of the PCM by preventing the withdrawal of the stored heat. The occurrence of supercooling is generally linked to the viscosity of the melt (as they have low diffusion coefficients) at the melting point.

- **Nucleation:** As mentioned above, supercooling can be a significant problem that prevents the full use of PCMs due to issues related to nucleation and/or nuclei. However, this issue can be prevented/reduced with the help of nucleating materials. This can be done by either homogenous nucleation (where no foreign materials are added) via methods such as ultrasonic waves, or by heterogeneous nucleation, where impurities among the melt plays the role of a catalyst for nucleation. In order for an impurity to act as a good nucleating agent, it should have various characteristics including having a melting point higher than the highest temperature reached by the energy storage material in the storage cycle, being water insoluble at all temperatures and being chemically inert with the hydrate.
- **Fire resistance:** The reaction of a PCM to fire creates a significant impact on the safety of the system. Some PCMs are flammable and create toxic gasses upon combustion, while others are non-flammable. However, embedment of flammable PCMs in layers of non-flammable coating, application of a fire-resistant treatment to the surface or incorporation of self-extinguishing products can be done to prevent any undesired outcome during operation. Additives are also being used to enhance the materials' response to fire and recently the use of sub-products such as flame retardants are also being studied for practical applications.
- **Compatibility:** Another important aspect of PCMs is to be compatible with the application and the encapsulant material in terms of operating temperature range as well as rate of expansion, toxicity and corrosion. Thus, it becomes imperative that the selected material fulfills the compatibility requirement of the desired application.
- **Cycle Life:** In order for PCMs to have practical use in thermal management applications, they need to display stable and predictable characteristics over long periods of time, especially since in most cases; it is cumbersome as well as expensive to repair/replace the PCM that is embedded in the system. Thus, it becomes imperative that the material do not have any changes in its aforementioned characteristics over repetitive cycles of melting and solidifying.
- **Encapsulation:** One of the most widely utilized methodologies to improve the heat transfer between the PCM and the source/sink and to protect it from the outside environment is through containment, such as the utilization of steel cans, plastic bottles/tubes and high-density polyethylene pipe. The container plays an important role in becoming an effective barrier to prevent oxygen penetration, the loss of material and exchange of water in and out of the PCM as well as in increasing the safety (mostly mechanical damage) and the effective heat transfer (when the encapsulating material is also a good heat conductor). More information on encapsulation will be provided in the following sections.

Once the aforementioned aspects are well understood, the next step is to find PCMs that display the desired characteristics for the considered application. Since no PCM demonstrates all the necessary traits, it is important to select the ones that satisfy the main criteria and mitigate the effects of undesired properties using methods that will be described in the next sections. The essential thermo-physical, kinetic and chemical, technical, economic and environmental properties that PCMs should incorporate in the design of a TMSs are listed in Table 3.2.

Finally, solid-liquid PCMs are generally categorized according to the type of heat exchange and the change of state they are designed to perform through their composition and application. They are divided into three main types, namely organic, inorganic

Table 3.2 Essential properties desired in the selection of the suitable PCM.

Characteristics	Expectation
Thermo-Physical Properties	Melting/solidifying temperature in the desired operating range
	High latent heat of fusion per unit volume that allows smaller containers
	High thermal conductivity to reduce charge/discharge duration
	High specific heat to allow the additional sensible heat
	Small volume changes on phase transformation and small vapor pressure at operating temperatures to reduce the containment problems
	Congruent melting to allow homogenous solid and liquid phases
	Thermally reliable to have stable melting temperatures and latent heat of fusion
Kinetic Properties	High nucleation rate to avoid super cooling in liquid phase
	High rate of crystal growth to meet demands of heat recovery
Chemical Properties	Virtually reversible freezing/melting cycle
	Chemical stability and compatibility with encapsulated materials
	No degradation after a large cycle life to assure long operating time
	Corrosion resistant to encapsulated materials
	Non-toxic, non-flammable and non-explosive to assure safety
Technical Properties	Simple to use, compact, viable and compatible with the considered application and reliable for long-term use
Economic Properties	Cost effective
	Large-scale (commercial) availability
Environmental Properties	Low overall environmental impact
	High recyclability

Source: (compiled from Sharma *et al.*, 2009; Memon *et al.*, 2014; Zhou *et al.*, 2012).

and eutectics, which are further divided based on their LHF and their melting point as shown in Figure 3.3. The blue boxes represent the areas which will be focused in the following sections.

Organic PCMs have plethora of useful characteristics that enables them to be used in various thermal management applications (such as chemical stability, congruent melting and recyclability). However, they have low thermal conductivity with large volume variation during phase-change process and are flammable. On the contrary, inorganic PCMs are non-flammable and have lower costs when compared to organic PCMs and are more widely available. However, they may undergo improper re-solidification, suffer from decomposition and supercooling affects their phase change properties. Finally, eutectics are mixtures of various compounds with melting points lower than the constituent compounds and therefore can be modified in order to acquire the desired specific melting points for the necessary application. They virtually always melt/solidify without segregation before the components have a chance to separate, however, all components should be miscible in the liquid phase in order to have congruent melting. These types of PCMs are further explained in the following sections.

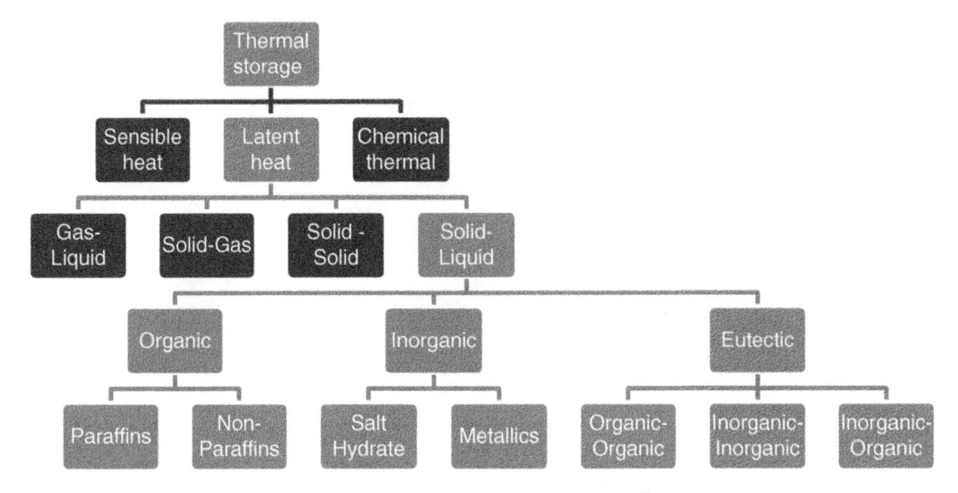

Figure 3.3 Types of solid-liquid phase change materials.

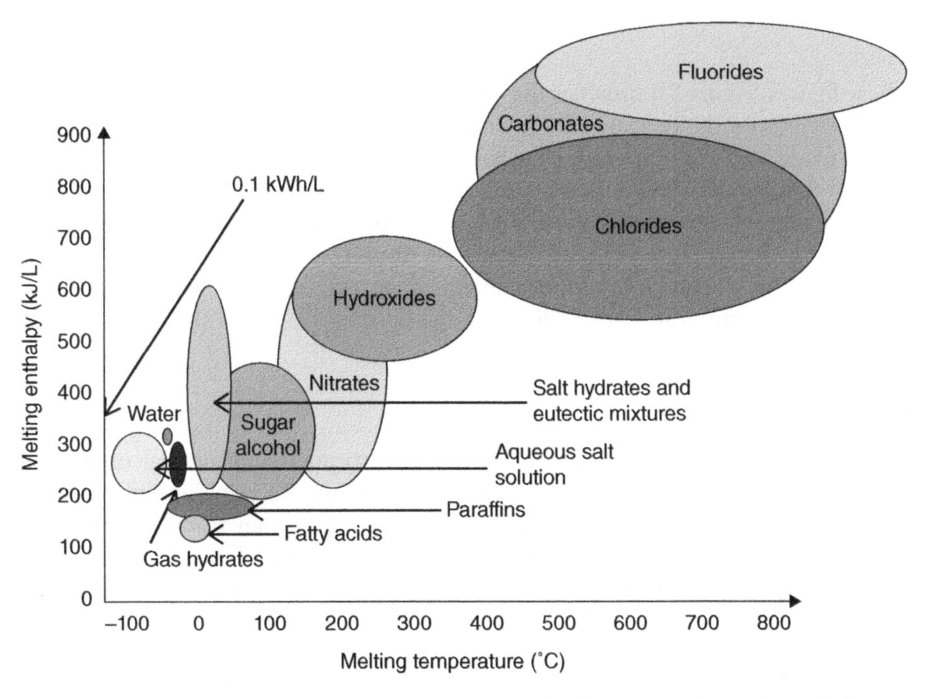

Figure 3.4 Melting temperature and phase change enthalpy for existing PCMs (adapted from Dieckmann, 2006).

At this point, it should be noted that, even though virtually all materials can be considered as PCMs, the melting/solidifying temperatures and enthalpies only allow a small portions of materials to be utilized effectively for the desired applications. The materials for the given temperature band and energy storage characteristics are illustrated in Figures 3.4 and 3.5.

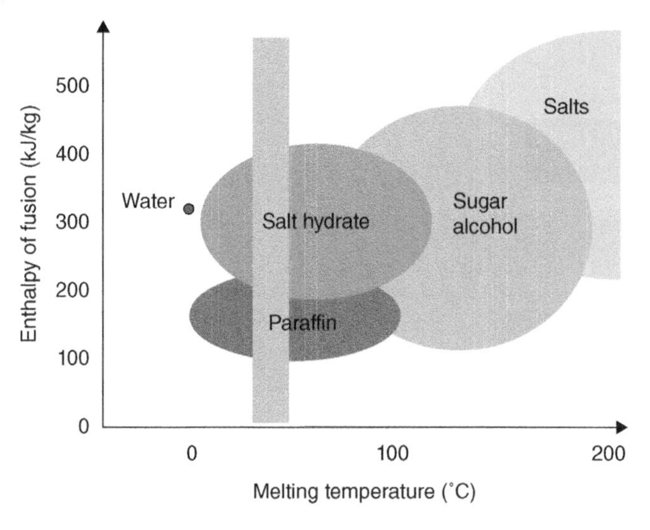

Figure 3.5 Temperature ranges and associated enthalpy of fusion of various PCMs that can be used in EV BTMS applications (adapted from Troxtechnik, 2015).

In these figures, Figure 3.4 provides the overall operating temperature band, whereas Figure 3.5 focuses on the typical melting temperature range that can be used in battery thermal management applications for the battery chemistries explained in Chapter 2. It can be seen that paraffins and salt hydrates are the most suitable to be utilized for various Li-ion batteries, whereas salts and sugar alcohols require chemistries with higher operating temperatures. However, the mixing two or more of these PCMs as well as the implementation of several heat transfer enhancement techniques (as will be discussed in the following chapters) are utilized to attain the optimum PCM for the H/EV BTMS.

3.2.1 Organic PCMs

Organic PCMs are generally composed of paraffins, fatty acids and sugar alcohols and encompass a melting point range of 0 to 200 °C. They are mostly stable at high temperatures due to their carbon and hydrogen contents. Moreover, they usually have lower densities (under 1 g cm^{-3}) compared to most inorganic PCMs, which cause them to have lower volumetric heats of fusion compared to their inorganic counterparts. A picture of a sample organic PCM is provided in Figure 3.6.

Figure 3.6 Picture of an organic PCM (courtesy of Rubitherm GmbH).

3.2.1.1 Paraffins

Paraffin is composed mostly of straight chain alkanes (hydrocarbons with single bond molecules) with 2-methyl branching groups near the end of the chains and have the chemical formula of C_nH_{2n+2}, where $n \leq 40$. For $n > 17$. Paraffin is usually solid at room temperature and if the number of carbon atoms is an even number, it is called n-paraffin, while an odd number denotes an iso-paraffin. The crystallization of the chain releases considerable latent heat. Thus, the melting point as well as the LHF increases as the carbon atoms or the chain length increase. This property along with their existence over a wide temperature range, makes them an exceptional thermal storage material. On the other hand, since pure paraffins are highly costly, technical grade paraffins (which are hydrocarbons mixtures with dissimilar chain lengths) are generally used. The pure paraffins contain 14–40 C-atoms, which results in melting temperatures 6 °C–80 °C (for pure paraffins), whereas technical (or commercial) paraffins, on the other hand contain only 8–15 C-numbers with significantly lower melting temperatures of 2 °C–45 °C. When the number of carbon atoms range in between 13–28 C-atoms, the melting temperature drops to −5 °C–60 °C range. Moreover, technical grade paraffins became largely popular among various applications due to being safe, non-corrosive, inexpensive, and widely available as well as having broad phase-transition temperatures (compared to pure paraffins) and freezing without supercooling. Besides having high heats of fusion, paraffin waxes (solid phase of paraffin) are chemically inert and stable with no phase segregation (below 500 °C), which makes them reliable and predictable in most applications they are being utilized. Furthermore, they exhibit only slight volume change during melting and have low vapor pressure in the melt form. On the other hand, they have mediocre thermal storage densities (~ 200 kJ/kg) low thermal conductivity (~ 0.2 W/m°C) and are non-compatible with plastic containers. In addition, in the case of pure paraffins, they are expensive and moderately flammable. However, due to their low vapor pressure, certain additional steps need to be taken to prevent any fire related hazard in the utilized system.

3.2.1.2 Non-Paraffins

The organic non-paraffins are the most widely available types of PCMs with considerably diverse characteristics, unlike paraffins which tend to have more common ones. Although there is a wide variety of non-paraffin PCMs with a wide range of qualities, they share numerous defining features with various advantages (such as are having large heats of fusion and no/limited supercooling for fatty acids) and disadvantages (such as inflammability, low thermal conductivity and instability at high temperatures). As opposed to paraffins, the temperature range of most non-paraffins (especially fatty acids) that can be used for heating applications is much smaller (usually between 20 °C–30 °C) (Hyun *et al.*, 2014).

Research associated with processing fat and oil derivatives are rapidly growing, which enables them to become a biomaterial substitute to paraffins and salt hydrates. Fatty acids are characterized by the chemical formula $CH_3(CH_2)_{2n}COOH$ and have comparable storage densities with paraffins. Moreover, like paraffins, their melting/solidifying temperatures are also highly correlated with their molecule length. Moreover, they have advantageous characteristic including melting congruency and low toxicity, large latent heats of transition, high specific heats, negligible supercooling and small volume change during melting/freezing. However, even though they have chemical stability upon cycling, they usually react with the surrounding environment due to being acidic by nature. Another type of non-paraffins, fatty alcohols, generally have even number of carbon atoms and can be biodegradable with a chain length up to C_{18}. Unlike paraffins, these bio-based PCMs are harmless if ingested and are mostly nontoxic and biocompatible.

3.2.2 Inorganic PCMs

Inorganic PCMs have considerably higher densities than organic PCMs. Thus, even though they generally exhibit similar melting enthalpies per mass, their enthalpies per volume are significantly higher. Moreover, they can have thermal conductivities as much as three times higher than organic materials. They also demonstrate sharp melting points and can cover wide temperature ranges. These properties enable them to be used in various applications, especially where corrosiveness is not an important consideration. Inorganic PCMs are generally sub-categorized as salt hydrates and metals. A picture of a sample inorganic PCM is provided in Figure 3.7.

3.2.2.1 Salt Hydrates

Salt hydrates can be considered as alloys of inorganic salts and water that form a crystalline solid of general formula AB $\cdot n\mathrm{H_2O}$, with examples of $CaCl_2 \cdot 6H_2O$, $LiNO_3 \cdot 3H_2O$ and $3KF \cdot 3H_2O$ and cover a melting temperature spectrum of 5°C–130°C. The solid-liquid transformation of salty hydrates is essentially the dehydration/hydration of a salt, which is comparable to the melting/freezing process in thermodynamics. At the melting point, the hydrate crystals breakup into a lower hydrate and water or anhydrous salt and water as shown in equations 3.2a and 3.2b, respectively (Sharma *et al.*, 2009).

$$AB \cdot nH_2O \rightarrow AB \cdot mH_2O + (n-m)H_2O \tag{3.2a}$$

or in its anhydrous form;

$$AB \cdot nH_2O \rightarrow AB + nH_2O \tag{3.2b}$$

Salt hydrates have high latent heat of fusion (per unit volume) and thermal conductivity (up to two times of the paraffin wax) as well as slight volume change during melting, which makes them great candidates to be utilized in latent-heat thermal-energy storage system applications. Moreover, they are suitable to be utilized with plastics, not very corrosive, and have very low toxicity. They usually have higher densities compared to paraffins, but are only marginally more effective on a per volume basis, in spite of having a slightly lower heat of fusion. On the other hand, salt hydrates usually have

Figure 3.7 Picture of an inorganic PCM (courtesy of Rubitherm GmbH).

lower specific heats in both the solid and liquid states compared to paraffins. Three types of behavior can be identified for melted salts; namely congruent, incongruent and semi-congruent melting. Congruent and incongruent melting occurs depending on the solubility of the anhydrous salt (in its water of hydration) to be complete or not complete. In semi-congruent melting, the solid and liquid phases are at different melting compositions at equilibrium due to transformation of hydrate to a lower hydrated material by losing water. Salt hydrates have significantly higher vapor pressures (increases with the degree of hydration), which causes considerable water loss, leading to deviations in the material's thermal behavior. These PCMs display change in chemical stability and experience long-term degradation by oxidization, hydrolysis, thermal decomposition and other reactions.

Most salt hydrates experience problems with congruent melting (melting/freezing repeatedly without phase separation) as n or $(n - m)$ moles of hydration water are not enough to dissolve anhydrous or lower hydrate salts; thus the solution becomes superheated at the melting point. Due to their relatively high storage density of salty hydrate materials, the anhydrous or lower hydrate salt settles at the bottom of the container, which impedes its recombination with water during the reverse process of freezing. Thus, the storage density becomes difficult to maintain and generally starts to decrease with cycling. However, the issues can be resolved/mitigated by mechanical stirring, encapsulating the PCM, by utilizing additional water to inhibit supersaturation or modification of the compound. Excess water is generally utilized as a thickening material to dissolve the entire anhydrous salt during melting. Thickening materials can reduce the magnitude of the phase separation in solid and liquid phases as well as averting nucleating agents from settling down due to their high densities (Ryu *et al.*, 1992). Moreover, various materials (especially graphite fiber or metal) can be embedded into the PCM to form a composite to enhance thermal/mechanical properties. Alternatively, PCM can be incorporated into the matrix of other materials, where it holds the PCM inside the pores even when it has melted (Kosny *et al.*, 2013). Furthermore, nucleating materials can also be utilized to reduce their undesired super-cooling properties. Finally, it should be noted considerable research is being conducted to enhance Glauber's salt, which is among the least expensive heat storage materials that are commonly utilized in construction and building applications. In this regard, Borax was able to be used as a nucleating agent to prevent subcooling and to improve thermal cycling of the mixture.

3.2.2.2 Metals

Metals are also another wide category of inorganic PCMs with properties of low heat of fusion per unit mass, high heats of fusion per unit volume and thermal conductivity as well as low specific heats and vapor pressures. Even though metals were not previously thought to be practical for PCM applications due to their weight and relatively high melting points, they are starting to be considerably more viable due to the recent advancements in the fabrication of nanomaterials. However, unlike most other PCMs, metals require special attention as they may pose certain engineering problems. Since these materials have high thermal conductivity and stable melting behaviors, they become viable candidates for numerous applications, particularly for absorptive heat sinks in electronic components. Moreover, with the accumulation of metallurgic knowledge, the development of metal PCM allows with regulated melting points and latent heat become possible. Currently, the most attracting metallic PCMs are Bi, PB, In, Sn, Cd (as well as their alloys) due to their low melting points.

In PCMs, significant work is being conducted to reduce the size of the PCM structure for virtually instant phase change of the material. The simplest method to produce tiny PCM structures is by using the top-down approach, where the particles are processed by performed PCMs or by using bottom-up approach where the molecular building blocks are crystallized (Hyun *et al.*, 2014). The bottom-up approach is an easy and effective technique to develop colloidal particles from inorganic PCMs. The top-down approach is composed of fabrication with microfluidic devices that consist of networks of small channels. This method has the advantage of independently varying the size, construction and configuration of particles. Another method is fabrication by replication non-wetting templates, which requires producing isolated structure without a residual layer an also have the advantage of independently controlling the particle size and composition. Finally, the last method is the utilization of a mechanical shear force, which stretches and breaks the big liquid droplets into smaller and uniform ones, enough to be elongated by the sheer force through mechanical agitation.

3.2.3 Eutectics

Eutectics are mixtures of dissimilar compounds (with lower melting points than the encompassed compounds) that melts/solidifies congruently and forms a mixture of component crystals during crystallization. These mixtures could contain only salts, salt hydrates, alkalis or mixture of two or all. Their sharp melting points allow melting and freezing of all components to occur simultaneously. Thus, they can be modified to develop specific melting temperatures for given applications and virtually always melt/solidify without segregation as they freeze to an intimate mixture of crystals, before the components have a chance to separate; both components liquefy simultaneously during melting. In order to have congruent melting, all components should be miscible in the liquid phase. It should be noted that even though eutectic materials do not have the drawback of incongruent melting. They generally have issues with non-homogenous expansion in volume (since they are mainly made from inorganic salts) that can be detrimental to the container and or the internal structure. As eutectics are developed with respect to particular specifications (for the desired application), they tend to be more costly compared to the rest of the PCMs that exhibit lower heat of fusion. In addition to the information provided above regarding different types of PCMs, a compact list of commonly used PCMs as well as the advantages/drawbacks of the type that belong to are tabulated in Table 3.3.

3.3 Measurement of Thermal Properties of PCMs

The effectiveness of a thermal management system is directly affiliated with the utilized PCM. Since the selection of the type and the quantity of utilized PCM is mainly dependent on the melting temperature, heat of fusion and thermal conductivity; accurate measurement of their thermal properties has paramount importance. Even though various measurement methods exists, differential scanning calorimetry (DSC) and differential thermal analysis (DTA) are the most widely utilized ones in the industry. DSC

Table 3.3 List of some commonly used PCMs.

Type		Materials	Melting Point (° C)	Heat of Fusion ($kJkg^{-1}$)	Advantages	Drawbacks
Organic	Paraffins	Paraffin C15	10	205	Availability in large temperatures	
		Paraffin C17	21.7	213	Chemically stable and thermally reliable over extended cycles	
		Paraffin C19	32	222		Low thermal conductivity
		Paraffin C23	47.5	232		
		Paraffin C26	56.3	256	Virtually no phase segregation	Relatively large volume change
		Paraffin C30	65.4	251	Low vapour pressure when melt	
		Paraffin C34	73.9	268		Non compatible with plastic containers
	Non-paraffins	Oleic acid	13.5–16.3	N/A	Non-corossive (except for some fatty-acids)	
		Acetic acid	16.7	184		
		1-dodecanol	24	215	Virtually no supercooling during freezing	Moderately Flammable
		Capric acid	36	152		May generate fumes on combustion
		1-tetradecanol	38	205	Innocuous (except for some non-paraffins)	
		Laurie acid	49	178		
		9-heptadecanone	51	213		
		Phenylacetic acid	76.7	102	Recyclable	
		Acetamide	81	241	High latent heat of fusion	

(Continued)

Table 3.3 (*Continued*)

Type		Materials	Melting Point (° C)	Heat of Fusion ($kJkg^{-1}$)	Advantages	Drawbacks
Inorganic	Salt Hydrates	$LiNO_3 \cdot 3H_2O$	30	189	High thermal conductivity	May undergo supercooling and improper re-solidification
		$Na_2(SO_4)_2 \cdot 10H_2O$	32	251		
		$LiBr_2 \cdot 2H_2O$	34	124	Availability and lower cost	Corrosiveness (to metals)
		$FeCl_3 \cdot 6H_2O$	37	223	Neither toxic nor flammable	Have high vapor pressure
		$CoSO_4 \cdot 7H_2O$	40.7	170	Sharp phase change	Long term degradation
		$Ca(NO_3) \cdot 4H_2O$	47	153	Low environmental impact	Variable chemical stability
		$NaAl(SO_4)_2 \cdot 10H_2O$	61	181		
		$Mg(NO_3)_2 \cdot 6H_2O$	89.9	167	Potentially Recyclable	High volume change
	Metals	Ga	30	80.3	Large heats of fusion per unit volume	High melting points
		Bi-Cd-In alloy	61	25		Low specific heats
		Bi-Pb-In alloy	70	29	High thermal conductivity	May pose engineering problems
		Bi-In alloy	72	25	Sharp, well defined melting point	
	Eutectics (Organic and Inorganic)	$34\%C_{14}H_{28}O_2 + 66\%C_{10}H_{20}O_2$	24	14.7		
		$50\%CaCl_2 + 50\%MgCl_2 \cdot 6H_2O$	25	95		
		Octadecane + docosane	25.5–27	203.8		
		Octadecane + heneicosane	25.8–26	173.98	High sharp melting temperatures	
		$50\%CH_3CONH_2 + 50\%NH_2CONH_2$	27	163	High volumetric thermal storage density (marginally above organic PCMs)	Limited available test data on their thermos-physical properties
		Ga	30	80.9		
		$47\%Ca(NO_3)_2 \cdot 4H2O + 53\%Mg(NO_3)_2 \cdot 6H2O$	30	136		
		$60\%Na(CH_3COO) \cdot 3H_2O + 40\%CO(NH_2)_2$	30–31.5	200.5–226		

Source: (compiled from Hyun *et al.*, 2014; Memon *et al.*, 2014).

testing is an analytical method developed in 1962 and measures the material temperature and heat flows as a function of time and temperature in a controlled test set up. In this method, the sample and a reference material with known properties (usually pure water) are maintained at virtually equal temperatures throughout the measurement process and many of the thermal properties of the sample is obtained by acquiring data on the heat transfer differences amount them. It provides both qualitative and quantitative data about physical/chemical changes that involve endothermic and exothermic processes (Memon *et al.*, 2014). This method is mainly utilized for analyzing the thermal properties of PCM wallboards that are used in building and construction applications. DTA testing, which was invented much earlier (in 1887), is a thermoanalytic technique that the heat (as opposed to the temperature) between the sample and the reference remains constant and the thermal properties are determined through the temperature difference between the two. The obtained plot that shows the differential temperature vs. temperature (or time) is commonly referred as DTA curve. However, due to the limitations of these methods, newer ones are also starting to be utilized, such as the T-history method, which enables the simultaneous determination of thermo-physical properties of numerous PCMs. These properties are acquired from temperature vs. time plots of the PCM and compared against that of the reference material. Thus, various properties such as the melting temperature, degree of supercooling, heat of fusion, specific heat and thermal conductivity of PCMs can be determined. Further details on experimental set up and testing procedures for these methods will be provided in Chapter 4.

3.4 Heat Transfer Enhancements

Even though PCMs have considerable advantages that can be applied in various applications as mentioned in the previous chapters, they still possess various drawbacks that impede their wide-spread commercialization. Among these, one of the main difficulty is having low thermal conductivity, which prolongs the time it takes for a PCM (mostly most paraffin waxes, hydrated salts as well as eutectics) to melt/solidify, limiting the performance of the thermal management system. In order to improve the thermal conductivity of these materials, there are various techniques implemented, such as:

- Impregnation of high conductivity particles (such as graphite, copper, silver etc.) and porous materials (such as copper and aluminum matrices)
- Adding metal matrix structures
- Using low density highly conductive materials
- Increasing the heat transfer area through finned tubes
- Applying various encapsulation techniques

Another significant design parameter in using PCMs is the phase change time, as premature state change can reduce the effectiveness of the used PCMs. In this regard, addition of aluminum additives into certain PCMs (especially paraffin wax) is shown to considerably reduce the phase change time in thermal management systems. However, this should be weighed against the increase weight and cost associated with these additives. Adding metal foams can also help with heat transfer enhancements as many studies determine that adding metal foams to heat exchangers can generate as much as one-third of the commercially available high end heat exchangers.

Moreover, the heat transfer area can also be extended to improve the thermal capabilities of the PCM in the utilized application. This can be done under numerous methods; most commonly used ones being modifying the shape to become more tubular or plate-like and/or adding fins to increase the heat transfer between the PCM and the transfer fluid. Generally, the fins are positioned on the side with the PCM since it has a lower heat transfer coefficient. In this regard, it becomes important to verify that using fins and/or enlarging the surface area will enhance the heat transfer rate as they may be reduced due to the corresponding increase in the wall resistance. Moreover, multiple types of PCMs can be used in different layers in order to make the temperature transition more uniform during the heat transfer period. In this regard, if $T_{m,aver}$ can be taken as the average melting temperature of the PCM located in the middle of a multiple PCM system. The optimal temperatures difference for the charging period between the melting temperature extremes (highest and lowest) are written as follows:

$$T_{MP,first} - T_{MP,last} = \frac{NTU}{1 + \frac{NTU}{2}}(T_{HTF,in} - T_{m,aver}) \tag{3.3}$$

where NTU represents the number of heat transfer units.

In addition, one of the most frequently used methods to enhance the heat transfer between the PCM and the source/sink is to enclose it by incorporating various materials and elements into it, which are most commonly done through direct incorporation, immersion and encapsulation. Among these, direct incorporation is the most straightforward, practical and inexpensive method where the PCM is directly mixed with the associated material. In the immersion technique, the elements are dipped into the liquid PCM and absorb it by capillary action. However, in this process, the PCM may not have a large cycle life due to leakage and the used PCM may have impact on the associated element. These two methods are commonly used in construction materials (concrete, bricks, wall boards, etc.). More importantly (especially for the scope of this book), encapsulation is also used in various thermal management applications as they prevent leaks and enable them to be manipulated while they are laid. Moreover, mixing PCMs with various additives during this procedure can improve various characteristics of the material such as storage capacity, thermal conductivity, congruent melting, stability (Pons *et al.*, 2014). Products containing encapsulated PCMs can be classified according to the method of encapsulation, their shape and size, substrate and construction process. Among these the encapsulation method plays the most important role and can be separated as micro encapsulation in which small particles are enclosed in a protective shell material, and macro encapsulation, where PCM is sealed in containers as shown in Table 3.4.

In microencapsulation technique, PCM particles are enclosed in a thin sealed and high molecular weight plymetric film (each less than 1 mm diameter) to prevent the PCM to distort its shape or leak during the phase change process by impeding possible interactions with its surrounding through keeping the material isolated, as shown in Figure 3.8. It is a considerably useful method to increase the heat transfer area due to establishing a large surface area and is mainly established when the PCM is in liquid phase. Their main advantages include increased flexibility during repeated phase change, improved the heat transfer rate and enhanced thermal/mechanical material stability as well as improved compatibility of hazardous materials by enabling them to be used in

Table 3.4 Types of encapsulation for PCMs.

Encapsulation	Capsules		PCM Substrate	Construction Process
	Shape	Size		
Micro-encapsulated	Spherical or cylindrical	Ø <1 mm	Paraffin	Embedded
Macro-encapsulated	Depending on the container	Ø >1 cm	Organics and inorganics	Wall and ceiling linings

Source: (compiled from Pons *et al.*, 2014).

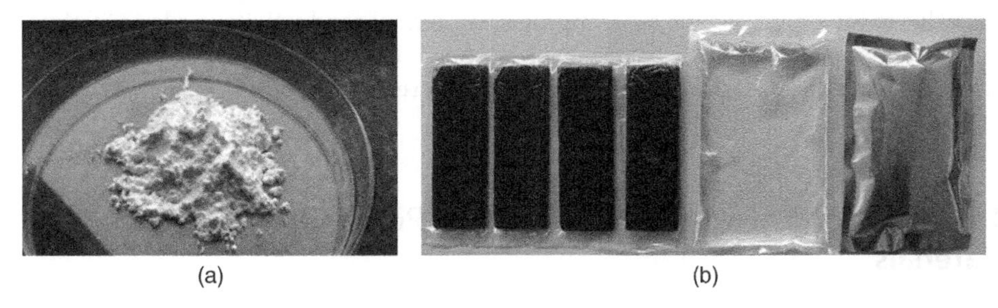

Figure 3.8 Picture of a microencapsulation of PCMs (courtesy of Rubitherm GmbH).

sensitive applications. They are widely used in plasterboards (mostly as building material) as well as suspension in hot/cold water to increased heat transfer capacity, but are mostly restricted by their flammability or cost.

Currently, numerous microencapsulation methods with various different benefits/drawbacks with respect to size, size dispersion, usable material, and are utilized in a broad range of industrial applications. Among these, air suspension technique uses a coating material which is sprayed onto PCM particles until the anticipated coating thickness is achieved. Coacervation method uses forced-precipitation of nanoparticles by creating alterations in the solution. Solvent evaporation adds an organic solution of a polymer as a coating material to an aqueous solution of the core material and mixing to the point of evaporation. In spray drying, emulsion is forcefully expulsed through an atomizer in a hot jet of air which creates microparticles and removes the solvent at the same time. In various studies, it was found that the microencapsulation efficiency of coacervation and spray-drying processes depend on core to coating ratio, emulsifying time and the amount of cross-linked agent (Memon *et al.*, 2014). Interfacial polymerization involves polycondensation at the interface of two-phase system and finally dip coating techniques directly pipets and aqueous solution of shell polymer into the core particles. Even though these aforementioned microencapsulation techniques are currently the most commonly used ones, due to the large number (that are not listed here) and adaptability of these methods, they are virtually unlimited in composition as well as application.

Macro encapsulation, which is usually done in units larger than 1 cm diameter, is the most commonly utilized type of PCM containment as it prevents large phase separations and increases the heat transfer. It is used to hold the liquid PCM, prevent any potential changes in its composition due to being in contact with its environment, thus improving the compatibility and reducing the volume change (Riffat *et al.*, 2015). Through this method, a considerable quantity of PCM can be packed in tubes, spheres and panels to be used for thermal management systems. The main advantages of macro encapsulation are ease of shipping and handling, ability to be modified to fit in various applications, and can reduce the external volume changes. However, it also has certain drawbacks such as relatively poor thermal conductivity, and potential solidification at the corners/edges that can impede effective heat transfer (Zalba *et al.*, 2003).

Moreover, high-density polyethylene, styrene and butadiene can be used to develop shape-stabilized PCMs. In this method, PCM and the supporting material are melted and mixed with each other at high temperatures and subsequently cooled below glass transition temperatures until it becomes solid. This method is also gaining wide acceptance due to their high specific heat, appropriate thermal conductivity and long lifetime and most importantly having a fixed/stabilized shape. This method also does not require a container and the mass proportion of PCM (compared to the supporting material) can be up to 80%.

3.5 Cost and Environmental Impact of Phase Change Materials

Due to their complex structures and material, PCM systems are significantly costly compared to sensible heat storage systems. As these materials are in relatively early stages of development and are just starting to be widely commercialized (as they are currently being mainly used in niche applications), they still don't have a fully developed market and/or high demands, causing them to have relatively high prices. Even though various PCMs are currently under research for a broad range of industrial applications, only building and construction applications are seemed to have a dominant market, mainly due to the recent environmental regulations for buildings. Even in these sectors, the time required to recover the initial investment in terms of thermal benefits can range from 10 to 50 years. However, considerable research is being conducted to use low cost materials as PCMs, such as residues, waste from industrial processes and natural oils. Thus, as similar environmental regulations are starting to be in place, the market potential for PCMs are expected to increase significantly in the future, reducing their unit price.

The cost of these materials is currently dominated by the raw PCM cost (in the range of range around \$10–\$55 per kWh, depending on its classification) and its encapsulation. The encapsulation costs can increase dramatically when expensive methods (that can avoid the use of heat exchanger surfaces) are utilized. It is estimated that the cost of macro encapsulation is ~20% of the total cost, whereas this number can reach to ~50% for micro encapsulation techniques (Hauer, 2011). Since commercial paraffins are byproducts from oil refineries, they are easily available with relatively cheaper prices, but the prices are highly correlated with the purity of the material. As an example, the price of technical grade eicosane is ~\$7/kg, whereas the pure laboratory grade is ~\$54/kg. It should be noted that these prices can vary depending on the season and the geopolitical scenarios. Moreover, the estimated cost of microencapsulation for such a paraffin would be around 55% of the total cost of the paraffin PCM. On the other hand, the price of low-cost paraffin alternatives can be as low as ~\$0.5/L, in the case of POLYWAX. In addition, even cheaper alternatives, such as salt hydrates and biobased PCMs are also starting to become viable options with the properties and improvements in the process technology (Kosny *et al.*, 2013). When other types of PCMs with lower costs are examined, the cost of fatty acids, such as stearic acid, palmitic acid, and oleic acid are $1.5/kg, $1.6/kg and $1.7/kg, respectively (based on the 2013 market prices).

In terms of environmental impact, even though PCMs generally have higher impact than the alternatives in terms of the used materials (especially in construction), this negative impact seem significantly lower when it is evaluated under the complete life cycle of the system (this will be covered in Chapter 6), as the impact in production is compensated by the reduction in the energy demand and the associated CO_2 emissions over the life of the product.

The impact varies according to the type of the used PCM and its estimated life cycle in the application. At the end of their useful life, the majority of PCMs can be recycled; as organic PCMs are biodegradable and inorganic ones are innocuous. However, since this technology and its widespread applications are at its infancy, further studies are still needed to be conducted to better understand the environmental implications of the use of PCMs, compared to their alternatives.

3.6 Applications of PCMs

Based on the aforementioned properties of PCMs, it is not surprising that they are starting to be used in a broad range of industrial applications. In general, the key points in using PCMs for different applications are the selection of the appropriate PCM with its melting point in the anticipated temperature range, a suitable heat exchange surface and a proper container that is compatible the utilized material. Most commonly utilized PCMs are water/ice, salt hydrates, and certain polymers where their usage for thermal management purposes dates back to late 1800s as seat warmers for British railroad cars. Other early applications of PCMs include 'eutectic plates' used for cold storage in trucking and railroad transportation applications. PCM were also being used in space missions, with NASA, where it was utilized on thermal management of electronic packages. However, the first experimental application of PCMs for cool storage occurred in the early 1970s using eutectic salts at the University of Delaware in the design and construction of a solar energy laboratory of eutectic salts. In 1982, Transphase Systems Inc. installed the first eutectic salt storage system for cool storage to serve a commercial or industrial building (Dincer and Rosen, 2010).

Today, PCMs are being utilized throughout a wide range of heat management applications, from conventional ones such as building and construction, HVAC, textiles, fixed refrigeration, to relatively new ones such as solar energy utilization, peak load shifting, thermal insulation for functional fibers as well as emerging ones such as general containers for temperature sensitive food, isothermal water bottles for cycling, catering products and medical devices (Hyun *et al.*, 2014). Pictures of most common types of PCMs used in thermal energy storage applications are provided in Figure 3.9.

Among these applications, construction and building holds a large portion of the market share, especially due to the recent environmental regulations for buildings. For these applications, the latent enthalpy per unit volume and flammability are the key properties that lead to the selection of the right type of PCMs. Materials with melting point ranges of 15–35°C are replaced low thermal mass light-weight building materials to efficiently prevent high temperatures variations in buildings. The PCMs are mostly being used in

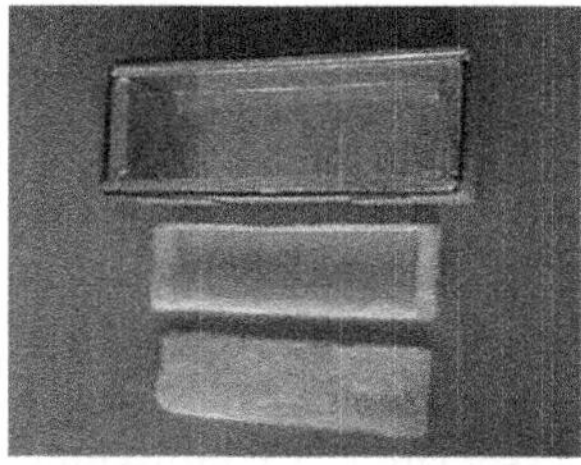
Moulded Rubber PCM Box

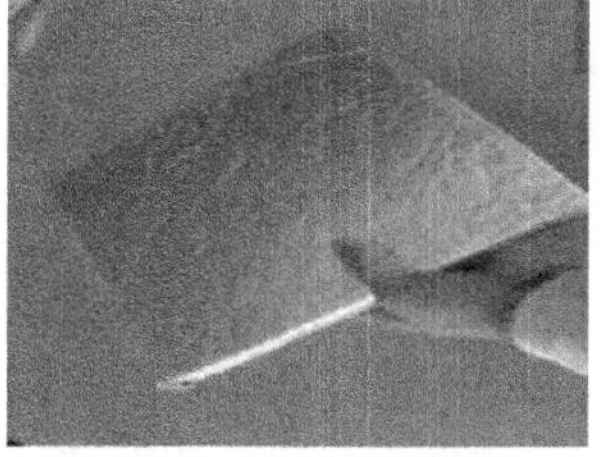
Aluminum Foil PCM Sheet

PCM Rubber Balls

Figure 3.9 Most common types of organic PCMs used in TES applications (courtesy of PCM Products).

wallboards as they are cheap and the extensive surface area of the wall enables maximal heat transfer with its surroundings. In addition, they are also highly utilized other energy storage system applications due to their high energy storage density (Sharma *et al.*, 2009). Today even though paraffins are mainly used in these applications due to being nontoxic, highly available and easy to microencapsulate; salt hydrates are starting to replace these materials with properties such as their higher latent heat per unit volume and non-flammability. However, before they could start to be widely utilized, the subcooling effect and the difficulty in microencapsulating must be overcome.

In the aforementioned applications, PCMs are mainly selected based on their melting temperature, as PCMs below 15°C are mostly used for air conditioning and above 90°C are used for absorption refrigeration. They are also starting to be used in various solar energy applications with the aim of storing the available solar energy and using it when needed to reduce the peak load. Significant research is also being conducted to effectively utilize them in various electronic devices, engines and even industrial chemical reactors.

As an example, PCMs are starting to be used in main circuit boards and processors as they generate significant heat during operation which negatively affects (or shuts down) the electronic systems significantly if not dissipated before reaching the maximum operating limit of the specific electronic system (generally 45°C). In these systems, the temperature of the enclosed system can be maintained as the heat entering the building (e.g., through sunlight) and/or dissipated by the electrical equipment is absorbed by the PCM, thus enabling the equipment to operate reliably without any auxiliary thermal management system as illustrated in Figure 3.10a. In general, organic solutions used in these systems are safe to contact any electrical component as they are both non-conductive and non-corrosive and can be molded into any shape. Furthermore, PCMs are also starting to be used for even larger volume energy storage options for an entire building. An illustration of some global chilled water cooling application examples is provided in Figure 3.10b.

Other commonly used applications include personalized items such as pocket heaters for mountain-rescue operations for winter and vests with stable temperatures to provide cooling during the summer. Moreover, PCMs are also being incorporated to various clothing items, such as gloves, shoes and even underwear. In addition, in the recent years, they have also been starting to be used in different areas such as drug delivery systems (biocompatible PCMs), where the temperature of the drug is regulated in order to control its diffusivity or its carrier, thus enabling the drug to be delivered to the target areas on demand. Significant research is also being conducted on the variances between

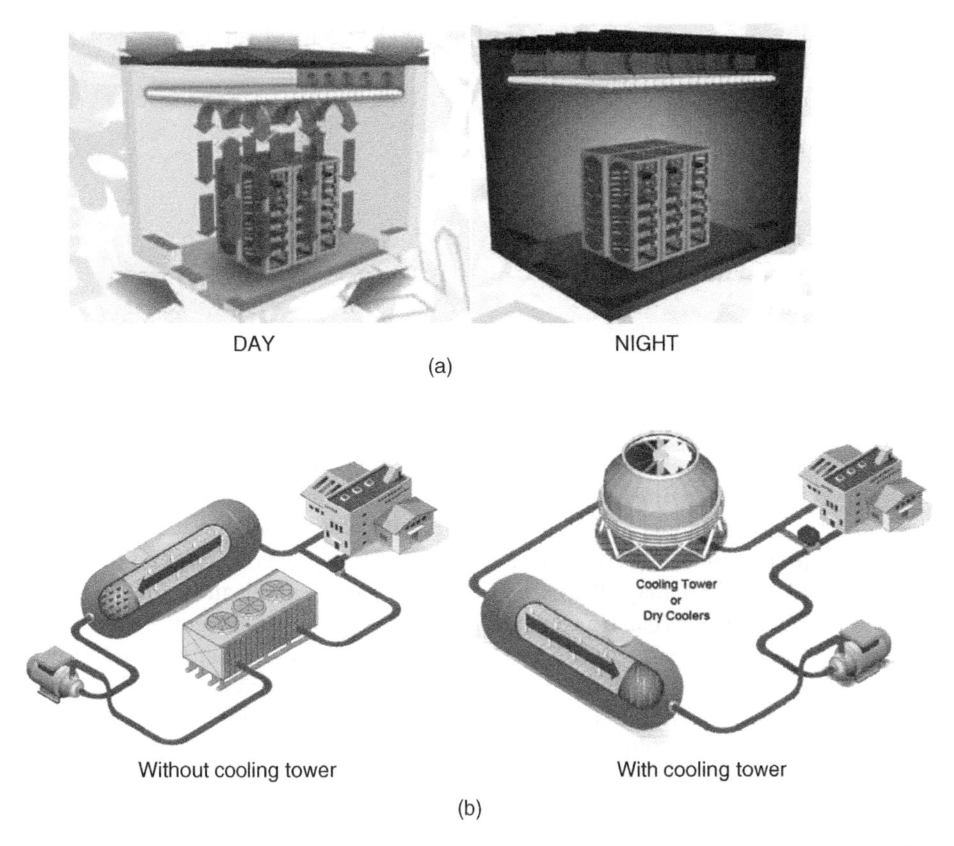

Figure 3.10 PCMs used in electrical equipment and chilled water cooling application (courtesy of PCM Products).

the amorphous and crystalline states that can assist with the progression of solid-state memory and optical-storage applications. Other interesting applications include being used as biomarkers in cancer detection (by using PCMs with slightly different melting points as thermally sensitive biomarkers) as well as barcoding (using unique differential scanning calorimeter fingerprints). However, the successful use of these applications is currently at its infancy and many outstanding issues need to be resolved before they can become widely available (Hyun *et al.*, 2014).

Finally, PCMs are also starting to be used in battery thermal management applications (especially for EVs and HEVs) as shown in Figure 3.11. As mentioned in Chapter 2, all battery chemistries dissipate heat during charging and discharging periods, which can both reduce the performance and the life of the battery. Thus, PCMs also provide significant advantages in the battery thermal management systems (that will be discussed in more detail in the following chapters). When applied appropriately, PCMs can shave the temperature peaks in the battery pack and eliminate or reduce the need of using any energy intensive active cooling systems. For these applications, it is imperative to select the PCM that has capable heat transfer properties, melts/solidifies within the operating temperature band and can withstand the cycle life of the battery. Moreover, for an

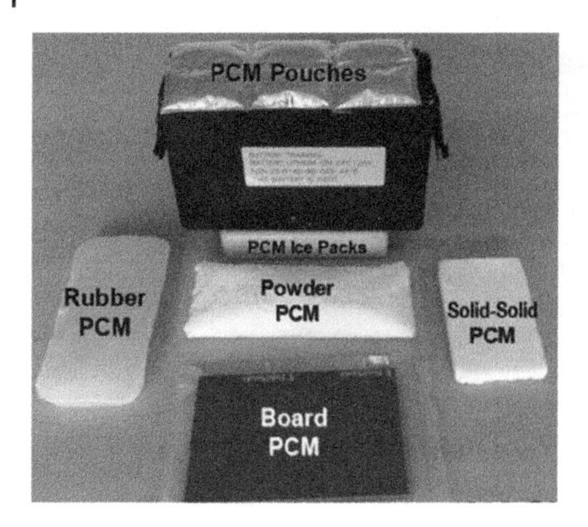

Figure 3.11 Example of various PCMs used for battery applications (courtesy of PCM Products).

effective TMS, the material also needs to have a small mushy phase and short temperature interval for melting. These materials are often contained in rugged, self-stacking, water impermeable containers made of a high-density polyethylene inside the battery pack. Thus, both the operating energy (e.g., due to parasitic power losses) and the associated cost to thermally manage the battery can be reduced significantly. Moreover, as the amount of moving parts (such as pumps and compressions) will be reduced or scaled down, the reliability and maintenance cost of the system can be further reduced. However, due to the current limitations of the PCMs and their encapsulation techniques, these materials are currently used as secondary TMSs that prevent the battery from reaching certain temperature limits when the vehicle goes through demanding driving and/or operating conditions. As the size, cost and the cycle life of the PCMs increase, they are expected fully replace these TMS that only uses sensible heat.

3.7 Case Study I: Heat Exchanger Design and Optimization Model for EV Batteries using PCMs

3.7.1 System Description and Parameters

3.7.1.1 Simplified System Diagram

In this case study, the aforementioned PCM selection criteria and heat transfer enhancement techniques will be utilized to develop a thermal management system for an electric vehicle battery application. In this system a Li-ion battery that is already being cooled by an active liquid thermal management system is being utilized to improve the system effectiveness with the integration of PCMs. The simplified diagram of the initial battery cooling system is provided in Figure 3.12 (Javani *et al.*, 2014).

In order to improve this system, a shell and tube heat exchanger is selected as latent heat thermal energy storage system to be installed in the cooling cycle. The main reasons for selecting shell and tube model of heat exchanger are as follows:

- Heat loss form shell and tube configurations is minimal
- Shell and tube heat exchangers require less charging and discharging time compared to its counterparts
- Shell and tube technology is relatively low cost and it is easy to manufacture

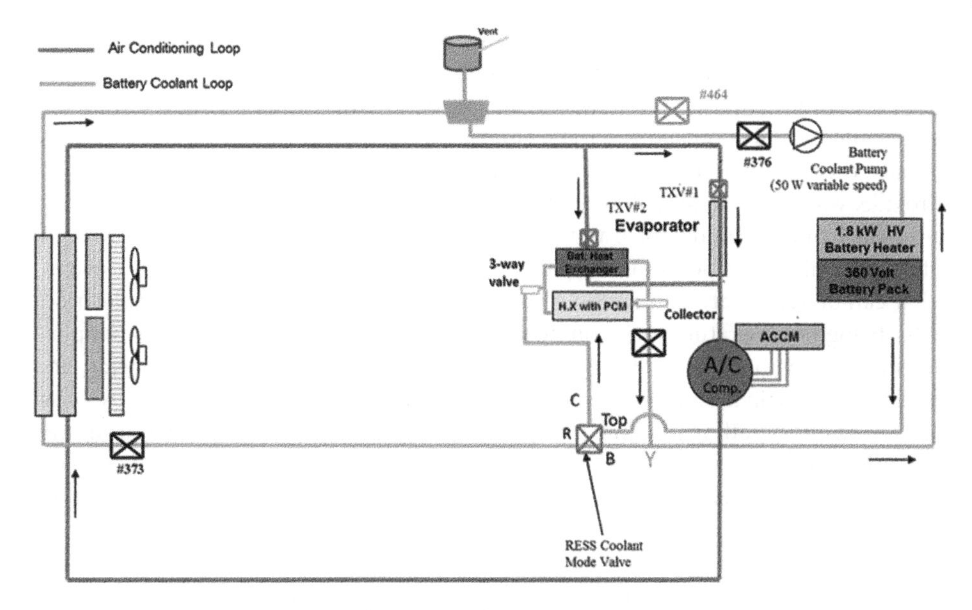

Figure 3.12 Simplified schematics of the analyzed HEV TMS.

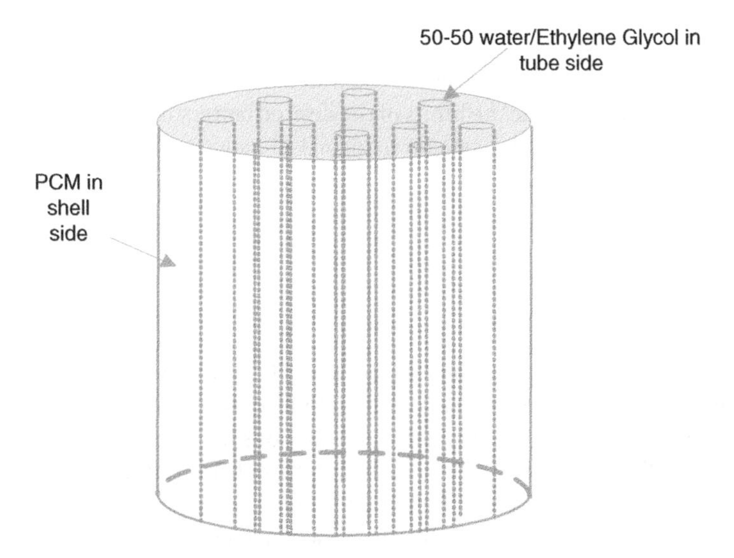

Figure 3.13 Physical model of heat exchanger.

As shown in Figure 3.13, the system is designed to have the coolant flow through the copper tubes and PCM is enclosed in the shell space. Since the outer surface of the tank is kept insulated, the heat exchanger effectiveness is able to be improved.

3.7.1.2 PCM Selection For the Application

In order for the most suitable PCM to be selected for this application, the associated competitive advantages and drawbacks are evaluated. In this regard, congruent melting and limited supercooling are the strong suites of organic PCMs. As previously mentioned, inorganic materials have a relatively high LHF and their density can be twice the density of organic materials. Even though this property gives the material the advantage of small volume and compactness, the incongruent melting, partial corrosion and toxicity make them unfavorable for battery applications. One of the example being sodium hydroxide, where the material has a satisfactory thermal conductivity and limited volume change, but it is highly toxic and corrosive. Most eutectics have the same problem as they are mainly composed of inorganic materials. When other materials are evaluated, it can be seen that, iso-paraffin would not be able to be used for this application due to its high temperature variation during freezing which puts the battery's temperature stability and uniformity in jeopardy. When non-paraffins, such as fatty acids, are considered, the main limitation becomes the cost of the system as they can cost twice as much as organic paraffins. When hard to obtain materials are also eliminated from the list, paraffin wax become the most optimal PCM to be used in this application, mostly due to its useful characteristics such as incongruent melting, high latent heat of fusion and non-corrosivity.

Among the paraffin wax, it can be seen that pure alkanes such as A18 or A20 have suitable melting temperatures to be used in conjunction with Li-ion batteries (around 37°C), but have relatively high prices. Among these two materials, A20 has higher latent heat of fusion. Even though it has low thermal conductivity, it can be enhanced by various aforementioned techniques, especially encapsulating or embedding the material. Finally, among these Alkanes, Eicosane is non-reactive and water soluble characteristics. However, its melting temperature is outside the limits of the desired battery temperature range. For this reason, a material in a similar category, normal-Octadecane is selected, as it has a melting temperature of 29.5°C, which is highly close to the ideal operating temperature for the selected battery.

3.7.1.3 Nano-Particles and PCM Mixture For Thermal Conductivity Enhancement

3.7.1.3.1 Analysis of the PCM-CNT Mixture The effectiveness of the PCM can be improved by adding/mixing nanotubes. The predicted thermal conductivity of this mixture is utilized as a designing parameter of the heat exchanger. The equations have been derived for the case where a carrying fluid *(f)* and PCM additives are utilized. Thus, in a similar way, it is assumed that PCM particles are carried by the CNT as it has replaced the flow. Note that the series configuration for PCM leads to parallel arrangement for CNTs. In this case, the following equations are obtained. c is defined as the mass concentration in the foam and PCM.

$$c = \frac{V_{pcm}}{V_{tot}} = \frac{V_{pcm}}{V_{pcm} + V_{cnt}} \tag{3.4}$$

$$k_{eff,parallel} = \left(\frac{c}{k_{pcm}} + \frac{1-c}{k_{cnt}}\right)^{-1} \tag{3.5}$$

$$k_{eff,series} = ck_{pcm} + (1-c)k_{cnt} \tag{3.6}$$

As defined previously, the ratio of thermal conductivities are given as follows:

$$\frac{k_{eff}}{k_{cnt,parallel}} = \left(c\frac{k_{cnt}}{k_{pcm}} + (1-c)\right)^{-1} \tag{3.7}$$

$$\frac{k_{eff}}{k_{cnt,\,series}} = c\frac{k_{pcm}}{k_{cnt}} + (1-c) \tag{3.8}$$

3.7.1.4 Thermal Modeling of Heat Exchanger

3.7.1.4.1 Heat Transfer Coefficients and Pressure Drops LMTD method is utilized here for predicting the heat exchanger performance where, its heat transfer rate is calculated as

$$Q = UA_{tot}\Delta T_{lm} = (\dot{m}c_p\Delta T)_h = (\dot{m}c_p\Delta T)_c \tag{3.9}$$

where ΔT_{lm} is the logarithmic mean temperature as

$$\Delta T_{lm} = \frac{(T_{h,i} - T_c) - (T_{h,o} - T_c)}{\ln((T_{h,i} - T_c)/(T_{h,o} - T_c))} \tag{3.10}$$

Here, h and c are subscripts of hot and cold stream. U is the overall heat transfer coefficient and A_{tot} is total heat transfer surface.

$$U = \frac{1}{\frac{A_o}{A_i}\frac{1}{h_i} + \frac{1}{\eta_o h_o} + \frac{R_{f,o}}{\eta_o} + \frac{A_o}{A_i}R_{f,i} + A_o R_w} \tag{3.11}$$

$$A_{tot} = A_b + N_f \times s_f \tag{3.12}$$

where A_b and A_i are outside base and internal heat transfer surface area defined as

$$A_b = \pi d_o N_f (s_f - t) \tag{3.13}$$

$$A_i = \pi d_i N_f s_f \tag{3.14}$$

Here, d_i, d_o, t, N_f and s_f are inside and outside tube diameters, fin thickness, number of fins and distance between the fins. Moreover η_o in Equation 3.15 is overall surface efficiency defined as follow and R_w is wall thermal resistance.

$$\eta_o = 1 - \frac{N_f A_f}{A_{tot}}(1 - \eta_f) \tag{3.15}$$

where η_f is the efficiency of a single fin. It is worth mentioning that in the absence of any fins, its efficiency is taken as unity. When the circular fin is considered for the external surface, its efficiency can be written as

$$\eta_f = \frac{C_2\left[K_1(mr_1)I_1(mr_{c2}) - I_1(mr_1)K_1(mr_{c2})\right]}{\left[I_0(mr_{r1})K_1(mr_{c2}) + K_0(mr_1)I_1(mr_{c2})\right]} \tag{3.16}$$

where I and K are modified Bessel function of 1st and 2nd kind. Moreover, C_2 and m are given as follows:

$$C_2 = \frac{2r}{m(r_{2c}^2 - r_1^2)} \tag{3.17}$$

$$m = \sqrt{\frac{2h_o}{k_w t}} \tag{3.18}$$

Here, convection heat transfer coefficient in tube side, h_i is calculated based on the corresponding Reynolds number as follows:

$$h_i = \frac{k_f}{d_i}\left(3.657 + \frac{0.0677\,(\mathrm{Re\ Pr}\ d_i/L)^{1.33}}{1 + 0.1\ \mathrm{Pr}\,(\mathrm{Re}\,d_i/L)^{0.3}}\right) \quad for \quad \mathrm{Re} \le 2300 \tag{3.19}$$

$$h_i = \frac{k_f}{d_i}\left\{\frac{\frac{f}{2} \times (\mathrm{Re} - 1000)\ \mathrm{Pr}}{1 + 12.7\sqrt{\frac{f}{2}}(\mathrm{Pr}^{0.67} - 1)}\right\} \quad for \quad 2300 < \mathrm{Re} \le 10000 \tag{3.20}$$

where friction factor for this case is $f = (1.58\log(\mathrm{Re}) - 3.28)^{-2}$

$$h_i = \frac{k_f}{d_i}\left\{\frac{\frac{f}{2} \times \mathrm{Re\ Pr}}{1.07 + \frac{900}{\mathrm{Re}} - \frac{0.63}{1 + 10\,\mathrm{Pr}} + 12.7\sqrt{\frac{f}{2}}(\mathrm{Pr}^{0.67} - 1)}\right\} \quad for \quad \mathrm{Re} > 10000 \tag{3.21}$$

Here, friction factor $f = 0.00128 + 0.1143(\mathrm{Re})^{-0.311}$ and *Re* is Reynolds number which is defined as

$$\mathrm{Re} = 4\dot{m}/(\pi d_i \mu N) \tag{3.22}$$

Here, N is the number of tubes. Furthermore, the pressure drop and outside convection heat transfer coefficient (h_o) are calculated as

$$\Delta P = 4NfL\dot{m}^2/(\rho\pi^2 d_i^2) \tag{3.23}$$

Once the heat transfer inside the tube is calculated, the heat transfer coefficient for the outer surface is required. To calculate this value, the outer surface Nusselt number is calculated first.

$$h_o = \frac{Nu \times k_f}{\Delta r_m} \tag{3.24}$$

Here, k_f, Δr_m and Nu are PCM conductivity, thickness of heat storage material and Nusselt number. Similar to internal surface of the tube, the equivalent Reynolds number is the criterion to select the proper equation

$$Nu = 0.28\left(\frac{\Delta r_m Ra}{L}\right)^{0.25} \quad for \quad Ra \ge 1000 \ \ and \ \ \Delta r_m \le 0.006 \tag{3.25a}$$

$$Nu = 1 \quad for \quad Ra < 1000 \ \ and \ \ \Delta r_m \le 0.006 \tag{3.25b}$$

$$Nu = 0.133(Ra)^{0.326}\left(\frac{\Delta r_m}{L}\right)^{0.0686} \quad for \quad \Delta r_m > 0.006 \tag{3.25c}$$

where R_a = Gr. Pr and Grashof number is $Gr = \dfrac{g\beta(T_{h,i} - T_c)\Delta r_m^{\ 3}}{\upsilon^2}$

$$D_s = 0.637 p_t \sqrt{(\pi N_t) CL/CTP} \tag{3.26}$$

Here, D_s, L and Ra are diameter of the heat exchanger, length of the tubes and Rayleigh number, respectively. Also, p_t tube is pitch and CL is tube layout constant that has a unit value for 45° and 90° tube arrangement and 0.87 for 30° and 60° tube arrangement. Also CTP is tube count constant which is 0.93, 0.9, 0.85 for single pass, two passes and three passes of tubes, respectively.

3.7.2 Design and Optimization of the Latent Heat Thermal Energy Storage System

Another option to utilize PCMs in the EV BTMSs is to use them as passive cooling systems installed on the battery pack. Thus, in this section, two tube configurations have been considered, including straight and helical tube heat exchangers. In addition, fins are studied as extended surface to investigate their effect. The optimization has been carried out based on the constraints, including limit volume and length of the heat exchanger due to the space restrictions in the vehicle.

3.7.2.1 Objective Functions, Design Parameters and Constraints

The main criteria in the heat exchanger are the occupied volume. The heat exchanger is considered as the objective function, and the design parameters considered are listed as follows:

- number of tubes
- tube inside and outside diameter
- shell diameter

When the PCMs are utilized as storage media, the length of the heat storage system exceeds the proposed limits, mainly due to their low thermal conductivity. In order to prevent this issue, nanoparticles have been introduced to increase the thermal conductivity and heat transfer rate in the PCM, which can improve the compactness of the system and enable the length to be within the criteria of objective function. Carbon Nanotubes and Graphene Nano-platelets have been added to PCM as described in the experimental section.

3.7.2.2 Effective Properties of the PCM and Nanotubes

The effective thermal conductivity largely depends on the direction of the nanotubes, where it can be increased considerably with series configuration, but not as much with parallel arrangements. The effective thermal conductivity for the parallel case is shown in Figure 3.14, which can be considered as the "worst case scenario". The label pointing to the zero concentration corresponds to the following effective thermal conductivity as:

$$\frac{k_{eff}}{k_{cnt\,parallel}} = 5.067 \times 10^{-5} \quad \text{Therefore: } k_{eff,parallel} = 5.067 \times 10^{-5} \times 3000 = 0.152\ \mathrm{W/m\,K} \tag{3.27}$$

The obtained value is virtually identical to thermal conductivity of pure PCM. On the other hand, the 'best case scenario' involves the carbon nanotubes to be in series configurations with the direction of temperature gradients as shown in Figure 3.15.

As an example, for 90% concentration for PCM, which is equal to 10% concentration of CNTs, we will have:

$$\mathrm{c} = \frac{V_{pcm}}{V_{tot}} = \frac{V_{pcm}}{V_{pcm} + V_{cnt}} \tag{3.28}$$

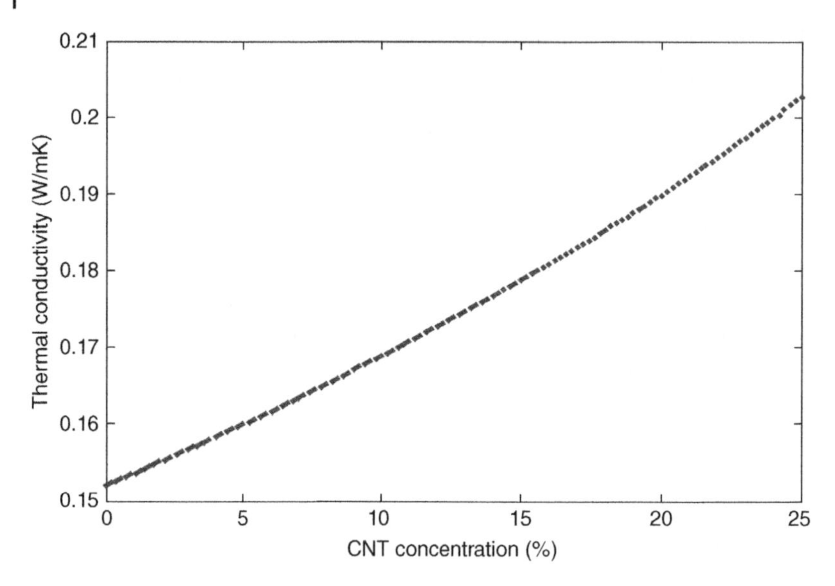

Figure 3.14 Effect of CNT concentration on the thermal conductivity of the mixture in parallel configuration.

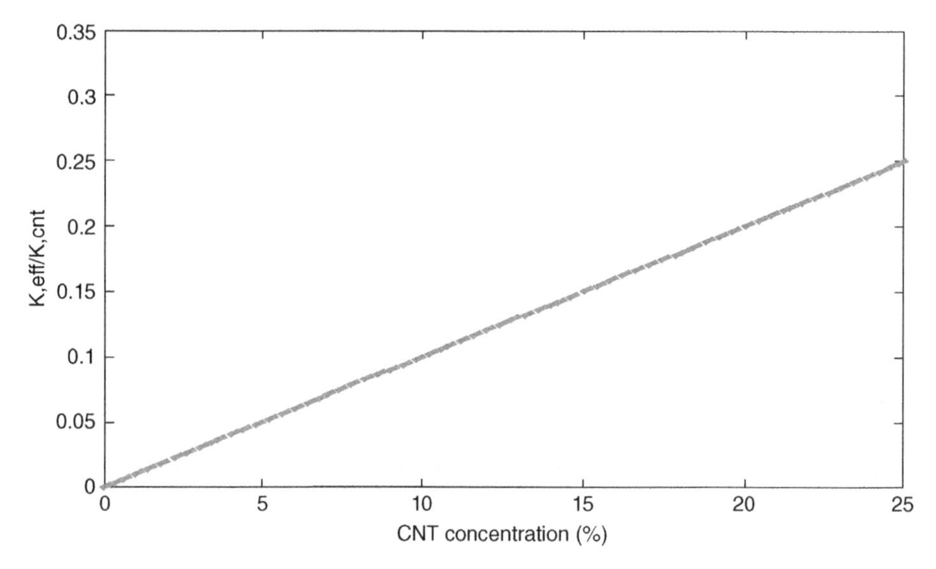

Figure 3.15 Thermal conductivity of the PCM and nanoparticles in series arrangement.

$$k_{eff}(parallel) = \left(\frac{c}{k_{pcm}} + \frac{1-c}{k_{cnt}}\right)^{-1} \tag{3.29}$$

$$k_{eff}(series) = ck_{pcm} + (1-c)k_{cnt} \tag{3.30a}$$

$$k_{eff} = k_{cnt} \times 0.1 = 3000 \times 0.1 = 300\, w/mK \tag{3.30b}$$

3.7.2.3 Combined Condition

The second approach is the distribution for the CNTs where the weighting criteria for the best and worst arrangements of CNTs in the mixture are utilized. If "P" denotes the probability of series arrangement (best case), then the following equations can are obtained:

$$k_{mix} = P \times k_{eff,s} + (1 - P) \times k_{eff,p} \tag{3.31}$$

where $k_{eff,s}$ and $k_{eff,s}$ represent thermal conductivity in series and parallel configurations.

$$k_{mix} = P\left(c.k_{pcm} + (1 - c)k_{cnt}\right) + (1 - P)\left(\frac{c}{k_{pcm}} + \frac{1 - c}{k_{cnt}}\right)^{-1} \tag{3.32}$$

which shows that the concentration of nanoparticles is an effective parameter in determining the mixtures thermal conductivity, as provided in Figure 3.16. Therefore, similar to previous section where the mass and volumetric concentration is defined for soaked foam, the same idea is valid for the mixture as follows:

$$c = \frac{V_{pcm}}{V_{tot}} = \frac{V_{pcm}}{V_{pcm} + V_{cnt}} \tag{3.33}$$

$$k_{eff}(parallel) = \left(\frac{c}{k_{pcm}} + \frac{1 - c}{k_{cnt}}\right)^{-1} \tag{3.34}$$

$$k_{eff}(series) = ck_{pcm} + (1 - c)k_{cnt} \tag{3.35}$$

$$k_{mix} = P\left(c.k_{pcm} + (1 - c)k_{cnt}\right) + (1 - P)\left(\frac{c}{k_{pcm}} + \frac{1 - c}{k_{cnt}}\right)^{-1} \tag{3.36}$$

3.7.2.4 Model Description

In terms of the operating conditions of heat exchanger; the hot water with the minimum mass flow rate of 0.02 kg/s enters in the tube side as hot stream. The PCM is positioned in the shell side to absorb the heat dissipated by the battery (300W). The selected PCM has a melting point of 28.5 °C. The liquid coolant is a 50/50 water-glycol mix that exits the tubes at 29.5 °C. Two types of tubes including the straight tube and helical tubes are studied, where both finned and unfinned structures are evaluated in the first case.

3.7.2.5 Sensitivity Analysis

The optimum value of effective thermal conductivity is determined as 34 W/mK for the case of without finned tube. Figure 3.17 displays the variation in the heat exchanger length with respect to the effective PCM thermal conductivity. It can be seen that increasing the effective thermal conductivity decreases the necessary heat exchanger lengths where the required heat transfer surface areas also decreases for the specific heat duty.

The heat exchanger length and shell diameter with respect to the standard tube are shown in Table 3.5, where it becomes apparent that they increase with increasing tube diameter. Thus, the minimum available tube diamater was used in the analysis. Thus, the minimum available tube diameter in the market is suitable in this case. Essentially, by increasing the tube diameter, the Reynods number decreases, and as a result, the inner convection heat transfer coefficient and overall heat transfer coefficient decreases which in turn increases the total needed heat transfer surface area (length of tubes).

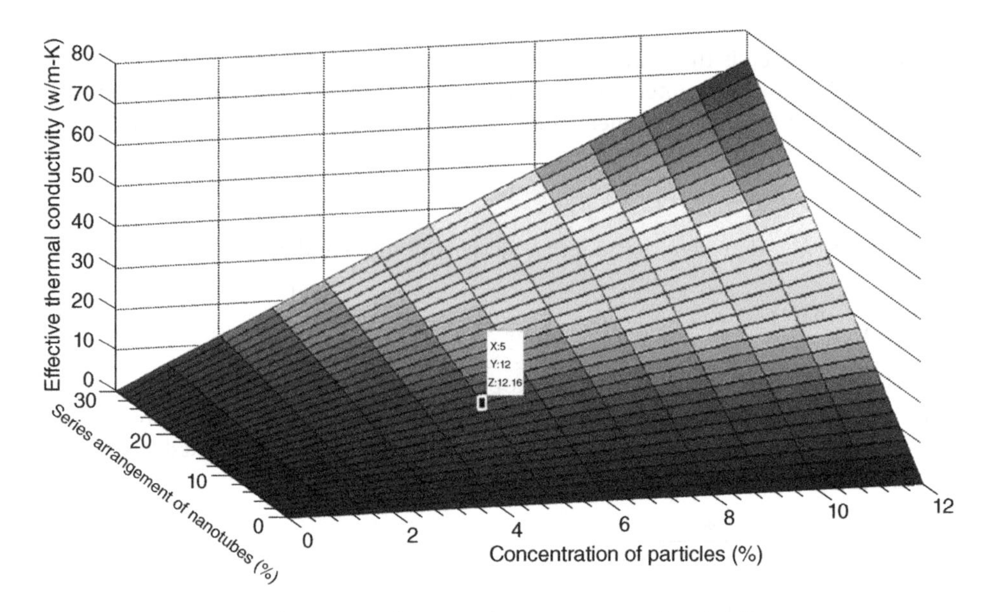

Figure 3.16 Thermal conductivity as a function of concentration and probability.

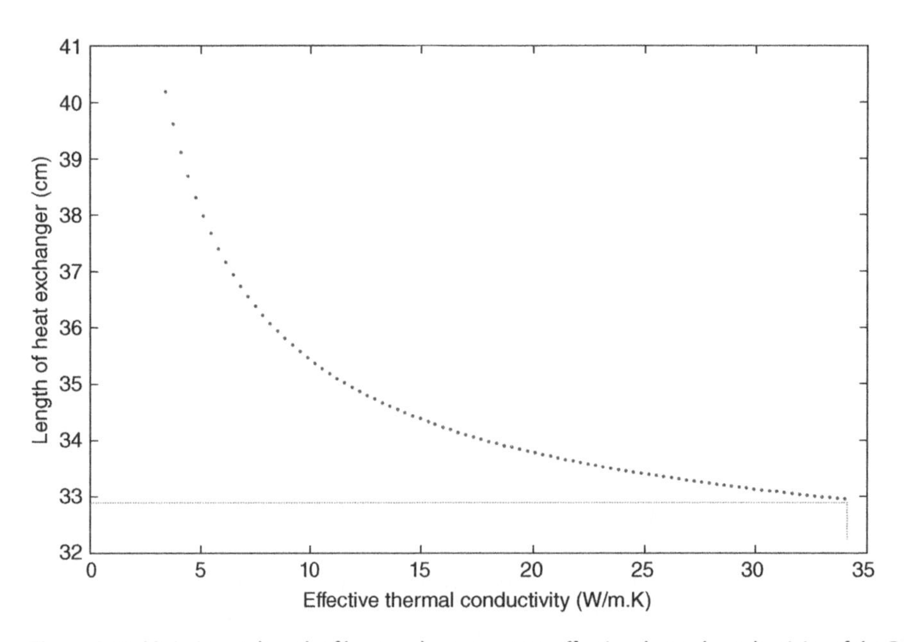

Figure 3.17 Variation on length of heat exchanger versus effective thermal conductivity of the PCM.

Table 3.5 Variations of heat exchanger length and shell diameter with respect to tube diameter.

Tube size	Index	d_i(mm)	d_o(mm)	L(mm)	D(mm)
1/16.	1	1.14	1.59	16.4	13.7
1/8.	2	1.65	3.18	16.72	13.93
3/16.	3	3.23	4.75	20.8	17.53
1/4.	4	4.83	6.35	24.7	20.74
5/16.	5	6.30	7.93	28.41	23.79
3/8.	6	7.90	9.53	32.1	26.77
7/16.	7	9.49	11.11	35.33	29.51
1/2.	8	11.13	12.70	39.33	32.64

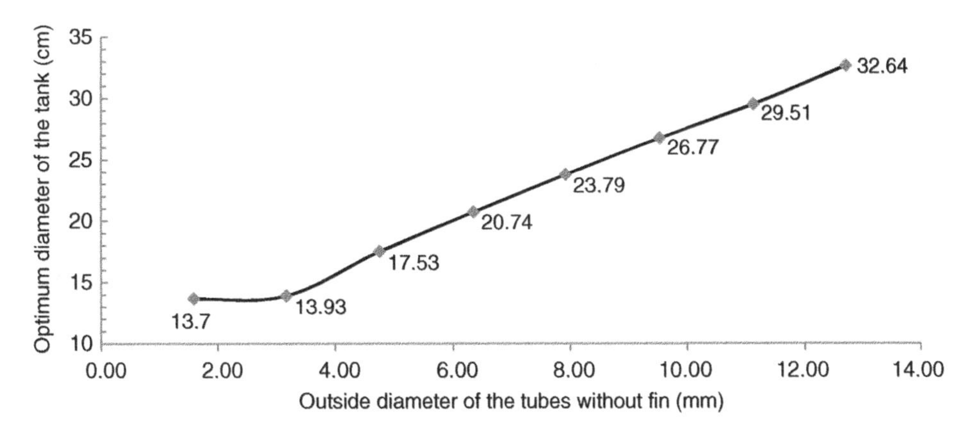

Figure 3.18 Variation of optimum shell (tank) diameter versus tube outside diameter in the case of without fin.

By increasing the tube index, the tube inside and outside dimensions are also increased and L/d_i and D/d_o decreases. Figures 3.18, 3.19 and 3.20 shows the variations of L/d_i and D/d_o versus the variations of tube index. It is determined that the rate of increment in the tube inside and outside diameters is higher than the ones in the tube length and shell diameter.

Figure 3.21 shows the relationship between the optimum value of tube length and tube inner diameter for various heat loads. As it can be seen, the optimum tube length increases by increasing the tube inner diameter with a constant slope as well as be increasing rate heat transfer rate.

The relationship of Reynolds number and rate of heat transfer with respect to "L/d" for various tube diameters is provided in Figure 3.22. It is shown that, the higher value of heat transfer needs the higher value of Reynolds number and L/d, respectively.

In addition, heat exchanger length with respect to CNT series probability and CNT concentration in optimum point (that reside within the problem constraints) are also shown in Figures 3.23 and 3.24. It is clear that the heat exchanger length decreases by

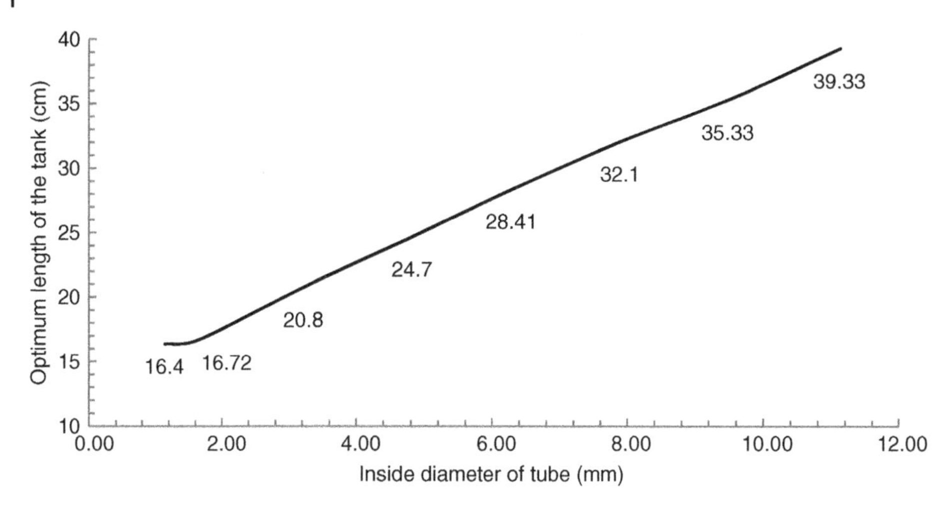

Figure 3.19 Variation of optimum tube length versus tube inside diameter in the case of without fin.

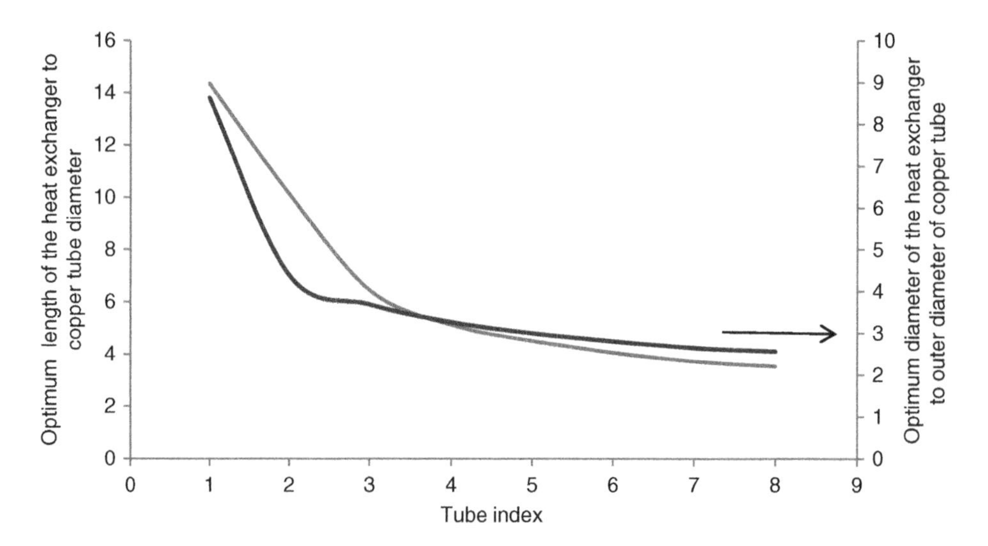

Figure 3.20 Variation of L/di and D/do with tube index.

increasing both CNT probability and concentration. Moreover, the optimal design is satisfied with zero CNT probability and concentration (pure PCM).

The optimal heat exchanger length is provided as a function of CNT concentration and series configuration probability in Figure 3.23. The contours show the regions where the constraints cannot be satisfied. In the figure, the bottom left corner corresponds to the pure PCM which provides the lengths that fail to meet the requirements and to satisfy the constraint.

The optimum lengths have been determined based on various heat generation and mass flow rates that result in a range of Reynolds numbers. As all evaluated tube diameters (from 1/16" to 1") with respect to the aforementioned criteria are investigated to be handled in through the heat exchanger diameter, the following relationship fits the

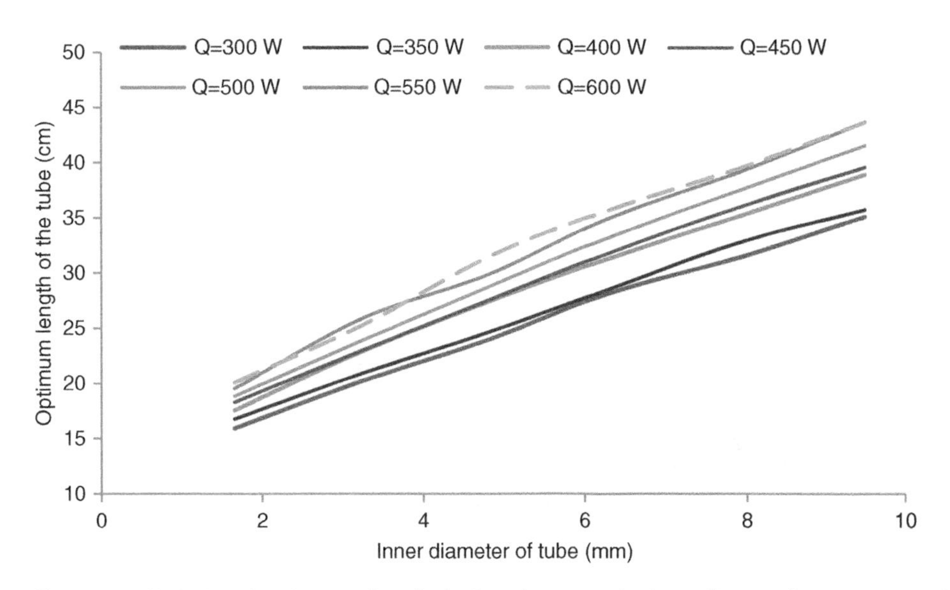

Figure 3.21 Variation of optimum value of tube length versus tube inner diameter for various rate of heat transfer.

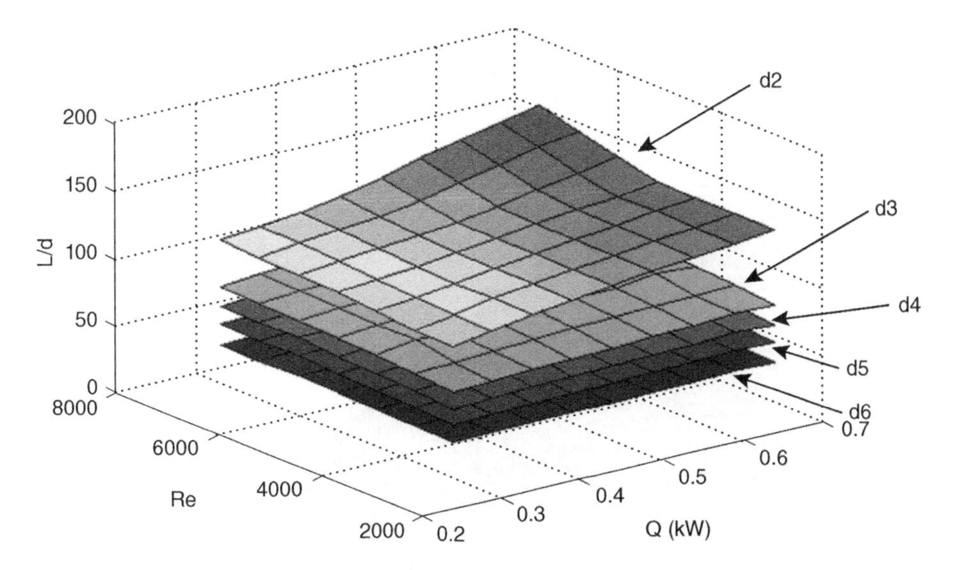

Figure 3.22 Dependency of Re, rate of heat transfer and L/d for various tube diameter.

set of diagrams with the least error.

$$\frac{LD}{d^2} = 471.55 + Re\left(1 + 381.28 \times \frac{Q}{L_{melt}\left(1 + \frac{Re}{893}\right)}\right) \tag{3.37}$$

"*L*" and "*D*" are optimum length and diameter of the tank. L_{melt} is latent heat of fusion for the phase change material. It is also worth mentioning that Re number is calculated

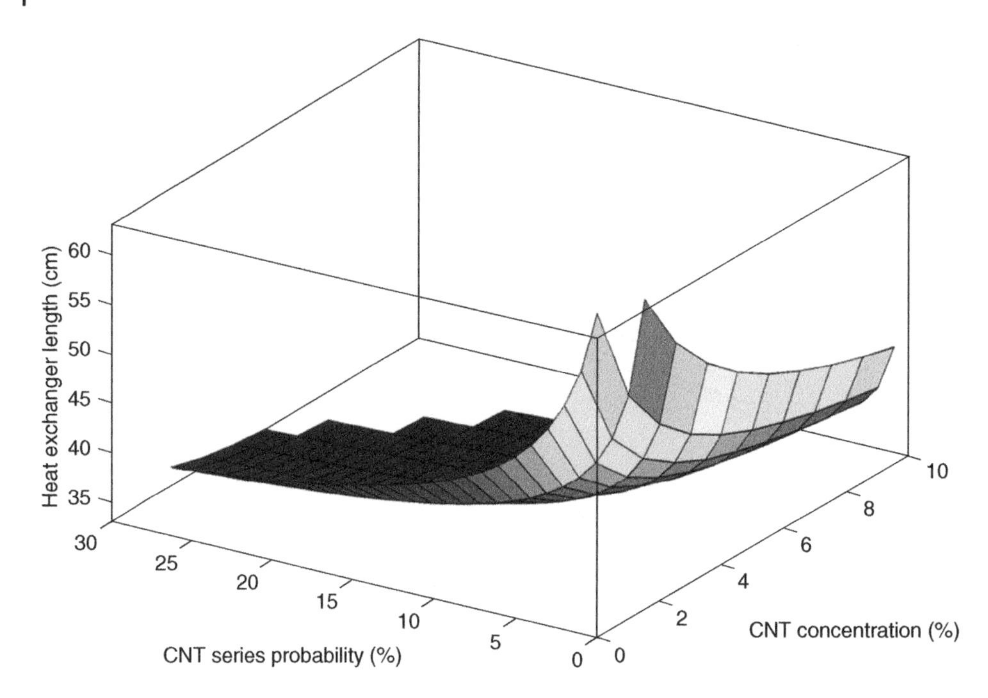

Figure 3.23 Variation of heat exchanger length with the probability of CNT in series configuration and concentration.

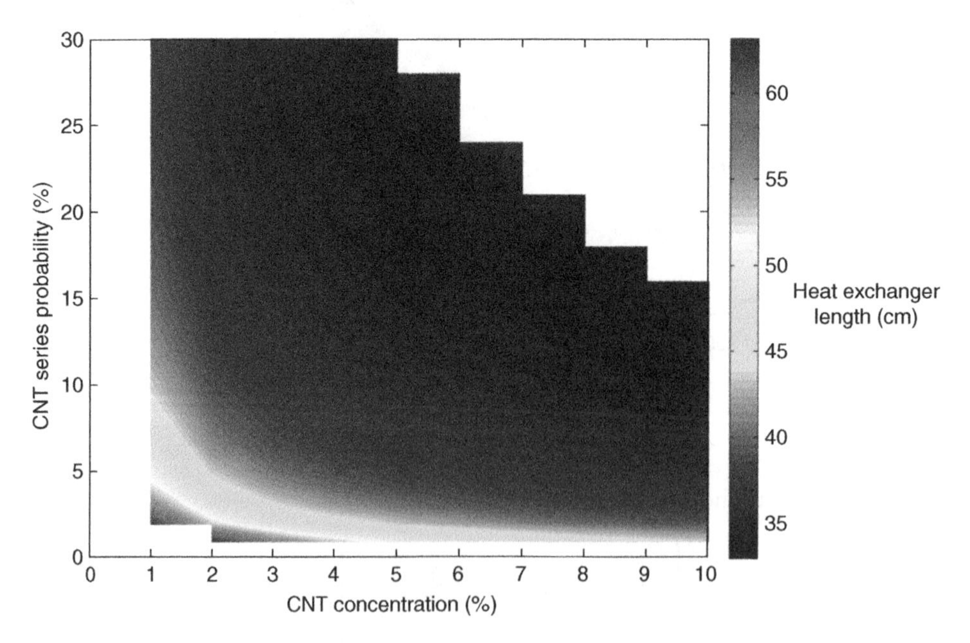

Figure 3.24 Contour of heat exchanger length versus CNT series probability and CNT concentration.

based on the total mass flow rate:

$$D_H = \frac{AA}{P} \tag{3.38}$$

where D_H is the hydraulic diameter. If "N" is the number of tubes in heat exchanger, then $D_H = d\sqrt{N}$. Also Reynolds number is calculated as $R_e = \frac{4\dot{m}_{total}}{\mu\pi D_H}$.

3.7.2.6 Helical Tube Heat Exchanger

Another alternative for the tube configuration is helical. Helical tubes increase the rate of heat transfer due to the curvature in the tube. At the same time, they apply specific geometrical limitation in the design. In order to minimize the heat exchanger length, seven design parameters are selected as shown in Table 3.6. Design tube schedules and corresponding tube outside diameter, tube thickness and tube fin length are listed in the Table 3.7.

The variations of heat exchanger length versus CNT series probability and CNT concentration in optimum point is shown in Figure 3.24. It should be mentioned that the heat exchanger lengths that lay outside of the aforementioned constraints are not illustrated in these figures. It is shown that the heat exchanger length can be reduce by increasing both CNT probability and concentration. Moreover, it can be seen that the pure PCM (with no CNT probability or concentration), the optimal design does not satisfy the constraint as it is the case for straight copper tubes.

Table 3.6 Design parameters and their range of variation in the case of helical tubes.

Parameter	Lower bound	Upper bound
Number of tubes	1	200
Index of tube	1	6
Tank diameter (m)	0	0.3
CNT concentration (%)	0	10
CNT series probability (%)	0	20
Helical radius	0.02	0.1
Aspect ratio	1.5	10

Table 3.7 Soft copper tube specifications for optimization.

Size (O.D, inch.)	Outer diameter (mm)	Inner diameter (mm)	Wall thickness (mm)
1/8"	3.16	1.65	0.762
3/16"	4.76	3.24	0.762
1/4"	6.35	4.83	0.762
5/16"	7.94	6.31	0.813
3/8"	9.53	7.90	0.813
1/2"	12.70	11.07	0.813

3.8 Case Study 2: Melting and Solidification of Paraffin in a Spherical Shell from Forced External Convection

In this section, a practical example is provided that incorporates the use of paraffins as a latent thermal energy storage system, where the charging and discharging processes inside an encapsulated-paraffin latent TES sphere is modelled. The results are validated with experimental data and several input and boundary conditions are considered to determine their impacts on various performance criteria. Finally, a detailed discussion of the results is provided along with a description of the importance of the data obtained (Dincer and Rosen, 2010).

The case simulated is similar to that reported by Ettouney *et al.* (2005). A paraffin blend is contained in a spherical, copper shell, and air is used as the heat transfer fluid. The paraffin is heated until it completely melts and then cooled to solidification using a fan, which blows air at 10 m/s through a vertical glass tube containing the copper sphere. The experimental apparatus is shown in Figure 3.25.

The glass column has a diameter of 20 cm and a height of 40 cm. The copper sphere, which is suspended by a thin supporting rod, has an outer diameter of 3 cm and a copper wall thickness of 1.2 mm, and is filled with a paraffin PCM. Thermocouples are located throughout the PCM to record temperature variations, and also to estimate the energy stored in the capsule during the charging and discharging processes. Due to the high velocity of the incoming air, heat transfer from the glass column to the ambient environment is neglected.

During charging of the latent store, the air is heated to a specified temperature above the melting point of the PCM, and the fan is used to force convective heat transfer to the sphere from the air. During discharging, the inlet air has a temperature below that of the PCM, so that heat is removed from it and solidification occurs.

The thermophysical properties used for the air and the PCM are taken for specified ranges from Ettouney *et al.* (2005), and in cases where variations exist between the solid and liquid states (e.g., density) the property is taken as the average of the two. Although this is not a precise approximation, its use as a simplifying assumption reduces the cost of computation. The impact on the results of this assumption is assessed subsequently.

The paraffin wax used as the PCM is thus assumed to have a constant density of 820 kg/m^3, a latent heat of fusion of 210 kJ/kg and a melting temperature of 48.51°C (321.66 K). Its specific heat and thermal conductivity, taken as the average between solid and liquid states, are taken to be 2.5 kJ/kg K and 0.195 W/m K, while the viscosity of the paraffin wax in the liquid state is taken to be 0.205 kg/m s. The air used as a heat transfer fluid has a density of 1.137 kg/m^3, a specific heat of 1.005 kJ/kg K, a thermal conductivity of 0.0249 W/m K and a dynamic viscosity of 2.15×10^{-5} kg/m s. The copper shell, in which the PCM is contained, has a FLUENT-defined density of 8978 kg/m^3, a thermal conductivity of 387.6 W/m K and a specific heat of 0.381 kJ/kg K.

An analysis of the experiment of Ettouney *et al.* (2005) reveals an axis of symmetry, which can be exploited to reduce the computational effort. Consequently, the cylindrical column in Figure 3.25 is split into four quadrants, reducing the computational cost of simulation by 75%. This simplification does not change the validity of the results, as shown subsequently, and lowers simulation times significantly. The quarter section considered of the physical domain, as created in GAMBIT, is shown in Figure 3.26.

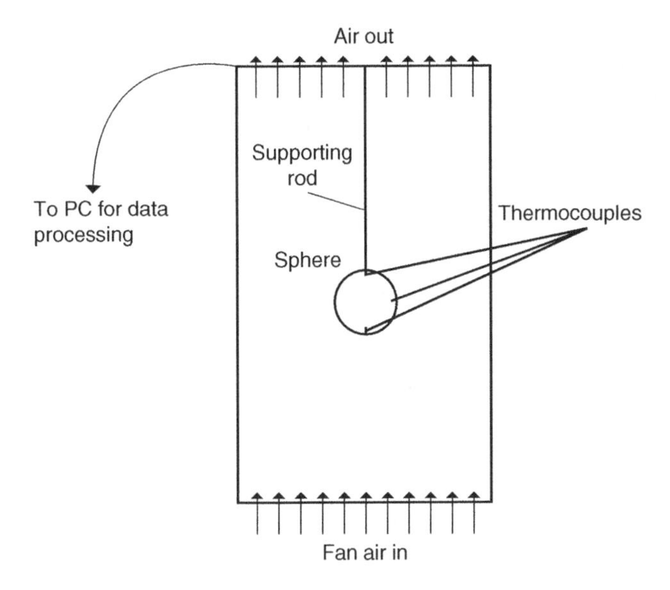

Figure 3.25 Experimental apparatus used for the charging and discharging of the latent store using a forced air flow, modified from Ettouney *et al.* (2005).

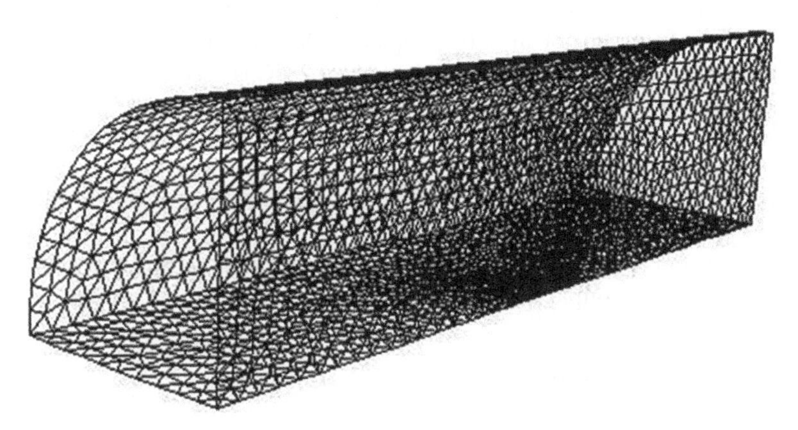

Figure 3.26 Wall grid volumes for the computational domain considered for a latent TES. The PCM capsule volumes are shown in the darker center region (adapted from Dincer and Rosen, 2010).

In addition to assuming axisymmetric properties along the cuts made to the original domain, a number of other assumptions are also introduced to facilitate calculations such as neglecting kinetic and potential energy effects, including buoyancy, and assuming a zero-shear wall for the symmetric walls in Figure 3.26. Moreover, the outer walls, while exhibiting no-slip wall conditions, are assumed to be adiabatic. Since the heat transfer fluid passes rapidly through the volume, this assumption reduces the complexity of the model considerably without sacrificing its accuracy or the reliability.

Once the geometric creation from GAMBIT is loaded into FLUENT, the remaining steps in the simulation set up are concerned with using the correct models, defining material properties, monitoring data for performance calculations, choosing a suitable time step and setting the residual tolerances. Since these simulations involve phase change, the solidification/melting model is turned on, with the material properties for air, copper and paraffin entered as previously described.

The case study investigates the effects of variations in the inlet air temperature on the performance of the system. In the study, four cases are considered: two charging and two discharging. For the charging (melting) cases, the initial temperature of the PCM is set to a sub-cooled temperature 5°C below its freezing point (i.e., at 316.66 K), while the inlet air temperature is set to either 10°C or 20°C above the melting point (i.e., to 331.66 K and 341.66 K). The resulting performance values help indicate which option is more efficient. For the discharging (solidification) cases, the initial temperature is set 5°C above the melting point (i.e., at 326.66 K) so that the PCM is completely melted. As with the charging case, the inlet air temperatures are set between 10°C and 20°C below the melting point (i.e., at 311.66 K and 301.66 K). The reference-environment temperature for exergy calculations in all cases is taken to be 23°C.

3.8.1 Validation of Numerical Model and Model Independence Testing

In order to validate the numerical model, conditions from Ettouney *et al.* (2005) are mirrored so that the resulting temperature variations on the inner portion of the shell can be monitored and compared. The experimental steps are as follows: Air enters the glass column at 60°C and with a velocity of 10 m/s. The PCM capsule, initially at a temperature of 23°C, receives heat from the hot air until complete melting is achieved after 1120 seconds. Then, the air heater is removed and ambient air at 23°C is used to re-solidify the PCM, which occurs after 760 seconds. The PCM temperatures are recorded with thermocouples located just inside the copper shell at three locations: facing the flow direction, away from the flow direction, and at the side of the sphere where the air moves the fastest. That is, the thermocouples are located at the top, bottom and side of the sphere as shown in Figure 3.27.

The FLUENT simulation is configured to match the above conditions, and the temperature profiles at the thermocouple locations are monitored by selecting a point in

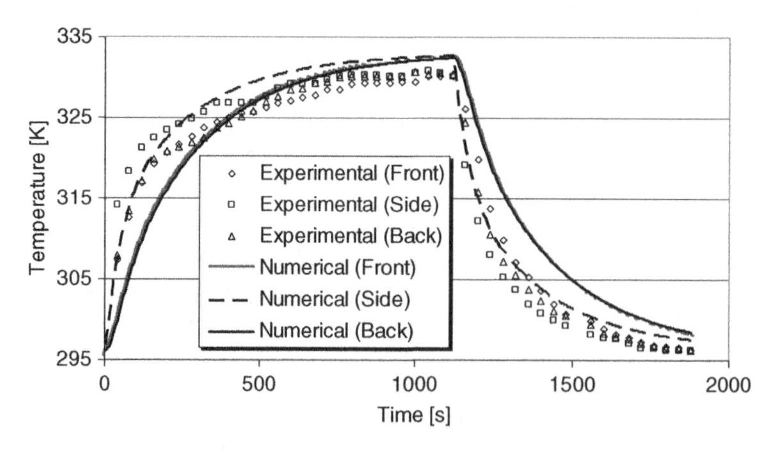

Figure 3.27 Comparison of numerical temperature profiles at three locations in the sphere with the experimental results from Ettouney *et. al.* (2005), for the model validation (adapted from Dincer and Rosen, 2010).

the PCM domain and recording the temperature in the corresponding volume or cell as time progresses. The resulting temperature profiles are shown in Figure 3.27.

The differences between the numerical and experimental temperature profiles are likely due to a number of factors, including contact resistance, effects caused by embedding thermocouples in the PCM and convection effects in the PCM. However, the solidification and melting times, as well as the overall heat transfer rates, are important to this investigation, the purpose of which is to determine heat transfer characteristics between the PCM and HTF and associated performance criteria.

The time required for the PCM to completely melt is similar with both the experimental and numerical approaches. Complete melting is detected in 600 seconds experimentally, while the numerical approach quite accurately predicts complete melting in 615 seconds. A duration of 280 seconds is required to completely solidify the PCM after solidification is initiated in the experimental unit. This value is again in good agreement (within 1.5%) with the numerically obtained value of 285 seconds.

The similarities in the numerical and experimental results are encouraging and well within normal acceptable errors when validating a numerical model such as the one considered here. With these results, the model can next be subject to grid size and time step validations, to improve confidence in the performance data.

The grid size performance independence tests are presented first. In any numerical model, if the results are to be taken as a realistic representation of actual behavior, the simulation results must be shown to be independent of small changes in the structure of the computational volume. Here, therefore, the mesh spacing is changed in the copper shell region, causing the spacing in the other connected regions of the domain to change correspondingly. The cell density throughout the region is varied so that more and fewer volumes are considered compared to the grid used in the base performance analysis. The cell distribution in all three cases is shown in Table 3.8.

In order to determine the dependence of the overall results on the small changes in cell distribution, simulations are performed using the same conditions as in the model validation tests in Figure 3.27, and the variation of the liquid fraction inside the PCM with time is monitored. The resulting liquid fraction variations are shown in Figure 3.28.

Since the liquid fraction is a good indicator of the heat transfer characteristics to and from the capsules, it is taken here to be a sufficient gauge of the overall grid performance. From Figure 3.28, it can be clearly observed that that the changes in grid size have little

Table 3.8 Cell distribution for the three cases considered in the grid independence tests for the sphere.

Cell data	Grid size		
	Small	Base	Large
Mesh spacing in copper shell (cm)	0.21	0.20	0.18
Mesh volumes in copper region	164	200	221
Mesh volumes in PCM region	1,523	2,048	2,313
Mesh volumes in air (HTF) region	27,775	31,864	33,238
Total volumes	29,462	32,312	35,772

Source: Dincer and Rosen (2010).

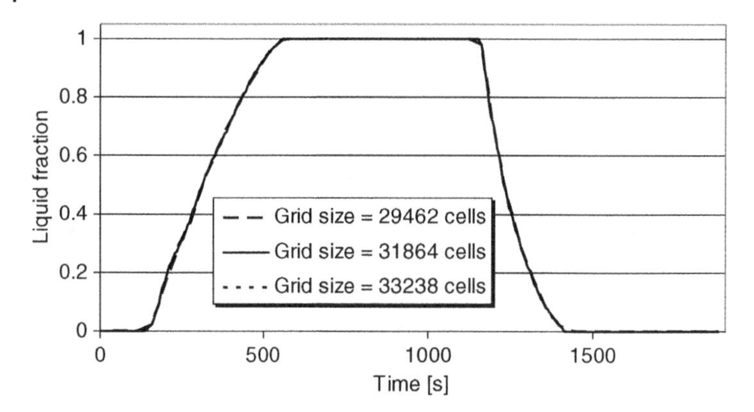

Figure 3.28 Liquid fraction as a function of time for the grid size independence tests for the sphere (adapted from Dincer and Rosen, (2010).

impact on the overall solutions. Hence the chosen grid size and orientation are deemed adequate. In general, care must be taken when initially choosing the cell distribution in a computational domain to ensure the model is not only able to model physical scenarios well, but also capable of passing grid independence tests.

The time step independence is now tested. In the present case, the time step is set to 1.0 second. The dependence of the simulation progression on changes in time step can be evaluated straightforwardly by altering the time step in the FLUENT controls, and monitoring the changes in liquid fraction for the overall process. To illustrate, alternative time steps of 0.1 seconds and 2 seconds are considered, and the resulting liquid fraction variations with time are shown in Figure 3.29.

The liquid fraction variations in Figure 3.29 correlate very well for all time steps considered, suggesting that the step of 1.0 second used in the initial analysis is adequate. Although the smaller time step of 0.1 second may permit a more accurate representation of the physical domain and phenomena, the computational cost is increased tenfold, increasing the simulation time greatly and, depending on the computer resources available, perhaps unreasonably. For the time step of 2 seconds, little difference is observed in the variation of liquid fraction while computational cost is severely decreased but,

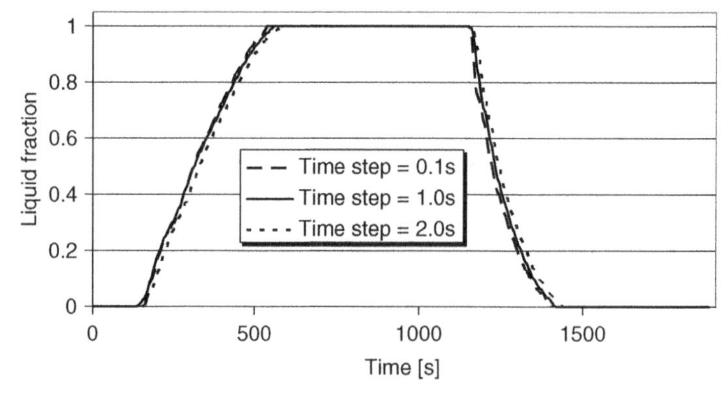

Figure 3.29 Liquid fraction as a function of time for the time step independence tests for the sphere (adapted from Dincer and Rosen, 2010).

although not shown here, it was found that increasing the time step beyond 2.0 seconds significantly influenced the validity of results. This test shows the importance of choosing the time step appropriately. The selection of time step and grid sizes can serve as a useful tool for determining the most computationally efficient scenario for any given problem.

With the model having been validated and passing time step and grid size independence testing, the remainder of the analysis focuses on the performance criteria.

3.8.2 Performance Criteria

Since kinetic and gravitational potential energy effects are neglected, the energy balance for a system undergoing either charging or discharging processes can be written as follows:

$$\Delta E_{sys} = E_{in} - E_{out} = U_{in} - U_{out} \tag{3.39}$$

The change in energy of the system itself is comprised of the changes in energy of each material within the domain, namely the air, copper and PCM (paraffin) regions:

$$\Delta E_{sys} = \Delta E_{air} + \Delta E_{copper} + \Delta E_{pcm} \tag{3.40}$$

The change in energy of the air and copper regions at any time can be calculated using the change in volume-averaged temperature within each region:

$$\Delta E_{air} = m_{air} C_{air} \left(\overline{T}_{air} - T_{air,ini} \right) \tag{3.41}$$

$$\Delta E_{copper} = m_{copper} C_{copper} \left(\overline{T}_{copper} - T_{copper,ini} \right) \tag{3.42}$$

Due to the complexities resulting from solid/liquid interfaces within the PCM, its energy change is evaluated by monitoring the heat transfer from the copper to the PCM on the inner surface of the copper shell:

$$\Delta E_{pcm} = \int_0^t Q_w dt = \sum_{i=0}^{t} Q_{w,t} \tag{3.43}$$

The efficiency of the system in general differs for each process. First, we consider energy efficiencies. For the charging process, the purpose is to add heat to the PCM region. In other words, the energy content of the PCM and copper shell regions is increased, while the energy input to achieve this objective is the change in enthalpy from inlet to outlet of a heat transport fluid (air in the present case).

$$\eta_{ch} = \frac{E_{prod}}{E_{input}} = \frac{\Delta E_{sys}}{H_{in} - H_{out}} = \frac{\Delta E_{sys}}{U_{in} - U_{out} + V(p_{in} - p_{out})} \tag{3.44}$$

The term $V(p_{in} - p_{out})$ can be written as follows:

$$V(p_{in} - p_{out}) = \frac{\dot{m}t}{\rho_{air}} \int_0^t (\overline{p}_{in} - \overline{p}_{out}) dt = \frac{\dot{m}t}{\rho_{air}} \sum_{i=0}^{t} (\overline{p}_{in,t} - \overline{p}_{out,t}) \tag{3.45}$$

During discharging, the purpose is to recover heat from the solidifying PCM, so that the total enthalpy received from the capsule is the product energy content. The total energy obtained is the change in energy of the PCM and its shell:

$$\eta_{dc} = \frac{E_{prod}}{E_{input}} = \frac{H_{in} - H_{out}}{\Delta E_{sys}} = \frac{U_{in} - U_{out} + V(p_{in} - p_{out})}{\Delta E_{sys}} \tag{3.46}$$

The change in energy of the system is negative during the discharging process, and there is a drop in pressure through the column. The energy efficiency varies between zero and 100% for both discharging and charging cases. Although the dependence of the evaluated parameters on the progression time of the simulation is not written explicitly here, the energy and exergy efficiencies are examined as the simulation time progresses.

The exergy performance criteria can be similarly examined. For charging or discharging processes, the exergy balance can be expressed as follows (more detailed information regarding the concept of exergy efficiency and its calculations will be provided in Chapter 5):

$$\Delta \Xi_{sys} = \in_{in} - \in_{out} - I \tag{3.47}$$

Here, I denotes exergy destruction and includes exergy destroyed due to heat transfer and viscous dissipation in the flowing fluid. The exergy balance excludes thermal exergy flows across system boundaries, since the walls are assumed adiabatic. The exergy destruction can be determined with an entropy balance:

$$I = T_{\infty}\Pi \tag{3.48}$$

where

$$\Delta S_{sys} = S_{in} - S_{out} + \Pi \tag{3.49}$$

The change in entropy of the system is calculated as the summation of the changes in entropy for each of its three components:

$$\Delta S_{sys} = \Delta S_{pcm} + \Delta S_{copper} + \Delta S_{air} \tag{3.50}$$

where

$$\Delta S_{pcm} = \frac{\Delta E_{pcm}}{T_{m,pcm}} = \sum_{i=0}^{t} \frac{Q_{w,t}}{T_{m,pcm}} \tag{3.51}$$

$$\Delta S_{copper} = m_{copper} C_{copper} \ln\left(\frac{\overline{T}_{m,copper}}{T_{ini}}\right) \tag{3.52}$$

$$\Delta S_{air} = m_{air} C_{air} \ln\left(\frac{\overline{T}_{m,air}}{T_{ini}}\right) \tag{3.53}$$

The above expression for ΔS_{pcm} presumes that no entropy generation occurs within the PCM. Further, it is assumed that heat transfer to the PCM occurs at its mean temperature. The expressions for ΔS_{copper} and ΔS_{air} assume incompressible fluid behaviour and constant specific heats.

The difference between inlet and outlet entropy is dependent on the flow temperatures, so that after time t the entropy difference is

$$S_{in} - S_{out} = \sum_{i=0}^{t} \dot{m} C_{air} \ln\left(\frac{T_{in}}{\overline{T}_{out,t}}\right) \tag{3.54}$$

To solve the exergy balance, the total exergy change of the system is needed at each time t. The exergy balance depends on the change in energy of the system as well as the entropy change:

$$\Delta\Xi_{sys} = \Delta E_{sys} - T_\infty \Delta S_{sys} \tag{3.55}$$

With this information, the charging and discharging exergy efficiencies can be written analogously to their energy counterparts:

$$\psi_{ch} = \frac{\Delta\Xi_{sys}}{\in_{in} - \in_{out}} \tag{3.56}$$

$$\psi_{dc} = \frac{\in_{in} - \in_{out}}{\Delta\Xi_{sys}} \tag{3.57}$$

The energy and exergy efficiencies can now be assessed while the simulation progresses.

It is instructive to examine the nature of the irreversibilities, which are a result of viscous dissipation within the fluid and heat transfer. The exergy destruction associated with viscous dissipation can be expressed as follows:

$$I_{dissipative} = T_\infty \Pi_{dissipative} \tag{3.58}$$

The exergy destroyed via heat transfer during the melting and solidification processes can be written as the difference between the overall exergy destruction I and the exergy destruction associated with viscous dissipation:

$$I_{ht} = I - I_{dissipative} \tag{3.59}$$

3.8.3 Results and Discussion

For the charging and discharging processes, many of the performance results as well as energy and exergy quantities are presented here as a function of a dimensionless time t^*, defined as

$$t^* = \frac{t}{t_{sf}} \tag{3.60}$$

where t_{sf} denotes to the time for total solidification or melting.

The liquid fractions for solidification and melting are plotted against this dimensionless time in Figure 3.30. In this figure, as noted previously, inlet air temperatures 10°C and 20°C above the melting point (i.e., at 331.66 K and 341.66 K) are considered for the charging (melting) cases, while inlet air temperatures 10°C and 20°C below the melting point (i.e., at 311.66 K and 301.66 K) are considered for the discharging (solidification) cases. It is observed in Figure 3.30 that inlet air temperatures which are further removed from the phase change temperature promote heat transfer more readily between the air and the copper capsule, causing the liquid fraction to decrease more rapidly for the solidification process, and increase more rapidly for the melting process.

In order to illustrate the energy-related processes during charging and discharging, energy quantities for each case are shown in Figures 3.31 and 3.32. These include the heat generated through viscous dissipation, the energy stored during charging and the energy recovered during discharging.

The heat generated through viscous dissipation for both the charging and discharging cases is small compared with the stored and recovered energy quantities. This result is

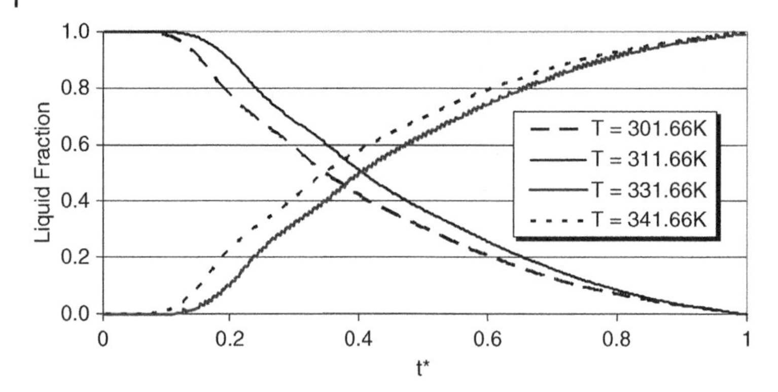

Figure 3.30 Variation of liquid fraction for the charging and discharging cases with dimensionless time t^*, for several inlet air temperatures (adapted from Dincer and Rosen, 2010).

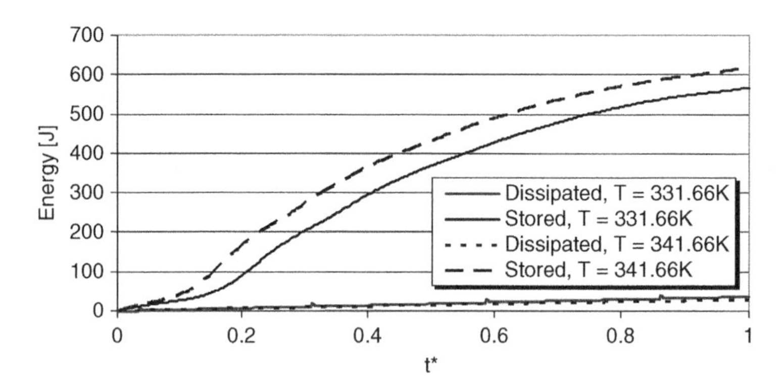

Figure 3.31 Variation of the energy stored and the heat generated through viscous dissipation for the two charging cases, as represented by the two inlet air temperatures considered (adapted from Dincer and Rosen, 2010).

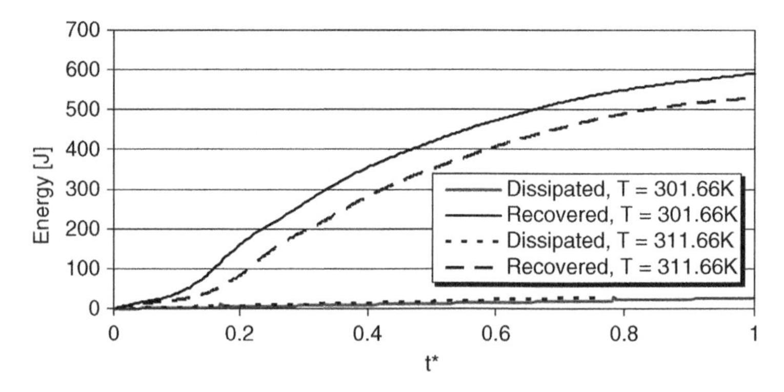

Figure 3.32 Variation of the energy recovered and the heat generated through viscous dissipation for the two discharging cases, as represented by the two inlet air temperatures considered (adapted from Dincer and Rosen, 2010).

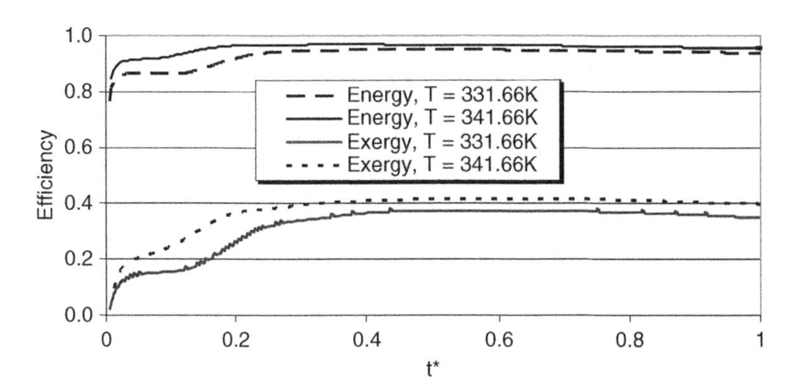

Figure 3.33 Variation of energy and exergy efficiencies with dimensionless time t^* for the charging process, for two inlet air temperatures (adapted from Dincer and Rosen, 2010).

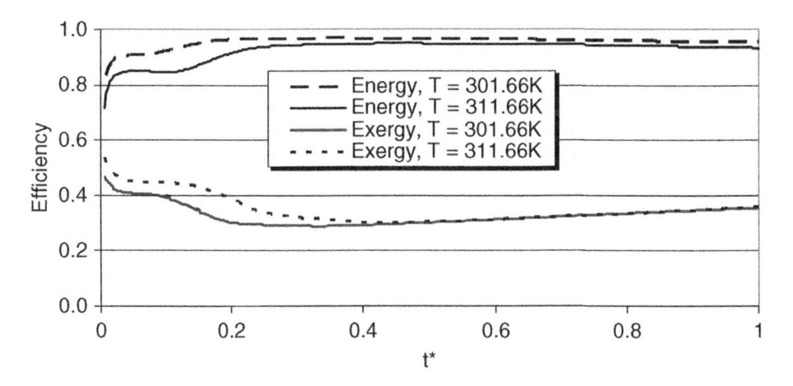

Figure 3.34 Variation of energy and exergy efficiencies with dimensionless time t^* for the discharging process, for two inlet air temperatures (adapted from Dincer and Rosen, 2010).

expected, since heat generation through viscous dissipation is usually small compared to the input and recovered thermal energy quantities in most TES systems. In this case study, the high velocity of the heat transfer fluid (air) causes the heat generation through viscous dissipation to be somewhat larger than usual.

Energy and exergy efficiencies are shown in Figure 3.33 and 3.34 for the charging and discharging cases, respectively. Note that there are no external heat losses from the system shown in Figure 3.25, so the energy efficiencies accounting for only external energy losses are always 100%. In this case study, the heat generated through viscous dissipation, which is relatively small, is assumed unrecoverable and thus treated as an energy loss. Hence the energy efficiencies are somewhat less than 100%. The exergy efficiencies in Figures 3.33 and 3.34 are discussed further after the various exergy quantities (destroyed, stored and recovered) are presented.

The energy efficiency varies with dimensionless time as observed because the stored and recovered energy quantities in the initial stages of the simulations are high, due to the good heat transfer characteristics of the copper shell. The shell's high thermal conductivity permits it to convey heat readily when $t^* < 0.1$. As t^* increases, the paraffin must be heated or cooled to its fusion temperature, so little heat storage or retrieval occurs until melting or solidification begins. Heat storage or retrieval further increases the energy efficiency until $t^* \approx 0.6$, where a maximum is realized (although the efficiency is fairly constant for $t^* > 0.3$). The existence of this energy efficiency maximum can be attributed to the superheating of the paraffin liquid state (or supercooling of the solid state), which reduces heat transfer significantly. Since the energy stored or retrieved is slightly curtailed, the energy efficiency decreases.

Another important observation in Figures 3.33 and 3.34 is that the energy efficiencies are higher for inlet air temperatures which lead to lower solidification or melting times. Inlet air temperatures which decrease the simulation time also decrease the energy loss, thereby increasing the energy efficiency.

In order to explain the behaviours of the exergy efficiencies, the quantities of exergy destroyed, stored and recovered are shown in Figures 3.35 and 3.36. Two primary exergy destruction mechanisms occur, and both are considered: exergy destruction as a result

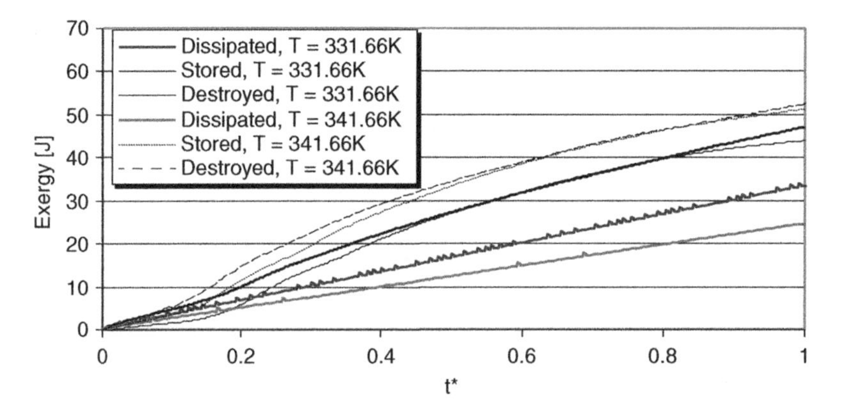

Figure 3.35 Variation with dimensionless time t^* of exergy stored, destroyed via viscous dissipation and destroyed via heat transfer during charging, for two inlet air temperatures (adapted from Dincer and Rosen, 2010).

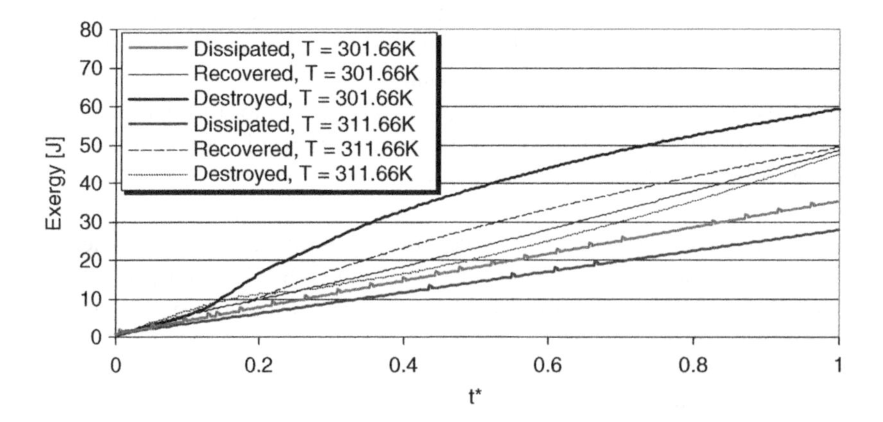

Figure 3.36 Variation with dimensionless time t^* of exergy recovered, destroyed via viscous dissipation and destroyed via heat transfer during discharging, for two inlet air temperatures (adapted from Dincer and Rosen, 2010).

of irreversibilities associated with viscous dissipation (referred to as "dissipated") and exergy destruction due to irreversibilities associated with heat transfer (referred to as "destroyed"). This use of "dissipated" and "destroyed" is followed throughout the remainder of this case study.

The variations in exergy quantities with time for the charging and discharging cases are noteworthy. A complex balance exists between the magnitudes and modes of exergy destroyed and recovered (or stored), which affects the exergy efficiency. The balance suggests multiple interpretations of the most efficient scenario. The main factors involved are the entropy production associated with the two irreversibilities: viscous dissipation and heat transfer.

In Figures 7.42 and 7.43, the largest exergy quantity is the destroyed exergy due to heat transfer, followed by the stored (recovered) exergy and then the exergy destruction due to viscous dissipation. This finding suggests that the temperature of the inlet air is an important factor when optimizing charging or discharging processes involving latent encapsulated PCMs. Note that when the inlet air temperature is further removed from the solidification temperature, the exergy destruction associated with heat transfer increases. However, since reducing the difference between the inlet air temperature and the PCM solidification temperature raises the charging/discharging time, the exergy destruction due to viscous dissipation is greatly increased in this instance, which reduces the exergy efficiency.

The destroyed exergy due to heat transfer and viscous dissipation, as well as the overall destroyed exergy, after solidification or melting is shown in Figure 3.37 for each air inlet temperature case.

An important inference from the results is that efficiencies increase as inlet temperatures are lowered for the discharging case, and are raised for the charging case. This result can be attributed to the increased viscous dissipation in the computational domain as the air speed used to charge or discharge the capsule rises. Note that in most cases the heat transfer fluid used in macro encapsulated paraffin PCMs is not air, but rather a glycol based refrigerant (for cold storage) or water (for warm storage). These fluids have advantageous thermophysical properties (e.g., relatively high specific heats), heat transfer characteristics and low costs. Thus, the results of the current case study are most meaningful when a high pressure head is present during charging or discharging.

In most other packed bed analyses, the irreversibility associated with heat transfer usually is responsible for the dominant exergy loss. For example, consider the detailed

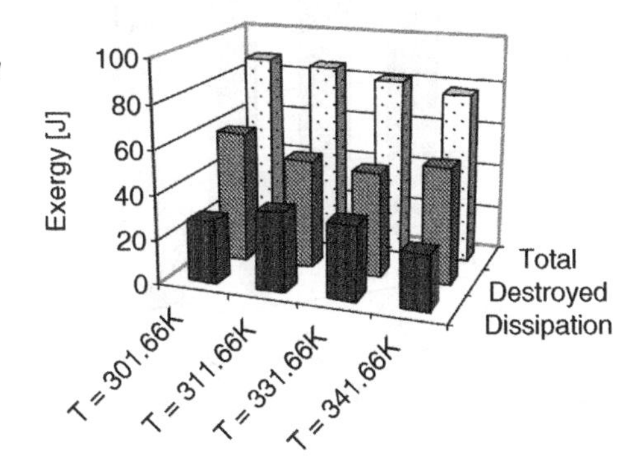

Figure 3.37 Exergy destroyed via viscous dissipation and heat transfer, and the total exergy destruction, after the overall storage process, for all inlet air temperature cases (adapted from Dincer and Rosen, 2010).

TES investigation by MacPhee (2008) of the charging and discharging efficiencies over a wide range of inlet temperatures, flow rates, and geometries, including spherical, cylindrical and slab capsules. The capsules are situated in a packed bed of similar capsules, and an ethylene glycol solution is the heat transfer fluid. The results indicate that all processes are more efficient at lower inlet velocities, which decreases the exergy destroyed due to viscous dissipation (rendering it almost negligible). As a result, the dominant exergy loss is the exergy destruction due to heat transfer, and this quantity decreases as inlet temperature approaches the solidification temperature.

For higher inlet velocities of the heat transfer fluid, the results suggest that it may be more efficient to consider inlet temperatures which achieve solidification or melting more quickly, to decrease exergy destruction via viscous dissipation. This result could be useful for understanding TES using micro encapsulated packed beds or any medium which normally experiences a high pressure drop in a flowing heat transfer fluid. Nonetheless, the irreversibility associated with heat transfer is likely to be of major importance for any space heating or cooling applications. This irreversibility is reduced when the two materials exchanging heat are at similar temperatures. When larger temperature differences exist, entropy generation and exergy destruction are greater. In most cases, like those reported by MacPhee (2008), it is beneficial in terms of efficiency to keep the inlet temperature close to the fusion temperature of the PCM.

The overall energy and exergy efficiencies after the solidification or melting processes have concluded are shown in Figure 3.38.

An important outcome of this case study is the demonstration provided of the usefulness and versatility of exergy analysis. From an energy perspective, only heat generation through viscous dissipation is considered for as a loss (and usually even this is not treated as a loss), resulting in very high overall efficiencies. In Figure 3.38, exergy efficiencies are contrasted with their energy counterparts. The lower magnitudes of the exergy efficiencies reflect more accurately and comprehensively the actual efficiency of the TES system.

Since exergy analysis provides a detailed understanding of system performance and efficiency as well as the exergy destroyed and its breakdown by cause, the method allows designers to test different options and determine performance trends, ultimately facilitating system enhancements and optimization.

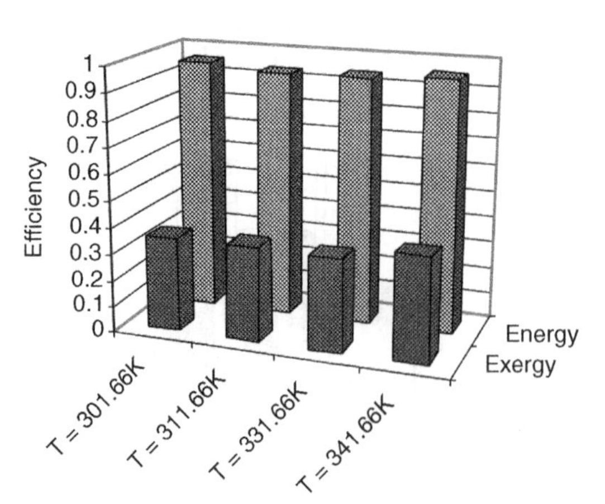

Figure 3.38 Energy and exergy efficiencies at the completion of the overall storage process, for all inlet air temperature cases (adapted from Dincer and Rosen, 2010).

3.9 Concluding Remarks

In this chapter, phase change materials are presented, their basic properties/types are examined and various methods of measuring and enhancing their heat transfer capabilities are provided for the readers. In addition, the current and near future utilization of this technology in different industries are shown. With over 50 different PCMs currently being commercially available, the global market for these materials was $460 million in 2013 (mostly utilized in Europe and North America) and is expected to reach $1.15 billion by 2018. Thus, it can be seen that PCMs would provide practical and effective solutions to a wide range of thermal management issues that currently exists in a wide range of area. However, in order for them to be widely available, they need to be changing phase within the useful temperature range of the application, melt congruently with minimum subcooling and volume expansion along with being long lasting, chemically stable and inexpensive.

Nomenclature

a	fraction
A	area (m^2)
c	volumetric concentration of CNT
C_p	specific heat (kJ/kgK)
CL	tube layout constant
d	tube diameter (m)
D	heat exchanger diameter (m)
f	friction factor
Gr	Grashof number
h	convection heat transfer coefficient
h_o	outside convection heat transfer
ΔH	enthalpy of reaction (kJ/mol)
k	thermal conductivity (W/mk)
L	heat exchanger length (m)
m	mass (kg)
N	number of tubes
N_f	number of fins
Nu	Nusselt number
P	pressure (Pa)
P	probability of series CNT configuration
Pr	Prandtle number
$\dot{Q}$	heat transfer rate (kW)
R	resistance (Ω)
Ra	Rayleigh numebr
Re	Reynolds number
r_m	thickness of heat storage material (m)
R_w	thermal resistance
s_f	distance between the fins
t	time (s) or fin thickness

t^*	dimensionless time
s	distance between the fins
T	temperature (K or °C)
U	overall heat transfer coefficient
V	velocity (m/s)

Greek Symbols

β	thermal expansion coefficient (1/K)
Δ	change in variable
Π	entropy production
Ξ	exergy (J)
$\in$	flow exergy (J)
η	energy efficiency, overall efficiency
ψ	exergy efficiency
Δ	change in variable
ρ	density (kg/m^3)

Subscripts

aver	average
b	base
c	cold
f	fin, final
h	hot
HTF	heat transfer fluid
i	initial, internal, inside
lm	logarithmic mean
m	melting
o	outside
P	parallel
q	heat
t	fin thickness
tot	total
w	wall

Acronyms

CNT	carbon nanotube
DSC	differential scanning calorimetry
DTA	differential thermal analysis
EV	electric vehicle
GHG	greenhouse gas
HEV	hybrid electric vehicle
HTF	heat transfer fluid
HVAC	heating ventilation and air-conditioning
ICE	internal combustion engine

LHTES	latent heat thermal energy storage
NTU	number of heat transfer units
PCM	phase change material
TES	thermal energy storage
TMS	thermal management system

Study Questions/Problems

3.1 Define what is commonly referred as PCMs and elaborate on the main properties and advantages that separate them from other materials.

3.2 What are different types of PCMs and the selection criteria for determining the most optimum PCM for a given thermal management application? Provide some examples where the use of PCMs would not be currently feasible/viable and state your reasons.

3.3 What are some of the techniques that are currently used to increase the heat transfer rate of PCMs? What are the drawbacks associated with these techniques?

3.4 What are some of the methods that can be used to assess and evaluate PCMs? Where do they stand in terms of cost and environmental impact compared to commonly used sensible heat storage systems?

3.5 What are the current niche applications of PCMs and which areas do you think they will dominate in the next decade? Explain your answer.

3.6 Based on the different EV and/or HEV configurations, drive cycles and various battery chemistries learned in the previous chapters, conceptualize an electric vehicle application. Which types of PCMs would you utilize in your selected battery pack and why? What would be your design limitations?

References

Dieckmann A. (2006). *Latent heat storage in concrete*. University of Kaiserslautern, Germany, Available at: http://www.eurosolar.de/ [Accessed August 2016].

Dincer I, Rosen MA. (2010). *Thermal Energy Storage: Systems and Applications*, Second Edition. John Wiley & Sons, Ltd, West Sussex, England.

Ettouney H, El-Dessouky H, Al-Ali A. (2005). Heat transfer during phase change of paraffin wax stored in spherical shells. *Journal of Solar Energy Engineering* **127**:357–365.

Hauer A (2011). *Storage Technology Issues and Opportunities, Committee on Energy Research and Technology (International Energy Agency), International Low-Carbon Energy Technology Platform, Strategic and Cross-Cutting Workshop "Energy Storage – Issues and Opporutnities" 15 February 2011*, Paris, France.

Hyun DC, Levinson NS, Jeong U, Xia Y. (2014). Emerging Applications of Phase-Change Materials (PCMs): Teaching an Old Dog New Tricks, *Angewandte Review* 3780–3795.

Ivancevic VG, Ivancevic TT. (2008). *Complex Nonlinearity: Chaos, Phase Transitions, Topology Change and Path Integrals*, Springer-Verlag Berlin Heidelberg.

Javani N, Dincer I, Naterer GF. (2014). New latent heat storage system with nanoparticles for thermal management of electric vehicles, *Journal of Power Sources* **268**:718–727.

Kosny J, Shukla N, Fallahi A. (2013). Cost Analysis of Simple Phase Change Material-Enhanced Building Envelopes in Southern U.S. Climates. US Department of Energy.

MacPhee D. (2008). *Performance investigation of various cold thermal energy storages*, M.A.Sc. Thesis, University of Ontario Institute of Technology, Oshawa, Ontario, Canada.

Memon SA. (2014). Phase change materials integrated in building walls: A state of the art review. *Renewable and Sustainable Energy Reviews* **31**:870–906.

PCM Products. (2016) Company Website Catalogue. Available at: http://www.pcmproducts.net/ Accessed August 2016].

Pons O, Aguado A, Fernández AI, Cabezac LF, Chimenos JM. (2014). Review of the use of phase change materials (PCMs) in buildings with reinforced concrete structures, *Materiales de Construcción* **64**:315–331.

Riffat S, Mempouo B, Fang W. (2015). Phase change material developments: a review. *International Journal of Ambient Energy* **36**:102–115.

Rubitherm. (2016) Company Website Catalogue. Available at: http://www.rubitherm.eu/ [Accessed August 2016].

Ryu HW, Woo SW, Shin BC, Kim SD. (1992). Prevention of Supercooling and Stabilization of Inorganic Salt Hydrates as Latent Heat-Storage Materials." *Solar Energy Materials and Solar Cells* **27**:161–172.

Sharma A, Tyagi VV, Chen CR, Buddhi D. (2009). Review on thermal energy storage with phase change materials and applications, *Renewable and Sustainable Energy Reviews* **13**:318–345.

Trox. (n.d.). Cooling naturally with phase change materials. Available at: www.troxtechnik.com [Accessed August 2016].

Zalba B, Marin JM, Cabeza LF, Mehling H. (2003). Review on thermal energy storage with phase change: materials, heat transfer analysis and applications, *Applied Thermal Engineering* **23**:251–283.

Zhou D, Zhao C Y, Tian Y. (2012). Review on thermal energy storage with phase change materials (PCMs) in building applications. *Applied Energy* **92**:93–605.

4

Simulation and Experimental Investigation of Battery TMSs

4.1 Introduction

In previous chapters, comprehensive information regarding the state-of-the-art on electric and hybrid electric vehicle technologies, battery chemistries and thermal management systems are provided and phase change materials are introduced to the readers. Moreover, in the next chapters, the methods and procedures for conducting energy, exergy, exergoeconomic and exergoenvironmental analyses as well as multi-objective optimizations for BTMSs will be illustrated under real life scenarios. Thus, before these detailed energy- and exergy-based efficiency, cost and environmental impact analyses can be conducted and system optimizations be performed; the battery and its thermal management system should be modelled through numerical simulations and the results should be verified experimentally to better understand the system components, their associated operating conditions and corresponding performance. This will enable the readers to have more in-depth information on the electrochemistry of the cell as well as the thermodynamic properties of the thermal management systems before rigorous performance assessments and improvement methods can be introduced.

Therefore, in this chapter, a walkthrough of the necessary steps are provided to develop representative models of the battery, achieve reliable simulations, form correct set ups of the data acquisition hardware and software, as well as use right procedures for the instrumentation of the battery and the vehicle in the experimental set up. The simulations mainly focus on the battery cell and submodules to provide the fundamental concepts behind the heat dissipation in the cell and heat propagation throughout modelling with and without BTMSs, and can be easily scaled up to represent the entire battery pack. The experimentations are shown both on the cell, test bench and a full size vehicle levels to familiarize the readers with the components and devices used in the experimentations and enable them to better understand the associated testing techniques and procedures.

In the following sections, a wide range simulations and experimentations from battery cell to the vehicle level is provided that can be utilized for simulating and testing several different thermal management systems, with the special focus on phase change materials.

Thermal Management of Electric Vehicle Battery Systems, First Edition.
İbrahim Dinçer, Halil S. Hamut and Nader Javani.

4.2 Numerical Model Development for Cell and Submodules

In order to conduct a representative and reliable heat transfer simulations of a BTMS, a thorough mathematical model needs to be developed, which should be able to represent the physics of the problem, input variables as well as the relationship between these parameters through the governing equations and their valid range of variation with initial and boundary conditions. Furthermore, such a model is required to achieve the determined objectives by assigning constraints for the problem and defining the corresponding methodology. Therefore, in this section, the steps that need to be taken to develop mathematical models for of the BTMS in cell and submodule levels will be provided. A commonly utilized software in this discipline, ANSYS FLUENT, will be used in this process.

For the simulations, the system physical domain is initially needed to be presented, followed by the assumptions made while developing the model. Furthermore, the governing conservation equations in the cell zone along with the PCM needs to be defined, which is discretized to be used by numerical methods. Finally, boundary and initial conditions required to complete the model formulation are also needed to be illustrated.

Developing a transient three dimensional model for the battery cell and module plays a vital part for better understanding the heat generation and propagation processes as well as the properties of the thermal material (i.e., PCM and/or coolant) that needs to be used to mitigate the unwanted temperature rise. Thus, in the next sections, a layout for a cell and the submodule will be introduced along the physical domain and the necessary boundary conditions. Existing assumptions and governing equations will form the model. Experiments are carried out to select the main elements of the submodule. Basic elements are the lithium-ion cell, cooling plate and the foam which will be applied to the system. The model can be applied to scale up to the entire battery pack to analyze the battery as a whole and its interactions with the vehicle.

4.2.1 Physical Model for Numerical Study of PCM Application

Developing involved battery models commonly start from a single Li-ion cell (and the surrounding PCM) as it is the most basic component in the system. Once the cell geometry, properties and heat dissipation is modelled and verified through experimentations; various other major items that exist in the modules, such as foams and cooling plates and other submodule components, are also needed to be added. Figure 4.1 illustrates the cell geometry and aspect ratio of the Li-ion cell, foam and cooling plate. The heat generation rate is assumed to be the only heat source in the cell zones in this stage. Submodules up to four cells are introduced to provide the relationships among the cells in the module.

Figure 4.2 shows a sample configuration for the submodule. In order to have meaningful values in the transient solution, first a steady state condition needs to be solved and

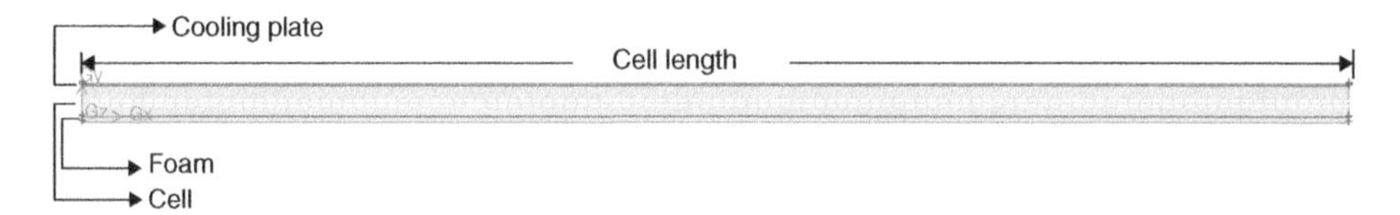

Figure 4.1 Relationship single cell model in the submodule.

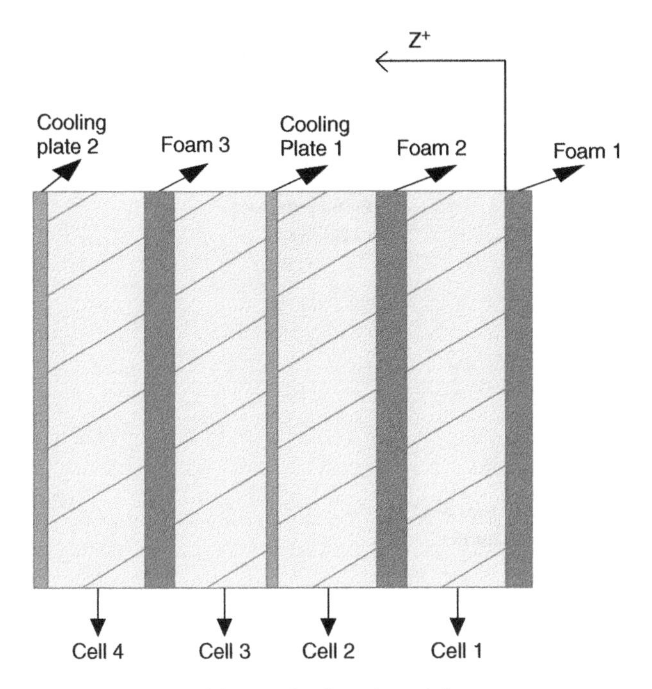

Figure 4.2 Physical domain for the submodule.

the results be utilized as initial conditions for the model, as this is the common practice used in ANSYS FLUENT software (Ansys Inc., 2009).

4.2.2 Initial and Boundary Conditions and Model Assumptions

In the layout for the submodule, the boundary conditions for the interface between the cells and the foam are usually defined as "wall boundary conditions". In addition, for the cell and cooling plate, "wall" boundary condition is considered for locations where there is no mass transport. The ends of the system need to be considered to attain free convection. The source term for heat generation rate is applied to the system for the zone condition. The upper/lower surfaces are also exposed to free convection. In order to model the initial simplified version of the cell, free convection is usually considered for all surfaces. In fact, a sample cell is shown in Figure 4.3.

The assumptions made for modeling the single cell and submodule are listed as follows:

- The volumetric heat generation rate: $\dot{q}_v = 63{,}970\ \mathrm{W/m^3}$ for 2C (C-rate) (4.45 W/cell) (in accordance with a lumped system analysis).
- The heat transfer from the terminal surfaces and ambient temperature is free convection.
- The boundary between the terminal and cell itself is "coupled" type boundary conditions which assures the continuity of the temperature across the defined boundary
- The initial temperature is considered to be 294.15 K equal to the ambient temperature.
- The cell has orthotropic thermal conductivity.

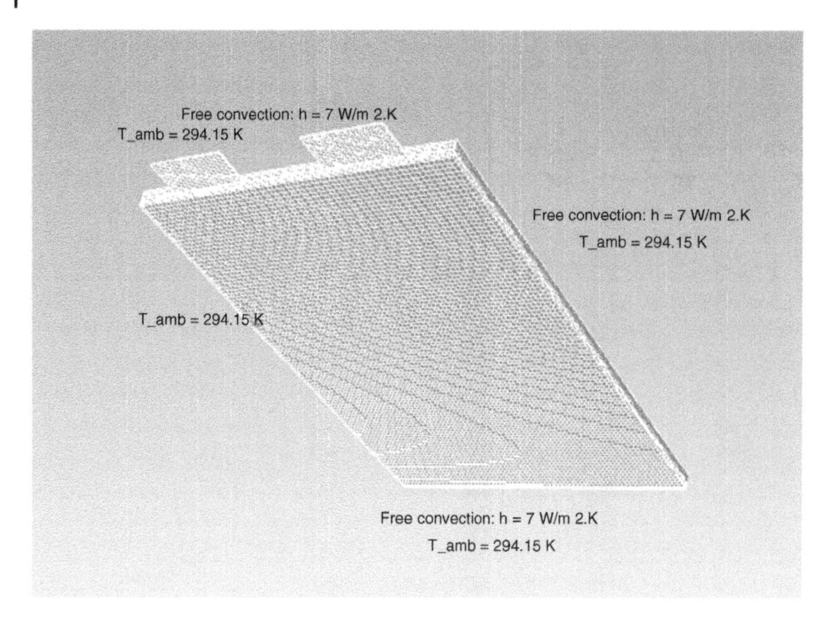

Figure 4.3 Boundary conditions for a sample Li-Ion cell.

- The specific heat for the cell, plate and fin is averaged.
- The thermal conductivity of foam is considered as the value for insulation foams.
- No flow is filed for the liquid phase of PCM.
- The radiation is not considered.

4.2.3 Material Properties and Model Input Parameters

In battery packages, the physical domain contains different zones, each one potentially having different materials and properties. Thus, the data for the cell, phase change material, foam, cooling plates and other components in the battery need to be represented in the simulations. The properties for these components and materials are shown in the next sections.

4.2.3.1 Li-ion Cell Properties

Chemical reactions within the cell have considerable impact on the spatial distribution of cell heat generation. In this developed model, various scenarios are analyzed with respect to cell heat generation rate under different operational conditions. Moreover, the option of applying of PCM around the cell is also selected due to the orthotropic behavior of the cell.

As mentioned in Chapter 2, the rate at which the current is discharged from the cell is described by the C-rate. The thermal stability as well as the heat generation inside the cell is commonly linked to the internal resistance, which is correlated with the performed C-rate. The new generations of Li-ion cells have significantly more uniform heat generation rates. This is in accordance with the assumption of uniform heat generation in cell zone. For the rest of this chapter, the following battery cell heat generation rates are used:

$\dot{q}_v = 6855\ \mathrm{W/m^3}$ by Standard US06
$\dot{q}_v = 22{,}800\ \mathrm{W/m^3}$ at 135 Amps (150 kW), 3.6 W/cell
$\dot{q}_v = 63{,}970\ \mathrm{W/m^3}$ at 2C, 4.45 W/cell
$\dot{q}_v = 200{,}000\ \mathrm{W/m^3}$ at full power, uphill condition

The heat generation can be stored as sensible and latent heat in the pack. In the case of the phase change process taking place, it can be described as

$$Q = m_{PCM}C_p(T_m - T_i) + m_{PCM}LHV \tag{4.1}$$

Cell heat generation can be determined using the calorimetric method or can be modeled mathematically by considering the associated inputs and assumptions. In order to fully understand the heat generation rate and/or the cell surface temperature, phase change inside the cell, electrochemical reactions, mixing effects and Joule heating need to be determined. The following expression developed by Bernardi *et al.* (1985) is commonly utilized to determine the heat generation in the cell:

$$\dot{q} = I\left(U - V - T\frac{dU}{dT}\right) = h_s A(T - T_a) + M_A C_p \frac{dT}{dt} \tag{4.2}$$

where U and V are open circuit and cell potentials, respectively and h_s is per-cell heat transfer coefficient. Term $I(U - V)$ is the heat generation due to cell polarization and $-IT\frac{dU}{dT}$ is entropy coefficient due to a reversible process in the cell. M_A is the cell mass per unit area and C_p is the weight-averaged value of cell heat capacity. When the temperature change is assumed to be uniform within the cell, Equation (4.2) can be rewritten as follows (Wu *et al.*, 2002):

$$\dot{q} = \frac{I}{V_b}\left[(E_0 - E) + T\frac{dE_0}{dT}\right] \tag{4.3}$$

Here, $\dot{q}$, V_b, E_0, E and I are heat generation rate per unit volume, battery volume, open circuit potential, cell potential and current respectively. If heat generation consists of reversible and irreversible effects, the reversible heat released or absorbed in the cell as a result of chemical reactions can be stated as (Selman *et al.*, 2001):

$$Q_{rev} = n_{Li}T\Delta S = T\left(\frac{\partial E_{eq}}{\partial T}\right)It_{dc} \tag{4.4}$$

In order to determine the heat generation with thermodynamic relations, the following Gibbs free energy equation can considered

$$\Delta G = -n \times F \times U \tag{4.5}$$

where F is Faraday‘s constant and n is the number of electrons.

4.2.3.2 Phase Change Material (PCM)

The selection of PCM material plays a vital role in developing an effective BTMS. In order to find the different specific heat values for the PCM with respect to the experimented time, they must be interpolated based on the curve fitting method. Thus, a methodology for the prediction of the specific heat behavior over time needs to be established.

Table 4.1 Effect of temperature variations on the specific heats.

Temperature (K)	$C_p (J/kg\,K)$	Normalized temperature	Normalized C_p
299.15	2150	0.983561	0.00881148
300.15	2150	0.986848	0.00881148
300.65	2150	0.988493	0.00881148
300.95	2150	0.989479	0.00881148
301.05	5000	0.989808	0.0204918
301.13	122000	0.990071	0.5
301.15	244000	0.990136	1
301.2	244000	0.990301	1
301.25	244000	0.990465	1
301.35	244000	0.990794	1
301.55	244000	0.991452	1
301.75	244000	0.992109	1
301.95	244000	0.992767	1
302.05	244000	0.993095	1
302.25	5000	0.993753	0.0204918
302.35	2180	0.994082	0.00893443
302.95	2180	0.996054	0.00893443
303.15	2180	0.996712	0.00893443
304.15	2180	1	0.00893443
302.25	5000	0.993753	0.0204918
302.35	2180	0.994082	0.00893443
302.95	2180	0.996055	0.00893443

In order to determine a closed form function for the experimental data set, a certain type of fitting algorithm needs to be utilized. Wide range of methods exit to obtain an accurate closed form function for fitting a data set; including Chebyshev polynomials, least squares, Levenberg-Marquardt and Gauss-Newton algorithms. First, Chebyshev polynomials (first kind) are illustrated in Table 4.1, due to its compatibility with these applications. The first kind Chebyshev polynomials are defined by the recurrence relation and are shown as:

$$T_0(x) = 1. \tag{4.6}$$

$$T_1(x) = x. \tag{4.7}$$

$$T_{n+1}(x) = 2xT_n(x) - T_{n-1}(x) \tag{4.8}$$

The conventional generating function for T_n is

$$\sum_{n=0}^{\infty} T_n(x)t^n = \frac{1 - tx}{1 - 2tx + t^2}. \tag{4.9}$$

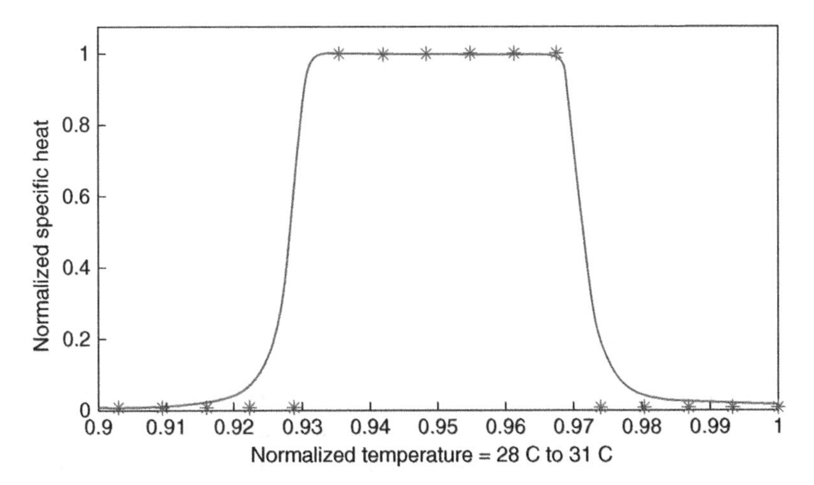

Figure 4.4 Chebyshev polynomial to interpolate specific heat.

The generating function relevant for 2-dimensional potential theory and multiple expansions is

$$\sum_{n=1}^{\infty} T_n(x)\frac{t^n}{n} = \ln \frac{1}{\sqrt{1-2tx+t^2}}. \tag{4.10}$$

Eventually, an explicit form is found for the system based on the superposition rule and Chebyshev polynomials. Using six orders of Chebyshev polynomials, the following functions are obtained (shown in Figure 4.4):

$$y_a = \frac{1-0.06}{\sqrt{1-0.09^2 T_6^2\left(\frac{1}{0.018}(x-0.95)\right)}}. \tag{4.11}$$

$$y_b = \frac{0.007}{\sqrt{1-0.09^2 T_6^2\left(\frac{1}{0.0125}(x-0.987)\right)}}. \tag{4.12}$$

$$y_c = 0.0065.$$

$$T_6(x) = 32x^6 - 48x^4 + 18x^2 - 1. \tag{4.13}$$

By combining the superposition rule and the above functions, an explicit form, that can predict the behavior of the system with substantial accuracy is determined. Figure 4.4 represents the superposition rule which has been utilized to obtain the closed form function for the objective problem.

$$y_{total} = 0.0065 + \frac{1-0.06}{\sqrt{1-0.09^2 T_6^2\left(\frac{1}{0.018}(x-0.95)\right)}} + \frac{0.007}{\sqrt{1-0.09^2 T_6^2\left(\frac{1}{0.0125}(x-0.987)\right)}}. \tag{4.14}$$

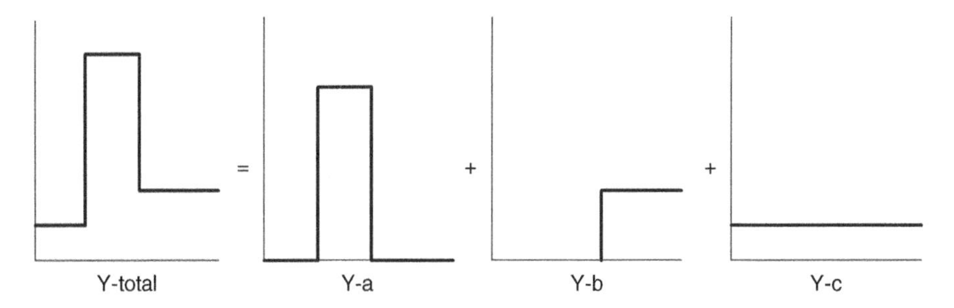

Figure 4.5 Representation of superposition rule for our system.

Figure 4.4 illustrates the curve fit for the variation of the normalized specific heat. Figure 4.5 shows the superposition method used to obtain the resultant piecewise function of specific heat with respect to the normalized temperature. The above-mentioned Chebyshev estimation is used to find the curve. ANSYS FLUENT can be utilized at this step to enter input data as discrete numbers and use the piecewise polynomial option. However, it should be noted that the abrupt increase of properties can lead to divergence (e.g., specific heat has a very low value of 2,150 J/kgK for solid phase and at phase change stage, can increase up to 244,000 J/kgK) and may cause instability in the simulation if goes unnoticed by the users.

As a second method, the Gauss function can be used. For the available system, Gauss function has the following form:

$$C_p = 0.0088 + Gauss(54.79 \times Gauss(1.628e - 7/ (T^4 - 3.687 \times T^3 + 5.0713 \times T^2 - 3.108 \times T + 0.7144))) \quad (4.15)$$

The values for the specific heat, thermal conductivity and density of the phase change material are given in the following form to be used in simulations:

$$C_p = \begin{Bmatrix} 2150\ J/(kgK) & T_{solidus} > T & \text{solid phase} \\ 0.255\ J/(kgK) & T_{solidus} < T < T_{liquidus} & \text{Mushy zone} \\ 0.152\ J/(kgK) & T > T_{liquidus} & \text{liquidus phase} \end{Bmatrix} \quad (4.16a)$$

$$k = \begin{Bmatrix} 0.358\ W/(mK) & T_{solidus} > T & \text{solid phase} \\ 0.255\ W/(mK) & T_{solidus} < T < T_{liquidus} & \text{Mushy zone} \\ 0.152\ W/(mK) & T > T_{liquidus} & \text{liquidus phase} \end{Bmatrix} \quad (4.16b)$$

$$\rho = \begin{Bmatrix} 814\ kg/m^3 & T_{solidus} > T & \text{solid phase} \\ 769\ kg/m^3 & T_{solidus} < T < T_{liquidus} & \text{Mushy zone} \\ 724\ kg/m^3 & T > T_{liquidus} & \text{liquidus phase} \end{Bmatrix} \quad (4.16c)$$

It should be noted that the selected PCMs are available in two categories: technical grade and pure PCM. It is essential to accurately know the latent heat of fusion of the PCM that is used in the system. Even though values are assigned for these materials in the product catalogues or in the literature, the readers are encouraged to measure the latent heat of fusion of their own material themselves if possible, as it may deviate significantly from the listed values. As an example, the latent heat of fusion value for the pure Octadecane is found to be 244 kJ/kg in most literature (Agyenim *et al.*, 2010). However, the carried DSC tests show a value of 225 kJ/kg. This difference usually occurs due to the purity of the material. The technical grade will have even lower values depending on the provided grade.

4.2.3.3 Foam Material

Foam material also plays an important role in the battery as it has a high thermal stability and acts as a separator once it is placed in between the cells. In most cases, they are urethane foams with a density that generally varies between 240 kg/m^3 to 400 kg/m^3. For the absorbed PCM in the foam, the following relation can be used to estimate effective specific heat and thermal conductivity. The time-dependent thermal conductivity is usually given as $k(T) = 0.02064 + 11.28 \times 10^{-5}$ T. Effective thermal and density (K_{eff} *and* ρ_{eff}) are defined similarly and used in the simulations. The porosity and the average specific heat of the foam can be formulated as below in the simulations.

$$Cp_{aver} = \frac{1}{m_{tot}} \times \sum m_i Cp_i = \frac{m_{pcm}}{m_{tot}} Cp_{pcm} + \frac{m_{foam}}{m_{foam}} Cp_{foam} \tag{4.17}$$

The porosity is defines as:

$$\varepsilon = \frac{V_{pcm}}{V_{tot}} \tag{4.18}$$

Therefore, the average specific heat can be formulated as

$$Cp_{aver} = \frac{\rho_{pcm}}{\rho_{tot}} \times \varepsilon \times Cp_{pcm} + \frac{\rho_{foam}}{\rho_{tot}} \times (1-\varepsilon) \times Cp_{foam} \tag{4.19}$$

4.2.3.4 Cooling Plate

The physical and thermophysical properties of the cooling plate material need to be accurately considered in the simulations. In today's BTMSs, the cooling plates are mainly made out of aliminum. However, the property of the material should not be directly used in the simulation without applying corrections for the thermal convection effect (Pesaran *et al.*, 1999). Table 4.2 shows the elements' properties used in the simulation. In order to have the exact heat generation rate, experimental methods such as calorimetry readings would provide more reliable data that increases the accuracy of the model.

4.2.4 Governing Equations and Constitutive Laws

Once all the material geometries and properties are inputted in the simulation, the governing equations and laws need to be defined to attain accurate interactions among

Table 4.2 Thermo-physical properties of materials for simulation.

Property/component	Cell	Cooling fin	Foam
Density (kg/m^3)	4035	2719	277
Specific heat (J/kgK)	1027	871	1500
Thermal conductivity (W/mK)	$K_{x,y} = 25$, $K_z = 1$	202.4	0.083
Heat generation rate (kW/m^3)	22.8	0	0

the components in the system. In these simulations, a specific partial differential equation is used to describe each phenomena which comprises a transient term, diffusion term, convection term and a source term. By taking these terms into account, a general equation can be derived as follows:

$$\frac{\partial}{\partial t}(\rho\Theta) + \nabla\cdot(\rho\vec{u}\Theta - \Gamma_{\Theta}\nabla\Theta) = S_{\Theta} \tag{4.20}$$

where Θ is 1, u, Y and h, in the continuity, momentum, species and energy equations, respectively. Γ_{Θ} and S_{Θ} are the diffusion coefficient and source terms which have consistent units. The main governing equation is energy equation to analyze the heat transfer in the model. In order to obtain the cell temperature distribution, an energy balance is applied. The expression for the conservation of energy is stated as:

The net rate of change for internal and kinetic energy = The net rate energy change by convection (fluid flow) + The net rate of energy change by heat by conduction (heat) + The net rate of work

Here, the rate of change for internal and kinetic energy of element and is represented by the following equation:

$$\frac{\partial}{\partial \mathrm{t}}\left[\rho\left(\hat{u} + \frac{V^2}{2}\right)\right] dx\,dy\,dz \tag{4.21}$$

where û is internal energy. The net rate of energy change by convection (fluid flow contribution) is

$$-\left\{\nabla.\left[\left(\hat{u} + \frac{V^2}{2}\right)\rho V\right]\right\} dx\,dy\,dz \tag{4.22}$$

The net rate of energy change by heat by conduction (heat contribution in the balance equation) is formulated as

$$-(\nabla.\ddot{q})\,dx\,dy\,dz \tag{4.23}$$

The net rate of work by element on the surroundings (works by body forces and surface forces) is

$$-\rho(V.g)dx\,dy - \left[\frac{\partial}{\partial x}(u\sigma_{xx} + v\tau_{xy} + w\tau_{xz}) + \frac{\partial}{\partial y}(u\tau_{yx} + v\sigma_{yy} + w\tau_{yz}) + \frac{\partial}{\partial z}(+u\tau_{zx} + v\tau_{zy} + w\sigma_{zz})\right] dx\,dy\,dz \tag{4.24}$$

Substituting all terms into the Equation 5.24 yields the following equation

$$\frac{\partial}{\partial t}\left[\rho\left(\hat{u} + \frac{V^2}{2}\right)\right] = -\nabla.\left[\left(\hat{u} + \frac{V^2}{2}\right)\rho V\right] - (\nabla.\ddot{q}) + \rho(V.g) + \left[\frac{\partial}{\partial x}(u\sigma_{xx} + v\tau_{xy} + w\tau_{xz}) + \frac{\partial}{\partial y}(u\tau_{yx} + v\sigma_{yy} + w\tau_{yz}) + \frac{\partial}{\partial z}(+u\tau_{zx} + v\tau_{zy} + w\sigma_{zz})\right] \tag{4.25}$$

Here, the stress tensor has 9 normal and shear stresses which have symmetry and can be extracted from the momentum equation to simplify this equation as

$$\rho\frac{D\hat{u}}{Dt} = -\nabla.\dot{q} + \left(\sigma_{xx}\frac{\partial u}{\partial x} + \tau_{yx}\frac{\partial u}{\partial y} + \tau_{zx}\frac{\partial u}{\partial z}\right) + \left(\tau_{xy}\frac{\partial v}{\partial x} + \sigma_{yy}\frac{\partial v}{\partial y} + \tau_{zy}\frac{\partial v}{\partial z}\right) + \left(\tau_{xz}\frac{\partial w}{\partial x} + \tau_{yz}\frac{\partial w}{\partial y} + \sigma_{zz}\frac{\partial w}{\partial z}\right) \quad (4.26)$$

In order to relate heat flux $\dot{q}$ to temperature field and eliminate normal and shear stresses, a constitutive equation is needed. For temperature fields, the constitutive equation is represented as Fourier's law:

$$\dot{q}_n = -k_n\frac{\partial T}{\partial n} \quad (4.27)$$

A Newtonian approximation can relate the stresses and velocity fields as the other set of constitutive equation becomes

$$\tau_{xy} = \tau_{yx} = \mu\left(\frac{\partial v}{\partial x} + \frac{\partial u}{\partial y}\right) \quad (4.28)$$

Using these two equations, the principle equation of energy conservation can be arranged to

$$-\rho\frac{D\hat{u}}{Dt} = -\nabla.k\nabla T - p\nabla.V + \mu\varphi \quad (4.29)$$

where enthalpy h is $\hat{u} + \frac{p}{\rho}$.

The differentiation of this equation and combining with the main equation leads to the energy equation in form of

$$\rho\frac{D\hat{h}}{Dt} = \nabla.k\nabla T + \frac{Dp}{Dt} + \mu\varphi \quad (4.30)$$

If the temperature is of interest, enthalpy can be replaced by temperature as follows:

$$D\hat{h} = C_p dT + \frac{1}{\rho}(1-\beta)dp \quad (4.31)$$

Here, β is coefficient of thermal expansion as

$$\beta = -\frac{1}{\rho}\left(\frac{\partial p}{\partial T}\right)_p \quad (4.32)$$

The resulting energy conservation equation is

$$\rho C_P\frac{DT}{Dt} = \nabla.k\nabla T + \beta T\frac{Dp}{Dt} + \mu\varphi \quad (4.33)$$

4.2.5 Model Development for Simulations

In order to better understand the thermal response of the battery from cell to pack level and discretize the aforementioned energy equations, Finite Volume Method (FVM) simulations are utilized. In these simulations, the computational domain is divided into a number of control volumes and the governing equations in the differentiable form are integrated over each control volume. Among the centroid of the cells, the variables are interpolated to have a profile for variations of the parameter such as pressure or temperature where the result becomes a discretized equation in the domain. Finally, a solver (along with various other numerical techniques) is used to converge the solution in the model. These fundamental steps are introduced in the next sections.

4.2.5.1 Mesh Generation

Generating a characteristic and reliable mesh plays an important part in having representative and accurate simulations. By defining the domain and zones that require detailed analysis, pre-processors such as Gambit and ICEM are commonly used to generate the mesh. Since the geometry of the cell and the module are relatively straightforward, structural meshes are usually preferred in these applications. In generating the meshes, the bottom-up method is commonly used where the "edges" are meshed first. Depending on a successive ratio, zones adjacent to the interface or boundaries (where the PCM is applied) is then meshed in detail. Then "face" mesh is generated by selecting the element of the mesh. The structured mesh requires specifications such as the mesh density in the zones of interest and other preliminary defined zones as well as boundary conditions. It should be noted that even though refining the mesh increases the number of grids which generally increases the accuracy of the solution, it also prolongs the calculation time and increases the cost. Thus, the mesh structure should be set by concentrating on the zones that have the largest impact in the system. In simulation software, three steps of pre-processing, solver and post-processing are used to complete the analysis. In the finite volume method, each node of the grid should be enclosed by a corresponding control volume. The grid network with attributed characteristics needs to be exported to the solver software.

4.2.5.2 Discretization Scheme

In order to convert a Partial Differential Equation (PDE) to an algebraic equation, a discretization technique needs to be used and solved by numerical methods. The control volume technique consists of integrating the transport equation about each control volume. In this way, the conservation law on a control-volume is expressed by a discrete equation. Discretization of a transient, convection, diffusion and source terms and general form of governing equations, can be demonstrated by the following equation written in integral form for an arbitrary control volume V:

$$\int_V \frac{\partial \rho\theta}{\partial t} dV + \oint \rho\theta\vec{v} d\vec{A} = \oint \Gamma_\theta \nabla\theta \, d\vec{A} + \int_V S_\theta dV \tag{4.34}$$

where $\rho, \vec{v}, \vec{A}, \Gamma_\theta$, $\nabla\theta$ and S_θ are density, velocity vector, surface area vector, diffusion coefficient, θ gradient and source of θ per unit volume, respectively. The above equation is applied to all cells or control volumes in the computational domain. A discretization of equation for a given cell becomes

$$\frac{\partial \rho\theta}{\partial t} V + \sum_f^{N_{force}} \rho_f \vec{v}_f \theta_f \vec{A}_f = \sum_f^{N_{force}} \Gamma_\theta \nabla\theta \, d\vec{A} + \int_V S_\theta dV \tag{4.35}$$

A common method to obtain temporal discretization for an arbitrary variable of β is

$$\frac{\partial \beta}{\partial t} = F(\beta) \tag{4.36}$$

Using backward differences as a first-order accurate method, a spatial discretization function of F can be written as

$$\frac{\beta^{n+1} - \beta^{n}}{\Delta t} = F(\beta) \tag{4.37}$$

Then, the first order implicit time integration becomes

$$\frac{\beta^{n+1} - \beta^{n}}{\Delta t} = F(\beta^{n+1}) \tag{4.38}$$

The final form to be solved at each time level iteratively is expressed as

$$\beta^{n+1} = \beta^{n} + \Delta t \, F(\beta^{n+1}) \tag{4.39}$$

The same methods are applicable to spatial discretization.

4.2.5.3 Under-Relaxation Scheme

The changes of a dependent or auxiliary variable need to have specific controlling possibility and constraints, where the under-relaxation can be considered in these scenarios. Thus, it is imperative to maintain the stability of the coupled, non-linear system of equations. In this regard, under-relaxation is a useful method to stabilize the solution to achieve convergence, especially for non-linear and stiff problems. When the under-relaxation is applied, the process divides in steps in a way that the under-relaxed variable does not reach its final value directly, but technically, under-relaxation factors make the solution take a lot longer to converge.

4.2.5.4 Convergence Criteria

There should be limiting criteria to control the variable, undertaking steps of iteration. In a numerical approach, this is used to stop the iterations. In order to control the convergence trend, a parameter called "residual" is defined to begin the solution procedure. When the residual sum for each of the variables is higher than a pre-defined value, iteration will continue until it reaches the defined amount.

4.3 Cell and Module Level Experimentation Set Up and Procedure

On the previous chapters, TMSs are analyzed from a holistic perspective with respect to their interactions with the rest of the TMS system components. Even though this approach has important benefits, it is imperative to understand the heat generation associated with the electrochemical energy conversion the in cell and the corresponding heat generation across the module. Thus, in this section, cell heat dissipation will be demonstrated and the experimental set up and procedures associated with instrumenting the battery cells, accurately gathering the necessary data and validating the results will be shown. In order for the covered experimental steps to be relevant to most readers, test set ups that are compatible with a wide range of thermal managements systems are selected.

4.3.1 Instrumentation of the Cell and Submodule

In order to understand the heat generation in the battery pack, it is imperative to accurately measure and control the temperature of each cell. For this reason, cells should be instrumented and monitored at all times to prevent any issues that can impact the performance and/or safety of the battery. In this regard, thermocouples (most commonly K-type) need to be connected through a thermal paste material to the surface of each cell and the collective information in the pack is gathered using data acquisition software. The data needs to be collected for all the cells in a submodule/module and be compiled to understand the temperature distribution associated with that region of the battery pack. Figures 4.6 through 4.8 shows the steps taken to instrument the cells, cooling plates and the submodule.

Once the cell is instrumented and its temperature curves have been monitored under different conditions and drive cycles, a compatible PCM needs to be selected based on the necessary melting temperature. To be able to efficiently utilize the PCM, it needs to be soaked in foam and placed around the cell. In order to select the foam, several experiments need to be conducted based on physical properties of the material. A list of most commonly used foams is presented in Table 4.3. The main criterion to select the right foam is its porosity that allows the effective absorption of the PCM (Javani *et al.*, 2014).

In the next step, the foam needs to be soaked by PCM where their absorption is assessed as shown in Figure 4.9. In these experiments, a temptronic device is usually used along with the heat chamber to control the test environment conditions and to melt the solid phase PCM in the testing process.

Figure 4.6 Connecting thermocouples on the surface of the Li-Ion cells.

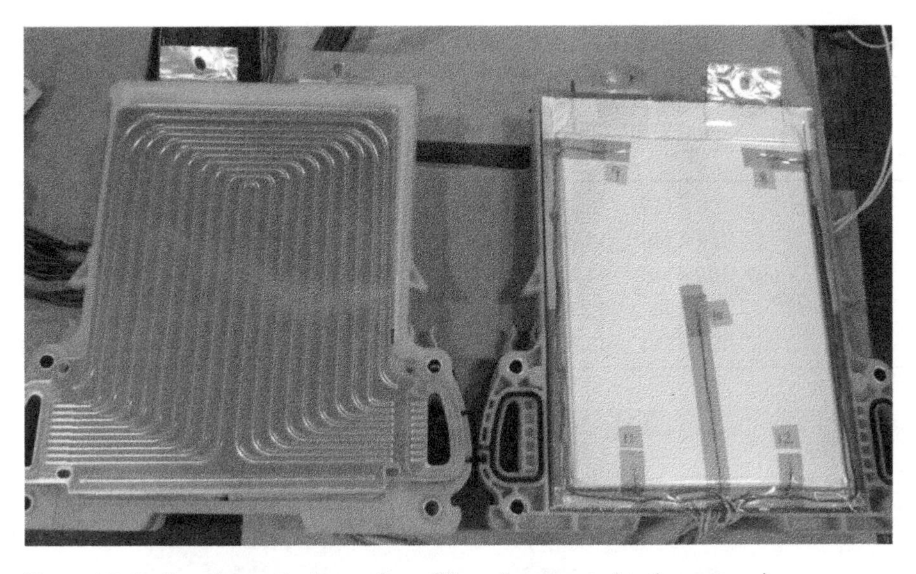

Figure 4.7 Cooling plate and other surface of the cells connected to thermocouples.

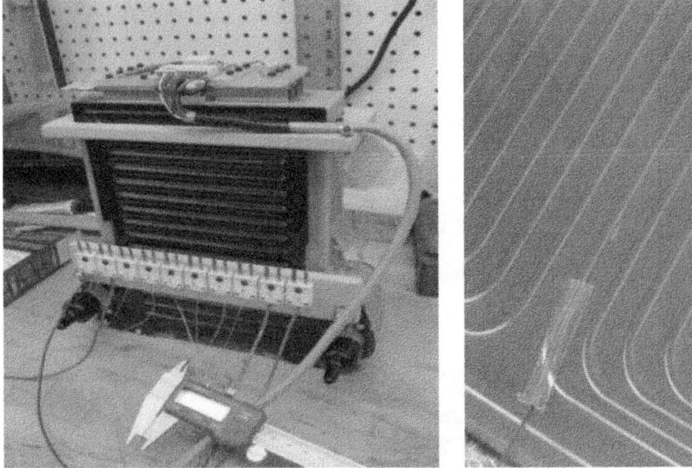

Figure 4.8 Submodule with three cells under the tests performed in Dr. Greg Rohrauer's Lab.

4.3.2 Instrumentation of the Heat Exchanger

As previously mentioned in Chapter 3, one alternative method of directly applying the PCM on the cell is to utilize a heat exchanger filled with PCM to condition the coolant temperature. In order to effectively utilize this heat exchanger, the temperature and the state of the PCM needs to be known accurately. Thus, several thermocouples need to be placed in the copper tubes of the heat exchanger. The melting process in the heat exchanger and location of thermocouples can be observed in Figures 4.10 and 4.11.

Table 4.3 Commonly used foams for BTMS applications.

Foam	Sample color	Sample size (mm^2)	Surface area (mm^2)	M1, dry mass (g)	M2 after soaking (g)	PCM, absorbed mass (g)	PCM density in foam (g/mm^2) × 1000
1	Yellowish	106 × 155	16,430	1.92	8.24	6.32	0.38
2	Black	107 × 140	14,980	4.15	5.68	1.53	0.10
3	Black (porous)	105 × 140	14,700	2.27	7.75	5.48	0.37
4	Blue (strip)	35 × 150	5,250	0.84	2.48	1.64	0.31
5	Blue	105 × 150	15,750	0.731	1.601	0.87	0.06

Figure 4.9 Foams after soaking in the PCM to assess their absorption.

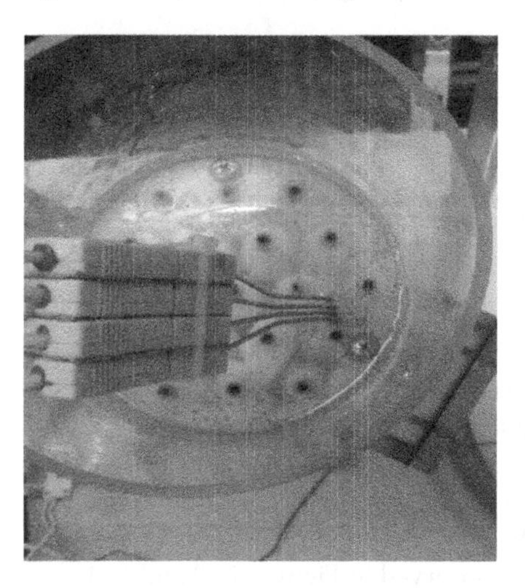

Figure 4.10 Position of thermocouples in equal distances from the copper tube.

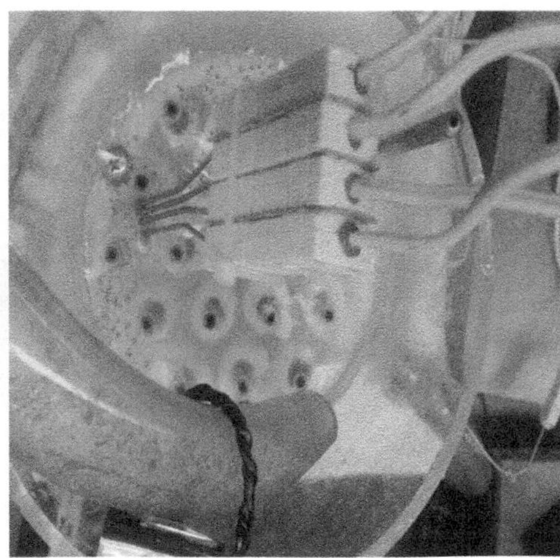

Figure 4.11 Position of thermocouples on the tube surface.

Figure 4.12 Manufactured heat exchanger with optimized dimensions.

Figure 4.12 shows a heat exchanger where the length and diameter of the tank, along with the number of copper tubes and their diameters are obtained through an optimization process (optimization procedures will be shown in Chapter 6). Different mass flow rates and heat transfer rates need to be considered in designing the heat exchanger. Figure 4.13 shows the PCM when it is in solidus phase.

4.3.3 Preparation of PCMs and Nano-Particle Mixtures

Different combinations of n-octadecane materials are used in the shell side of the heat exchanger. Pure PCM (99% purity) and technical grade (90.8%) are usually the ones considered for these applications (Figures 4.14a,b). An ultrasonic unit is used for

Figure 4.13 Solid phase of the PCM in the heat exchanger and test set up.

Figure 4.14 Preparation of samples using ultrasonic unit in Dr. Greg Rohrauer's Lab.

the preparation of the CNT and PCM. Ultrasonic waves are utilized to disperse the nanotubes in the PCM properly (Figure 4.14c). The advantage of the ultrasonic method is the prevention of agglomeration of nanoparticles through the phase change process. Moreover, by increasing the elapsed usage time of the ultrasonic device, thermal conductivity of the mixture will be increased (Amrollahi *et al.*, 2008). The prepared PCM and nanoparticle mixture is used for DSC tests and thermal conductivity analyses. In addition, these samples (Figure 4.14d) are also widely used in optical microscopic applications.

The nanoparticles used in these investigations are usually carbon nanotubes and graphene nano-platelets. The properties of graphene nano platelets are listed as follows:

- Purity: 97 wt%
- Diameter: 2 μm
- Grade 3
- Ash: <1.5 wt%

For the carbon nanotubes (CNT), the corresponding properties are

- Outer diameter: 8–15 nm
- Length: 10–50 μm
- Purity: >95 wt %

The graphene platelets and CNTs are needed to be mixed with technical grade and pure octadecane in different mass concentrations. It should be noted that, in concentrations greater than 10%, the mixture becomes less viscous and the latent heat of fusion for the mixture decreases. This resulting lower specific heat may create significant disadvantages the system. This also shows the thermal stability of the mixture of n-octadecane and nano-tube after high number of cycles. It is mainly due to the characteristics of organic materials and paraffin waxes. These mixtures are prepared and are analyzed with the differential scanning calorimetry (DSC) method and optical microscopic images which is briefly explained in the next section.

4.3.4 Improving Surface Arrangements of Particles

Once the cells and/or heat exchanger(s) are instrumented and the PCM and nano-particle mixture is prepared, the next step is to acquire a deeper understanding of the surface arrangements of the particles in the mixture to ensure the high thermal conductivity, especially after several cycles of melting and solidification. One of the most effective ways of investigating the dispersion of nanoparticles in the PCM mixture is using opto-imaging techniques.

In the reflection microscopic method, the surface of the PCM and nanoparticles during the phase change is observed to understand the surface arrangement of the particles. Using transmission microscopic images, the samples can be studied during the phase change process, and the rate of agglomeration for graphene nano-plates and carbon nanotubes can be further examined. Moreover, the fine extended surface and other methods can also be studied to investigate their effect on preventing convection of particles in the mixture.

On the other hand, transmission electron microscopy (TEM) is a microscopy technique whereby a beam of electrons is transmitted through an ultra-thin specimen,

Figure 4.15 The developed stainless steel micro-mesh with mesh size of 20 μm.

interacting with the specimen as it passes through. An image is formed from this interaction of the electrons transmitted through the specimen; the image is magnified and focused onto an imaging device, such as a fluorescent screen, on a layer of photographic film, or to be detected by a sensor such as a CCD camera. TEMs are capable of imaging at a significantly higher resolution than light microscopes. In a reflection microscope, light is reflected off a sample. Image contrast can arise in different ways.

One of the main drawbacks of a PCM and nano-particle mixture is the agglomeration of nanoparticles when the mixture gets melted, which decreased the thermal conductivity of the mixture. As a method to prevent this phenomenon, a very fine mesh of stainless steel is used in the mixture and experiments are done through reflection microscopy imaging and transmission electron microscopy. Figure 4.15 illustrates the micro-mesh employed in the mixture for this purpose.

4.3.5 Setting up the Test Bench

After each component that will be involved in the experimentation process is prepared, the test bench needs to be set up to conduct the experimentations required to acquire the necessary information for the BTMS. The test bench enables the engineers to examine the components of the system and the relationships within different circuits. It is important to include or simulate all the components and factors that have a significant direct or indirect impact on the system to ensure the accuracy and reliability of the results. Parametric studies and sensitivity analyses become very crucial to better understand the impact of each parameter on the system and the components that need to be improved. A test bench with all the components of refrigerant and coolant loops, and simulators of the engine (via an auxiliary bench) and an actual battery pack is shown in Figure 4.16.

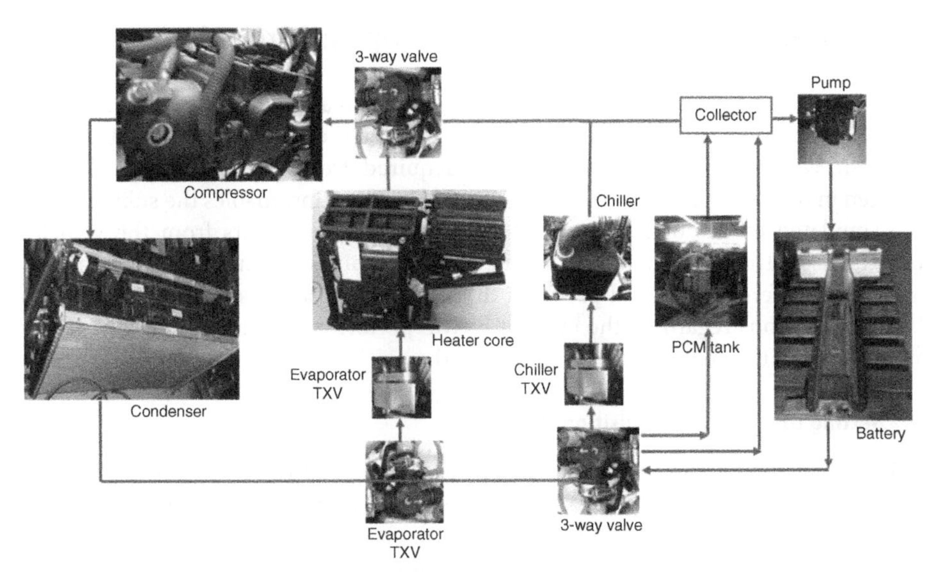

Figure 4.16 Schematic of the test bench refrigerant loop used.

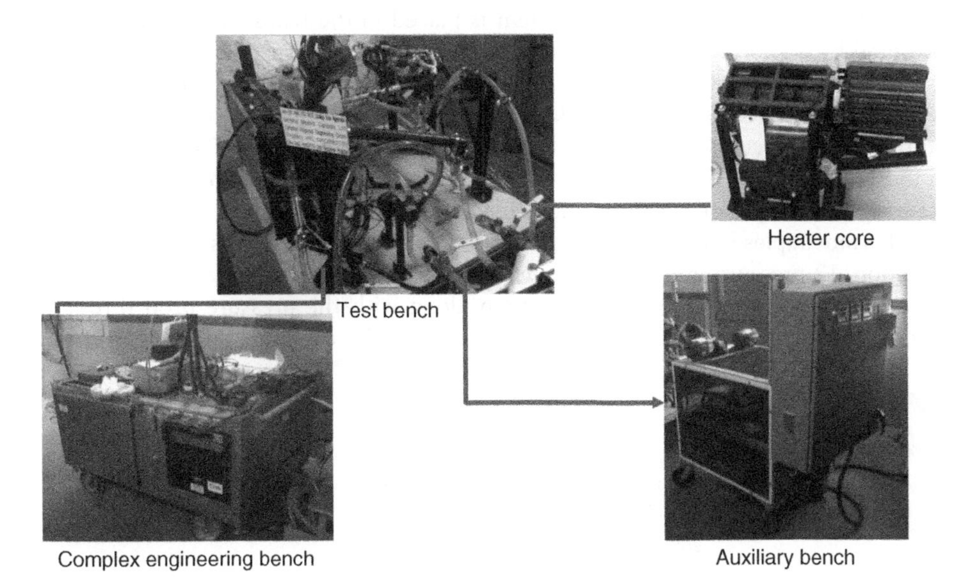

Figure 4.17 Schematic of the experimental set up.

It is important to initially inspect the plumbing for leaks and filled with cooling media. Temperature sensors and pressure gauges are placed at critical locations and mass flow rates were measured in points where there is a significant change in the flow. The test bench unit is then connected to the complex engineering bench in order to acquire data and power the electronics used in the test bench. The schematic of the experimental set up is provided in Figure 4.17.

4.4 Vehicle Level Experimentation Set Up and Procedure

In order to have a comprehensive understanding of the analyzed BTMS under real-life scenarios, it should also be tested in the vehicle after the aforementioned steps are completed. When combined with the data acquired from the experimentation is conducted in the cell and module level; the vehicle level testing enables the substitution of the simulated effects of auxiliary test benches with real inputs from the vehicle and/or the user. Thus, it is an important step that combines the information gathered from various different tests and provides the user with a larger spectrum of data to make final decisions regarding the current BTMS performance and areas that require improvements in the system (Hamut *et al.*, 2014).

4.4.1 Setting Up the Data Acquisition Hardware

In order to be able to conduct accurate and reliable tests in the test vehicle, it needs to be instrumented with a data acquisition system and numerous sensors and gauges. In the following steps, IPETRONIK data acquisition system is utilized. The simplified schematics on how the data acquisition system works is shown in Figure 4.18.

It is also crucial to select compatible hardware to make accurate and reliable measurements in the vehicle. In this step, M-Series hardware is used which include M-THERMO, M-SENS and M-FRQ. This modular system is placed in the trunk of the vehicle and is powered by a 12 V power supply which draws its power form the vehicle. All the sensors, gauges and flow meters that are placed in the thermal management system are wired through the vehicle to the trunk and are labeled with respect to their type and position. The data acquisition system in the test vehicle can be seen in Figure 4.19.

In the TMS, 82 M-THERMO K-type (16 Channel ANSI) thermocouples are used for measuring the temperature before and after every major component in the vehicle. These thermocouples have 16-bit analog converter and can measure as low as −60 °C. In addition, 12 pressure transducers are used in the TMS lines in order to determine the associated pressure values in the system along with the pressure drop through the components. These pressure transducers have 0.25% accuracy and temperature

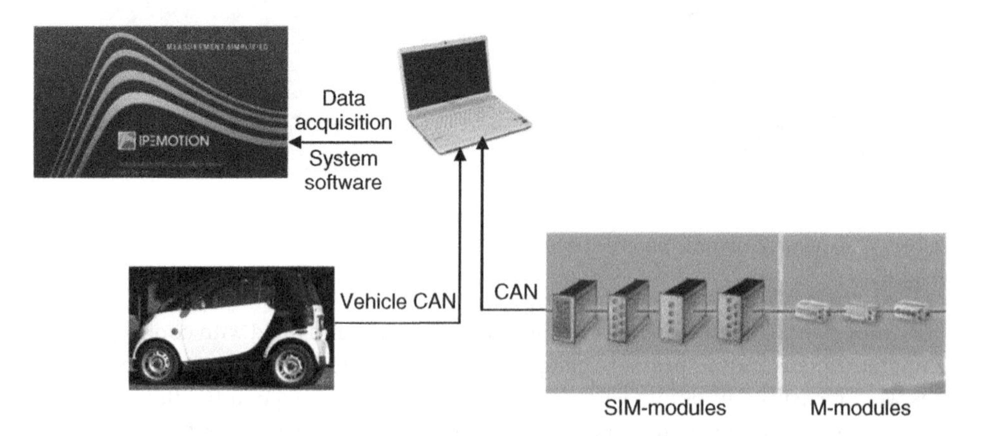

Figure 4.18 Application of data acquisition software in the vehicle (modified from IPETRONIK, 2009).

Figure 4.19 Data acquisition system installed in the trunk of an electric vehicle.

compensation and can operate between −20 °C and 80 °C temperature ranges, which cover the majority of the temperatures reached in the experimentation. The accuracy of the transducers decreases as they deviate from these temperature ranges. Moreover, M-Sens 8 (8 channel) voltage/current sensors are utilized in the experimentation in order to determine the corresponding mass flow rates associated with the refrigerant and coolant in the system. These sensors have 11 voltage and 2 current measuring ranges and work on a high speed CAN bus. Furthermore, M-FRQs are used which have 4 signal inputs with adjustable ON and OFF threshold and anemometers are placed on the condenser to determine the amount of air flow to the system. They have the measurement modes of frequency from period duration, plus duration, pause duration and duty cycle and can measure data output to CAN bus (high speed). The M-FRQ's have 4 inputs with the ranges of 14 V in 250 mV steps and 1±40 V in 200 mV steps (IPETRONIK, 2009). The sensors used are provided in Figure 4.20.

Furthermore, in order to record the flow rate in the system, 5 electromagnetic transmitters were placed in the vehicle. The electromagnetic flow meter with 3/4″ line size was installed to the vehicle, the sensor and converter is grounded and isolated from any source of vibrational and magnetic noise for the system to operate correctly. When started, the measured lines are completely filled with the associated cooling media and ensured that there is no flow in order to calibrate the equipment as well as that a compatible sample rate is selected for each device. The flow readings were acquired through the flow transmitter as well as the corresponding software package. A sample picture of one of the flow transceivers used is shown in Figure 4.21.

In order to acquire thorough and meaningful data from the experimentations, the sensors and gauges (for temperature, pressure and mass flow rates) need to be placed at every crucial point of the thermal management system. The list of these instrumentations and their most commonly utilized locations are provided in Table 4.4.

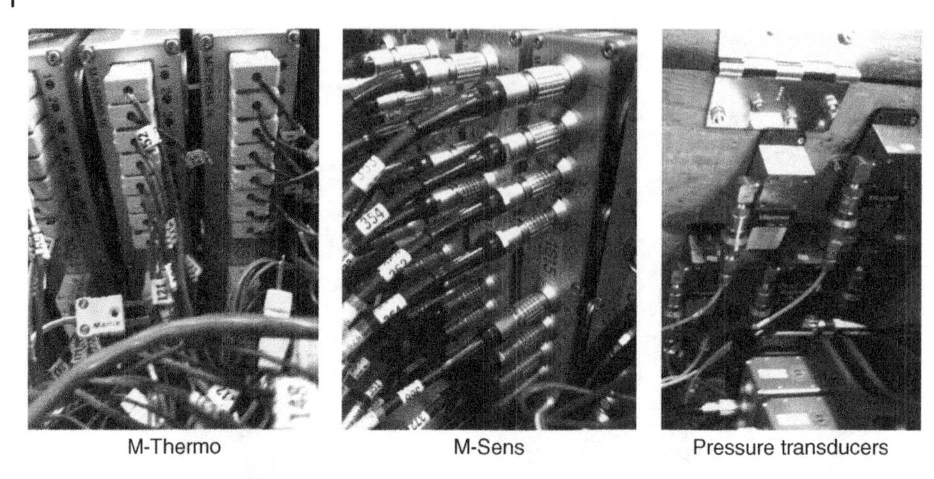

Figure 4.20 Sensors used in the data acquisition system.

Figure 4.21 A commonly utilized flow transceiver used in the experimentations.

4.4.2 Setting Up the Data Acquisition Software

In order to be able to read and process the data acquired from the instrumentation shown in Figure 4.20, a compatible software package (IPEmotion Developer Version 01.03 is used in this analysis) needs to be used (IPEmotion Manual, 2010). This software package allows configuring, displaying, measuring and storing the acquisition data. The signals are acquired by using manufacturer application layer which is a plug-in component as made of several dynamic link library files along with description files in XML format.

In this step, it is essential to label all the hardware according to their location and use in the vehicle and assign corresponding names in the data acquisition software. As the number of measurement devices increase, it becomes very easy to lose track and make wrong pairings between the devices and the readings. This creates significant issues, especially for sensors and gauges in hard to reach locations such as inside the

Table 4.4 Sample instrumentation list for general TMS experimentations.

Channel Name	Channel Description
RadInCool	Radiator Inlet Coolant - °C
RadOutCool	Radiator Outlet Coolant - °C
TrnAuxOilCoolAl	Transmission Aux Oil Cooler Air Inlet - °C
TrnAuxOilCoolAO	Transmission Aux Oil Cooler Air Outlet - °C
TrnAuxOilCoolFl	Transmission Aux Oil Cooler Air Fluid - °C
TrnAuxOilCoolFO	Transmission Aux Oil Cooler Air Fluid - °C
GrilleOATSens	Grille at AOT Sensor - °C
CowlAl	Cowl Inlet Air - °C
FrtBlwrAO	Front Blower Outlet Air - °C
CompOut	Compressor Outlet Stinger - °C
CondOut	Condenser Outlet Stinger - °C
FrtEvapInPipe	Front Evaporator Inlet Pipe Stinger - °C
FrtEvapOutlet	Front Evaporator Outlet Stinger - °C
Compln	Compressor Inlet Stinger- °C
RrEvapLnPipe	Rear Chiller / Evaporator Inlet Pipe Stinger- °C
RrEvapout	Rear Chiller / Evaporator Outlet Stinger- °C
FrtHtCorFl	Front Heater Core Inlet Fluid - °C
FrtHtCorFO	Front Heater Core Outlet Fluid - °C
Cond_Aln_Grid_1 through 12	Condenser Air In Grid # 1 - # 12 °C
Cond _AOutGrid 1 through 12	Condenser Air Out Grid #1 - #12 °C
Rad_AlnGrid_1 through 5	Radiator Air In Grid # 1 - # 5 °C
Rad_AlnGrid_1 through 5	Radiator Air Out Grid # 1 - # 5 °C
FrtEvapInGrid_1 through 9	Front Evaporator Air In Grid # 1 - # 9 °C
FriEvapAOGrid_1 through 9	Front Evaporator Out Grid # 1 - # 9 °C
FtHtrCoreAlGrd_1 through 6	Front Heater Core Air in Grid # 1 - # 6 °C
FtHtrCoreAOGrd_1 through 6	Front Heater Core Air out Grid # 1 - # 6 °C
Comp_Out_P	Compressor Outlet (0–500 psig) kPa
Cond_Out_P	Condenser Outler (0–500 psig) kPa
Evap_Frt_In_P	Front Evaporator Inlet (0-1000 psig) kPa
Evap_Frt_Out_P	Front Evaporator Outlet (0-100 psig) kPa
Comp_In_P	Compressor Inlet (0-100 psig) kPa
Evap_Rr_In_P	Rear Chiller/Evaporator Inlet (0-100 psig) kPa
Evap_Rr_Out_P	Rear Chiller/ Evaporator Outlet (0-1000 psig) kPa
Trans CoolIn_P	Transmission Cooler Inlet – kPa
Trans CoolOut_P	Transmission Cooler Outler – kPa
Rad_In_P	Radiator Inlet kPa
Rad_Out_P	Radiator Outlet kPa
FrHeatCoreln_P	Front Heater Core Inlet – kPa
FrHeatCoreOut_P	Front Heater Core Outlet – kPa

(Continued)

Table 4.4 (Continued)

Channel Name	Channel Description
Frt_Blower_V	Left/Main Cooling Fan – V
CoolingFan_Lt_A	Left/Main Cooling Fan – A
CoolingFan_Rt_V	Right Cooling Fan – V
CoolingFan_rt_A	Right Cooling Fan A
TransOilCool_lpm	Transmission Oil Cooler (3/4" Turbine) - 1pm
Radiator_lpm	Radiator – (1 - 1/2" Magnetic) 1pm
TPIMJpm	TPIM_Coolant – (3/4" Magnetic) 1pm
FrtHeatCore_Return_lpm	Front Heater Core Return to Engine (3/4" Magnetic) - 1pm
Cond_Fan_Freq_1 through 12	Condenser Anemometer # 1 – # 12 FREQ

Source: Hamut *et al.* (2014).

battery modules and behind the radiator, and can prolong the time it takes to complete the experimentations. Once all the measurement devices are placed, it is important to verify that the readings can be accessed in the software and fall within the estimated ranges. The noise and interferences must be kept to a minimum by frequent calibration and isolation of the devices.

In the software, the configuration is defined using the project properties. For signals, sampling rates of 5 Hz is generally used, as it is the recommended sampling rate (due to its optimal accuracy and frequency) for these applications. Subsequently, the corresponding units are selected for the channels. Next, the maximum and minimum displaying ranges of the acquired values need to be defined. Using the scaling calculator, the voltages need to be accurately converted to the corresponding measurement units for the set up. The limits for each value should be determined and logged in to the software. Then limit violations must be recorded and reset when the signal returns to the set range by passing a high level of hysteresis (generally, 2% for these applications).

The next step is to use the data manager main navigation to manage and analyze the acquired data. Loaded acquisition data sets are then converted to excel format through the export function. Finally, the analysis tab is used to visualize the data by using the software charts as shown in Figure 4.22. Once the data is acquired and stored in the software, it is then used to evaluate the vehicle performance.

Moreover, the high-speed and medium-speed CAN busses in the vehicle need to be monitored in order to log the associated signals. NeoVI RED device with Vehicle Spy 3 software package is used for this application. The list of most commonly obtained data is given in Table 4.5.

Finally, numerous experiments under different scenarios need to be conducted in order to gather a wide range of data of the vehicle thermal management system from both CAN busses. The experimentation procedure in this step is defined similar to the ones set by most OEMS and experimental analyses conducted in the literature for ease of comparison for the readers. Key parameters are varied systematically in order to record the associated changes in the system and new tests are conducted once the system reaches back to its steady state. These obtained data are used to re-validate the numerical analysis and further improve it by creating a more accurate representation

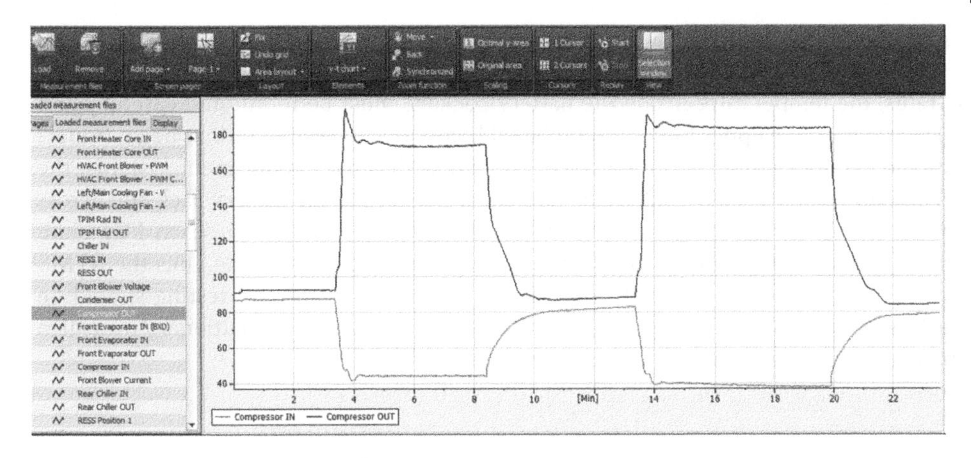

Figure 4.22 Sample screenshot from the data manager main navigation screen.

Table 4.5 List of commonly used medium-speed CAN bus signals received from the vehicle.

Channel Name	**Channel Description**
OAT	Outside Ambient Temperature - °C
HVBat_Max_Temp	Battery Maximum Temperature - °C
HVBat_Min_Temp	Battery Minimum Temperature - °C
RadInCool	Compressor High Side Pressure – kPa
HVBat SOC	Battery State of Charge - %
HVBat_Proc_Voltage	Battery Processed Voltage – V
HVBat_Proc_Current	Battery Processed Voltage – A
Comp_Volt	Compressor Voltage – V
Comp_Current	Compressor Current – A
Comp_Pwr	Compressor Power – kW
Comp_Speed	Compressor Speed – RPM

Source: Hamut *et al.* (2014)

of the actual vehicle system. A portion of the recorded sample data set is provided in Section 5.4.5 for reference and is obtained through the following scenario:

- The vehicle is turned on and data acquisition started.
- The heater and fan are fully turned on for t_1 seconds.
- The heater and fan are turned off until the parameters return to the initial state.
- The air conditioner and fans are fully turned on for t_2 seconds.
- The air conditioner and fans are turned off until the parameters are returned to the initial state.
- The data acquisition system is stopped.

It should be noted that, it is imperative to calibrate all the necessary instrumentation used in the vehicle before testing, to check for compatibility of the units used and place redundancies in critical measurement locations to ensure the reliability of the readings. Since most of the work is done around the high voltage battery, appropriate safety measures should be taken and all power must be disengaged before interfering with the electric flow of the battery. In scenarios where the battery is discharged at high C-rates, the temperature inside the battery package must be monitored carefully and restrictions on the upper limits should be set before the experimentations begin (unless destructive tests are being conducted). For liquid thermal management systems, the sealing of the pipes is essential as the coolant travels inside the battery pack and across the vehicle, passing through several electronic systems. In addition, when PCMs are used, significant degradation and volume change can occur over time, thus the BTMS should be checked for any leaks or deformations periodically. All testing should be performed by qualified personnel and all necessary safety and precautionary measures must be taken before the test procedures begin.

4.5 Illustrative Example: Simulations and Experimentations on the Liquid Battery Thermal Management System Using PCMs

In order to illustrate the applications and the outputs of the aforementioned steps and procedures, an experimental analysis of BTMS using both liquid cooling and PCMs are demonstrated on an electric test vehicle. The utilized vehicle is a hybrid electric vehicle (in series configuration) with 16 kW-h lithium-ion electric battery. The battery incorporated prismatic lithium-ion manganese-spinel ($LiMn_2O_4$) cells arranged in a number of modules. The cells are covered with PCMs with various different thicknesses and DEX-Cool (50/50 water glycol mix) is used on the battery coolant loop to keep the battery operating within the ideal temperature range. In the refrigerant loop, R134a is used to provide air conditioning to the cabin and remove the heat from the coolant loop when necessary. Moreover, an engine loop is used to keep the engine cool by the mixture of water and anti-freeze pumped into the engine block in order to draw the excess heat away from the crucial areas. Finally, a power electronics loop was used for cooling the battery charger and the power inverter module to ensure the main under-hood electronics do not overheat during usage. The schematic of the thermal management system loops in the test vehicle are provided in Figure 4.23. In the figure, the numbers represent the locations where the temperature sensors (°C), pressure gauges (kPa) or anemometers and flow meters (kg/s) are placed. Moreover, voltages (V) and currents (A) are measured where the power output was necessary.

In addition to the current thermal management system of the vehicle, PCMs are also applied internally (around the cells). For the PCMs, the solidus and liquidus temperatures are considered as 35.5 °C and 38.6 °C, which show a three degree interval for the segregation temperature due to impurities. Mass flow rates of 1.2 liter/min to 7 liter/min and heat transfer of 350 W to 700 W are considered in the heat exchanger design. The graphene platelets and CNTs are mixed with technical grade and pure octadecane in such concentrations as 1.25%, 3%, 6% and 9%.

The BMTS is first investigated at cell level with and without PCMs and followed by system level investigation using with liquid coolant running around the battery. The heat generation in the battery is determined both through finite volume-based methods and using a calorimeter. Fluent computer modelling is initially utilized the determined

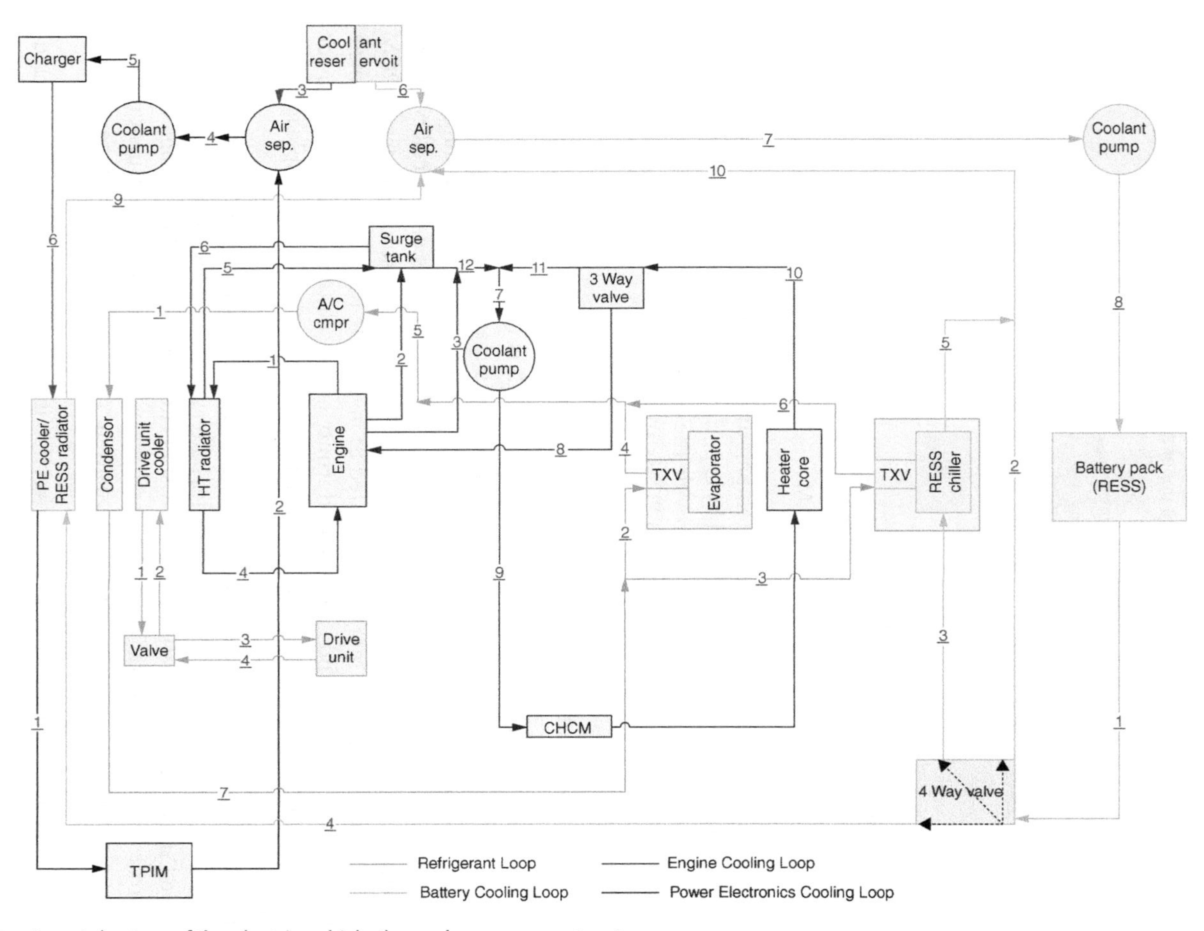

Figure 4.23 Experimental set up of the electric vehicle thermal management system.

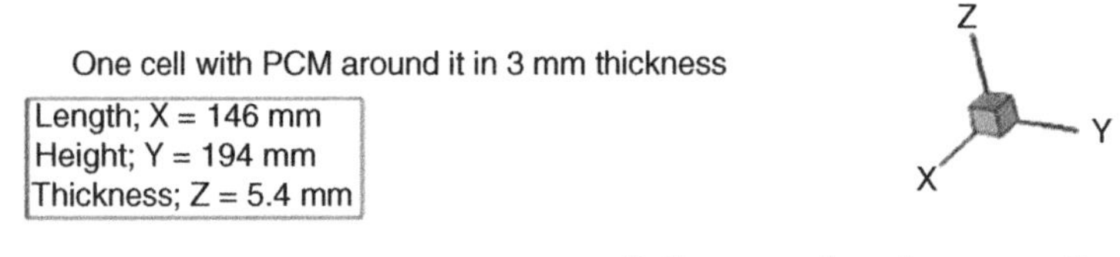

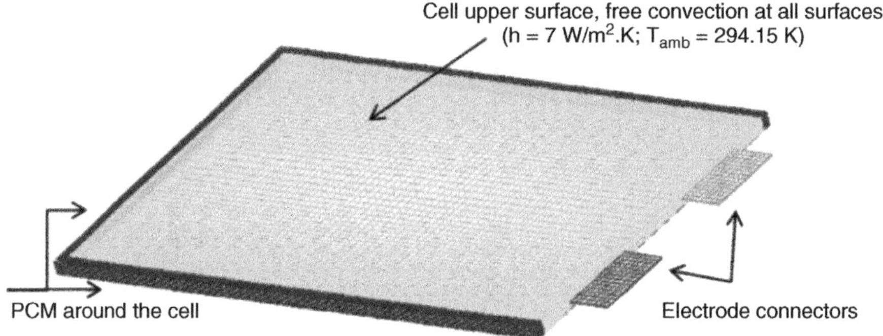

Figure 4.24 Schematic of cell and surrounding PCM.

the trends in average and maximum cell temperature increase and cooling performance of the battery, and the results are verified by experimental outcomes. It should be noted that the main goal of the analysis is to provide the methodology and necessary steps to replace battery active cooling TMSs with PCMs and reduce the number and cost of the cooling components in the system.

4.5.1 Simulations and Experimentations on Cell Level

In this section, the heat transfer with PCM in BTMS is examined in cell level using finite volume-based simulations. Significant heat dissipation can occur in the cells during charging and discharging of the battery, especially under demanding driving patterns. The distribution of the generated heat mainly depends on the cell type and geometry. Thus, various cell dimensions are analyzed in this case study. 4 different PCM thicknesses (3, 6, 9 and 12 mm) are evaluated to investigate the effect of the thickness of the PCM on the magnitude and the uniformity of the cell temperature and provide the readers with different options to be used in their applications (an example is shown in Figure 4.24).

It should be noted that the selection of PCM thicknesses is based on the practical applicability of the PCM in electrical vehicles battery pack. PCM with organic materials are used since they have relatively low corrosion effect on the PCM container, they are non-toxic and environmentally friendly, and have chemical stability, congruent melting (phase segregation), no chemical decomposition, and small/negligible super-cooling effect.

For the initial analysis of the cell, free convection is considered for all surfaces, including the cell and PCM interface with the ambient air where heat transfer coefficient is $h = 7\,W/m^2\,K$ and $T_{amb} = 294.15\,K$. The cell heat generation rate is calculated by the calorimeter tests. The assumptions and initial boundary conditions considered in the analyses are listed as follows:

- The heat transfer from the terminal surfaces is dominated by free convection.

- The boundary between the terminal and cell itself is a "coupled" boundary condition to assure the continuity of the temperature and other properties across the boundary.
- The initial cell temperature is considered to be 294.15 K (same as the ambient temperature).
- The cell has orthotropic thermal conductivity.
- The average specific heat for the cell and PCM is utilized.
- The spatial distribution of heat generation sources are assumed to depend solely on the cell chemical reactions
- The heat generation in the cell is considered to be uniform. (This assumption is in accordance with a lumped system analysis, which is applicable for these elements based on the insignificance of their thickness (Pals and Newman, 1995).
- No flow case exists for the liquid phase of PCM.
- The effect of radiation is considered to be negligible.

In the analyses, different scenarios have been considered for the cell heat generation rates depending on the operational conditions. The planar (or surface) thermal conductivity of the cell is considered to be 25 W/mK, and the value for the direction normal to the cell surface is considered as 1 W/mK. The specific heat value for the cell is taken as 1027 J/kgK. For most commonly used Li-ion batteries, when the state of the charge (SOC) equal to 50% at a discharge rate of C/1; the heat generation rate for the cylindrical cells is estimated to be 20 W/Liter (Al Hallaj *et al.*, 1999). For the case of C/1 and 2C, the heat generation rate is considered to be 1.33 W/cell and 4.45 W/cell, respectively. The C-rate determines the rate at which the current is discharged from the cell. The generated heat is mainly due to the internal resistance, which, in turn, depends on the C-rate of the battery. Thermal stability of the cell depends strongly on the internal resistance

In the examined cell, two terminals are symmetrically located. The heat generation is considered minimal in the terminals since they behave like a cooling fin in the cell as a very thin layer of metal (due to the compressed cell electrolyte) exists in the space between the cell margin and the terminals which decreases the heat transfer area between the cell and terminals.

In terms of the utilized foam, the average conductivity is considered as 0.083 W/m K. Measured mass of the foam in the experiments is 4.15 grams. The foam dimensions are 107 mm × 140 mm × 1 mm with specific density of 277 kg/m^3, which are in accordance with the standard values. Specific heat of the foam is considered as 1500 J/kg K from the data provided by the supplier. Moreover, the specific density, specific heat and thermal conductivity of the cooling plate used in the case study are 2719 kg/m^3, 871 J/kgK, K = 202.4 W/mK, respectively.

On the model is developed and the associated meshing is structured; the grid independence needs to be established before being able to make use of the simulations for the developed model.

4.5.1.1 Grid Independence Tests

Before interpreting the outcomes of any meshed analysis, first the grid independence test needs to be carried out to confirm that the mesh size has no effect on the simulation results. Normally, a point inside the domain is monitored to investigate the convergence history of a property such as temperature. In this case study, a rake crossing the cell is monitored as convergence criteria. Temperature distribution along a constant

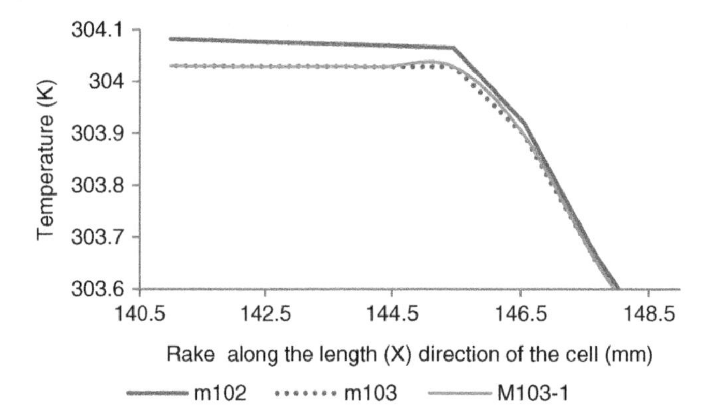

Figure 4.25 Grid independence investigation of the cell with PCM.

rake in the cell is given for three meshes with different sizes. Figure 4.25 shows the convergence history for the rake in each case. These three meshes are considered for grid independence test. For the first mesh, the grid size is 97 × 73. The successive ratio of the cell in the x and y directions are 1.02 and 1.05, respectively. The present mesh sizes are given as follows:

- M102: 87,700 volume elements
- M103: 196,300 volume elements
- M103-1: 300,800 volume elements

For the PCM thickness around the cell, the grid number in an x-y plane is 9 × 9. Based on Figure 4.25, the mesh with 196,300 elements is then selected.

4.5.1.2 Effect of Contact Resistance on Heat Transfer Rate

Once the grid independency is reached, the contact resistance (thermal barrier/resistance) needs to be examined on the cell before the PCM is placed as it can have an impact on the heat transfer rate. The contact resistance can be effective in any location with two or more different layers due to the surface roughness effect. In the current mode, the electrolyte in the Li-ion cell is encapsulated by a thin metal sheet where the terminals have been connected. There can be contact resistance in this interface that affects heat transfer rate. In addition, when the PCM is placed around the cell, an enclosure is required to prevent the leakage of melted PCM. If a thin layer of aluminum cover is considered, the contact resistance (R) can be written in the following form:

$$R_{tot} = \frac{\Delta T}{\dot{q}} = R_{thickness} + R_h + R_{contact} = \frac{L_{Al}}{k * A} + \frac{1}{hA} + R_{contact} \tag{4.40}$$

$$R_{contact} = \frac{T_{cell} - T_{terminal}}{Thickness} \tag{4.41}$$

where h is the heat transfer coefficient between the cell surface and ambient and L is the thickness of the casing. For the case of aluminum interface (10 micrometer surface roughness, 10^5 N/m^2 with air as interfacial fluid) the contact resistance is 2.75×10^{-4} m^2 K/W.

Based on the values for thermal conductivity and other dimensions, $R_{thickness}$, R_h , and $R_{contact}$ are calculated to be 1.7×10^{-5}, 3.531×10^1 and 7.8×10^{-6} ohms, respectively. The thermal conductivity is considered as 202 W/mK. Thus, after the calculations, it is determined that the magnitude of contact resistance is negligible compared to the coefficient of heat transfer. It should be noted that the main reason behind the contact resistance is the high temperature differences and heat flux in the interface. At micro-scale levels, this becomes more dominant. The phonons in the micro-scale level determine the contact resistance more effectively. Conduction takes place through the phonon radiative transport across the contact resistance. In order to examine the phonon intensity behavior across the interface, equivalent equilibrium temperature is introduced. This temperature is the analog of the usual thermodynamic temperature defined in the diffusive limit in any medium. It represents the average energy of all phonons around a local point and it is equivalent to the equilibrium temperature of phonons when they redistribute adiabatically to an equilibrium state. Furthermore, the contact resistance is small enough in boundary that can be neglected without a significant loss of accuracy in the simulations. This is mainly due to the high temperature gradients in the direction of heat transfer.

4.5.1.3 Simulation Results For Li-ion cell Without PCM in Steady State and Transient Response

In order to provide a baseline for the readers to compare the effects of applying PCM in various thicknesses, the cell without the PCM is introduced first. Initially, steady-state simulations are examined, followed by the transient response. Based on the conducted analysis under the conditions mentioned in the previous sections, the temperature contours shown in Figure 4.26 are obtained. For the considered boundary conditions and heat transfer coefficient, the maximum volume, averaged and minimum temperatures of the cell are compared with various other configurations.

Three rakes, 1, 2 and 3, are defined in three different locations in the cell along with vertical rake along the cell height. Once the thermal conductivity is not constant in all directions, the heat transfer rate will be different in the cell surfaces. This phenomenon

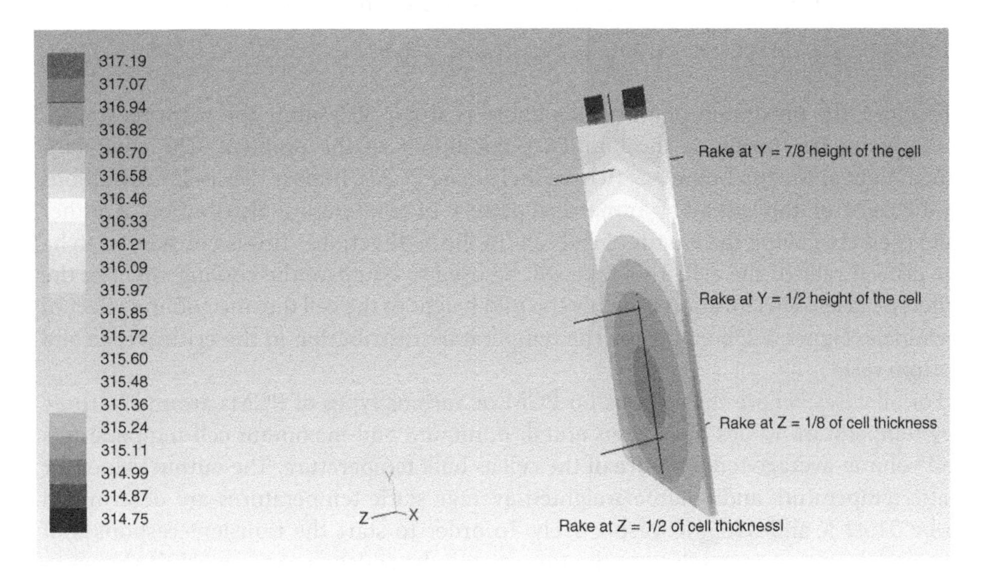

Figure 4.26 Temperature contours in cell without the application of PCMs.

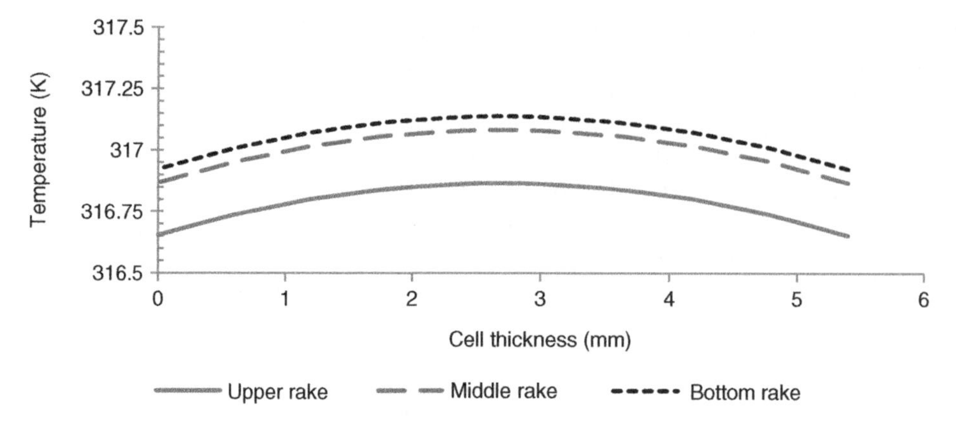

Figure 4.27 Temperature distribution along the horizontal rakes.

will be investigated under the transient case in the next section. The locations of rakes are shown in Figure 4.26, and the values are listed as follows:

- Rake 01: X = 73 mm, Y = 127.75 mm, Z ∈ [0 5.4] mm
- Rake 02: X = 73 mm, Y = 97 mm, Z ∈ [0 5.4] mm
- Rake 03: X = 73 mm, Y = 24.25 mm, Z ∈ [0 5.4] mm
- Rake 04: X = 73 mm, Y ∈ [0 194], Z = 2.7 mm

The temperature distribution in the created rakes is presented in Figure 4.27. It can be seen that the temperature increases to its peak value in the middle portion of the cell, which is due to the symmetrical boundary conditions. The vertical rake is created and the temperature distribution along this rake is shown in Figure 4.28a, in order to find the location for the maximum temperature. The curve fitting gives the following for the temperature distribution (with a correlation coefficient of nearly unity):

$$T = 2\times10^{-8}y^3 - 7\times10^{-5}y^2 + 6.9\times10^{-3}y + 317 \tag{4.42}$$

Therefore, the maximum point of this figure is at y = 50.37mm. The main reason for this asymmetric profile is the boundary conditions of the problem. The connector attachment at the top boundary (where there is no PCM), impose dissimilar conditions to the cell and imposes an asymmetrical pattern of temperature distribution. Another rake is created along the length of the cell (in the X direction). This is considered to be the critical rake in the cell. This rake will be used to compare the cooling effect of the different cases. This location is below the mid height of the cell due the cooling effect of terminals. Figure 4.28b compares the temperature distribution in the critical rake and bottom rake.

For all cases, where the cell has no PCM or various types of PCMs around it, three key temperature values have been noted: minimum and maximum cell temperatures and volume-average temperature of the cell as bulk temperature. The output values for static temperature and volume-weighted average static temperatures are determined to be 315.2 K and 316.7 K, respectively. In order to start the transient responses of the models in transient solution, the time-step independency of the mesh is shown in Figure 4.29.

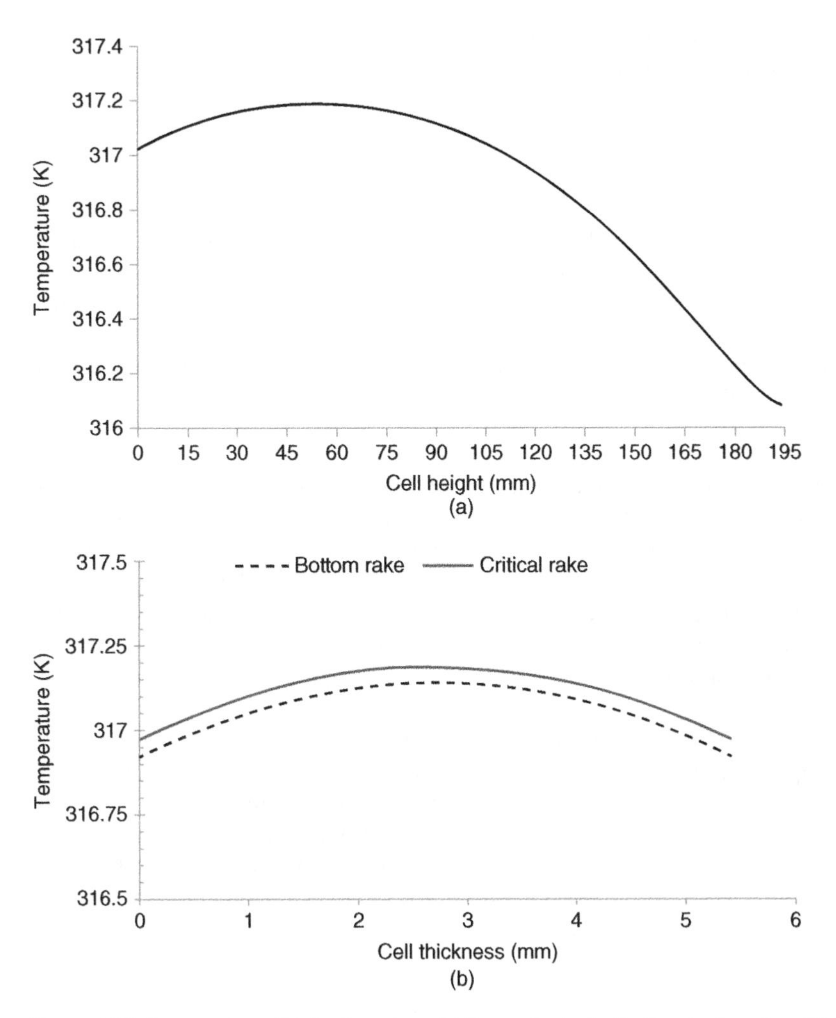

Figure 4.28 Temperature distribution along (a) the vertical rake for the single cell and (b) the critical rake compared to the bottom rake.

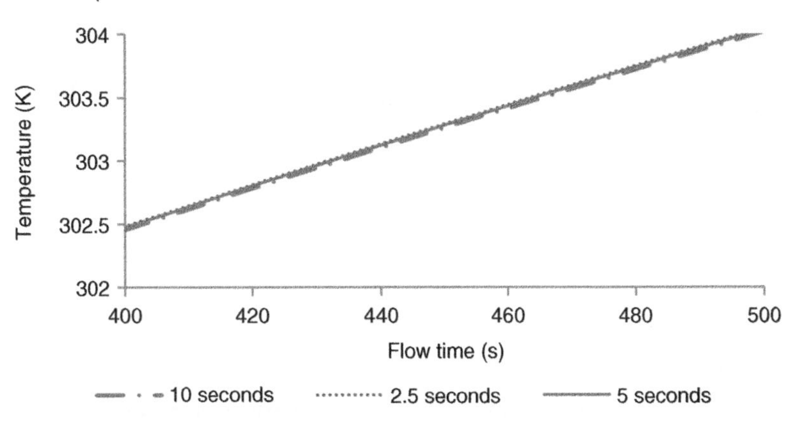

Figure 4.29 Time step independence study of the mesh.

The time steps of 2.5, 5 and 10 seconds for the time are considered. The percent difference is used instead of the percent error in the analysis.

$$\text{Percent difference} = \frac{|(T_{5s} - T_{2.5s})|}{\left(\frac{T_{5s} + T_{2.5s}}{2}\right)} * 100 \tag{4.43}$$

For the time equal to 500 seconds, time steps of 2.5 and 5 seconds result in temperatures of 304.05 K and 304.04 K in the cell. The percent error in this case is 0.003%. The time step of 5 seconds is mostly used in transient analysis unless higher heat generation rates are considered. Such a scenario is dominant for the heat generation rate of 200 kW/m^3, which may not last more than a few minutes. In these operational conditions, smaller time steps are used in the simulations. The results for the transient solution are used as comparison criteria with other configurations.

4.5.1.4 Simulation Results For PCM in Steady-State and Transient Conditions

Once the cells are examined without any means of thermal management to understand its heat dissipation pattern under natural convection, it is compared against applying PCMs in various (3, 6, 9 and 12 mm) thicknesses. Figure 4.30a shows the cell with 3 mm of PCM. In this case, the maximum cell squishiness is 0.076, which allows better geometrical mesh stability. The steady-state and transient responses of the system are compared with the previous case where there was no PCM around the cell. The effect of PCM around the cell gives a shift in the location of maximum temperature in the cell, as it can be seen in Figure 4.30b.

Moreover, Figure 4.31 shows the shift in the location of the critical rake in the cell, caused by the application of the PCM. The curve fit method gives the critical height of the cell to be $y_{critical} = 72.8$ mm; therefore, the new location is shifted 22.5 mm toward the cell interior. The dotted lines in the figure shows that the application of 3 mm PCM around the cell can reduce the maximum cell temperature and can provide better temperature uniformity, which are the two main criteria when evaluating battery thermal management systems.

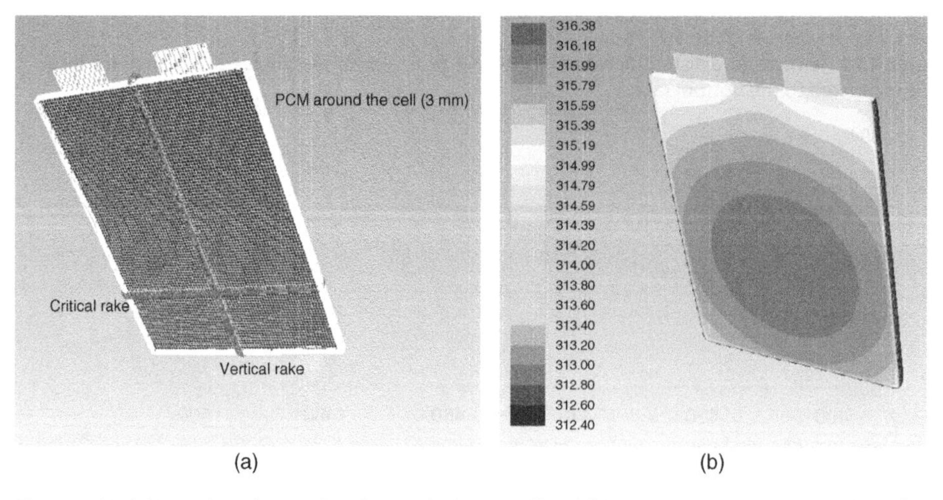

Figure 4.30 (a) Location of vertical and critical rakes in cell and (b) temperature contours in the cell with PCM (3 mm) around cell.

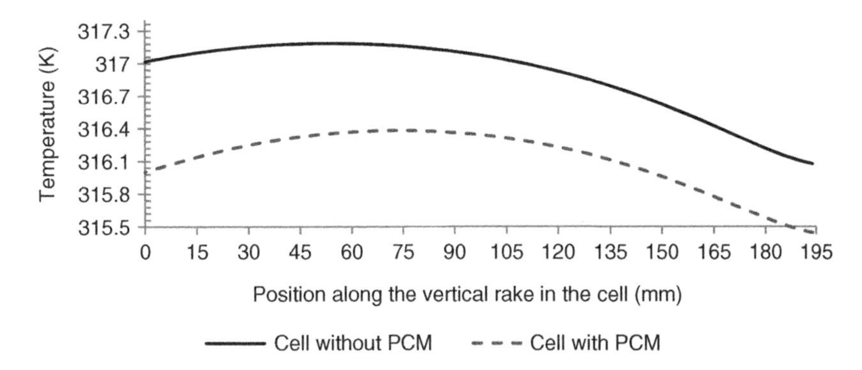

Figure 4.31 The change in the location of maximum temperature point in the vertical rake in cell.

It should be noted that the nature of the current BTMS analysis mostly relies on the transient behavior of the system. In steady-state solution, the PCM has already been melted and the liquid thermal conductivity will determine to be effective in the solution. The mathematical difference between the steady state and transient temperatures has been replaced by the quasi steady-state and steady state temperature differences in the cell. This means that when the average temperature (volume-average) in the transient conditions reaches a quasi-value (commonly around 85%), it is considered as the steady state response.

In order to monitor the temperature in the transient solution, a horizontal rake, created in the location of maximum temperature in the cell, is used with results to be compared with the other cases. Figure 4.32 compares the temperature rise in two critical rakes for the case with no PCM and when there is a 3 mm thickness PCM. It can be seen that after 10 and 20 minutes, the PCM has prevented temperature increase in the cell. Furthermore, 20 minutes elapsed time shows the PCM to be significantly more effective than the ones for 10 minutes. The temperature difference between the cell with and without PCM nearly doubles after 20 minutes, compared to 10 minutes. Figure 4.33 also shows the effects of PCM on the temperature of a specified point in

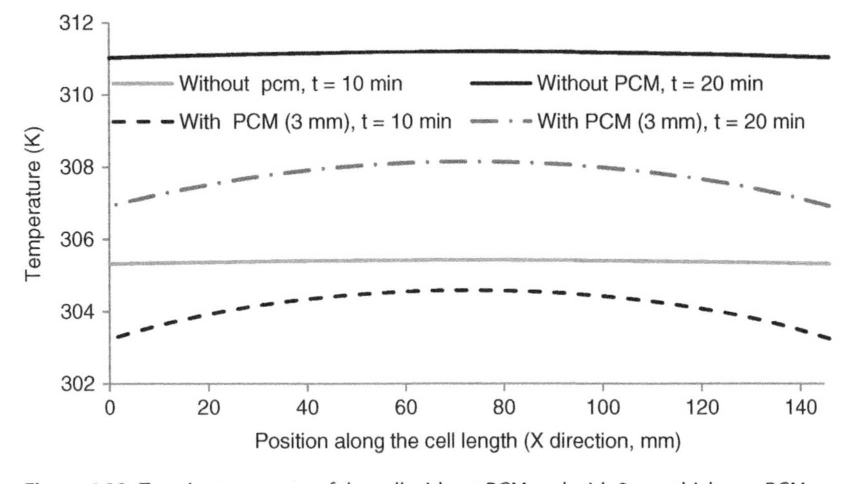

Figure 4.32 Transient response of the cell without PCM and with 3 mm thickness PCM.

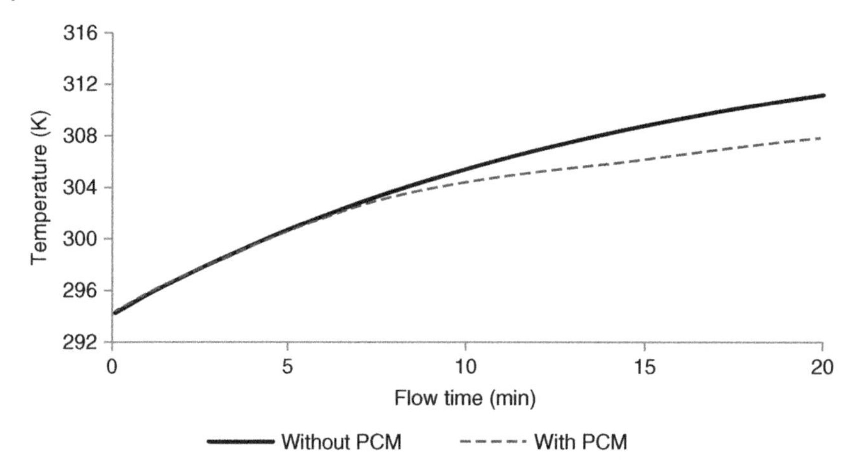

Figure 4.33 Effect of the PCM in preventing a temperature increase in the cell.

the cell. The figure indicates that the effects of flow time on the maximum temperature difference can be seen in both cases. Moreover, it also shows that the maximum cell temperature is reduced with the application of the PCM.

In order to illustrate the effects of PCM thickness, the same method is applied for the case with 6 mm PCM around the cell. In this case, the critical height for the highest cell temperature is calculated to be Y = 87.86 mm (critical height for vertical rake). Along the cell length (X-direction) and at the new height, the horizontal rake is created to monitor the critical cell temperatures. The results are recorded for the flow times of 10 minutes and 20 minutes. The new position is found to be Y = 84.37 mm, which may be considered close enough to the previous location of 87.86 mm. Temperature distribution in the cell is shown in Figure 4.34a, along with the shift of maximum temperature toward the upper side of the cell.

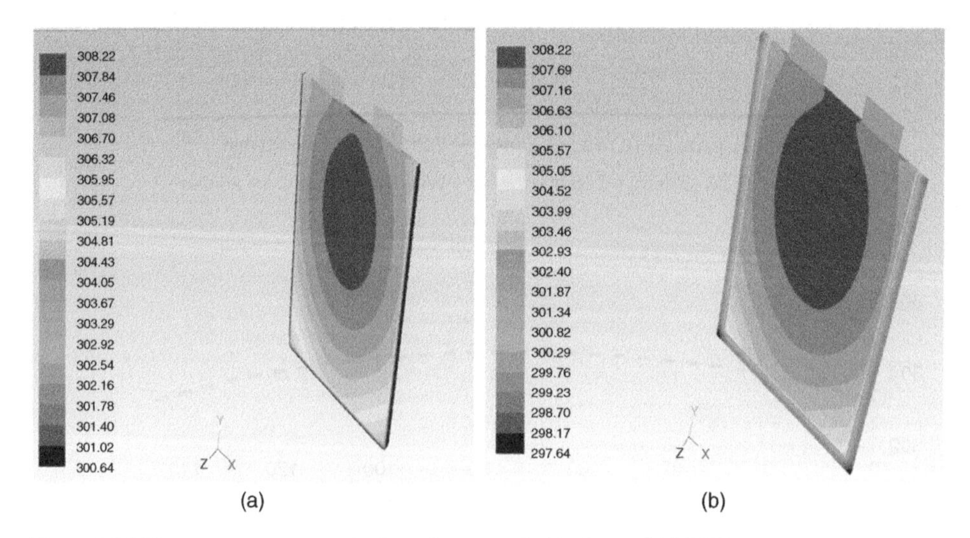

Figure 4.34 Temperature contours in the cell surrounded by 9 mm thick PCM.

In the steady-state situation, the vertical temperature distribution equation is

$$T = 7 \times 10^{-8} y^3 - 10^{-4}\, y^2 + 0.0158y + 314.79 \tag{4.44}$$

The maximum temperature along the rake occurs at the height of y = 86.94 mm. The temperature contours the uniform cell temperature distribution. The shift in the location of maximum temperature for 12 mm thicknesses is seen in Figure 4.34b. The PCM is melted by increasing the temperature. The amount of the PCM used extends the time interval that temperature is absorbed in the PCM and prevents temperature increase.

The overall results for the cell without the PCM are presented in Figure 4.35 using different thicknesses. The shift in maximum temperature can be clearly observed in the figure. Time dependent responses of different thicknesses are given in Figure 4.36 after 20 min. The higher temperature of 305.43 K refers to the case where there is no PCM thermal management in the cell.

A similar comparison is shown in Figure 4.37 for the elapsed time of 20 minutes. In order to investigate the effects of PCM thickness on the temperature distribution along the horizontal rake, the temperature range has been modified in Figure 4.36

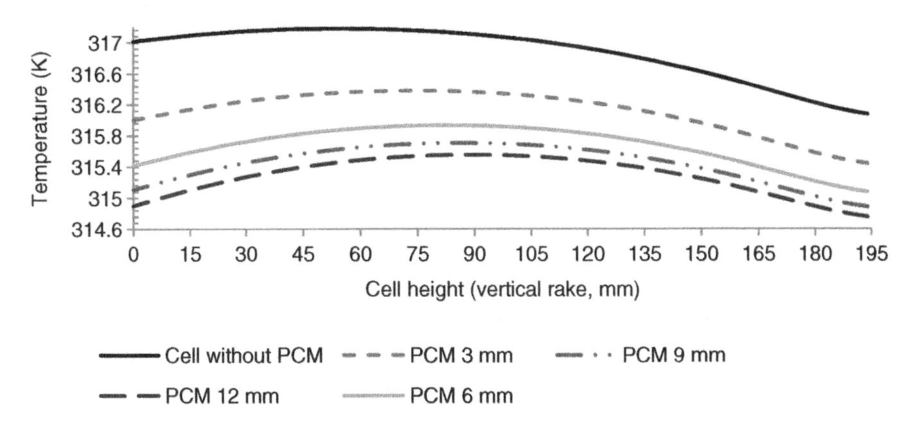

Figure 4.35 Steady-state temperature distributions along the vertical rake in cell.

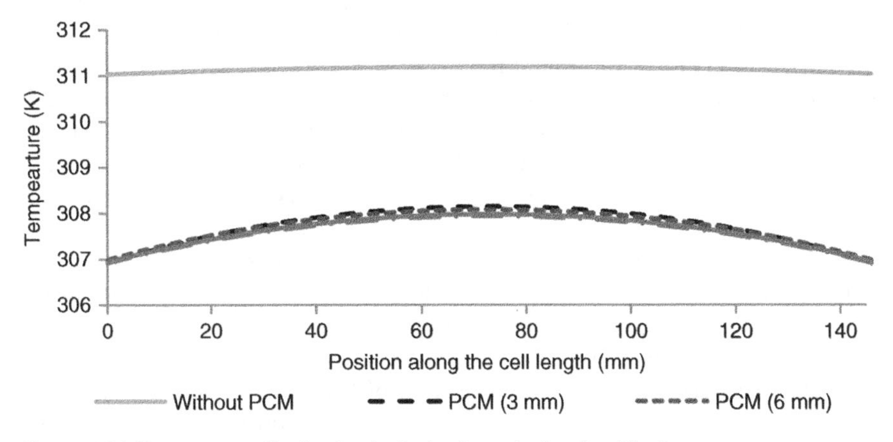

Figure 4.36 Temperature distribution in the horizontal rake after 20 minutes.

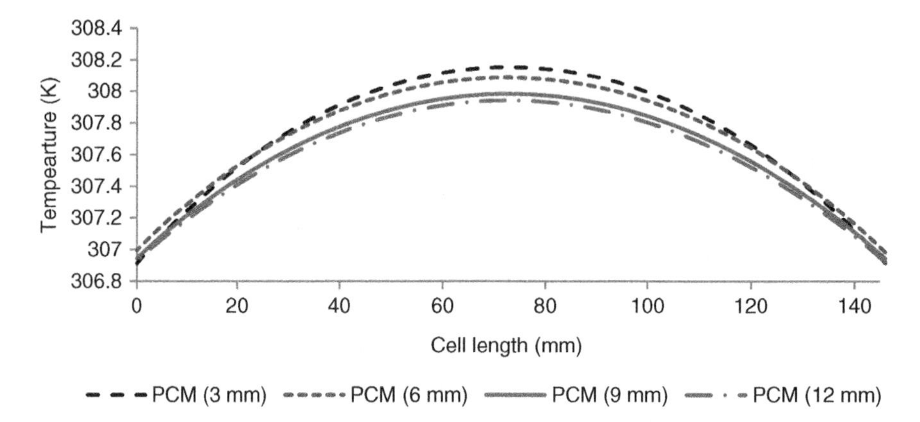

Figure 4.37 Temperature distribution along the horizontal rake in the cell after 20 minutes.

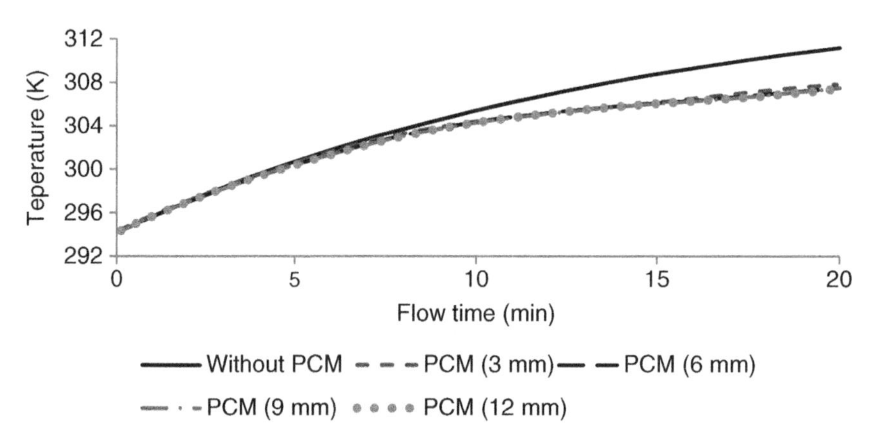

Figure 4.38 Transient response for maximum cell temperature for different thicknesses.

and reveals that PCM with 12 mm thickness has the greatest thermal management effect on the cell. However, it should be noted that there is a geometrical restriction in putting thicker layers of PCM around the pack. Moreover, considering the melted PCM, the thicker layer will impose higher cell thermal resistance, which is detrimental to thermal management of the cell. With the heat generation rate of 63.970 kW/m^3, in less than 7 minutes the effect of PCM around the cells will dominate in ameliorating the temperature increase in the cell (Figure 4.38). As the maximum cell temperature might not represent the entirety of the physics behind the problem, the temperature distribution along the rake is then compared for different thicknesses.

As Figure 4.38 shows, the temperature values along the rake result in a slower rate of increase for higher PCM thicknesses, which prevent the maximum temperatures in the short time. In order to compare the temperatures in the different models, minimum, volume average and maximum temperature values of the cell are measured and are shown in Table 4.6. The effect of PCM on average cell temperature with different PCM thicknesses is then illustrated in Figure 4.39.

Table 4.6 Critical temperatures in the cell for different models.

Configuration	Cell minimum temperature (K)		Average temperature of the cell (K)		Cell maximum temperature (K)	
	Flow time 10 min	Flow time 20 min	Flow time 10 min	Flow time 20 min	Flow time 10 min	Flow time 20 min
Cell without PCM around	304.54	309.82	305.24	310.89	305.43	311.20
Cell with PCM (3 mm)	302.37	305.05	303.98	307.58	304.70	308.43
Cell with PCM (6 mm)	302.48	304.72	303.97	307.37	304.64	308.30
Cell with PCM (9 mm)	302.34	304.63	303.92	307.34	304.61	308.22
Cell with PCM (12 mm)	302.35	304.58	303.92	307.28	304.60	308.16

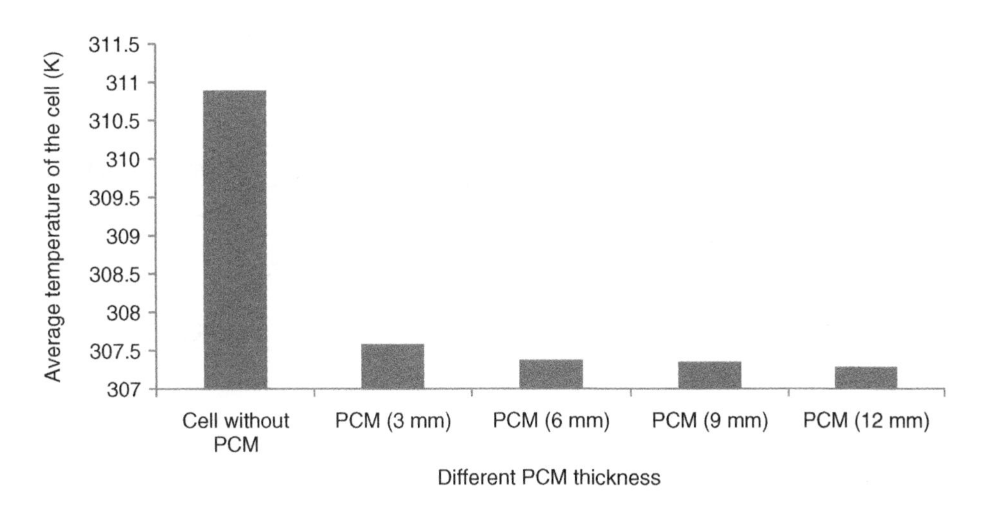

Figure 4.39 Average cell temperature with different PCM thicknesses after 20 minutes.

4.5.1.5 Cooling Effectiveness In the Cell

In order to investigate the cooling effectiveness, the following dimensionless parameter is defined for temperature variation along the critical horizontal rake in all five configurations (cell without PCM and with PCM in different thicknesses).

$$Cooling\ Effectiveness = \frac{T - T_{bulk,gen}}{T_{max,gen} - T_{bulk,gen}} \tag{4.45}$$

where $T_{bulk,gen}$ is the bulk temperature of all models and $T_{max,gen}$ is the maximum temperature of all models. Figure 4.40 shows the cooling effectiveness in the tested models. The other parameter used to assess the cooling effect of the PCM on the cell is defined as the relative temperature ratio. Two temperature coefficients are defined as follows

Overall temperature coefficient:

$$\theta_1 = \frac{T - T_{max,local}}{T_{max,gen} - T_{max,local}} \tag{4.46}$$

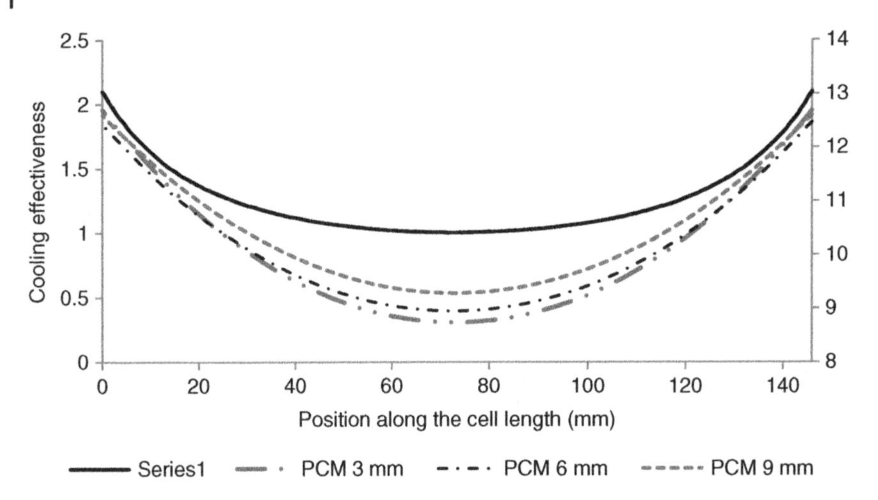

Figure 4.40 Cooling effectiveness for different PCM thicknesses.

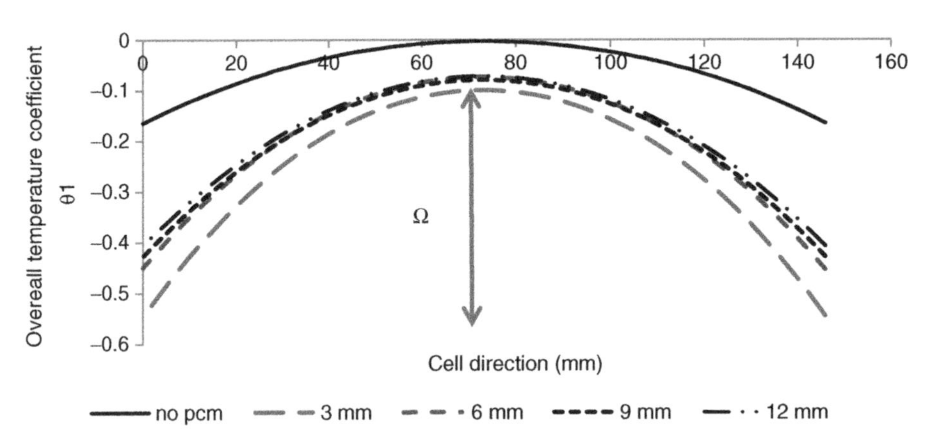

Figure 4.41 Overall temperature coefficients along the horizontal rake in the cell for different PCM thicknesses.

The local temperature coefficient:

$$\theta_2 = \frac{T - T_{bulk,local}}{T_{bulk,gen} - T_{bulk,local}} \tag{4.47}$$

Figure 4.41 shows the variations of temperature coefficient along the rake location for different thicknesses. Effectiveness Index (Ω) is defined as the depth of the curvature in the figure. A large PCM thickness around the cell will have larger depth in curvature which, in turn, provides better cooling in the cell. In this case, the maximum temperature in the cell attains low values and the difference between the maximum and the minimum temperature becomes small across the cell.

4.5.2 Simulation and Experimentations Between the Cells in the Submodule

Once the cell heat dissipation rate and the corresponding cooling load is determined individually, the next step is to analyze the cell with the components around it to better

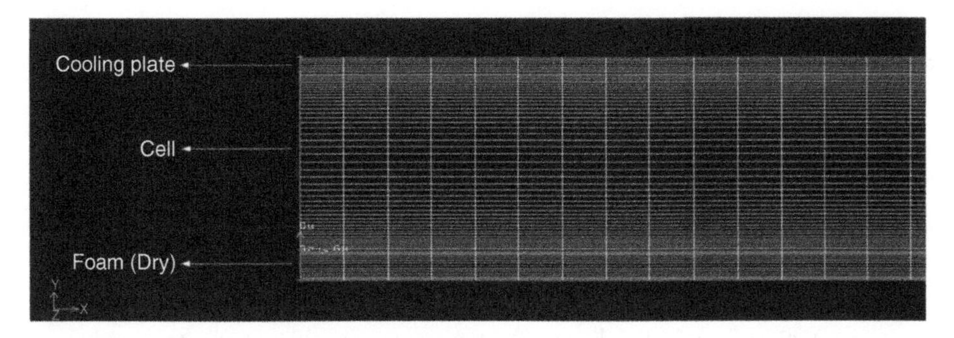

Figure 4.42 Li-Ion cell, cooling fin and foam mesh.

understand its thermal interaction with its surroundings. Thus, in this section, the cell is placed in between the cooling fin and foam, where the PCM starts to melt and gets absorbed by the foam as cell heat generation increases. In this regard, the amount of absorption along with its thermal stability and strength should be considered when selecting an appropriate foam for this application. Once the cell temperature increases enough to melt the PCM, the heat will be transferred to the foam, and in turn, will delay the temperature increase in the cells. A sample mesh is shown the battery cell, foam and cooling fin in Figure 4.42.

4.5.2.1 Effective Properties of Soaked Foam

For the absorbed PCM in the foam, the following relations can be used to estimate the effective properties of the wet foam.

$$Cp_{aver} = \frac{1}{m_{tot}} \times \sum m_i Cp_i = \frac{m_{pcm}}{m_{tot}} Cp_{pcm} + \frac{m_{foam}}{m_{foam}} Cp_{foam} \tag{4.48}$$

The mass concentration is defined as

$$c = \frac{M_{pcm}}{M_{total}} \tag{4.49}$$

$$M_{total} = M_{pcm} + M_{foam} \tag{4.50}$$

Therefore, if the mass concentration is of interest then

$$Cp_{ave} = c \times Cp_{pcm} + (1 - c) \times Cp_{foam} \tag{4.51}$$

For the foam, if the porosity is defined in the following way, then volumetric porosity will act like volume concentration and can be written as

$$\varepsilon = \frac{V_{pcm}}{V_{\mathrm{tot}}} \rightarrow m_{pcm} = \rho_{pcm} \times V_{pcm} = \rho_{pcm} \times \varepsilon \times V_{tot} \tag{4.52a}$$

Therefore, the specific heat can be calculated as:

$$Cp_{aver} = \frac{\rho_{pcm}}{\rho_{tot}} \times \varepsilon \times Cp_{pcm} + \frac{\rho_{foam}}{\rho_{tot}} \times (1 - \varepsilon) \times Cp_{foam} \tag{4.52b}$$

In specific applications, the volumetric concentration should be used in terms of mass concentration. The relationship between these two parameters is found to be as follows:

$$c = \frac{M_{pcm}}{M_{total}} = \frac{\rho_{pcm} V_{pcm}}{\rho_{pcm} V_{pcm} + \rho_{foam} V_{foam}} \tag{4.53}$$

By defining density ratios and volume ratios as constitutive relations in form of density and volume ratios of foam and PCM

$$\mathrm{r} = \frac{\rho_{pcm}}{\rho_{foam}} \tag{4.54}$$

$$x = \frac{V_{pcm}}{V_{foam}} \tag{4.55}$$

$$c = \frac{V_{pcm}}{V_{pcm} + \frac{1}{r} \times V_{foam}} = \frac{1}{\frac{1}{R} \times \frac{1}{x} + 1} \rightarrow x = \frac{c}{r(1-c)} \tag{4.56}$$

The porosity definition gives $\frac{1}{\varepsilon} = \frac{1}{x} + 1$, which is used to give the relation between mass and volume concentrations as follows:

$$\varepsilon = \frac{c}{c + r(1-c)} \tag{4.57}$$

In the simulation, the mass concentration is C=65.8 which is found the experimental measurements. The properties of normal Octadecane can be found in the literature, product catalogue or acquired after conduction appropriate tests in the lab. Based on this information, the following effective properties are calculated for the wet foam (foam with the absorbed PCM) which is going to be placed in between the cells in the submodule.

$$C_{p,wet\,foam} = \begin{Bmatrix} 1928 \;\; J/kgK & \mathrm{T}_{\mathrm{solidus}} > T & \text{solid phase} \\ 148560 \;\; J/kgK & \mathrm{T}_{\mathrm{solidus}} < T < T_{\mathrm{liquidus}} & \text{Mushy zone} \\ 1947 \;\; J/kgK & T > T_{\mathrm{liquidus}} & \text{liquidus phase} \end{Bmatrix} \tag{4.58a}$$

$$k_{wet\,foam} = \begin{Bmatrix} 0.264 \;\; W/mK & \mathrm{T}_{\mathrm{solidus}} > T & \text{solid phase} \\ 0.196 \;\; W/mK & \mathrm{T}_{\mathrm{solidus}} < T < T_{\mathrm{liquidus}} & \text{Mushy zone} \\ 0.128 \;\; W/mK & T > T_{\mathrm{liquidus}} & \text{liquidus phase} \end{Bmatrix} \tag{4.58b}$$

$$\rho_{wet\,foam} = \begin{Bmatrix} 630 \;\; kg/\mathrm{m}^3 & \mathrm{T}_{\mathrm{solidus}} > T & \text{solid phase} \\ 600 \;\; kg/\mathrm{m}^3 & \mathrm{T}_{\mathrm{solidus}} < T < T_{\mathrm{liquidus}} & \text{Mushy zone} \\ 571 \;\; kg/\mathrm{m}^3 & T > T_{\mathrm{liquidus}} & \text{liquidus phase} \end{Bmatrix} \tag{4.58c}$$

In this relations, the solidus and liquidus temperatures are considered as 301.15 K and 303.15 K, respectively.

4.5.2.2 Steady State Response of the Cells in the Submodule

When steady state response of the cells are examined, it becomes clear that the cells on the most inner locations of the submodule has the highest temperatures due to lower heat transfer rates of the inner components since the free convection with the ambient is not available. Even though the temperature difference between the adjacent cells are relatively low, in an hybrid/electric vehicle battery pack with hundreds of cells, this temperature difference can easily accumulate and reduce the performance of the vehicle and can even cause safety issues when not thermally managed properly. Thus, in Table 4.7, the temperature peaks in the submodule with and without PCM is compared.

Similar to the previous analyses, the region with higher temperature is selected and a horizontal rake has been placed along the submodule thickness to include

Table 4.7 Comparison between temperatures with and without PCM in between the cells.

Zone name	Maximum temp.(K) with PCM	Maximum temp.(K) without PCM
Foam 1	319.12	319.31
Cell 1	319.93	320.19
Foam 2	320.48	321.00
Cell 2	320.73	321.28
Cooling plate 1	320.72	321.27
Cell 3	320.72	321.27
Foam 3	320.40	320.89
Cell 4	319.75	319.90
Cooling plate 2	319.02	319.16
Terminals of cell 2	318.40	318.91

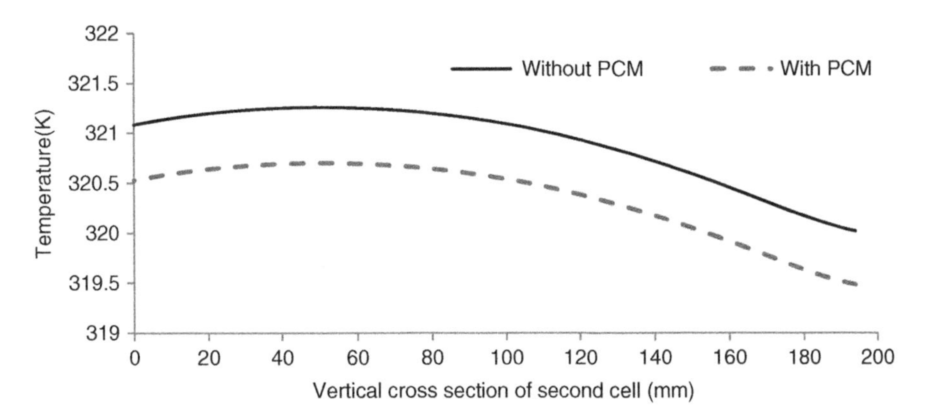

Figure 4.43 Temperature comparison in vertical direction in second cell with and without PCM.

all the 9 different zones and compare the effect of PCM on the system overall temperature. Figure 4.43 displays the effect of phase change materials in decreasing of the temperature in the submodule.

4.5.2.3 Transient Response of the Submodule

In this section, the transient response of the submodule is assessed under various time intervals throughout the PCM melting process and the results are compared to understand the effects of flow time. In addition to the previously created rakes, 5 points have been defined on the surface of the second cell (the critical cell in the sub module) to compare the simulation and experimental results. The PCM-soaked foam is applied to the submodule, where the locations of the measured points in the submodule are shown in Figure 4.44.

These values are compared with baseline where the foams are dry (without octadecane). It should be mentioned that points 1, 3 and 5 have been monitored due to the symmetry, which is an accurate description when the heat generation rate inside the

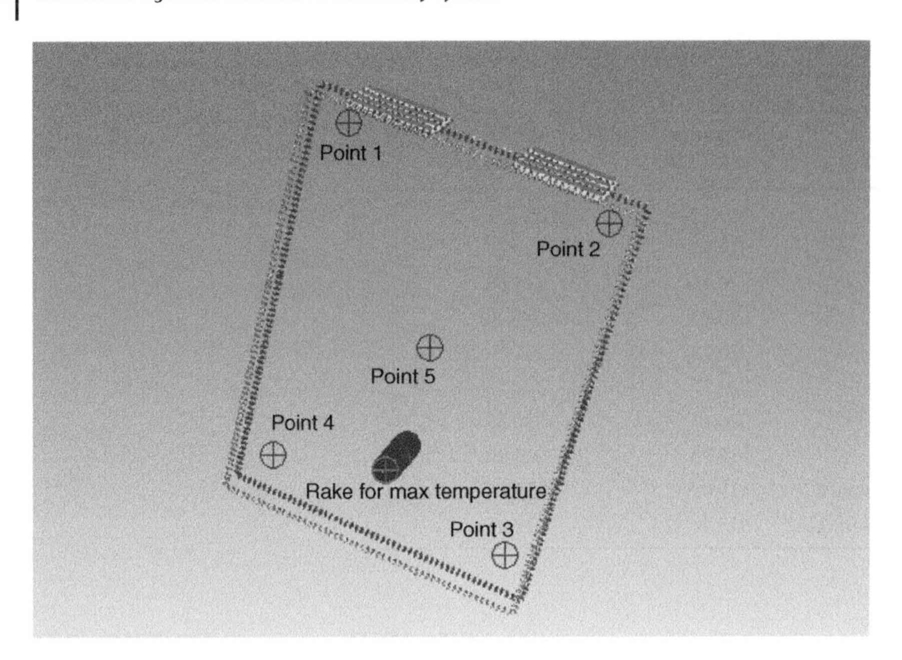

Figure 4.44 Location of points on the cell surface and the rake through submodule.

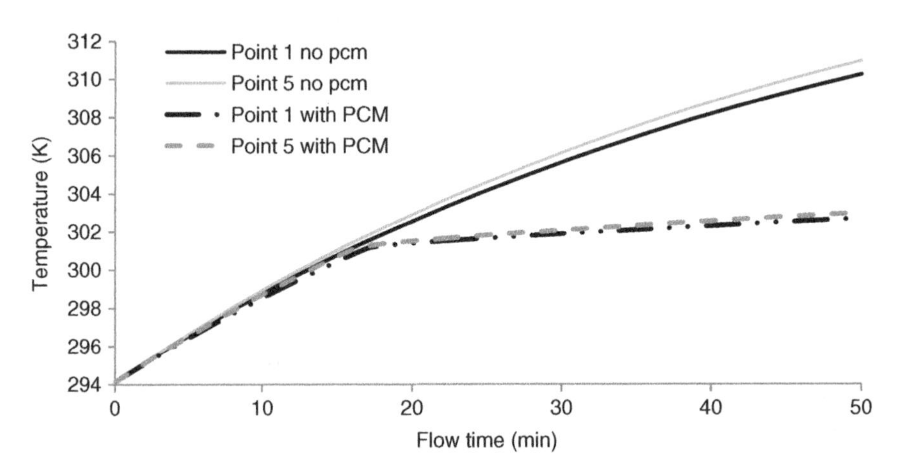

Figure 4.45 Effect of PCM on temperature increase on the cell surface.

cell is relatively uniform. For the other cases, the user defined functions should be integrated with the solver in the software to take into account the local heat generation rates, specially the higher rates close to the connectors, where the electrons are exchanged.

The temperature distribution in the rake after 50 min through the submodule can be seen in Figure 4.45. The PCM absorbs the generated heat by the battery and causes an abrupt decrease in the temperature trend of the submodule. When it is compared with the baseline (the case with no PCM), a temperature reduction of 8.1 K is observed.

The negative values in x axis of Figure 4.46 are simply due to the selected coordinate system. Thus, the value of −1 mm is the thickness of the first foam in the sub module.

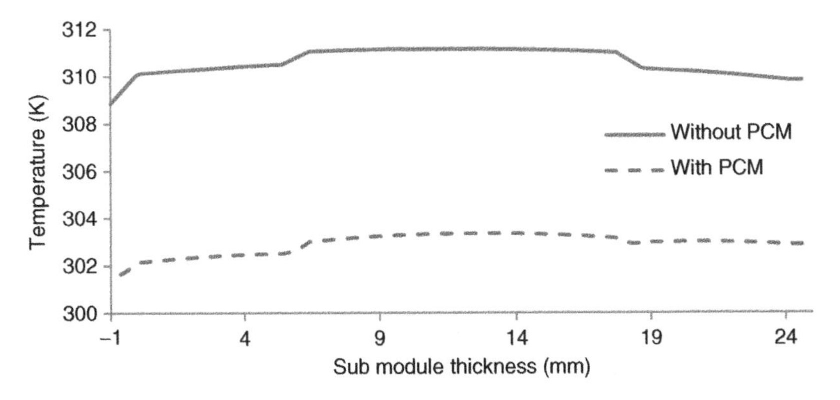

Figure 4.46 Temperature distribution along the thickness of the submodule after 50 minutes.

Table 4.8 Temperatures after 50 minutes for different zones.

Model	**Submodule with dry foam**			**Submodule with PCM in the foam**		
Zone name	**Volume average temp. (K)**	**Max. temperature (K)**	**Min. temperature (K)**	**Volume average temperature (K)**	**Max. temperature (K)**	**Min. temperature (K)**
Cell 1	310.05	310.51	308.98	302.30	302.52	301.85
Cell 2	310.87	311.16	309.85	303.13	303.85	302.65
Cell 3	310.85	311.15	309.84	303.20	303.36	302.73
Cell 4	309.88	310.30	308.98	302.88	303.00	302.44

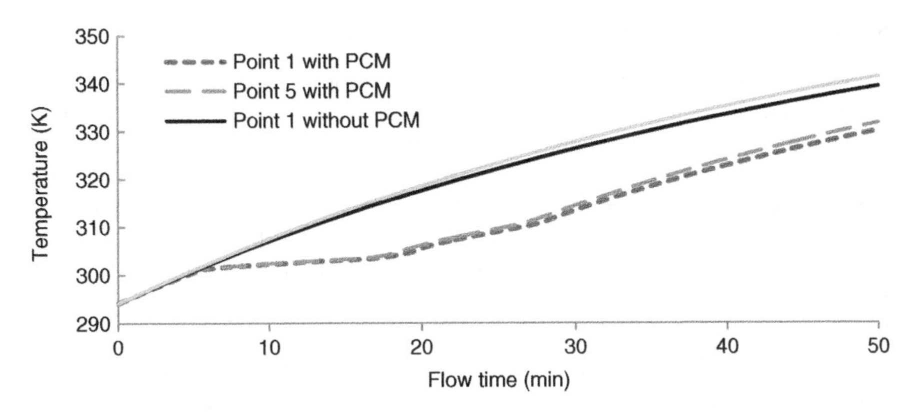

Figure 4.47 Effect of PCM on cell temperature increase (under a heat-generating rate of 63.97 kW/m^3).

Table 4.8 compares the temperature changes with respect to the time and their location in both dry and wet foam submodule.

4.5.2.4 Submodule with Dry and Wet Foam at Higher Heat Generation Rates

As Figure 4.47 indicates, the presence of PCM can be successfully utilized to shave the temperature peaks in the cells. However, it is also important to determine the minimum

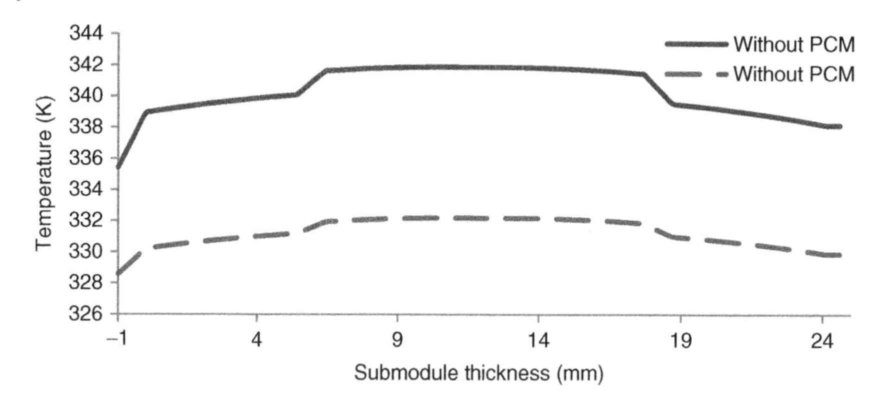

Figure 4.48 Temperature difference in submodule thickness with and without PCM under a heat generation rate of 63.97 kW/m^3.

volume of PCM that can absorb sufficient cell heat in order to maintain an effective thermal management system. Among the conducted analyses, for the heat generation rate of 22.8 kW/m^3, the utilized PCM melted after 16 minutes, and continues to be effective until 50 minutes of the flow time. For discharge rate of C-2, 63.97 kW/m^3 will be generated and the corresponding transient response of the system to this high heat generation rate is shown in Figure 4.47. This figure shows that after 5 minutes, the PCM acts as passive thermal management system. In this condition, PCM is effective to control the temperature for period of 15 minutes, before being completely melted. The same heat generation has been considered to find the temperature variations along the submodule thickness (Figure 4.48).

4.5.3 Simulations and Experimentations on a Submodule Level

In the previous sections, the PCM was applied at a cell level, around all the cells as well as in between the cells, to reduce the cell temperature rise under high discharge rates that is generally caused by demanding driving conditions. In this section, submodules are taken as the smallest denomination in the conducted analyses in order to understand the heat propagation in between the cells and the underlying reasons for any potential temperatures differences among the submodule. Since the battery pack performance is as good as its weakest cell, the temperature uniformity among the cells is crucial in BTMSs. Even though the sensitivity to the batteries performance varies depending on the battery chemistry, age, driving cycle and environmental conditions, generally, temperature differences less than 3 °C is aimed across the battery pack to prevent any thermal-related issues in the vehicle. Thus, analyzing the submodules enable the readers to see the thermal interactions of the cells and provide them with the information to build their battery architecture accordingly.

In the analysis, PCM is applied around the submodule by using enclosures and soaking into a material such as high stability foam and applying it all around the submodule as a cooling jacket. Figure 4.49 shows the configuration of the cells and the foam with the cooling plate. A unit of cells, foam and cooling plates are simulated only for a four-cell submodule to conduct high accuracy results in reasonable calculation times. As shown in the figure, cell 1 is in contact with this foam and the second foam

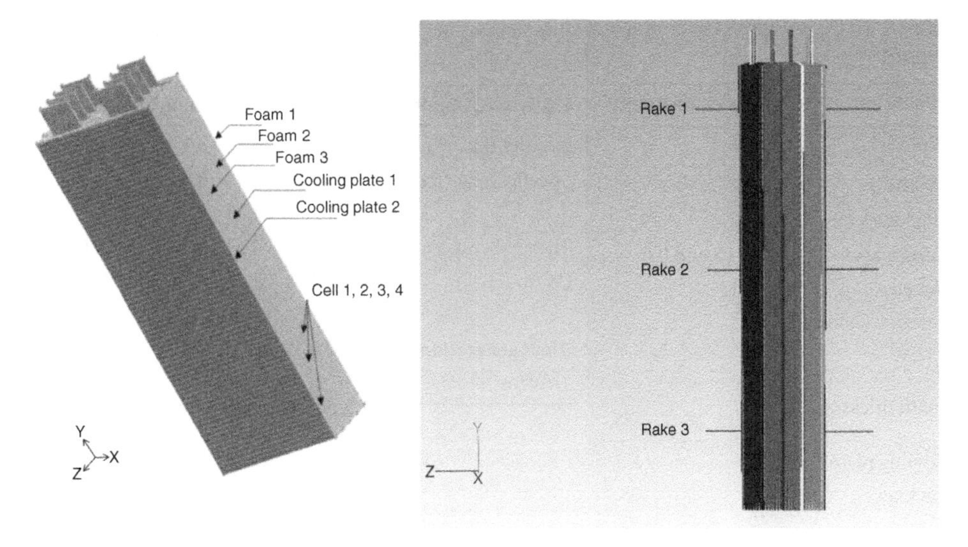

Figure 4.49 Configuration of cells, PCM sheets and cooling plates in the submodule with the monitoring line location on the right.

acts as a separator between cell 1 and cell 2. The 50-50 water-ethylene glycol coolant, is designed to flows through the plate and separate cell 2 and cell 3. Finally, rakes are created to determine the temperature uniformity across the surfaces of each cell.

As shown in Figure 4.49, there are 4 cells, 3 separating foams and 2 cooling plates, which form 9 zones in total in the system. In addition, there are also terminals that have been replaced symmetrically in the cells. Dimensions and properties of all components including cells and foam and cooling plate used in this simulation are listed in Table 4.9a and Table 4.9b respectively.

Using the model above, the finite volume code is employed to examine the steady-state and transient response of the submodule. A total of 261,000 cells are created in the mesh. Sensitivity analyses have been carried out and mesh quality has been enhanced by using the Gambit mesh generator. Temperature excursion along with the maximum temperature in the submodule is analyzed and compared for cases with and without PCM around the submodule. Free convection is considered for upper and lower surfaces and both ends of the system. For the zone condition, the source term for heat generation rate has been applied to the system.

4.5.3.1 Steady-State Response of the Submodule Without PCMs

Initially, a steady-state solution of the system is analyzed with free convection on the surfaces and volumetric heat generation rate of 22.8 kW/m^3. The cells have orthotropic thermal conductivity properties and the initial conditions are the results of the steady-state outputs. The initial temperature in the domain is set to 294.15 K (21 °C).

The high temperature contours are within the vicinity of submodule base (away from the terminals). Temperature contours for the submodule are shown in Figure 4.50. Numerical calculations give the steady-state maximum temperatures for each zone of the submodule. The results show the maximum and minimum submodules temperatures to be 321.28 K and 315.72 K.

Table 4.9 (a) Dimensions of submodule (without applying the PCM jackets) and (b) the properties of components.

Dimension	Magnitude (mm)
Wide (x)	146
Height (y)	196
Thickness (z)	25.6
Terminals placed symmetrically, L,H,W	35, 15 and 0.6
Foam thickness	1
Cooling plate thickness	0.5

(a)

Property/Component	Cell	Cooling fin	Foam
Density (kg/m^3)	4035	2719	277
Specific heat (J/kg.K)	1027	871	1500
Thermal conductivity (W/m.K)	$K_{x,y} = 25$ and $K_z = 1$	202.4	0.083
Heat generation rate (kW/m^3)	22.8	0	0

(b)

Due to the cooling effect of two terminals connected to the cell, temperature distribution is skewed from the center of cell. As shown in Table 4.10, maximum temperature belongs to cell 2. As the next step, a rake is created in to determine the temperature along the height of the cell with the highest temperature (Cell 2) in order to find location of the critical domain. Figure 4.51 shows the temperature profile along the vertical rake (cell height) where the maximum temperature is calculated to be within the vicinity of the cell base. In order to find the location for the highest temperature, the curve fit yields y = 43.6 mm for the highest temperature in the cell.

A new horizontal rake is created in this height and the results are compared with the other rakes. The locations of the rake are shown in Table 4.11.

The approximate location of the rakes is shown in Figure 4.52 in order to demonstrate the submodule geometry. If this rake is defined as the critical horizontal rake in

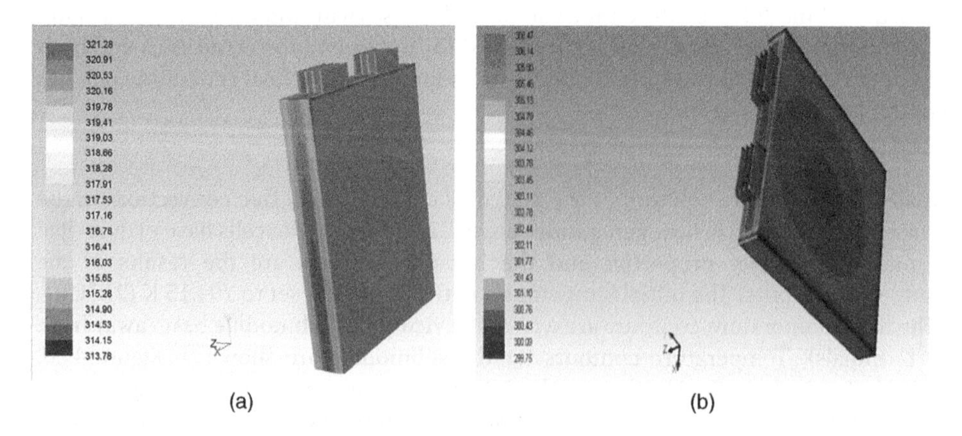

Figure 4.50 Temperature contours in the submodule (a) without PCM and (b) with PCM.

Table 4.10 Maximum temperature in different zones in the submodule with and without the PCM.

Zone name	Maximum temperature (K) with PCM	Maximum temperature (K) without PCM
Foam 1	318.67	319.31
Cell 1	319.52	320.19
Foam 2	320.34	321.00
Cell 2	320.61	321.28
Cooling plate 1	320.60	321.27
Cell 3	320.61	321.27
Foam 3	320.23	320.89
Cell 4	319.24	319.90
Cooling plate 2	318.50	319.16
Terminals of cell 2	316.94	318.91
PCM around the submodule	319.90	319.31

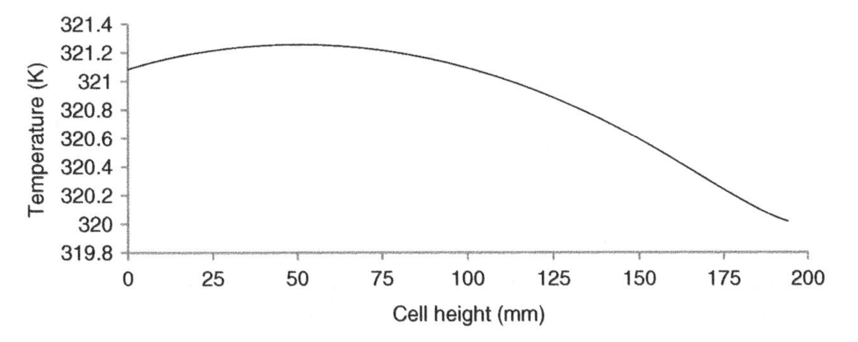

Figure 4.51 Temperature distribution in cell 2 height.

Table 4.11 Rakes locations in submodule to monitor the temperature distribution.

Monitoring surface	X (Length)	Y (Height)	Z (Thickness)
Rake 1	73 mm	1/8 height form top	Cell thickness
Rake 2	73 mm	Half of the height	Cell thickness
Rake 3	73 mm	1/8 height from bottom	Cell thickness
Vertical Rake	73 mm	43.6 mm	Cell thickness

the submodule, it can be considered as an assessment location for thermal management effect of the PCM. Figure 4.52, compares the temperature distribution in submodule thickness to verify the temperature peak in the critical rake. The negative value for the thickness in the figure is due to the coordinate location which has been set on the corner of cell 1.

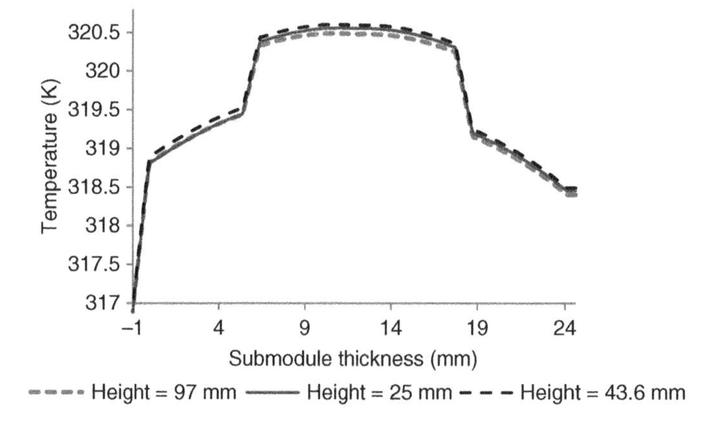

Figure 4.52 Comparison of temperature distribution along submodule thickness.

4.5.3.2 Steady-State Results of the Submodule with PCMs

For this analysis, a PCM thickness of 6 mm is selected to gain in depth information on the state-steady results. The used PCM foam encloses the submodules entirely, except for the north face. The PCM absorbs the heat dissipated by the submodules (through sensible heat storage) until the PCM reaches to the phase change temperature. As the PCM melts, the amounts of heat absorption increases considerably (latent heat storage), which mitigates the temperature rise in the submodule. Figure 4.53 shows the temperature contours in the submodule with the PCM. Once again, the rakes are created at the critical locations inside the submodule to determine the internal behavior of the submodule. In the steady state, when the PCM completely liquidifies, the lower thermal conductivity of the PCM will lose its effectiveness in extracting the heat from the submodule and behave as a thermal resistance. In the current condition, the PCM can operate in parallel to the conventional vapor compression cycle, where the coolant can extract the heat from the battery pack. Thus, the PCM, either in encapsulated form or integrated with proper foam in the form of cooling jacket, can be cooled down by an

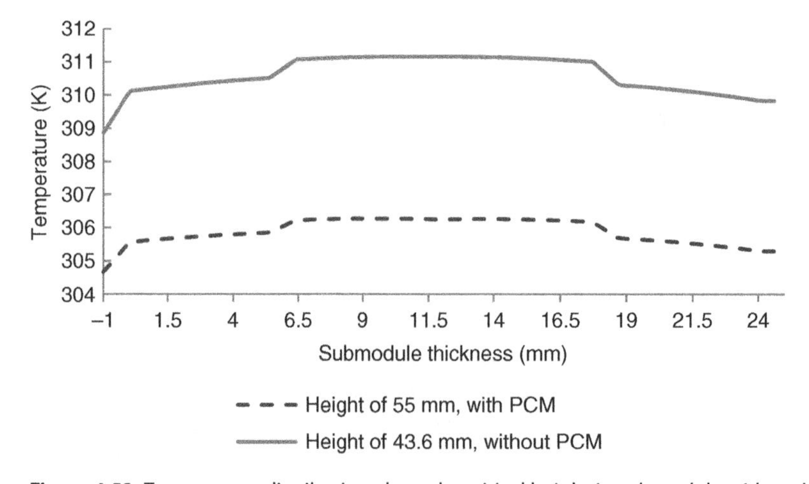

Figure 4.53 Temperature distribution along the critical height in submodule with and without PCM.

auxiliary cooling system such as heat pipe or part of the cooling loop returning from the cabin air.

When the temperatures in the submodule are evaluated in the presence and absence of PCMs, it is determined that Cell 2 has the highest temperature in the submodules (320.61 K) and the PCM jacket has the lowest (305.21 K). A second foam is then placed in between two cells as a separator to prevent direct contact between the cells. Temperature in the vertical alignment of the second cell is changed through the height of the cell. The value of zero represents the bottom of the cell while the $y = 194$ mm is the top of the cell, where the terminals are connected to the cell. Temperature is increasing in this direction and reaches a maximum value. Curve fitting of the temperature profile gives the maximum temperature location (similar to the previous case where there was no PCM).

A vertical position of 55 mm provides the maximum temperature which is located in the cell 2. A horizontal rake is created in Z-direction which includes all the 9 zones of the submodule. The value of −1 mm depicts the foam 1 zone since the mesh coordinates starts at cell 1 ($z = 0$ indicates the corner of cell 1). The temperature distributions for critical rake in cases of submodule with and without PCM are compared in Figure 4.53.

4.5.3.3 Transient Response of the Submodule

Thermal management systems using liquid and/or air cooling system are generally operated in a steady-state condition when the heat transfer rate relatively constant and/or the cooling demand is not demanding. However, during excessive charging/discharging of the vehicle and/or other operating condition where the cooling demand is higher than usual (higher battery C-rates), the compressor generally operates with higher speeds in short durations to compensate the cooling load, which makes the transient response becomes and its time-dependency more important than usual. In this regard, time independency test is carried out for the submodule mesh by selecting three different time step sizes (1, 2.5 and 5 seconds) with same total flow duration to understand its time dependency. Temperature in the middle of cell 2 has been monitored as a complementary convergence criterion. As shown in Figure 4.54, time-independency as different time steps provides negligible differences.

It should be noted that the governing boundary conditions also have impact on the submodule temperature distribution. In order to better understand this phenomenon, the submodule under the specific heat generation rate and boundary condition is evaluated in the presence and absence of PCM, where cell 2 midpoint is selected as the evaluation point. Figure 4.55 shows the impact of PCM on the temperature of this point. It can be seen that, the temperature is reduced significantly as the PCM reaches the phase change temperature. For further transient response of these models, Figure 4.56 compares the temperature distribution along the rake in height of 43.6 mm in two different elapsed times. It is seen that even though only negligible temperature difference is observed (0.5 K) after 10 minutes of operation, the difference becomes considerably more prominent after 50 minutes (7 K). Moreover, as the rake moves closer the boundaries where the PCM has been applied, it impact becomes considerably more visible in the system.

In both locations for the rake, the temperature distribution is seen to be more uniform with the integration of PCM in the submodule as seen in Figure 4.56. The melting front can be seen in Figure 4.57.

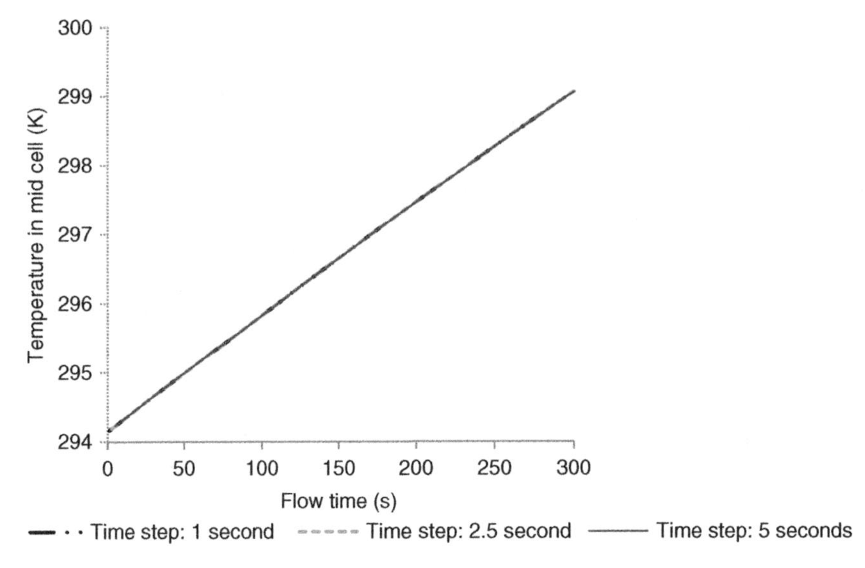

Figure 4.54 Transient response of submodule in different time steps.

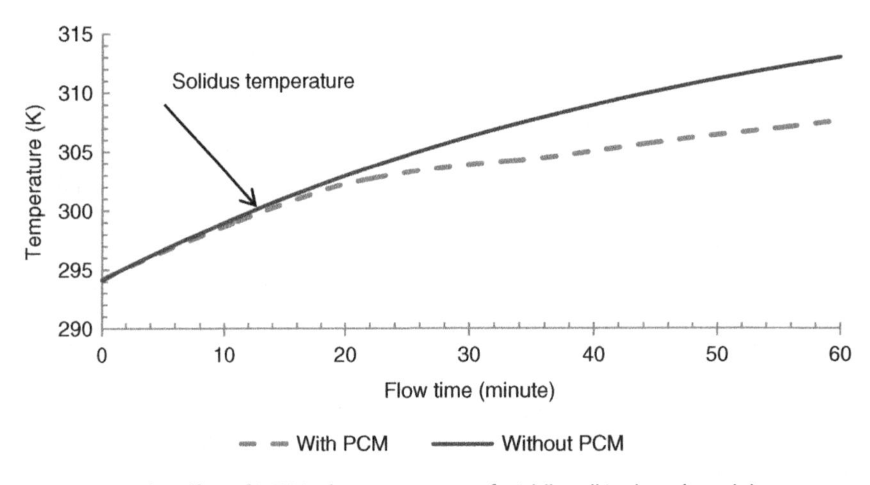

Figure 4.55 The effect of PCM in the temperature of middle cell in the submodule.

4.5.3.4 Quasi-Steady Response of the Submodule

Even though the steady-state and transient responses play an important role in determining the heat generation rates and the amount of cooling required in the battery; quasi steady-state temperature is also needed to be calculated to determine the critical intervals in the system, especially when the discharge rates and/or environmental conditions change drastically. In this case study, the duration before the system reaches its quasi steady-state becomes the main area of concentration. In the examined submodule, the steady-state temperature in cell 2 center is determined to be 320.58 K. If 85% of the steady-state temperature is defined as the quasi steady-state situation and initiating temperature considered to be 294.15 K, then the time it is required for the submodule to reach the temperature of 22.46 °C is quasi steady-state time period

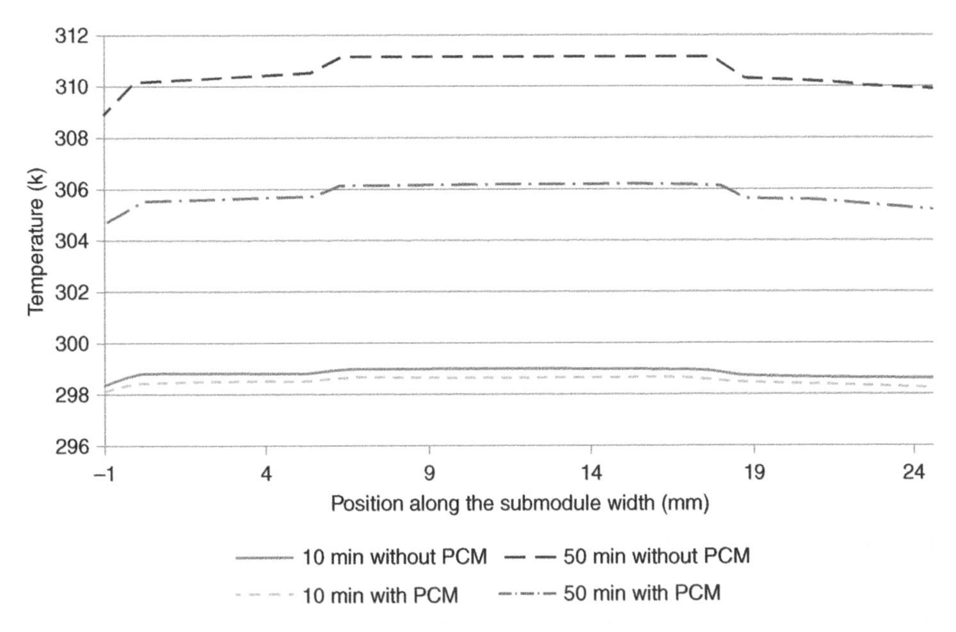

Figure 4.56 Time-dependent temperature behavior of the middle cell in the submodule.

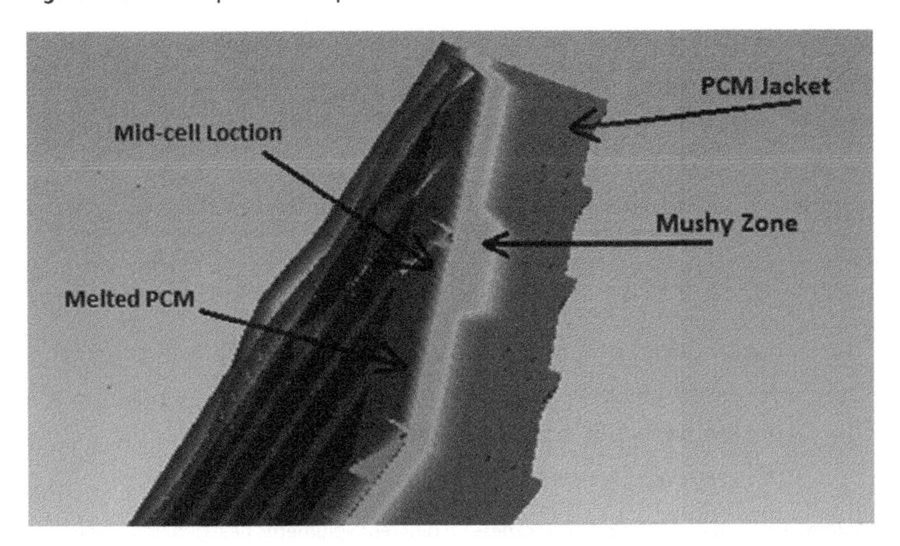

Figure 4.57 Transient melting behavior of PCM around submodule.

(316.61 K). As Figure 4.58 shows, the quasi steady-state temperature is satisfied within 3 hours of monitored flow time.

Figure 4.59 shows the final temperature of the submodule under various heat generation rates. Commonly, a heat generation of 6,866 W/m^3 is being considered for the Li-ion cells, where final temperature is not sufficient to change the phase of the PCM in the analyzed submodule. There are two reasons why one should not select a different PCM with lower melting temperature. The first reason is the thermal runaway problem in the battery pack. Secondly, in 2-C or higher discharge rates, which is

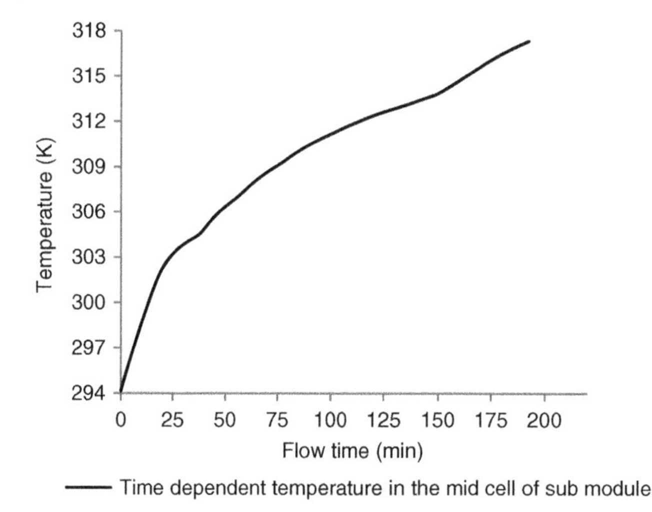

Figure 4.58 Quasi steady-state temperature dependence of submodule for heat generation of 22,800 W/m^3.

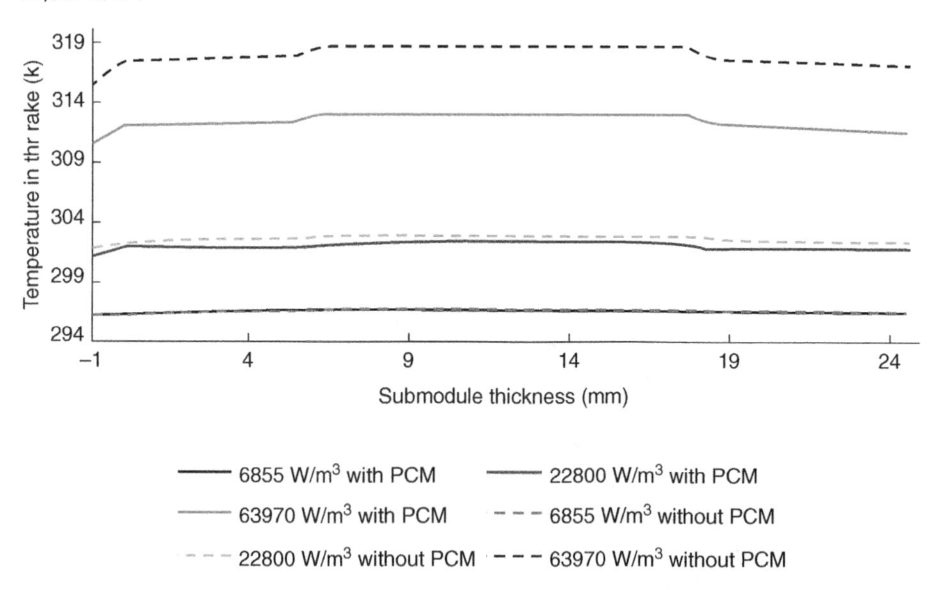

Figure 4.59 Temperature along submodule thickness for different volumetric heat generation rates.

possible in acceleration period, the 6855 W/m^3 will cause the submodule to attain temporarily high temperatures which can keep the temperature levels higher than the predicted temperature even after getting to its standard operating condition. It is necessary to mention that in higher heat generation rates, the PCM has an incredible effect on cooling down or preventing temperature rise in the system. In the various simulated heat generation rates, the results show that for the rake along the submodule thickness, the temperature will decrease more in higher heat generation.

The temperature rise in cell 2 mid-point with respect to flow time under various heat generation rates are shown in Figure 4.60. It can be seen that, the cell temperatures can

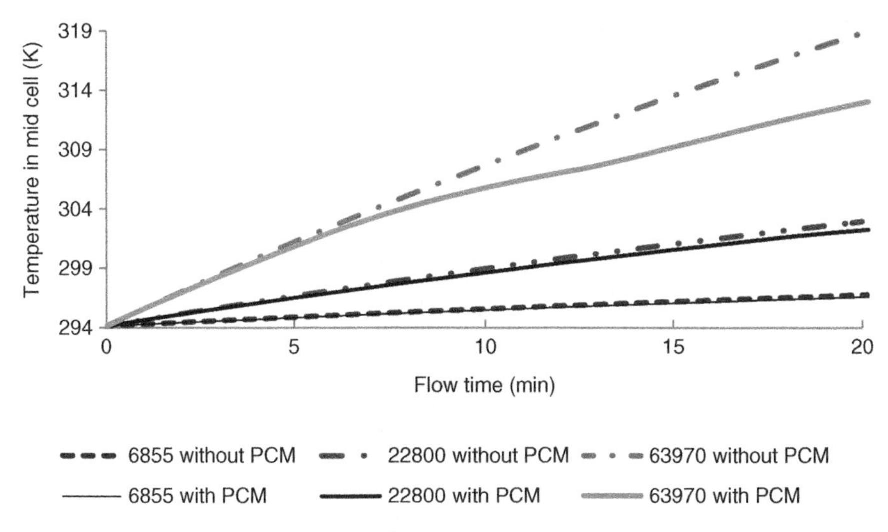

Figure 4.60 Temperature increase in mid-cell under various heat generation rates.

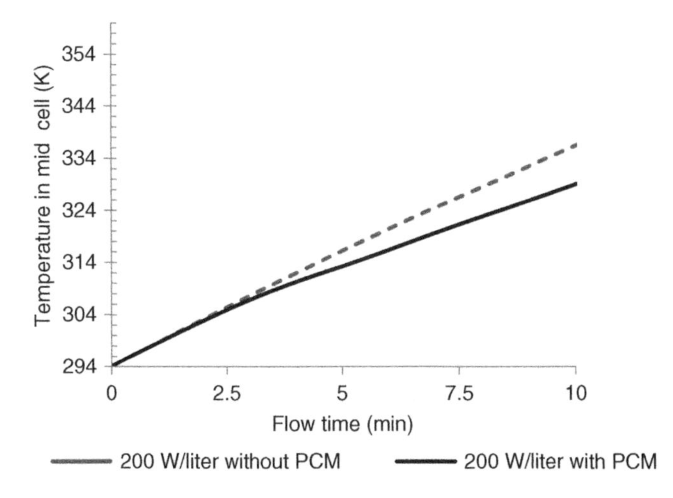

Figure 4.61 Submodule response for the higher heat generation rate in battery pack (200 kW/m^3).

rise significantly (especially in the absence of PCM) when exposed to excess battery heat generation rates (200,000 W/m^3) during rapid battery discharge under brief periods of time. Figure 4.61 compares the response of the cell 2 to similar heat generation rates in transient situation.

4.5.3.5 Model Validation

Once the numerical models are developed and the simulations are conducted, the next important step is to validate these outputs by comparing with the data acquired from the experimentations by the experimental set up and procedures described in Section 5.2.

In order to assess the temperature increase on the cell surface, battery cycler is used for the charging voltage and current extraction (in C/1 –Rate) to the Li-ion cells. Figure 4.62

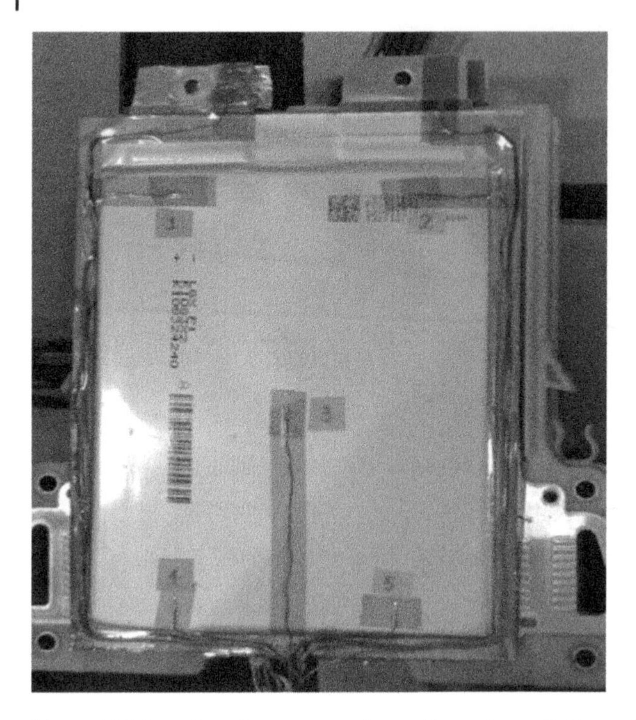

Figure 4.62 Location of thermocouples on the surface of Li-Ion cells.

shows the five locations where the thermocouples are placed for measurement purpose. Transient response of the cells was analyzed in two situations. First, the submodule was simulated without the PCM in the foam. In this case, the foams in the submodule act as a separator to prevent cell surfaces to contact. Then the foam was soaked in the PCM and tested as a submodule under the same operating conditions.

Figure 4.63 shows the all temperature profiles in the 10 locations on the cell surfaces after filtering the noise in the data. Figure 4.64 reveals the temperature variations in the specified locations. Due to the symmetrical position of the thermocouples, points 3 and 5 are selected to simplify the monitoring process of their transient temperature increase during the testing period.

The quasi steady-state has been considered and the elapsed time has been selected to be 3 hours. The same C-rate is applied for the case of dry foam and the foam soaked with the PCM. As it can be seen from Figure 4.63 and based on the simulation results, in both points, the surface temperature of the Li-ion cell decreases by replacing the dry foam with the foam soaked in PCM.

When the acquired experimental data is compared with the results obtained from the conducted simulations, it can be seen that the heat generation inside the cell, propagation in the submodule and the melting process of the PCM is predicted with a high accuracy. This allows for conducting further simulations in scenarios where experimental testing is not feasible, expensive and/or takes long time. Thus, having simulations and experiments in this parallel structure enables the decision makers with the right tools to obtain the necessary data to find ways of improving their BTMSs.

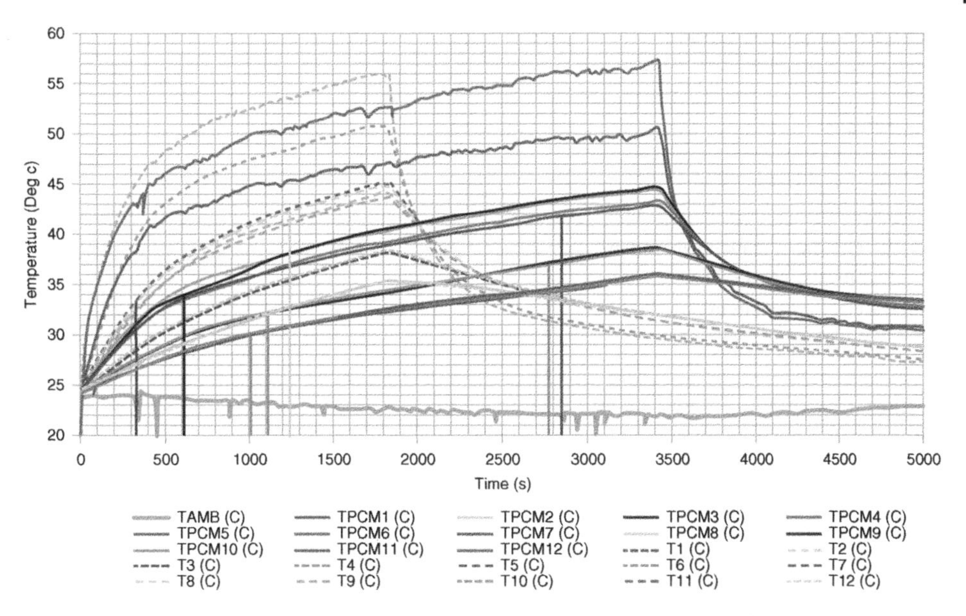

Figure 4.63 Temperature variations for all 10 points on both sides of the cell with and without the PCM.

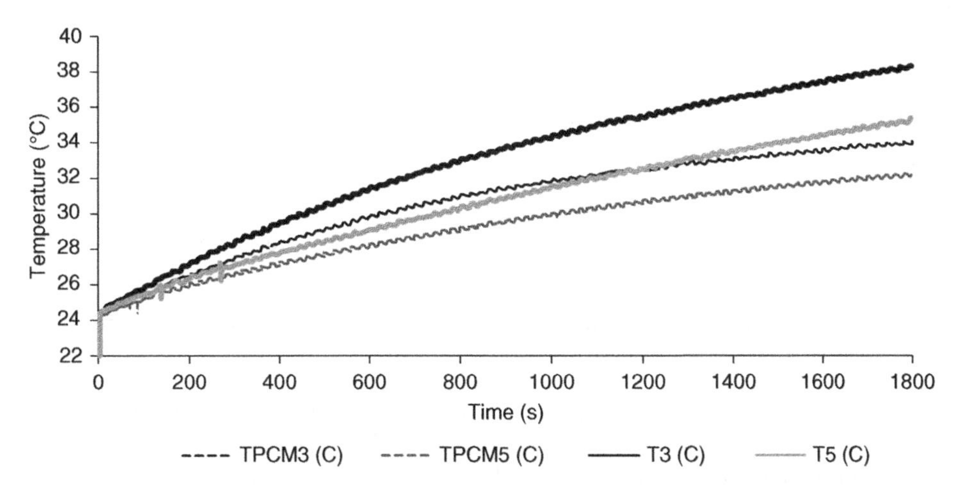

Figure 4.64 Experimentally measured cell temperature with and without PCM in between the cells.

4.5.4 Optical Observations

4.5.4.1 Thermal Conductivity Enhancement by Nanoparticles

In this section the effects of adding carbon nanotubes and grapheme nano-platelets will be presented using differential scanning calorimetry (DSC) tests for the readers to provide them with information on increasing the thermal conductivity of their phase change materials. At the same time, a method is developed to calculate and compare the effect CNT and grapheme additives in the PCM based on the DSC data. DSC method for the 4 samples has been carried out. Table 4.12 shows these tested and analyzed samples.

Table 4.12 Samples of PCM and nano-particles prepared for the tests.

No.	Sample
1	Technical Grade PCM
2	Pure PCM
3	6% Graphene platelets mixed in technical grade PCM
4	6% CNT mixed in in pure PCM

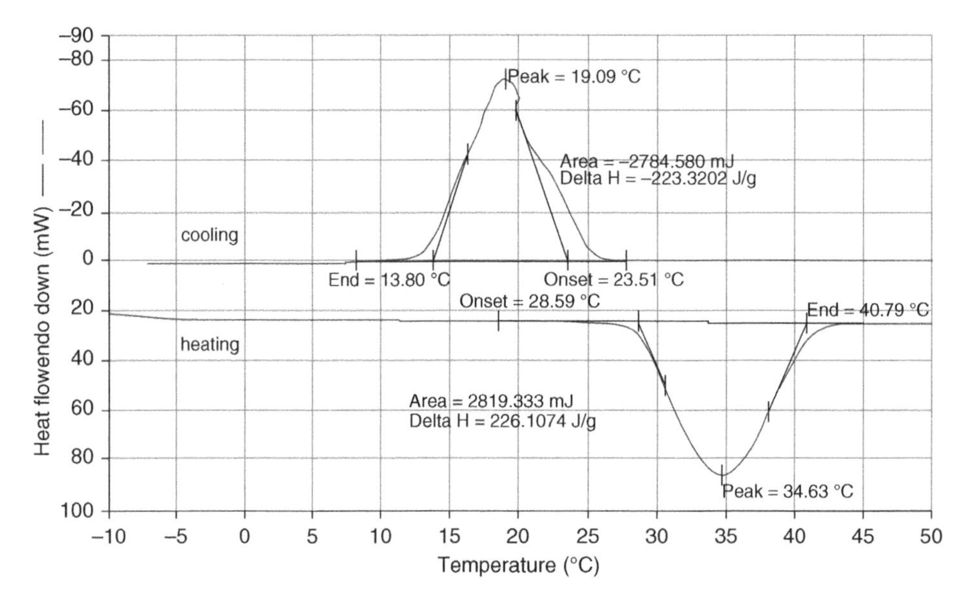

Figure 4.65 Differential scanning calorimetry results for pure n-octadecan (99%) in heating and cooling periods.

The properties of CNT and grapheme nano-platelets are briefly described in the previous chapters. The used Carbon Nano-tubes (CNT) are 8–10 nm in diameter and 10–50 micrometer in length and used octadecane can be either pure or technical grade, (with the purity of 99.66% and 90.8%, respectively in the case study). Cooling of the samples starts from 55 °C to −10 °C at 10 °C/min temperature ramp. Figures 4.65 and 4.66 show the heating and cooling DSC results for pure and technical grade octadecane PCMs, respectively.

Nanoparticles are added to the pure and technical paraffin as previously described. The same methodology is applied in order to obtain the DSC diagrams of these mixtures of PCM and nanoparticles. The related DSC graphs are given in Figure 4.67 and Figure 4.68. One result that can be deduced comparing these DSC graphs is that the reduced latent heat of fusion of the pure material by increasing the impurities. This value for pure PCM is 226,107 J/kg, which has been reduced to 187,322 J/kg for technical grade PCM as shown by the area under the power curve and horizontal axis.

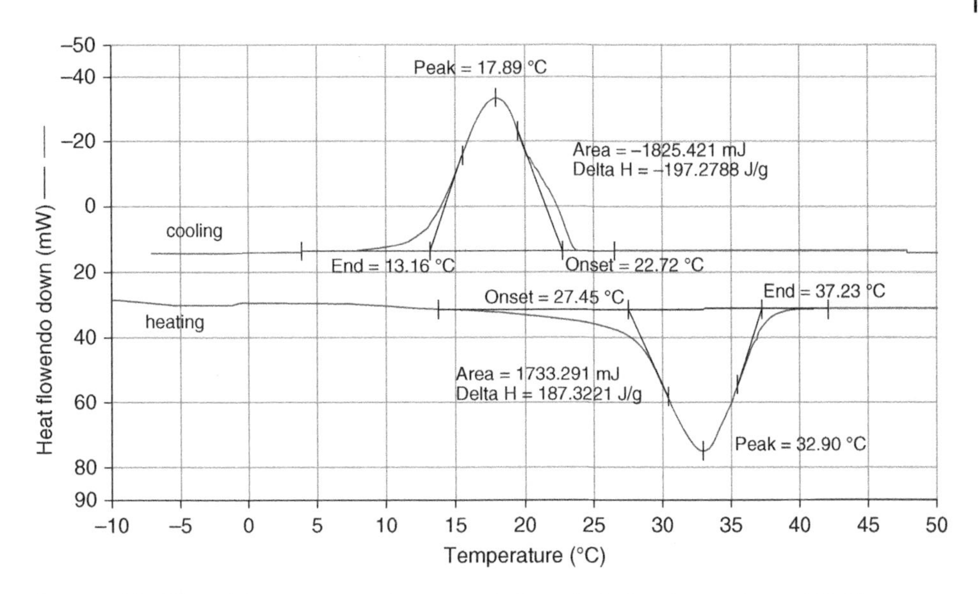

Figure 4.66 Differential scanning calorimetry results for technical grade octadecane (90.8%).

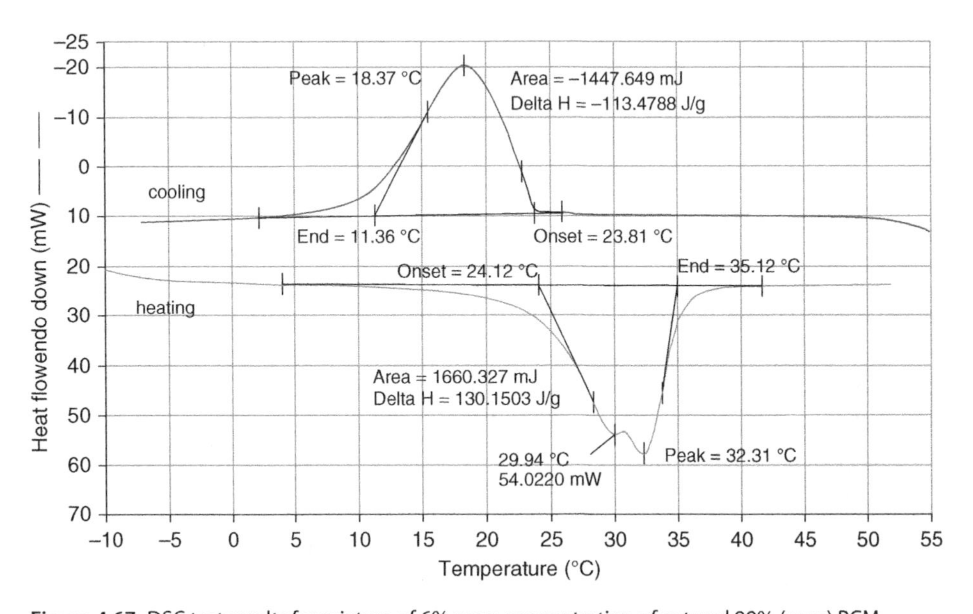

Figure 4.67 DSC test results for mixture of 6% mass concentration of cnt and 99% (pure) PCM.

In order to calculate the thermal conductivity based on the DSC test for specific heat, the following relations are utilized:

The relationship below can be obtained by solving the heat differential equation (Camirand, 2004):

$$\frac{d(\Delta P)}{dT} = \frac{2}{R} \tag{4.59}$$

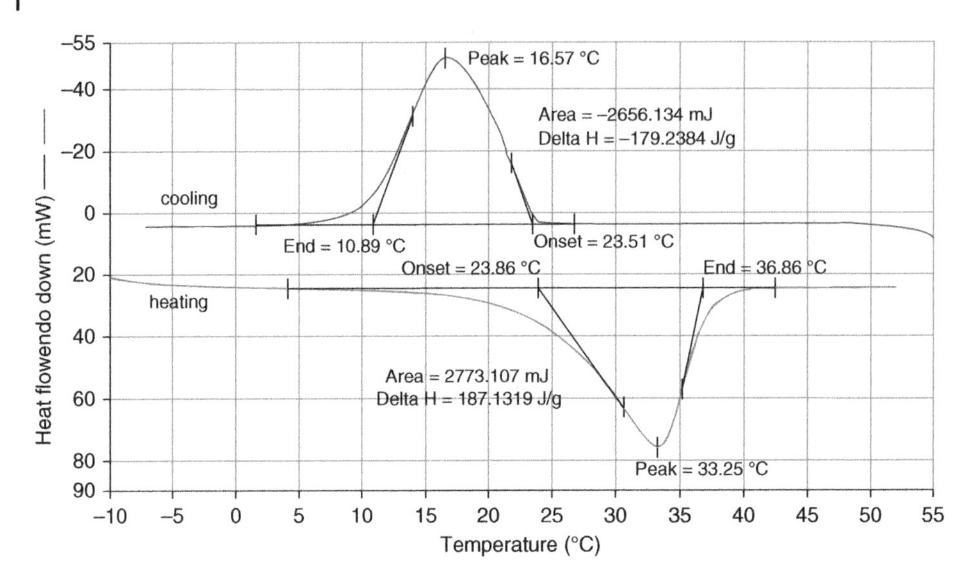

Figure 4.68 DSC test results for the mixture of 6% graphene platelets mixed with technical-grade PCM.

The slope is considered in general form as it can be seen from DSC diagrams for the samples. In general, it can be written in the following form:

$$\frac{d(\Delta P)}{dT} = \frac{C_1}{R} \tag{4.60}$$

where ΔP represents the differential power of calorimeter and R is the total thermal resistance.

$$R = R_1 + R_2 + R_s \tag{4.61}$$

Here, R_1, R_2 and R_3 are thermal contact resistances between the sample and the sample furnace, the thermal contact resistance between the sample and the calibration substance, and thermal resistance of the sample, respectively.

The device related resistance C_2 is defined as $C_2 = R_1 + R_2$

$$\frac{d(\Delta P)}{dT} = \frac{C_1}{R_1 + R_2 + R_3} = \frac{C_1}{C_2 + R_s} \tag{4.62}$$

The thermal conductivity can be calculated from the following thermal contact resistance relationship:

$$R_s = \frac{L_s}{k_s A_s} \tag{4.63}$$

where L_s, k_s, *and* A_s are the height of the sample, thermal conductivity of the sample and a horizontal cross-section area; respectively. By integrating $C_3 = \frac{L_S}{A_S}$ into the previous equation,

$$\frac{d(\Delta P)}{dT} = \frac{C_1}{C_2 + \frac{C_3}{k_s}} \tag{4.64}$$

$$\frac{C_3}{k_s} = \frac{C_1}{\frac{d(\Delta P)}{dT}} - C_2 \tag{4.65}$$

By arranging and re-writing the corresponding values for PCM and mixture of PCM and CNT:

$$k_{s}, {}_{PCM} = \frac{C_3 \frac{d(\Delta P)_{PCM}}{dT}}{C_1 - C_2 \frac{d(\Delta P)_{PCM}}{dT}} \tag{4.66}$$

$$k_{s}, {}_{PCM+CNT} = k_{s,mix} = \frac{C_3 \frac{d(\Delta P)_{mix}}{dT}}{C_1 - C_2 \frac{d(\Delta P)_{mix}}{dT}} \tag{4.67}$$

The ratio of thermal conductivities is

$$k_{ratio} = \frac{\frac{d(\Delta P)_{mix}}{dT}}{\frac{d(\Delta P)_{PCM}}{dT}} \cdot \frac{C_1 - C_2 d\left(\frac{(\Delta P)}{dT}\right)_{PCM}}{C_1 - C_2 d\left(\frac{(\Delta P)}{dT}\right)_{mix}} \tag{4.68}$$

It can be noticed that the following ratio is always greater than unit.

$$\frac{C_1 - C_2 d\left(\frac{\Delta P}{dT}\right)_{PCM}}{C_1 - C_2 d\left(\frac{\Delta P}{dT}\right)_{mix}} > 1 \tag{4.69}$$

If the terms C_1 and C_2 are considered, then, it can be observed that inequalities in these equations are always satisfied as explained below.

$$C_1 - C_2 d\left(\frac{\Delta P}{dT}\right) > 0 \tag{4.70}$$

$$\left(\frac{\Delta P}{dT}\right)_{mix} > \left(\frac{\Delta P}{dT}\right)_{PCM} \tag{4.71}$$

In order to have a positive thermal conductivity, inequality in this equation should be positive. The second inequality is true since the results of the experiments also show that the slope of power to temperature for sample with CNT is higher than that of pure PCM and indicates that carbon additives will lead to higher thermal conductivity, which implies that the ratio will be positive. Based on the given calculations, the ratio of effective thermal conductivity of 99% CNT to technical grade (90.8%) octadecane is shown in Figure 4.69.

Comparing 6% mass concentration CNT mixed with PCM and the same concentration of graphene platelet mixed with PCM is shown in Figure 4.70. It should be noticed that the values needs modification in terms of multiplication in a coefficient. As it was mentioned, this coefficient is a positive number and greater than one. This means that even if the ratio become near 1, after multiplication by this coefficient, it will be improved.

Dependency of this coefficient to other parameters is an important concept. It is worth to mention that the ratios for other temperatures are not shown in Figure 4.70, since this ratio is only for the phase change region. Therefore, for the solid part, which is expected to have higher thermal conductivity due to CNT presence, other methods are needed to be used. Figure 4.70 also shows that the thermal conductivity of CNT is approximately

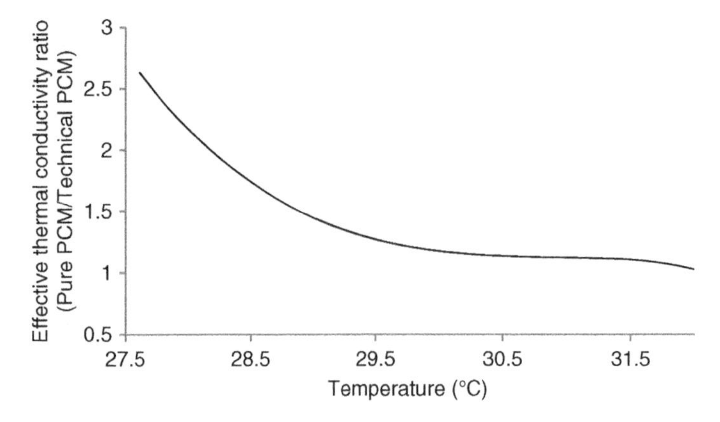

Figure 4.69 Ratio of effective thermal conductivity of 99% CNT to technical grade (90.8%) octadecane.

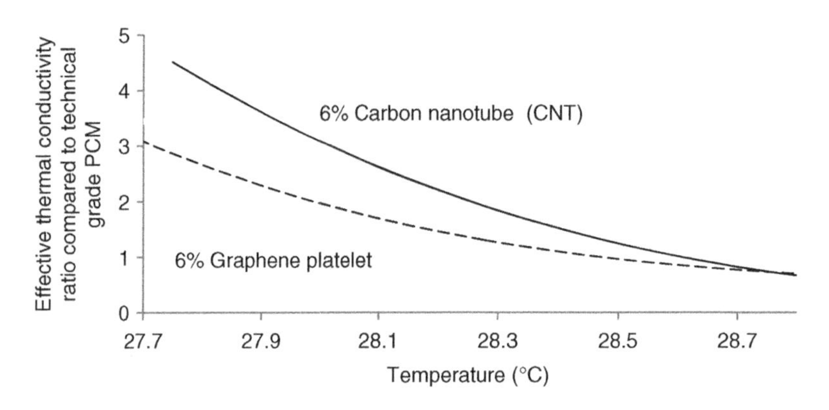

Figure 4.70 Comparing CNT and platelets of graphene-effective thermal conductivity.

1.6 higher than graphene platelets at 27.9 °C. Moreover, the thermal conductivity can be up to five times greater than that for technical PCM, as seen by the test results.

4.5.4.2 Data For the Case of Pure PCM (99% Purity)

Temperature in different locations in the pure PCM is shown in Figure 4.71. The time starts from the moment that hot water enters to the heat exchanger.

In case of pure n-octadecane, phase changing process takes shorter time than that for technical grade PCM. The pure PCM behaves differently in melting process. While the technical grade (Figure 4.69) takes longer to absorb, the pure PCM shows a sharper increase. This can also indicate that melting effect of the pure PCM and interaction of neighborhood tubes carrying the hot flow is faster than the case for technical grade. The experimental results presented in this case study also shows higher thermal conductivity for pure PCM compared to technical PCM as shown in (Figure 4.72).

4.5.4.3 Optical Microscopy Analysis of the PCM and Nanoparticle Mixture

In the previous sections, simulations observing the thermal properties of applying PCM around the cell, in between cells and around the submodule are analyzed. Even though

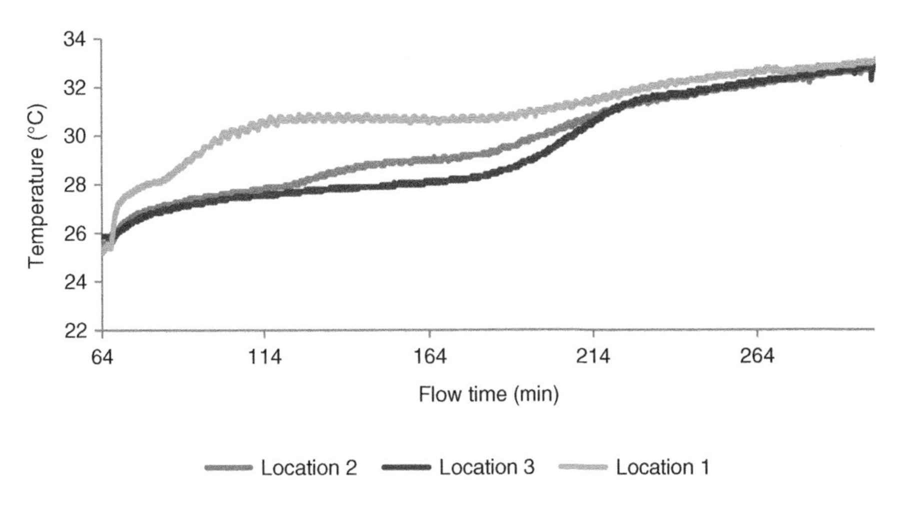

Figure 4.71 Temperature in locations 1, 2 and 3 in the case of pure PCM.

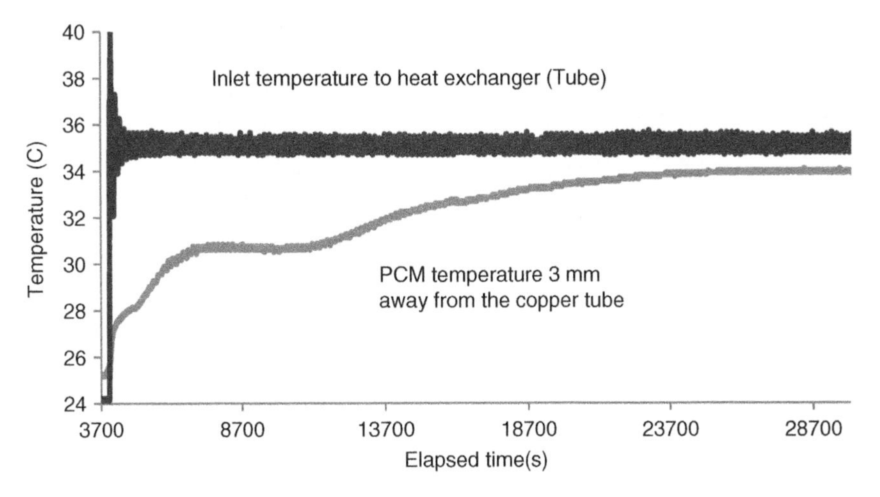

Figure 4.72 Temperature increase in the pure PCM compared to the inlet temperature.

determining the rate of heat dissipated from the cells and the volume of PCM needed to reduce the temperature play an important role in designing effective BTMSs, other approaches such as optical microscopy can also add significant benefits to the analysis by looking at the system from the PCM point of view, as the optical microscopic method can provide important insights of the CNT configuration in the PCM mixture. This is a type of light microscopy where the lights pass through the source and then incidence onto the lens. Optics properties of PCM and mixture along with their morphological characteristics are obtained through this method. CNT and graphene platelets mixed with pure and technical grade PCM, are prepared and studied.

Figure 4.73 shows an optical image of pure PCM (99% purity). Prior to melting, structural morphology demonstrates granulated and smooth surfaces. The granulated part is formed during the solidification at a high cooling rate, which indicates the initial solidification region in the PCM. It should be noted that initial temperature difference

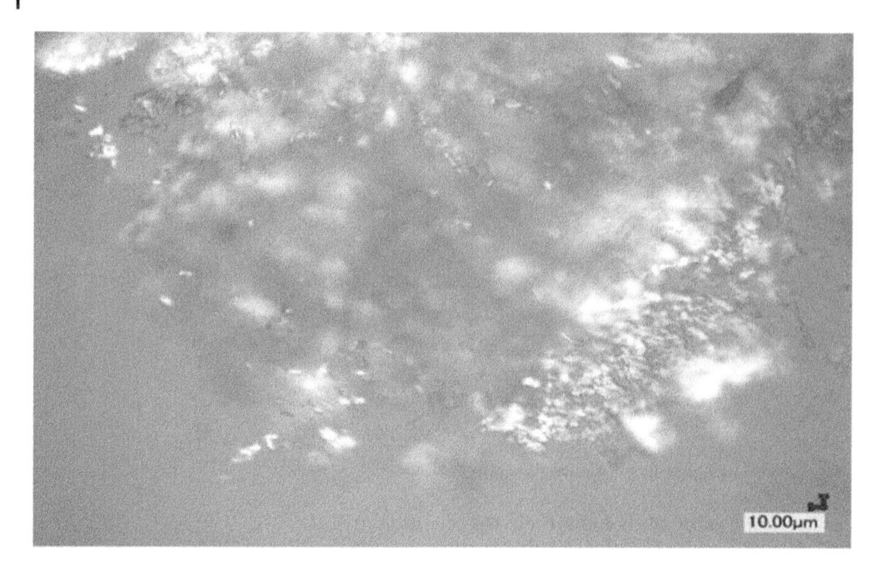

Figure 4.73 Optical image of pure PCM.

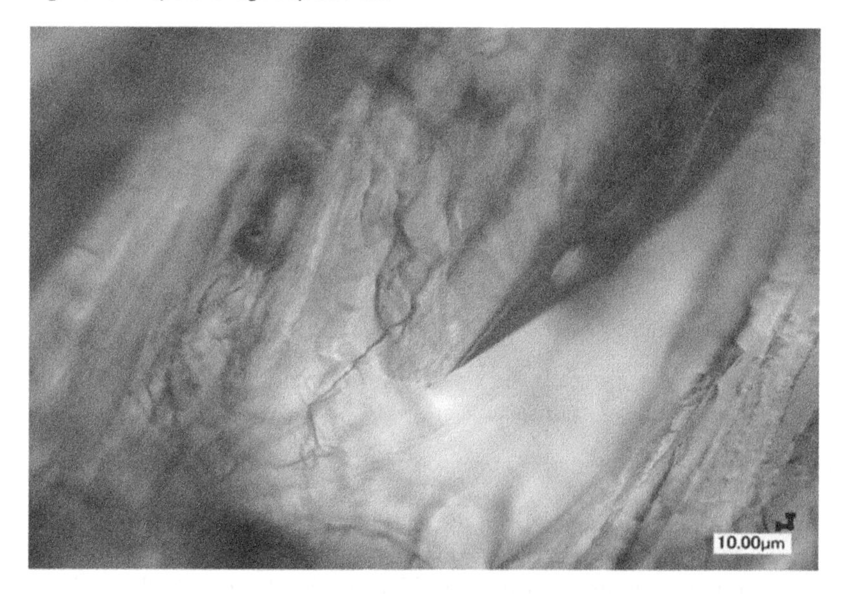

Figure 4.74 Structure of technical-grade PCM with x500 magnification using optical microscope.

in the PCM should be sufficiently high to initiate the solidification process in line with the solidification characteristic (Bejan, 2013).

Figure 4.74 shows an optical micrograph of technical grade PCM in solid phase. The surface morphology dictates the presence of irregular structure, closely situated at the surface. The distinguishing feature appears to be the boundaries in the heterogeneous structure. These boundaries can act as thermal boundary resistance suppressing the heat flow across the structure similar to grain boundaries in solids. Therefore, the melting initiation and competition is longer than that of the case observed for pure PCM.

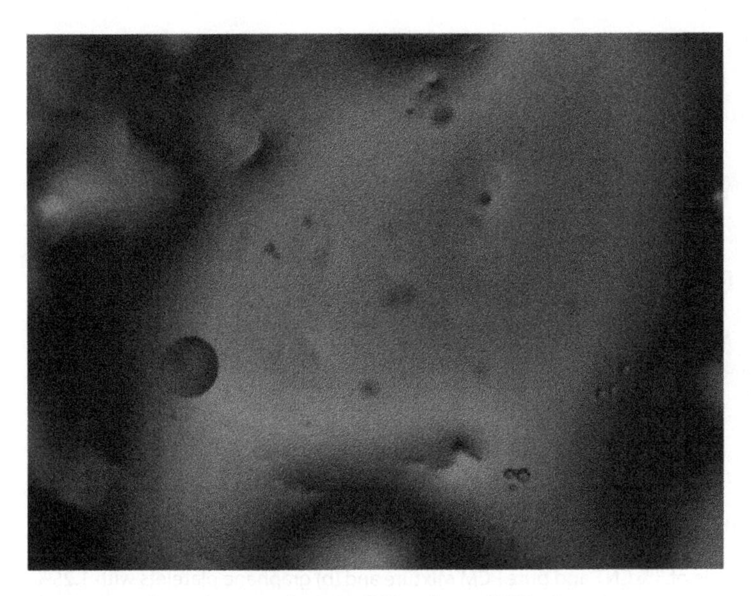

Figure 4.75 Optical image of 1.25% CNT and pure PCM mixture.

Figure 4.75 shows an optical image of 1.25% of CNT and pure PCM mixture onset of melting. The morphology of the structure reveals the minor and local scattered agglomeration of carbon nanotubes. The coverage area of agglomeration is only a small fraction of total are, shown in Figure 4.75. Agglomerated CNT appears to be dark inclusions with round appearance as it can be seen from the figure. It should be noted that the maximum magnification of the lens used in the optical microscope is not capable of capturing the images of the nano-sized particles. The image presented has the resolution of 10 micrometer per 0.5 cm on the captured picture. This only allows observing the agglomerated nano carbon tubes or graphine pellets on the image. Figure 4.76a shows 3% CNT and pure PCM mixture. The morphology of the image shows locally agglomerated CNT with circular appearance at the surface. In addition, partially dissolved solid structure is also visible in the photographic image. The size of CNT agglomerations is different at the surface.

However, for all sizes, spacing between inclusions is not close enough to merge forming a large size clustered CNT regions in the PCM. The morphology in Figure 4.76b indicates small size and large number of agglomerated platelets at the solid surface. This figure shows a 1.25% of graphene platelet and technical PCM mixture. The presence of small size agglomeration is associated with the small Van der Waals and other small forces such as surface tension forces. However their increased number in the PCM is expected to improve the thermal conductivity.

Figure 4.77 shows 6% of CNT and pure PCM mixture onset of melting. The radiation source is provided to melt the PCM during the microphotography. Since the absorption characteristic of CNT and PCM are different for the incident radiation, the topology of the image shows fine hilly like structures. In this case, the volume concentration of the CNT is high and agglomerated size of the CNT becomes large.

This, in turn, absorbs more radiation and generates more heat in these regions (compared to the corresponding PCM). Consequently, the local melting in regions with

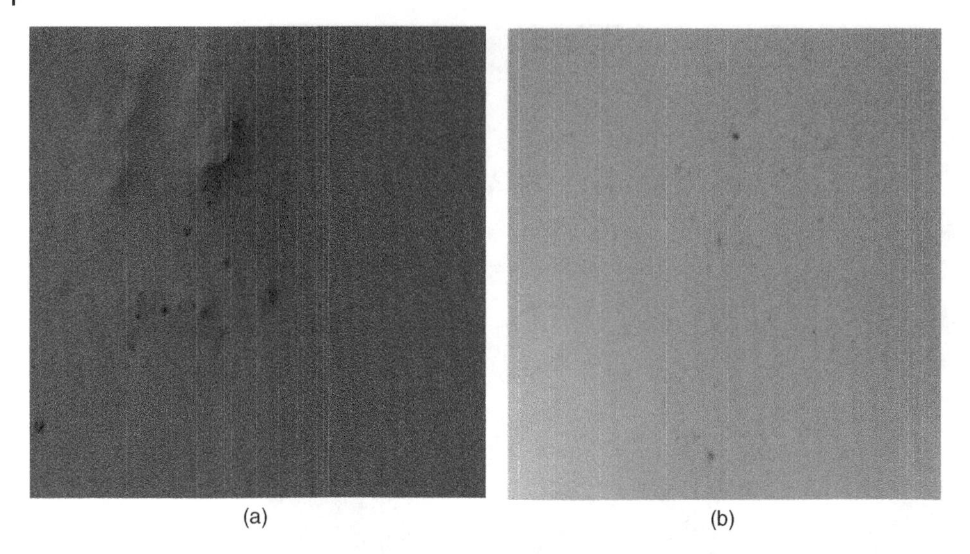

Figure 4.76 (a) Optical image of 3% CNT and pure PCM mixture and (b) graphene platelets with 1.25% mass fraction mixed with technical PCM.

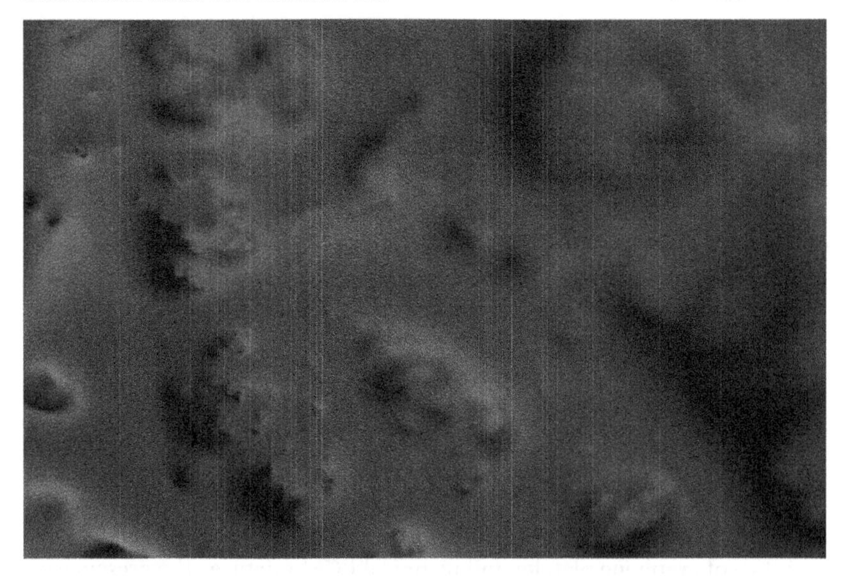

Figure 4.77 Optical image of 6% CNT and pure PCM mixture.

high PCM concentration forms the valley-like structures in the topology. Nevertheless, this structure is scattered without interconnecting at the surface, while indicating no overall merging of agglomerated CNT sites.

Figure 4.78 shows transmission optical image of 1.25% of CNT and pure PCM when it is partially melted. The melted regions appear to be bright color due to higher transmittance characteristics of molten PCM than solid PCM because of incident radiation emanating from optical microscope. The image shows the convection current forming streamline flow characteristics in the molten phase. This shows a feather-like appearance in the bright region of the image.

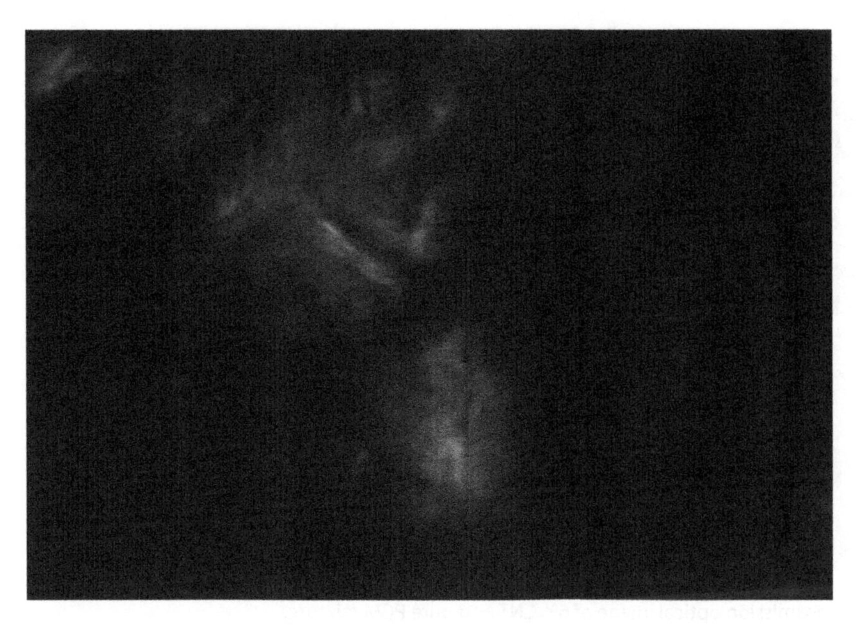

Figure 4.78 Transmission optical image of the 1.25% CNT and pure PCM mixture.

The convection current is formed due to the temperature and density variations in the molten phase of PCM. It is also possible that melted PCM (n-octadecane) molecules can enter into the nanotube while modifying the buoyancy forces. The molecular diameter (length of the molecular chain) is in the order of 1 nm (Sun *et al.*, 2007) which is smaller than the carbon nanotube (CNT) inner diameter (10-30 nm). The reduced buoyancy force facilitates CNT participating in convection current. This enhances the optical image at regions where the convection current is developed. Moreover, streamlined flow enhances the concentration of CNT in this region which further increases the absorption of incidence radiation emanating from the optical microscope. Hence, temperature variations in this region further enhances the convection current intensity. Figure 4.79 shows transmission image of 6% CNT and pure PCM for partially molten mixture. The image consists of dark and bright regions. Dark regions represent solid phases whereas the bright regions correspond to the liquid ones.

In the bright regions, scattered patterns of radiation are emitted from the microscope. This scattered pattern is circular and its origin is associated with the presence of CNT agglomeration sites. Concentrated CNT regions undergo early melting because of high rate of absorption of incident radiation, emanating from the microscope. These regions melt sooner, therefore, the dark regions in the vicinity of the bright region indicates the presence of less concentration of CNT sites. This behaviour inherently indicates that high concentration of CNT causes: i) reduction in the latent heat of melting, and ii) enhancement of the thermal conductivity. Figure 4.80 shows the effect of stainless steel mesh with 20 μm mesh size. Observations show that the mesh decreases the convection flow of the graphene platelets and CNTs. The agglomerated masses of nanoparticles are seen, beneath the mesh in the figure.

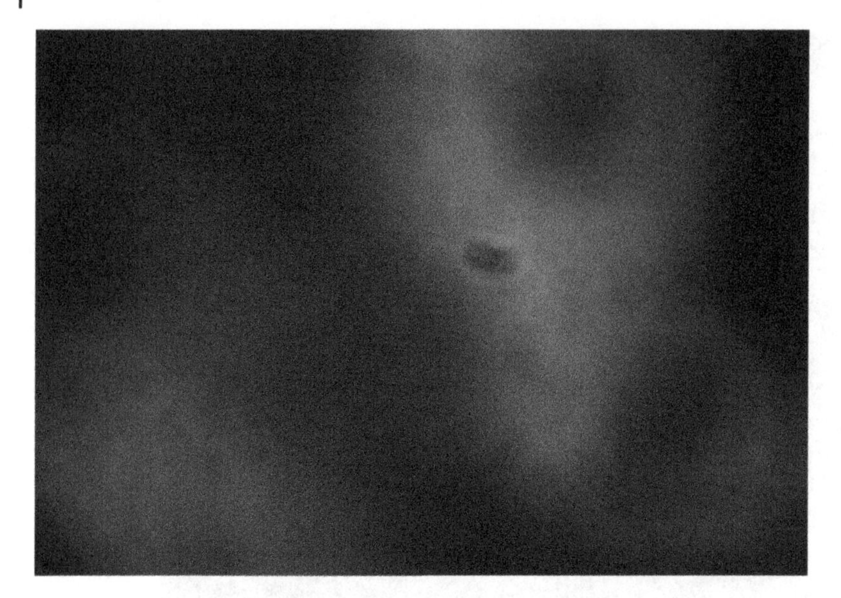

Figure 4.79 Transmission optical image of 6% CNT and pure PCM mixture.

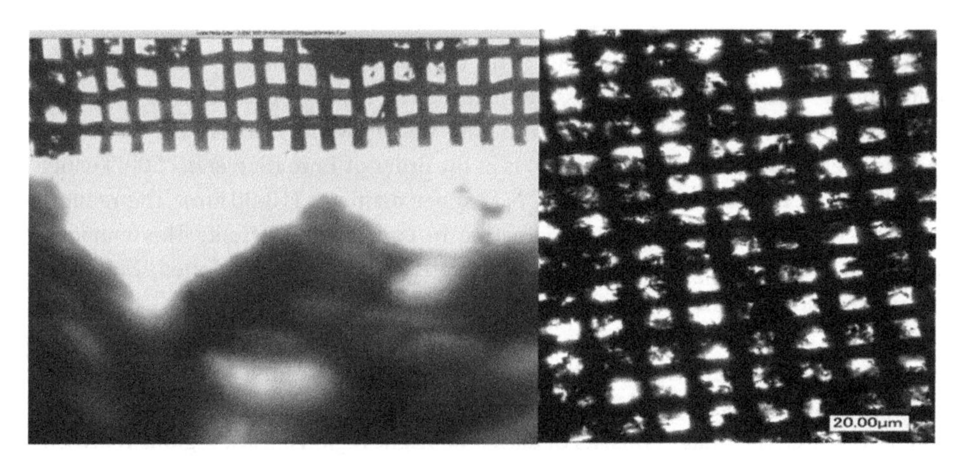

Figure 4.80 Effect of metal micro-mesh on agglomeration of nano-particles.

4.5.5 Vehicle Level Experimentations

Once the test bench and vehicle is set up and established using the aforementioned methods, procedures and data points using the data acquisition software and hardware described in Section 4.3 and the test scenarios provided in Section 4.4; experimental data is gathered to verify the accuracy and reliability of the conducted simulations in vehicle level. The specific values and trends in the collected data provide valuable insights on the inner workings of the vehicle under different driving conditions and operating environment and the corresponding response and performance of the BTMS. The results of some of the experimentations conducted on the test bench and the test vehicle are shown in Figures 4.81–4.90 and Figures 4.91–4.106, respectively.

4.5.5.1 Test Bench Experimentations

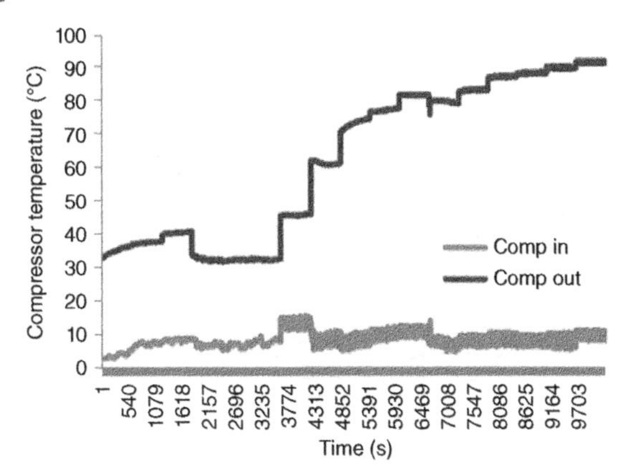

Figure 4.81 Refrigerant temperature before and after the compressor.

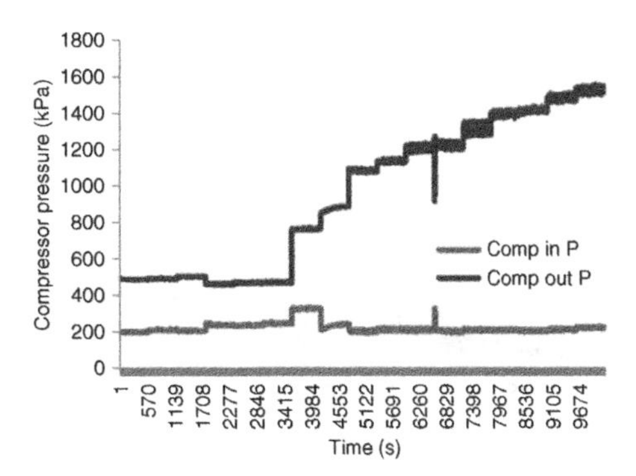

Figure 4.82 Refrigerant pressure before and after the compressor.

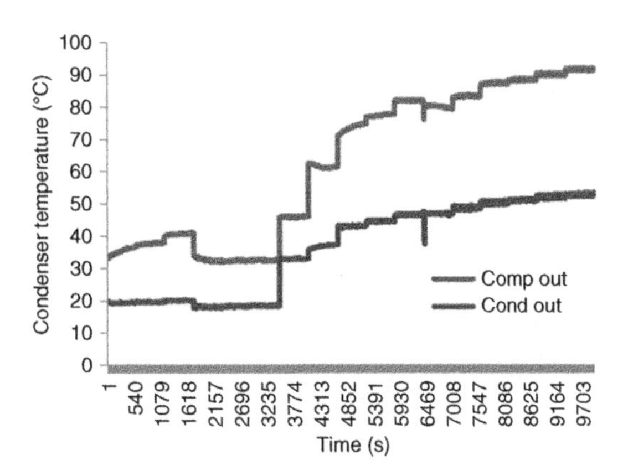

Figure 4.83 Refrigerant temperature before and after the condenser.

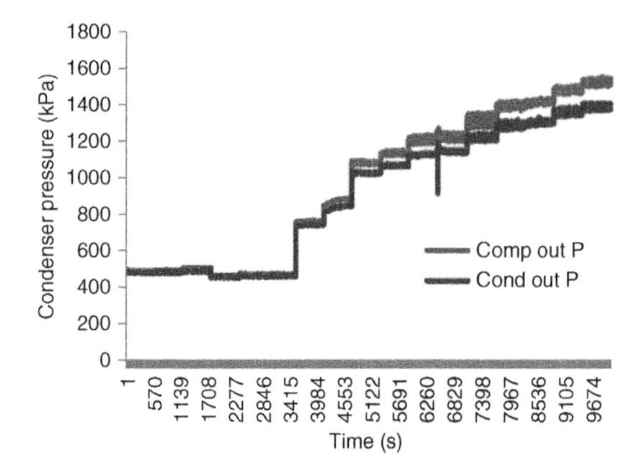

Figure 4.84 Refrigerant pressure before and after the condenser.

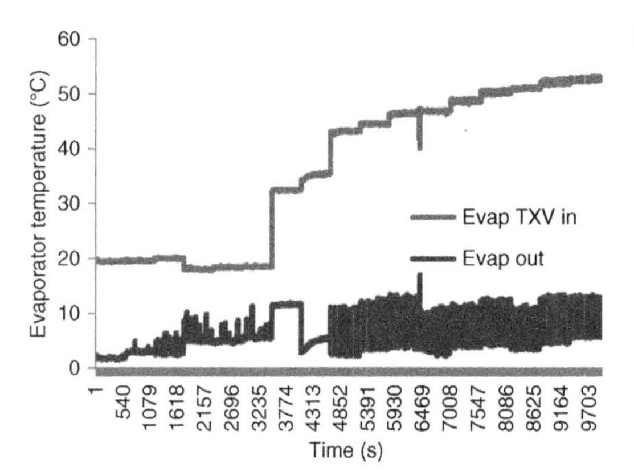

Figure 4.85 Refrigerant temperature before and after the evaporator.

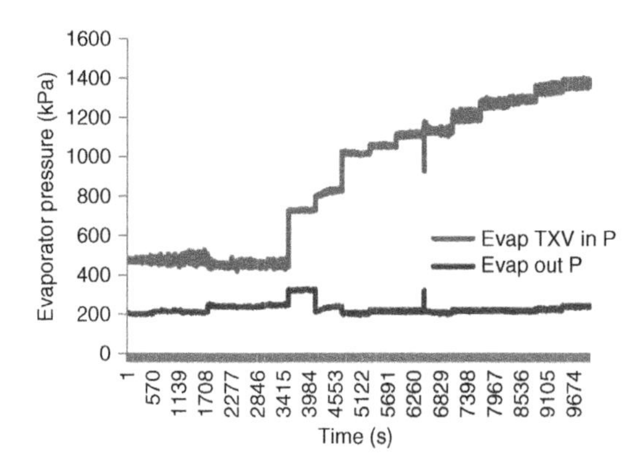

Figure 4.86 Refrigerant pressure before and after the evaporator.

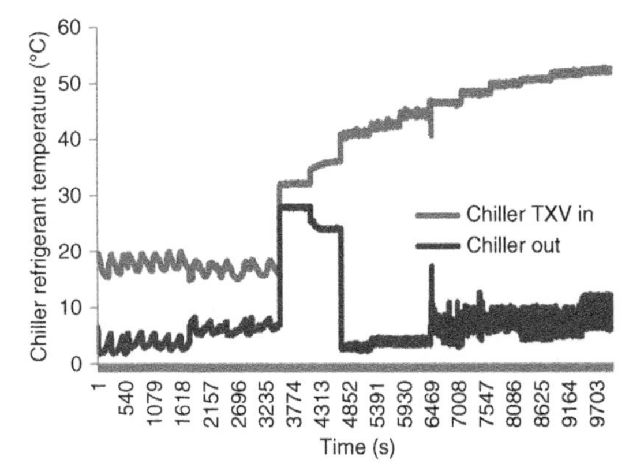

Figure 4.87 Refrigerant temperature before and after the chiller.

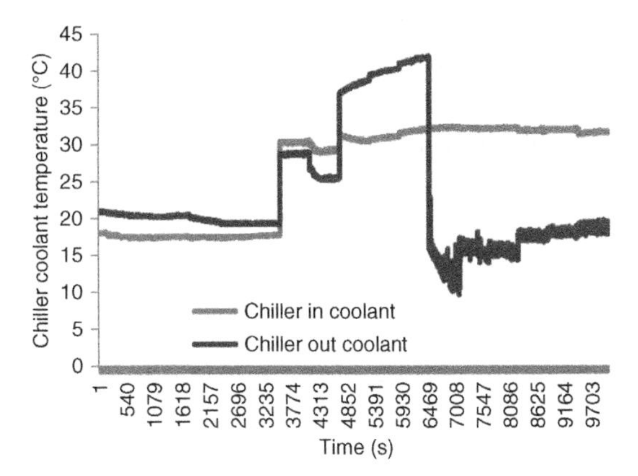

Figure 4.88 Coolant temperature before and after the chiller.

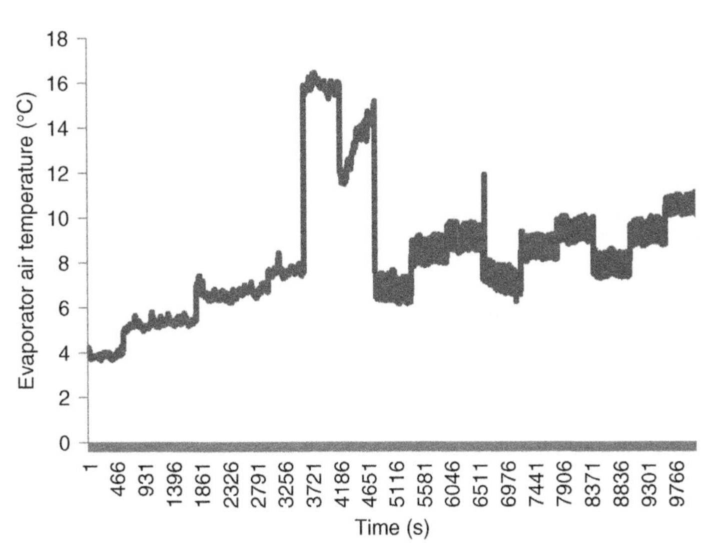

Figure 4.89 Refrigerant temperature before and after the compressor.

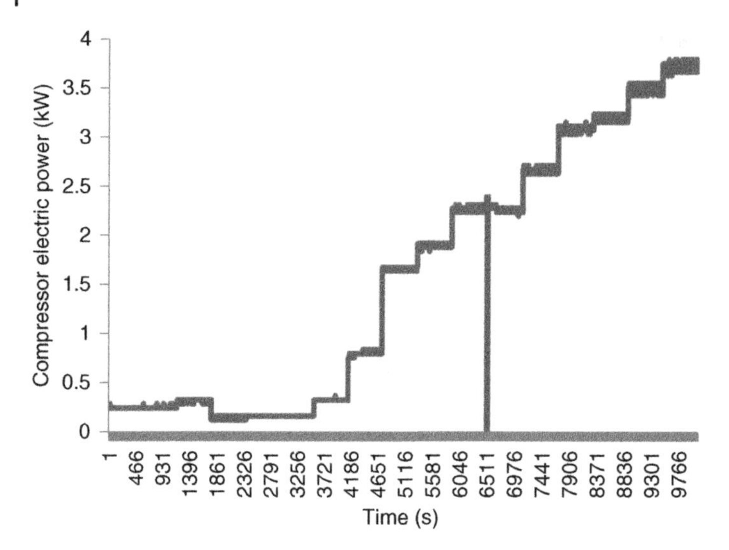

Figure 4.90 Compressor electric power.

4.5.5.2 Test Vehicle Experimentations

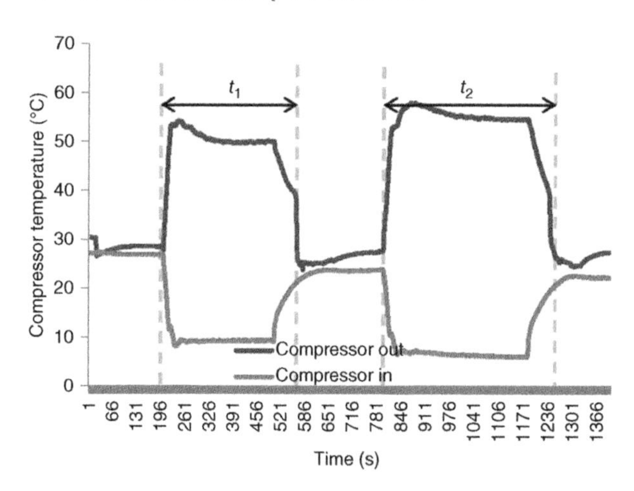

Figure 4.91 Refrigerant temperature before and after the compressor.

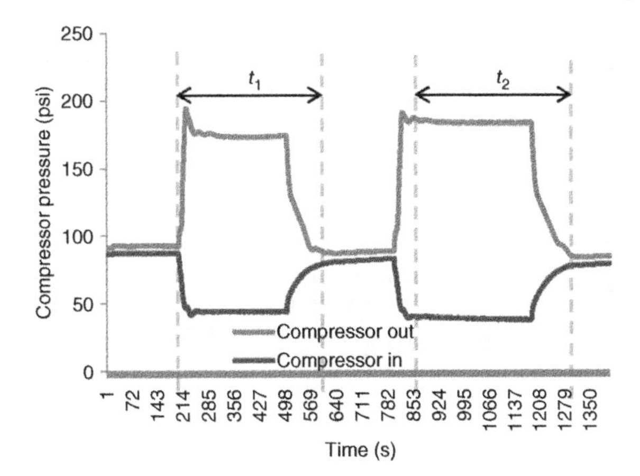

Figure 4.92 Refrigerant pressure before and after the compressor.

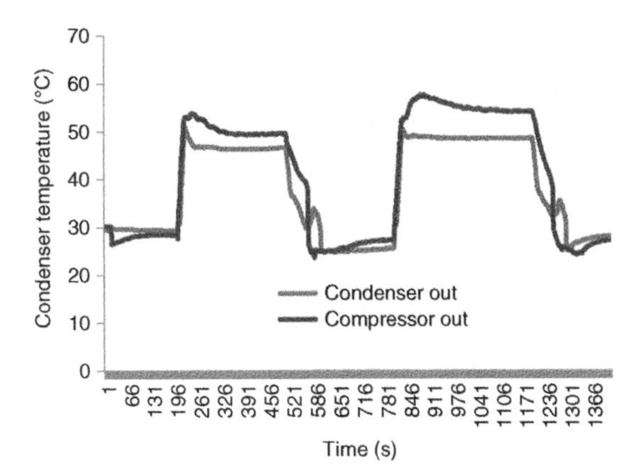

Figure 4.93 Refrigerant temperature before and after the condenser.

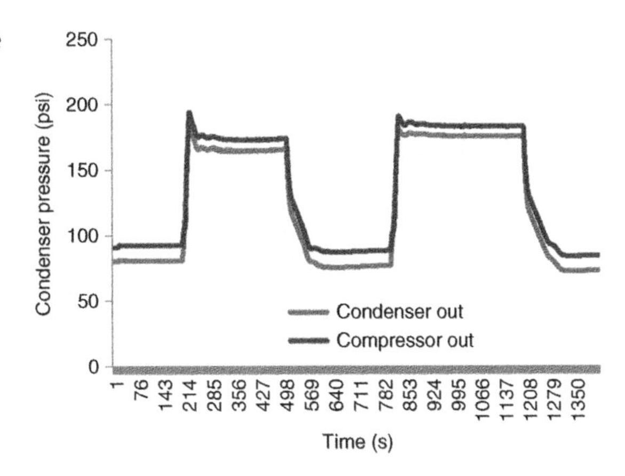

Figure 4.94 Refrigerant pressure before and after the condenser.

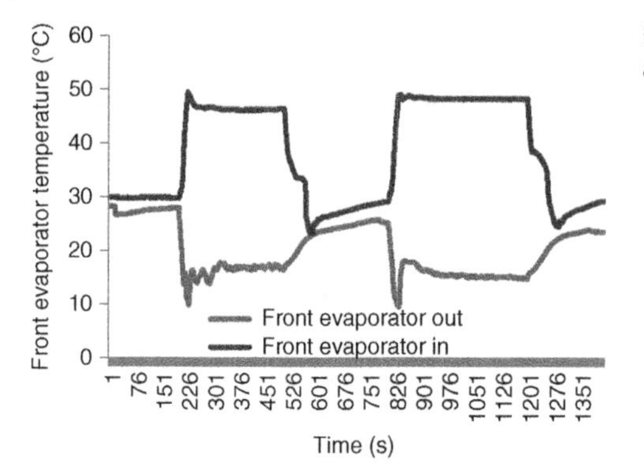

Figure 4.95 Refrigerant temperature before and after the evaporator.

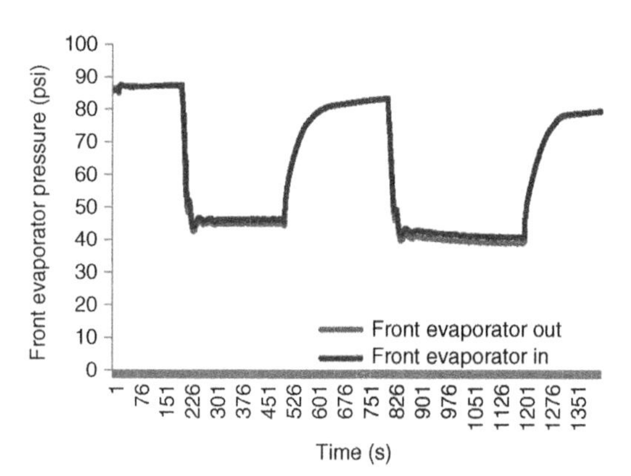

Figure 4.96 Refrigerant pressure before and after the evaporator.

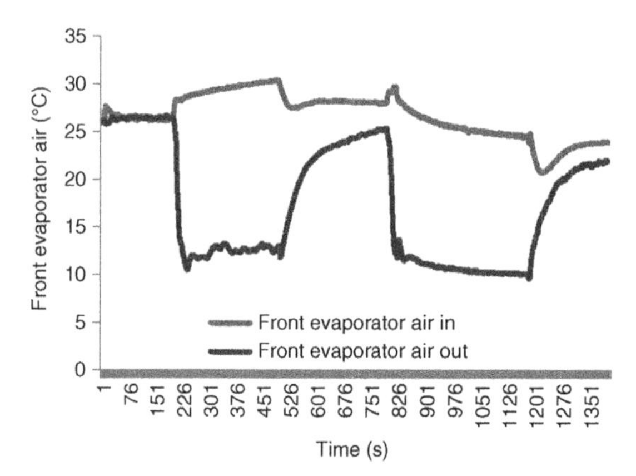

Figure 4.97 Air temperature before and after the evaporator.

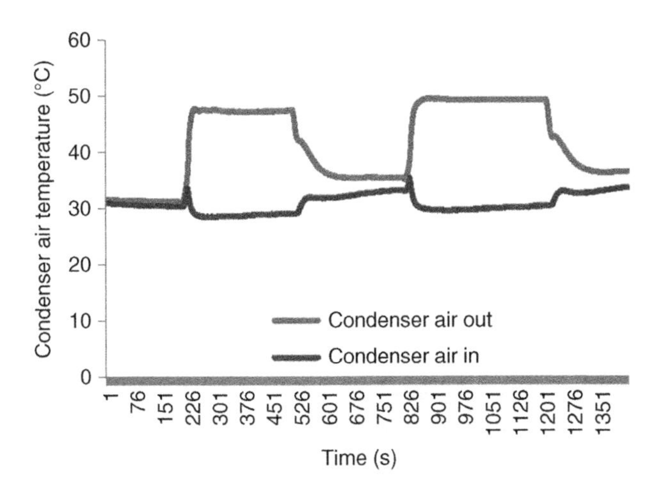

Figure 4.98 Air temperature before and after the condenser.

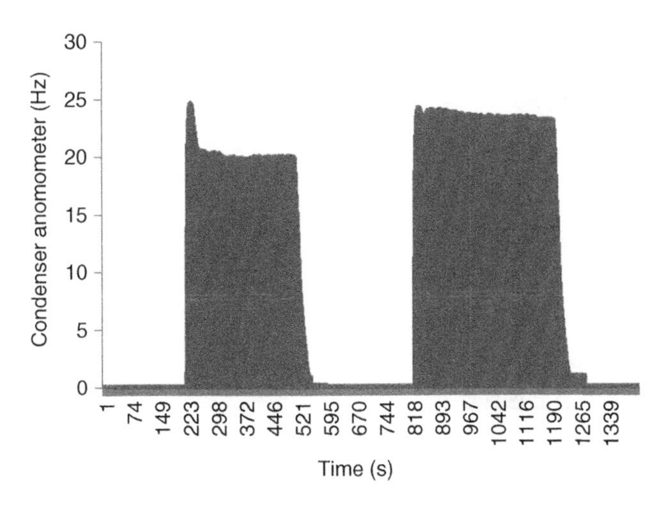

Figure 4.99 Anemometer readings of condenser fan.

Figure 4.100 Air temperature before and after the radiator.

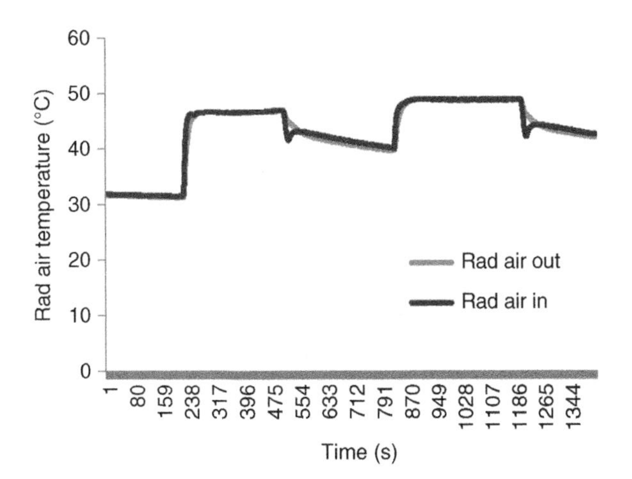

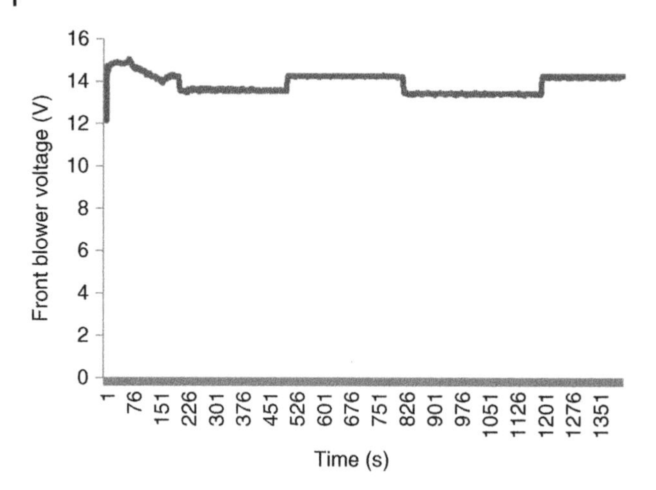

Figure 4.101 Front blower voltage.

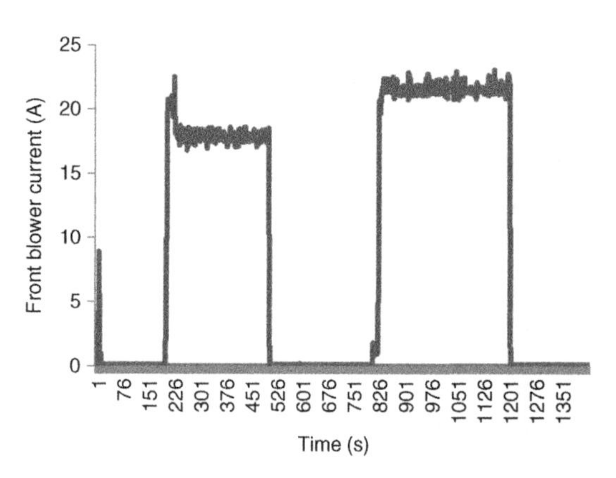

Figure 4.102 Front blower current.

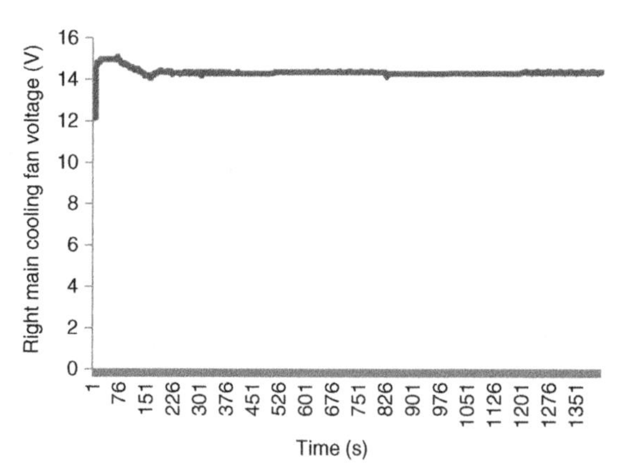

Figure 4.103 Right main cooling fan voltage.

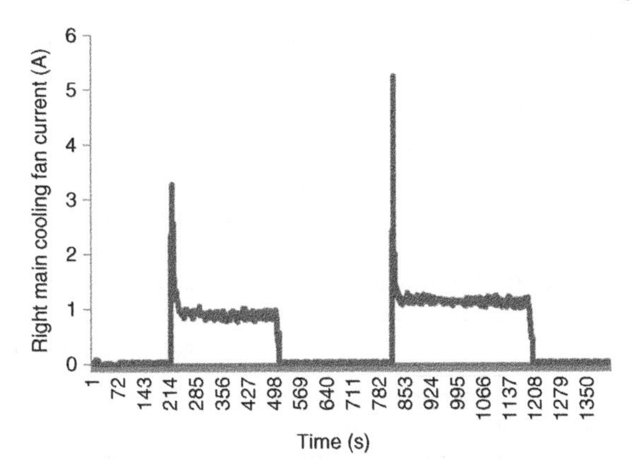

Figure 4.104 Right main cooling fan current.

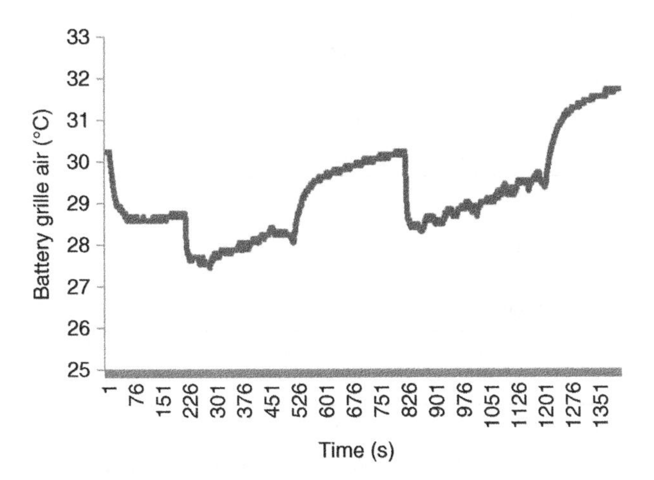

Figure 4.105 Battery grille air temperature.

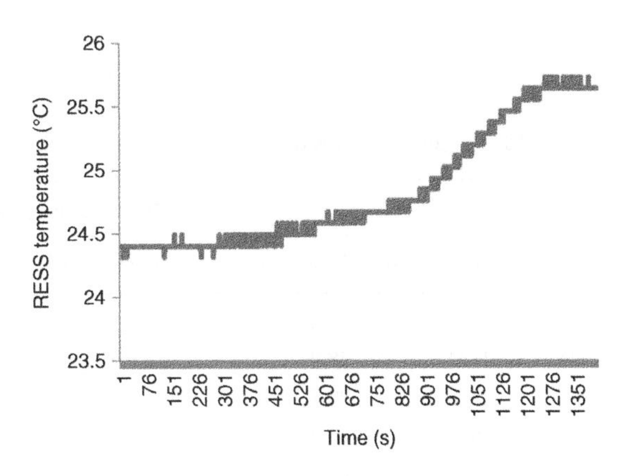

Figure 4.106 RESS temperature.

In order to provide a comparison, decision variable values for the model are retrieved from the experimental results taken from the data acquisition system and vehicle CAN bus; and the corresponding results calculated from the model are used to predict the remaining experimental outputs. The following data points provided in Table 4.13, which are taken from various different states representing a wide range of operation parameters in the experimentation, are selected to compare with the baseline model.

Table 4.13a Refrigerant temperatures used to validate the model.

	Refrigerant Temperature (°C)						
Time (s)	Compressor IN	Compressor OUT	Condenser OUT	Front Evap. IN	Front Evap. OUT	Rear Evap. IN	Rear Evap. OUT
207	26.8	28.2	29.6	29.9	27.3	27.7	26.8
224	12.1	53.0	50.8	46.3	11.7	46.0	12.3
376	9.2	49.7	46.5	46.1	16.8	46.2	9.0
693	23.5	26.3	25.1	27.5	24.8	25.1	23.7
857	7.3	56.4	49.1	48.7	17.9	48.8	7.3
1092	6.2	54.4	48.7	48.2	15.3	48.3	6.1

Table 4.13b Refrigerant pressures used to validate the model.

	Refrigerant Pressure (kPa)						
Time (s)	Compressor IN	Compressor OUT	Condenser OUT	Front Evap. IN	Front Evap. OUT	Rear Evap. IN	Rear Chiller OUT
207	87.4	92.6	80.7	87.1	87.2	87.2	87.2
224	47.6	193.8	183.6	52.2	49.3	49.6	49.1
376	304.7	1193.2	1135.6	320.1	311.8	313.8	311.8
693	560.9	602.6	522.2	566.0	564.8	563.4	564.0
857	40.4	186.7	177.8	43.3	41.4	41.8	41.4
1092	262.7	1263.3	1211.2	284.1	270.6	272.0	269.2

Table 4.13c Air temperatures used to validate the model.

	Air Temperature (°C)				
Time (s)	Ambient Air	Condenser IN	Condenser OUT	Front Evap. IN	Front Evap. OUT
207	26.1	26.5	26.6	30.7	31.5
224	25.8	28.2	13z.3	32.3	40.8
376	27.5	29.7	12.3	28.9	47.2
693	28.2	28.2	24.0	32.6	35.5
857	28.3	27.5	11.5	30.0	49.2
1092	27.5	24.9	10.2	30.2	49.3

Table 4.14 Comparison of results between the experimentation and the model.

Time (s)	ΔP_{cond} (kPa)		ΔP_{evap} (kPa)		θ		W_{comp} (kW)		Q_{evap} (kPa)	
207	38.22	34.11	37.89	35.35	1.43	1.41	0.89	0.84	0.37	0.38
224	70.34	71.53	31.71	27.54	4.07	4.05	1.21	1.10	2.63	2.65
376	57.67	50.54	15.33	20.67	3.92	3.87	2.42	2.32	3.08	3.12
693	80.38	75.12	5.07	3.56	1.07	1.01	0.91	0.88	0.74	0.76
857	61.66	57.76	20.16	25.5	4.63	4.42	1.45	1.41	2.83	2.87
1092	52.09	54.62	21.41	17.43	4.81	4.76	1.52	1.47	2.60	2.63

Initially, the temperatures and mass flow rates of the air entering the heat exchanger are used from the experimental data along with the evaporating and condensing temperatures of the refrigerant in the refrigeration cycle. In addition, the data acquired from the vehicle medium speed CAN bus is used to determine the battery heat dissipation rates to the coolant loop.

Based on these parameters, the thermodynamic states of the thermal management system, compression ratio and total work of the compressor as well as the heat load of the evaporator and chiller are predicted for each selected point in the experiment. The experimental results along with the ones developed from the model are provided in Table 4.14. In the table above, the first sub-columns in each parameter are obtained from the experimentations and the second sub-columns are calculated with respect to the developed model.

Based on Table 4.14, it can be seen that the calculated and experimentally obtained results are very similar with a relatively small percentage of error that comes from some of the aforementioned assumptions and simplifications done in the model.

4.5.6 Case Study Conclusions

In this case study, the numerical and experimental approaches of analyzing BTMSs are provided. The first part of the study included numerical simulation of the system with ANSYS FLUENT version 12.0.1 for a Li-ion battery cell with different thicknesses (3, 6, 9 and 12 mm) of PCM. Next, a submodule consisting of four Li-ion cells is modeled with the application of PCM around the cells. Furthermore, the separating foams in the battery pack are replaced by foam soaked in the PCM and placed between cells. Temperature variations and distribution are calculated and compared to understand the effect of PCMs in the system and the amount and application methods needed to develop the most effective thermal management in cell and submodules levels. These results are compared with cell level experimental data to ensure their accuracy and reliability.

Moreover, since some PCMs may face the problem of low thermal conductivity, carbon nano-tubes (CNT) and graphene nano-platelets are added to the pure and technical grade octadecane to improve the thermal conductivity of the mixture. Optical microscopic methods are implemented to study the PCM and nano-particles mixture, specifically to monitor the agglomeration of nanoparticles. Transmission and reflection optical microscopy methods are carried out in order to study the prepared samples. Furthermore, CNT and graphene platelets mixed with pure and technical grade PCM, are studied through differential scanning calorimetry (DSC) in order to compare the thermal conductivity of the mixtures.

Finally, the effects of the battery cells/submodule in the performance of the overall vehicle are demonstrated using a test bench/vehicle under different scenarios. The calculated and experimentally obtained results are compared to validate the accuracy of the developed model as well as to improve the simulations in order to increase the effectiveness of the BTMS.

The main findings of the case study are summarized as follows:

- When the PCM is applied in a 3 mm thickness around the Li-ion cell, the temperature distribution becomes 10% more uniform. Results show that phase change material with 12 mm thickness decreases the maximum temperature by 3.04 K. The corresponding value for thinner layers of 3 mm, 6 mm and 9 mm are 2.77 K, 2.89 K and 2.98 K, respectively. The results are calculated for the time of 20 minutes in transient conditions.
- For the submodule with PCM around it, more than a 7.7 K decrease in volume-average temperature occurs. Difference between the maximum and minimum temperatures in the submodule with PCM around it are decreased 0.17 K, 0.68 K, 5.80 K and 13.33 K for the volumetric heat generation rates of 6.885, 22.8, 63.97 and 200 kW/m^3, respectively.
- Experimental study of five different foams shows that the maximum mass concentration of PCM possible in this medium is 65.8%.
- Maximum temperature of the submodule with dry foams decreases from 310.87 K to 303.13 K when PCM is applied in the foam in between the cell. Moreover, the critical cell shows reduced temperature variation (9% less temperature excursion).
- The numerical results on temperature distribution of cells in the submodule are found to be in accordance with experimentally measured data in both cases with and without PCM.

Contact resistance in the cell and connectors boundary is 7.8×10^{-6} m^2K/W, which is smaller than air film resistance by an order of 10^{-4} and therefore is not considered in simulations. This is mainly due to the high temperature gradients in the direction of heat transfer.

- The agglomeration effect can be challenging in practical applications. Using a micro-mesh reduces CNT agglomeration in the mixture by reducing the convection in the melting process.
- The optical observations of technical PCMs show boundaries in their structure. These boundaries can act as a source of thermal boundary resistance, suppressing the heat flow across the structure similar to grain boundaries in solids.

4.6 Concluding Remarks

In this chapter, the necessary steps for developing the battery model from cell level to module is illustrated and the procedures for instrumenting the vehicle components and BTMSs and best practices for acquiring and verifying experimental data is demonstrated for the readers. The numerical simulations are tried to be performed as general as possible to be compatible with various different battery types and chemistries. Moreover, they are developed in scalable submodule units to reduce the complexity and the computation time of the models, while providing the fundamentals information and tools to transform it into the whole battery pack. In these models, application of PCMs has been considered around the cells, in between them and over the entire submodule.

Furthermore, a case study is conducted to provide detailed applications of the aforementioned methods and procedures and examine a BTMS both through numerical model and experimentation using ANSYS FLUENT numerical simulation tool, IPETRONIKS data acquisition hardware, IPEMOTION data acquisition software, and optical imaging instrumentations. The simulation and experimental results are provided (from cell to vehicle level) and are compared to validate the accuracy and reliability to the demonstrated models.

Nomenclature

A	area (m^2)
Cp	specific heat coefficient (J/kg K)
E	potential (V)
E_0	open circuit potential (V)
F	Faraday's constant
G	Gibbs free energy (J)
h	specific enthalpy (kJ/kg)
H	total enthalpy content
h_s	convective heat transfer coefficient (W/m^2 K)
I	current (A)
k	thermal conductivity of a cell (W/m K)
L	length scale (m), latent heat of fusion (J/kg)
n	number of electrons
$\dot{q}$	heat transfer rate (kW)
R	resistance (Ω)
S	entropy (J/K)
S_h	source term for heat generation rate (kW/m^3)
t	time (s)
T	temperature (K)
T_0	ambient temperature (K or °C)
U	open circuit potential (V)
u	velocity in x-direction (m/s)
v	velocity in y-direction (m/s)
V	cell potential (V)
w	velocity in z-direction(m/s)

Greek Symbols

α	coefficient of volumetric expansion (1/K)
β	liquid fraction
Δ	change in variable

θ_1	overall temperature coefficient
θ_1	local temperature coefficient
Θ	parameter in general equation
ρ	density (kg/m^3)
Ω	effectiveness index
Γ	diffusion coefficient
Θ	unspecified variable

Subscripts

0,amb	ambients
b	battery
cell	cell
gen	generation
G	generatio
m	melting

Acronyms

CFD	computational fluid dynamics
DSC	differential scanning calorimetry
EV	electric vehicle
FVM	finite volume method
GHG	greenhouse gas
HEV	hybrid electric vehicle
PCM	phase change material
ICE	internal combustion engine
PCM	phase change material
SOC	state of charge
TMS	thermal management system

Study Questions/Problems

4.1 Please define the critical safety measures that need to be implemented before conducting any experimentation involving high voltage batteries?

4.2 What are the rules of thumb when instrumenting the battery or the vehicle?

4.3 What are some of the reasons for discrepancy in the readings and white noise in data acquisition?

4.4 What are the critical inputs that are needs to be measured in BTMSs? Where are the optimum locations to measure them?

4.5 What is grid independence and how can it be obtained in simulations?

4.6 Please name some methods to improve the thermal conductivity of phase change materials and briefly describe the associated procedures to resolve the thermal conduct.

References

Agyenim F, Hewitt N, Eames P, Smyth, M. (2010). A Review of Materials, Heat Transfer and Phase Change Problem Formulation for Latent Heat Thermal Energy Storage Systems (Lhtess). *Renewable and Sustainable Energy Reviews* **14**:615–628.

Al Hallaj S, Maleki H, Hong J, Selman J. (1999). Thermal Modeling and Design Considerations of Lithium-Ion Batteries. *Journal of Power Sources* **83**:1–8.

Ansys Inc. (2009). Ansys Fluent 12.0.1 User's Guide. Available at http://users.ugent.be [Accessed August 2016].

Amrollahi A, Hamidi A, Rashidi A. (2008). The Effects of Temperature, Volume Fraction and Vibration Time on the Thermo-Physical Properties of a Carbon Nanotube Suspension (Carbon Nanofluid). *Nanotechnology* **19**(31):5701.

Bejan A, Lorente S, Yilbas B, Sahin A. (DATE?). Why Solidification Has an S-Shaped History. Available at: http://www.nature.com/srep/2013/130424/srep01711/full/srep01711.html [Accessed August 2016].

Bernardi D, Pawlikowski E, Newman J. (1985). A General Energy Balance for Battery Systems. *Journal of the Electrochemical Society* **132**:5–12.

Camirand CP. (2004). Measurement of Thermal Conductivity by Differential Scanning Calorimetry. *Thermochimica Acta* **417**:1–4.

Hamut HS, Dincer I, Naterer GF. (2014). Experimental and Theoretical Efficiency Investigation of Hybrid Electric Vehicle Battery Thermal Management Systems. *Journal of Energy Resources Technology* **136**(1):011202–1:13.

IPEmotion Manual. (2010). Ipetronik Gmbh & Co. KG, Jaegerweg 1, 76532 Baden-Baden, Germany.

IPETRONIK Manual. (2009). Ipetronik Gmbh & Co. KG, Jaegerweg 1, 76532 Baden-Baden, Germany.

Javani N, Dincer I, Naterer GF, Rohrauer GL. (2014). Modeling of passive thermal management for electric vehicle battery packs with PCM between cells. *Applied Thermal Engineering* **73**:307–316.

Pals CR, Newman J. (1995). Thermal Modeling of the Lithium/Polymer Battery. *I. Discharge Behavior of a Single Cell. Journal of the Electrochemical Society* **142**:3274–3281.

Pesaran AA, Burch S, Keyser M. (1999). An Approach for Designing Thermal Management Systems for Electric and Hybrid Vehicle Battery Packs. Proceedings of the 4th Vehicle Thermal Management Systems 24–27.

Selman JR, Al Hallaj S, Uchida I, Hirano Y. (2001). Cooperative Research on Safety Fundamentals of Lithium Batteries. *Journal of Power Sources* **97**:726–732.

Sun L, Siepmann JI, Schure MR. (2007). Monte Carlo Simulations of an Isolated N-Octadecane Chain Solvated in Water-Acetonitrile Mixtures. *Journal of Chemical Theory and Computation* **3**:350–357.

Wu MS, Liu K, Wang YY, Wan CC. (2002). Heat Dissipation Design for Lithium-Ion Batteries. *Journal of Power Sources* **109**:160–166.

5

Energy and Exergy Analyses of Battery TMSs

5.1 Introduction

As elaborated in the previous chapters, battery thermal management has a vital role in improving the life cycle, performance, safety and reliability of electric vehicles. Therefore, it is imperative to develop a capable battery thermal management system that can keep the battery cells operating within the ideal operating temperature range and uniformity, while minimizing the associated energy consumption and cost. In order to achieve these goals, engineers involved in electric vehicle battery development are required to answer the questions, such as:

- Which thermal management method and heat transfer medium should be selected for a given electric vehicle and how should the parts be arranged for the best outcome?
- What are the best process parameters and operating conditions of the thermal management system under different operating and ambient conditions?
- How much each component contributes to the operation inefficiencies of the overall thermal management system and how, where and why does the system performance degrades?

Thus, in this chapter various types of state of the art thermal management systems are examined and assessed for electric vehicle battery systems. Subsequently, step-by-step thermodynamic modelling of a real TMS is conducted and the major system components are evaluated under various parameters and real life constraints with respect to energy and exergy criteria to provide the readers with the methods as well as the corresponding results of the analyses. The procedures are explained in each step and generic and widely used parameters are utilized as much as possible to enable the readers to incorporate these analyses to the systems they are working on. A simplified version of the code for the conducted thermodynamic analysis is provided in the Engineering Equation Solver (EES) in the appendices section.

In the following sections, cabin cooling (via air), active moderate cooling (via refrigerant) and active cooling (via water/glycol mix) are examined and compared for a given battery system with respect to battery temperature increase, uniformity and entropy generation. Subsequently, active thermal management system is analyzed with respect to energy and exergy efficiencies under various system parameters. The major thermal management system components are modeled in detail, the heat transfer coefficients in the heat exchangers are calculated, and hence the pressure drops are determined with respect to the Reynolds Number correlations to provide a more accurate representation

Thermal Management of Electric Vehicle Battery Systems, First Edition.
İbrahim Dinçer, Halil S. Hamut and Nader Javani.

of the system. The thermodynamic analyses of the system are conducted under various operating conditions including evaporating and condensing temperatures, subcooling and superheating, compressor speed, heat exchanger pressure drop and battery heat dissipation rates to demonstrate the effects of these parameters on the thermal management system. Moreover, various other refrigerants are investigated in terms of compatibility with the existing system to provide alternatives to existing cooling mediums. In addition, an enhanced exergy analysis is conducted where the avoidable and unavoidable as well as endogenous and exogenous parts of exergy destructions are determined in order to enhance the readers' understanding of the interactions among the TMS components, establish priorities on which components should be improved first and assist in further optimization of the overall system. Consequently, recommendations are provided to improve the exergy efficiencies of the components as well as the overall system under the studied operating conditions and parameters. Finally, a case study is provided which shows a true application of the methods provided in the chapter and calculates the efficiencies of a transcritical CO_2 based BTMS.

5.2 TMS Comparison

The performance of thermal management systems in EVs and HEVs has great importance due to the limited supply of available energy onboard as well as the overall impact on vehicle performance in which efficiency is the main indicator. Thus, it is vital to have a good understanding of the efficiencies associated with the system and its components. In order to be able to accurately assess and further improve the efficiency of energy resource utilization, a clear and thorough approach that effectively characterizes all energy forms without the ambiguities and shortcoming of conventional energy analysis would be necessary. Especially with ever-increasing global competition and stricter environmental regulations, governments and large organizations are incorporating exergy into their energy analysis methods and utilizing the second-law principles to reduce the cost and environmental impact of their product and processes. Thus, exergy analysis has been used increasingly during the last several decades in the performance analysis of energy systems in many economic sectors including residential, commercial, institutional, industrial, utility, agricultural and transportation, especially with the purpose of enhancing energy sustainability. Exergy analysis is a powerful thermodynamic technique to acquire an in-depth understanding of the performance of processes, devices and systems and it is a valuable tool for the design, analysis, assessment and optimization of energy conversion systems and chemical processes to enhance the efficiency, reduce the cost and minimize the environmental impacts and to provide a sustainable development.

Energy-based efficiencies may lead to inadequate and misleading conclusions, since all energy forms are taken to be equal and the ambient environment is not taken into consideration. The second law of thermodynamics defines the energy conversion limits of this available energy based on irregularities between different forms of energies. The quality of the energy is highly correlated to the reference environment as well as the success level of this conversion capacity, and needs to be considered to prevent any incomplete and/or incorrect results. An analysis for examining the work potentials of the initial and final stages of a system can give an evaluation criterion for the quality of the energy. Such analysis is called "exergy analysis", which represents the amount of energy that may be totally converted to work and will be used to determine the efficiencies of the battery TMSs (Arcaklioglu *et al.*, 2005).

Exergy (also called available energy or availability) of a system is the "maximum shaft work that can be done by the composite of the system and a specified reference environment" (Dincer and Rosen, 2013). In every thermal management system, heat transfer within the system, or between the system and surrounding environment, occurs at a finite temperature difference, which is a key contributor to irreversibilities for the system. All real processes, including natural events are irreversible and the system performance degrades as a result of these irreversibilities in each individual thermodynamic process that makes up the system. The work potential is reduced by the irreversibilities and the corresponding amount of energy becomes unusable. Entropy generation measures the effect of these irreversibilities in a system during a process and helps compare each component in the system based on how much they contribute to the operation inefficiencies of the overall system. Therefore, entropy generation associated with each process needs to be evaluated to determine the overall system efficiency. Even though energy analysis is the most commonly used method for examining thermal systems, it is only concerned with the conservation of energy, which neither takes the corresponding environmental conditions into account, nor provides how, where and why the system performance degrades. Consequently, the energy analysis only measures the quantity of energy and does not reveal the full efficiencies of the system (Yumrutas *et al.*, 2002). Thus, in this chapter, the thermal management system will be examined with respect to exergy analysis in order to better understand the true efficiencies of the components by determining the irreversibilities in each cycle, as well as the overall system and how nearly the respective performances approach ideal conditions. By analyzing both the quality (usefulness) and the quantity of the energy, the true magnitude of losses, and their causes and locations are identified by investigating the sites of exergy destruction in order to improve the individual components and overall system.

5.2.1 Thermodynamic Analysis

In this section, before the detailed results of component level exergy analyses is provided, a high level comparison of the cabin air, refrigerant and liquid based thermal management systems is introduced based on thermodynamic and heat transfer analyses. The properties of air, refrigerant and the coolant are calculated in their associated circuits and used to conduct the exergy analysis based on the aforementioned balance equations. The energetic and exergetic COPs of each system are calculated with respect to Table 5.1. The same ambient and refrigerant circuit properties, as well as evaporator heat load are used in all TMSs in order to perform a consistent analysis.

In passive cabin air cooling, the battery is cooled by the conditioned ambient air that is transferred through the evaporator into the cabin as shown in Figure 5.1. By using the battery fans, some of this air is used in order to cool the battery. Since the battery is cooled by using the available cabin air, it does not require any additional compressor work other than the work used to provide thermal management to the vehicle cabin. Moreover, the system is highly compatible due to the optimum cabin and battery temperatures are generally very close to each other. However, since the battery cooling solely relies on the cabin temperature, it can be significantly affected when cabin temperatures

Table 5.1 Energetic and exergetic COP equations used in the analysis.

TMS	Coefficient of performance (COP)	
	Energetic	Exergetic
Passive cabin air cooling	$\frac{\dot{Q}_{Evap}+\dot{Q}_{bat}}{\dot{W}_{Comp}}$	$\frac{\dot{E}x_{QEvap}+\dot{E}x_{bat}}{\dot{W}_{Comp}}$
Active refrigerant cooling	$\frac{\dot{Q}_{Evap}+\dot{Q}_{bat\ evap}}{\dot{W}_{Comp}}$	$\frac{\dot{E}x_{QEvap}+\dot{E}x_{bat\ evap}}{\dot{W}_{Comp}}$
Active liquid cooling	$\frac{\dot{Q}_{Evap}+\dot{Q}_{chil}}{\dot{W}_{Comp}+\dot{W}_{Pump}}$	$\frac{\dot{E}x_{QEvap}+\dot{E}x_{chil}}{\dot{W}_{Comp}+\dot{W}_{Pump}}$

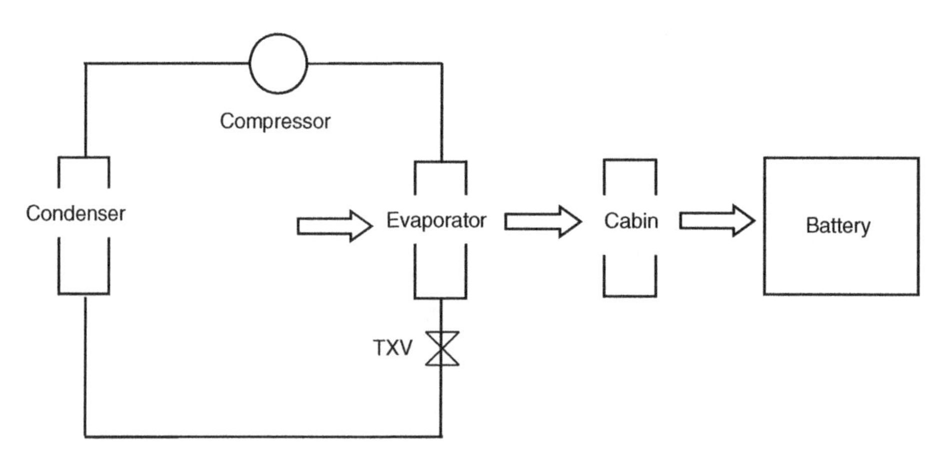

Figure 5.1 General schematic diagram of cabin air TMS.

are high. Moreover, a large fraction of the air flow rate is lost through its transition to the cabin. Therefore, significant fan power is required to increase the amount of air flow for the battery, which can increase the fan power consumption and noise level inside the cabin. This issue can be resolved by implementing independent air cooling with the help of a separate battery evaporator, but the trade-off will be the additional compressor power to flow the refrigerant through this evaporator.

The efficiency of this system depends on the heat load applied to the evaporator and the battery heat dissipation rate. For relatively low battery heat dissipation rates, the TMS is highly effective due to usage of already available air in the vehicle cabin. However, due to the low exergy and flow rate of the cabin air, it would not be sufficient when the battery is operating in a harsh operating environment and extreme duty cycles. Therefore, this system is mainly utilized when the battery heat dissipation is low and ambient air conditions are within the desired battery operating ranges.

In active refrigerant cooling, the battery is cooled with the additional evaporator utilized specifically for the battery as shown in Figure 5.2. The exergy associated with the battery cooling is higher due to the use of the evaporator with a refrigerant, as opposed

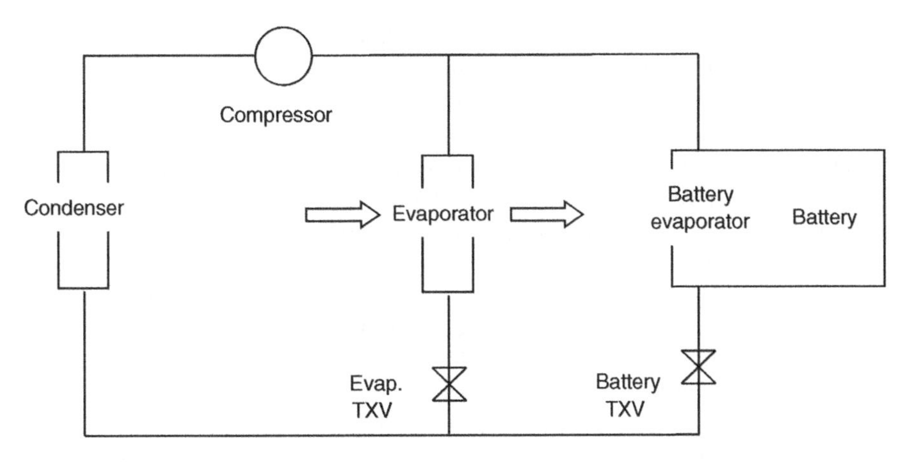

Figure 5.2 General schematic of refrigerant-based TMS.

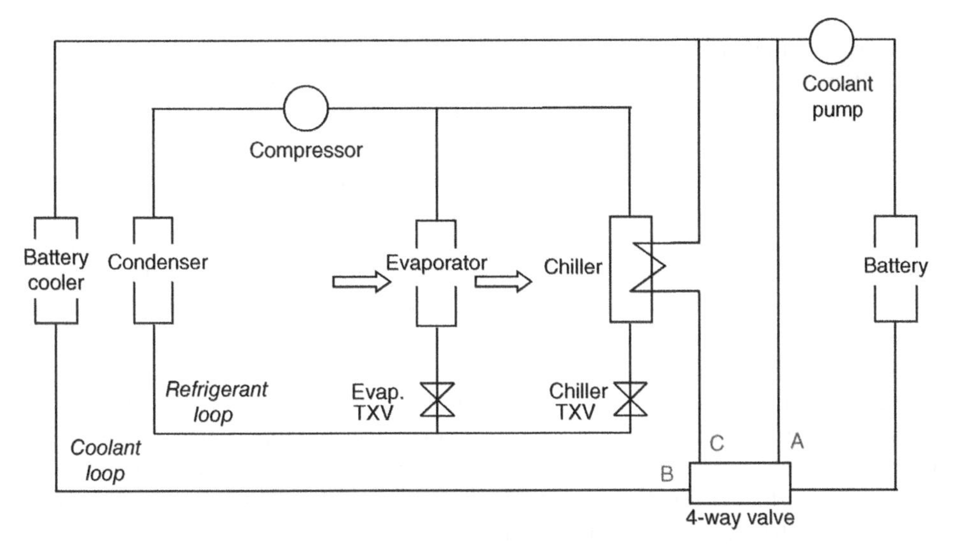

Figure 5.3 General schematic of liquid-based TMS (A: bypass route, B: battery cooler route, C: chiller route).

to just air flow in the previous TMS, but at the expense of the additional compressor work to pump the refrigerant to the battery evaporator. As a result, the total exergy efficiency of the system is calculated to be lower than the cabin air cooling TMS, since extra compressor work is needed to cool the battery via the refrigerant.

The active liquid cooling system on the other hand, incorporates the advantages of both the air cooling and refrigerant based cooling with the help of the additional battery cooler and chiller as shown in Figure 5.3. This additional cooling loop is kept cool via different procedures depending on the cooling load and ambient conditions as it will be discussed further in Section 5.4.

In this system, the method of keeping the battery cooling medium at a low temperature depends on several factors such as the amount of heat generated in the battery, cabin heat load and ambient temperature. In the baseline model, since the ambient temperature is higher than the battery desired temperature target (21.5°C), the battery is cooled solely with the help of the chiller which transfers heat from the coolant circuit to the refrigerant circuit. The coolant used (water/glycol mix) has high thermal capabilities, and the chiller is highly effective at cooling the battery to the optimal operating temperatures at high drive cycles. However, similar to the previous system, it uses additional compressor work associated with the chiller; therefore, the exergy efficiency of the system is calculated to be higher than the aforementioned active refrigerant cooling system. Another important advantage of this system is the flexibility under various required cooling rates as well as battery and ambient temperatures. This can be seen when the ambient temperature is reduced below 20°C. Below this temperature, some portion of the hot coolant that leaves the battery could be cooled with the help of the battery cooler that utilizes the ambient air flow, especially when the vehicle is travelling at high speeds. This cooling would be achieved by using a coolant pump in this circuit, which would consume negligible power compared to the compressor and thus increases the overall exergy efficiency of the system.

Even though the baseline refrigerant circuit model with different thermal management systems provides significant benefits in understanding the effectiveness of these systems based on the exergy analysis, the model neglects certain aspects which can have an important role when comparing these TMS. Among these aspects, the cost associated with manufacturing and maintenance has a significant role in selecting the appropriate thermal management system. Because of relatively recent widespread commercialization of these technologies in passenger vehicles, it is difficult to consistently compare each TMS based on cost. The passive air cooling TMS is in general considerably cheaper to install and maintain than active liquid cooling systems due to a significantly simpler design with less components and potential for leaks. Furthermore, passive air cooling systems utilize the already available cabin air to cool the battery and therefore, unlike most active liquid cooling TMSs, do not require any significant operating cost other than the power required for fan work. However, most passive air cooling TMSs are used on batteries that have lower cooling rate requirements than Li-ion and these batteries generally operate at higher temperatures. Since higher operating temperatures reduce the cycle life of the battery over the long term, especially under high drive cycles, the cost associated with replacing the battery increases the overall cost of the TMS drastically. Therefore, utilizing a TMS that is compatible with the desired driving cycle, cooling load and operating conditions is a key factor in selecting an appropriate TMS to have a low cost over the long term.

In addition, another important factor is the entropy generation associated with each different TMS due to cooling the battery and/or the hot coolant. Entropy is generated due to the finite temperature differences as well as fluid friction associated with the TMS. In these systems, the majority of entropy generation is based on the corresponding heat transfer and frictional effects. The heat transfer effects are correlated to the difference of the coolant inlet and outlet temperatures. For the analysis, the average of the minimum and maximum inlet – outlet temperature differences are used. The same inlet temperatures are used in all systems in order to provide a consistent comparison. The second terms are taken with respect to the difference in the inlet and outlet pressure due to the irreversible losses caused by the fluid friction inside the tubes. For the cabin air cooling system, the entropy generation for cooling the battery is represented by the heat transfer irreversibility in external flow as given below (Bejan, 1996):

$$\dot{S}_{gen,\,external} = \left(\frac{(T_b - T_{c,in})}{T_{c,in}}\right)^2 \overline{h}A + \frac{1}{T_{c,in}} A_{tube} \Delta P \, V_c \tag{5.1}$$

The fluid velocity is calculated from the mass flow rate and the associated area of the tube. ΔP is the pressure drop of the flow inside the battery. For the refrigerant system, refrigerant is utilized in the system to cool the battery. The respective equation becomes

$$\dot{S}_{gen,refrigerant} = \frac{\dot{m}_r h_{fg}}{T_{evap}} + \frac{\dot{m}_c \Delta P}{\rho_{c,in} T_{c,in}} \tag{5.2}$$

For the refrigerant cooling system, the refrigerant goes through phase change when cooling the battery, and the heat transfer occurs from latent heat of vaporization (h_{fg}), where the temperature of the refrigerant remains constant. Therefore, the enthalpy difference is used for calculating the thermal entropy generation rate, which is significantly higher due to the phase change in the refrigerant. The liquid cooling system goes through a similar procedure in terms of entropy generation rate, except that the coolant does not change phase through the battery. Therefore, the thermal entropy generation is based on the temperature difference of the coolant as given below:

$$\dot{S}_{gen,coolant} = \frac{Q\,(T_{c,out} - T_{c,in})}{T_b^{\,2}} + \frac{\dot{m}_c \Delta P}{\rho_{c,in} T_{c,in}} \tag{5.3}$$

5.2.2 Battery Heat Transfer Analysis

Heat transfer analysis is a critical method of analyzing and determining temperature distributions and heat transfer rates which are of great significance in thermal management of batteries in EVs and HEVs. In this section, we provide specific details and methodologies.

5.2.2.1 Battery Temperature Distribution

In the TMS, the rate of heat transfer between the walls of the module and the fluid depends on various properties of the transfer medium. Air cooling is used for batteries that operate in relatively uniform operating conditions that do not require significant cooling. They are in general simpler and cheaper than liquid cooling with less components and potential for leaks. However, they have a significantly lower heat capacity and heat transfer coefficient. Direct contact fluids have relatively high heat transfer rates, especially compared to air, due to their boundary layer and thermal conductivity. Furthermore, they have the best packaging densities among the aforementioned TMSs. Indirect contact water also has significantly higher heat transfer coefficients because of its relatively low viscosity and thermal conductivity. Moreover, they do not require fans or air ducts in the vehicle or occupy a large space for proper cell arrangement. However, their effectiveness can decrease significantly as a result of the added thermal resistance such as a jacket wall or air gaps. As a result, even if the heat removal rates from the cells to the coolant are the same in the different thermal management systems, liquid coolants such as water (or water/glycol mix) would not be heated as fast as air due to a higher heat capacity. Moreover, the difference between the coolant mean temperature and the cell surface temperature would be significantly lower in liquid cooling systems due to their larger heat transfer coefficient, which reduces both the maximum battery temperature and the temperature difference among the cells in the pack (Kim and Pesaran, 2006).

In order to calculate how fast each thermal management system cools the battery, first the battery heat generation needs to be determined. The heat in the battery is generated due to the internal resistance of the battery as follows:

$$Q_b = I^2R \tag{5.4}$$

Here, I is the current and R is the resistance associated with the battery cells. The internal resistance is based on several factors such as ohmic resistance and kinetic and diffusion polarization losses in the cell and the electrical collector system. In the considered battery, 80A current and 66 mΩ resistance is incorporated, which corresponds to 0.35 kW of heat dissipation. Without any active heat dissipation, over time, this performance and cycle life of the battery will be reduced significantly. According to the Arrhenius approximation of the temperature dependence on the battery life, a 10–15 K increase in a Li-ion battery can result in 30% to 50% reduction in the battery's life endurance (Kuper *et al.*, 2009). Without any cooling system, the battery heat would be dissipated by natural convection as follows:

$$m_b C_{P,b} \frac{dT}{dt} = \bar{h}A(T_b - T_0) \tag{5.5}$$

which can be solved in terms of time (t) as follows:

$$t = \frac{m_b C_{P,b}}{\bar{h}A} \, ln\left(\frac{T_b - T_0}{T_f - T_0}\right) \tag{5.6}$$

where m_b is the thermal mass of the battery of 0.45 kg for each cell and $C_{P,b}$ is the heat capacity of the battery of 795 J/kg-K, $\bar{h}$ is the effective heat transfer coefficient of 6.4 W/m^2-K and A is the battery surface area that is assumed to be 8 m^2. Based on the battery characteristics assumed for the model and an ambient temperature of 25°C, the time required for the battery temperature to decrease from 55°C to 30°C by natural convection is determined to be approximately 1 hour. Since this time is unacceptable for most cycles in the battery, the need for a better designed TMS becomes evident. With the utilization of a TMS, the time it takes for the battery temperature to reach optimal levels while used in the vehicle can be calculated by the following equation below:

$$m_b C_{P,b} \frac{\partial T}{\partial t} = I^2R - \bar{h}A(T_b - T_0) - \dot{m}_c C_{P,c}(T_{c,out} - T_{c,in}) \tag{5.7}$$

For the majority of TMS, the natural convection term is usually negligible compared to the cooling provided by the system. In order for the model to be more representative of the actual case, the internal heat generation of the battery, natural convection and cooling rates are written as a function of time. The above differential equation is solved and sample results of the battery temperature with respect to time can be obtained for each thermal management system, as shown in Figure 5.4. For the refrigerant cooling system, the temperature rise varies significantly based on the mass flow rate of the refrigerant and is limited by the cost associated with the compressor work, and therefore a specific value is not provided in the figure. In order to provide an adequate comparison, the battery temperature rise is based on natural convection alone and it is also provided in Figure 5.4.

The heat generation and cooling rates are assumed to be linear for the analysis. From Figure 5.4, it can be seen that natural convection has the maximum temperature rise in the battery, significantly higher than the liquid cooling system. Thus, when the different TMSs are examined solely with respect to the minimum battery temperature rise, the liquid cooling system provides significantly better cooling than the cabin air system.

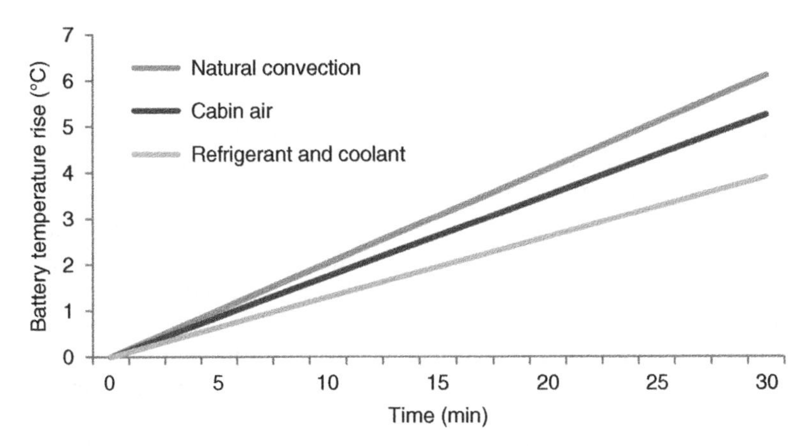

Figure 5.4 Temperature rise in the battery with time based on natural convection and various thermal management systems.

5.2.2.2 Battery Temperature Uniformity

The uniformity of the battery cells being cooled is just as important (if not more) as the maximum cell temperature, since it is one of the major causes of thermal runaway. Temperature variation between cells in the battery pack may result from ambient temperatures differences among the battery pack surface, non-uniform impedance distribution and heat transfer efficiency differences among cells in the pack (Al-Hallaj and Selman, 2002). Non-uniform impedance can result from defects in quality control or due to differences in the local heat transfer rate. Heat transfer efficiency differences are significantly related to the pack configuration since the cells along the edges are cooled by heat transfer to the environment while the ones in the center accumulate heat, which can magnify capacity differences among cells. The resultant excessive local temperatures rise in the cells, when not cooled down, may result in accelerating capacity fading and even thermal runaway in the battery pack. Even though the melting temperature for the battery is significantly high (e.g., 180°C for lithium-ion), if one or more internal cells in the stack is short-circuited, significant heat sources will exist locally, which is capable of raising the battery temperature from room temperature to above melting point of the battery in less than a minute. Since most battery packs are closely packed in order to exploit the energy and power densities of the battery (especially Li-ion), thermal runaway of a single cell can propagate and cause an entire battery to fail violently. Therefore, the uniformity of the battery cells has a significant role when comparing different TMSs and can be calculated for the battery used in the model. It is directly related to the temperature difference of the cooling medium before and after cooling the battery. In order to keep the cell temperature differences within tolerable limits, the coolant temperature difference needs to be small (less than 3°C in most cases). In thermal management systems, this is constraint by the specific heat of the medium as well as the mass flow rate, and it is shown by the following relationship:

$$\dot{Q}_b = \dot{m}_c C_{P,c}(T_{out} - T_{in}) \tag{5.8}$$

The model assumes the heat dissipation to be constant at 0.35 kW. In a cabin air cooling system, the maximum flow rate from the cabin to the battery is limited with respect to the cabin comfort and tolerable noise levels. It is typically between 1.9 kg/min and 4.8 kg/min (Kuper *et al.*, 2009). This results in a cooling temperature difference variation between 4.5°C and 11°C, approximately. In the refrigerant-based cooling, however, the cooling is provided with respect to latent heat instead, and the mass flow rate of the refrigerant depends on various factors, including the utilized compressor, and the cooling load of the evaporator and the battery, as is taken to be 0.07 kg/min to 0.7 kg/min. Moreover, the battery temperature uniformity will vary significantly based on the cooling parameters. In the liquid cooling system, the water flow rate is usually regulated around 1 kg/min to 10 kg/min which provides a cooling temperature difference of between 0.48°C and 4.57°C. Thus, when the different TMSs are examined solely with respect to their ability to cool the battery cells without major temperature differences among them, the liquid cooling system provides significantly better cooling than a cabin air system. This is due to indirect-contact heat transfer liquids (such as water) having higher specific heat and thermal conductivity than air, resulting in higher heat transfer coefficients. Moreover, generally the mass flow rates of the coolants (such as water) are significantly higher than the mass flow rates of the refrigerants (such as R134a) since the cost associated with the electricity consumption of the pump is significantly lower than the compressor. However, the decrease in indirect contact effectiveness is also a significant factor, since the heat must be conducted primarily through the walls of the jacket/container.

Before going through explaining the necessary steps of the exergy analyses, the system configuration is first introduced, and the system parameters are defined for the readers. The model of the battery thermal management systems is provided in the following section.

5.3 Modeling of Major TMS Components

Hybrid electric vehicle thermal management systems (HEV TMSs) are significantly different systems with unique requirements with respect to their commercial and industrial counterparts such as conventional vehicle and residential building air conditioning systems. The TMS needs to handle significant thermal load variations and provide comfort under highly fluctuating conditions, as well as be compact and efficient, and last several years without any significant maintenance. Moreover, the airflow volume, velocity and temperature must be adjustable over a wide range of ambient temperatures and drive cycles without having a significant impact on the all-electric vehicle performance characteristics. Furthermore, due to the limited time spent in the vehicles compared to buildings, along with the competing energy requirements between the cabin and the battery, the thermal management systems must be capable of conditioning the air in the passenger cabin quickly and quietly, while keeping the vehicle components operating under ideal operating temperature ranges (especially the electric battery) to prolong their lifetime, increase the fuel efficiency and all electric range. Thus, special attention needs to be given to hybrid electric vehicle TMSs (e.g., Jabardo *et al.*, 2002; Wang *et al.*, 2005).

A simplified thermal management system of an electric vehicle with liquid battery cooling is considered in Figure 5.5 as a baseline model. The system is composed of two loops, namely a refrigerant and battery coolant loop. The refrigerant loop enables air conditioning of the vehicle cabin, while the coolant loop keeps the electric battery operating within its ideal temperature range. These two loops are connected via a chiller,

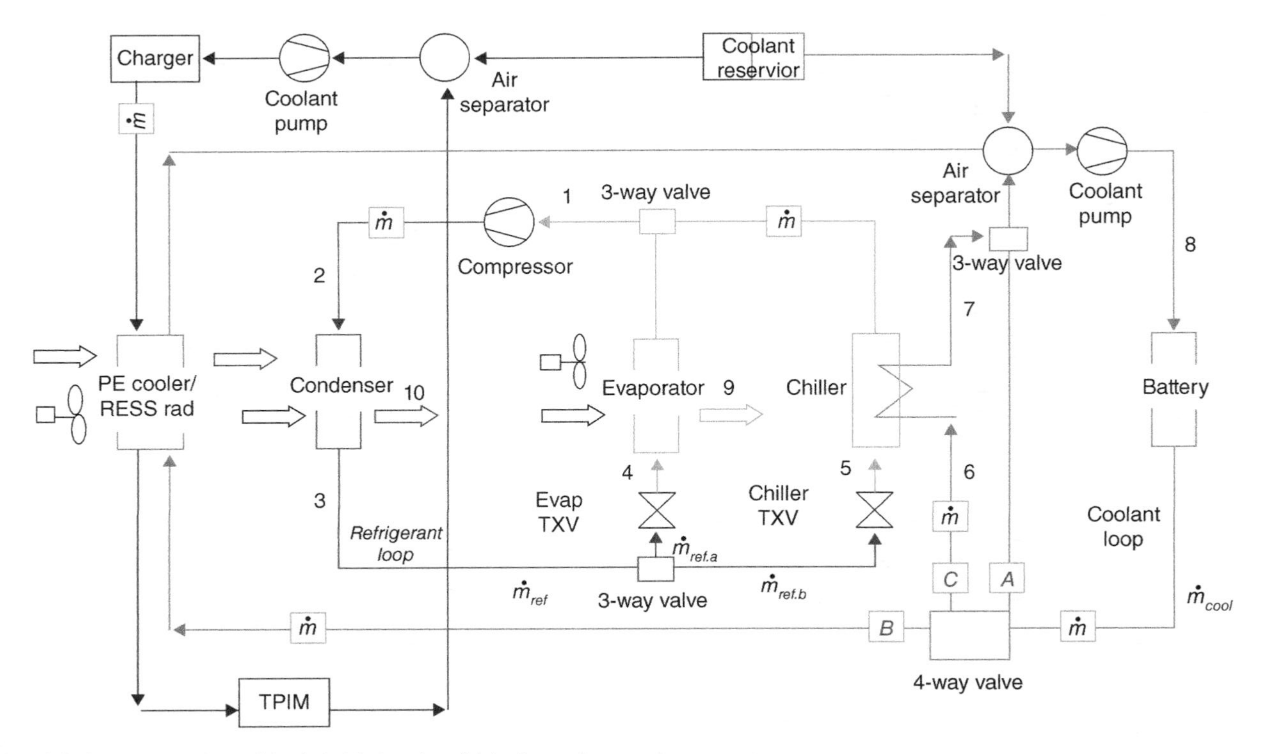

Figure 5.5 Simplified representation of the hybrid electric vehicle thermal management system.

which enables heat exchange among the loops to provide supercooling to the battery cooling as it passes through the chiller unit. This increases the efficiency of the system significantly since cooling via refrigeration circuit would consume more energy than operating the battery coolant circuit due to the need of the air compressor in the first case.

The thermal management system incorporates the advantages of both the air cooling and refrigerant based cooling with the help of the additional battery cooler and chiller. The additional cooling loop is kept cool via different procedures depending on the cooling load and ambient conditions. If the battery coolant circuit has stable temperatures within the ideal range, then it bypasses the thermal management systems and only re-circulates before getting pumped into the battery (Route A as shown in Figure 5.5). This loop permits temperature stability by controlling cell temperatures through pump control. When the battery temperature is high and the ambient temperature is lower than the desired temperature of the battery, the ambient air flow in the battery cooler is used to keep this coolant circuit cool (Route B). If the battery temperature is significantly higher and the ambient temperature is higher than the desired battery temperature, then by operating the electric air conditioning (A/C) compressor, R134a refrigerant is throttled by the thermal expansion valve (TXV) to permit super-cooling of the battery coolant as it passes through the chiller unit (Route C). This increases the efficiency of the system significantly since cooling via a cooling circuit would consume more energy than operating the battery coolant circuit due to the need of the air compressor in the first case (Behr, 2012).

When these systems are subdivided into its components, they are mainly composed of a compressor, heat exchangers, thermal expansion valves, pump and the battery. These components are described in more detail in the next section. The coolant pump is not described further due to its relatively negligible impact on the overall system.

5.3.1 Compressor

The compressor is a main component of the air conditioning system. A magnetic clutch is located at the front of the compressor and used to engage it when power is provided to the system. In the analysis, a scroll type compressor is used and modeled with respect to the isentropic efficiency correlation as follows (Brown *et al.*, 2002):

$$\eta_s = 0.85 - 0.046667\,\theta \tag{5.9}$$

Moreover, by using the ideal polytrophic equation for adiabatic and isentropic compression and with the assumption of ideal behavior for the compressor suction port gas, the following relationship can be obtained between the discharge and suction temperatures (Bhatti, 1999).

$$\frac{T_{discharge}}{T_{suction}} = 1 + \frac{1}{\eta_s}\left[\theta^{1-(1/\gamma)} - 1\right] \tag{5.10}$$

The above equations show that the refrigerants with lower pressure ratios ultimately result in a higher compression efficiency which then increases the COP of the system. Also, for a given pressure ratio, higher isentropic efficiencies lead to a lower compressor discharge temperature which results in lower compressor work, and consequently higher COP of the system.

5.3.2 Heat Exchangers

There are three heat exchangers in most of today's TMSs, namely the condenser, evaporator and chiller. The condenser is located in front of the radiator and the evaporator is located adjacent to the passenger compartment to condition the cabin. The chiller is placed somewhere between the air conditioner and battery loops and has coolant on one side and the refrigerant on the other. In order to be able to simply the heat exchanger models, the following assumptions are usually made in determining the heat transfer coefficients and pressure drops:

- The heat exchangers operate under steady-state conditions.
- The heat losses to surroundings are negligible.
- The changes in kinetic and potential energies of the fluid stream are negligible.
- There is no fouling.
- The temperature of the fluid is uniform over the flow cross section.
- There are no thermal energy sources and sinks in the heat exchanger walls or fluids.
- The velocity and temperature at the entrance of the heat exchanger on each fluid side are uniform.
- The overall heat exchanger surface efficiency is assumed uniform and constant throughout.

The overall heat transfer coefficient can be determined as follows:

$$\frac{1}{UA} = \frac{1}{\eta_i A_i \bar{h}_i} + \frac{\delta_w}{k_w A_w} + \frac{1}{\eta_o \bar{h}_o A_o} \tag{5.11}$$

Since the wall thickness of the tube is generally small and the thermal conductivity of the tube material is high, the thermal resistance of the tube is often negligible and therefore the equation can be simplified as follows:

$$\frac{1}{U} = \frac{A_o}{\eta_i A_i \bar{h}_i} + \frac{1}{\eta_o \bar{h}_o} \tag{5.12}$$

where the internal efficiency (η_i) is set equal to 1 since the channels with smooth internal surfaces are assumed in the model. The finned heat transfer surface efficiency (η_o) is given by the equation below:

$$\eta_o = 1 - \left(\frac{A_{fin}}{A_o}\right)(1 - \eta_{fin}) \tag{5.13}$$

In addition, the internal heat transfer coefficients ($\bar{h}$) associated with the refrigerant in the heat exchangers can be determined based on the correlation below (Dittus and Boelter, 1930):

$$\bar{h}_{evap,i} = \frac{3.6568k}{D_i} \quad for\ 0 < Re < 2000 \tag{5.14a}$$

$$\bar{h}_{cond,i} = \frac{4.3636k}{D_i} \quad for\ 0 < Re < 2000 \tag{5.14b}$$

$$\bar{h}_{evap\ or\ cond,i} = \frac{k}{D_i} \cdot \frac{0.5f\ (Re - 1000)Pr}{1 + 12.7\ (0.5f)^{0.5}\left(Pr^{\frac{2}{3}} - 1\right)} \tag{5.14c}$$

where

$$f = 0.054 + 2.3 \times 10^{-8}\, Re^{3/2} \;\; for\; 2300 < Re < 4000 \tag{5.14d}$$

$$f = 1.28 \times 10^{-3} + 0.1143\, Re^{-0.311} \;\; for\; 4000 < Re < 5 \times 10^{6} \tag{5.14e}$$

Here, Pr is the Prandtl number. On the air side, the heat transfer coefficients ($\bar{h}_o$) for the condenser and evaporator can be calculated for forced convection by the correlation below (Churchill and Chu, 1975):

$$\bar{h}_0 = \frac{k}{D_o} \cdot \left[0.6 + \frac{0.387\, Ra^{\frac{1}{6}}}{\left(1 + (0.559/Pr)^{\frac{9}{16}}\right)^{\frac{8}{27}}} \right]^2 \tag{5.15}$$

where Ra is the Rayleigh Number:

$$Ra = \Pr\, g\, \beta\, (T_0 - T_w) D_0^3 / v^2 \tag{5.16}$$

Here, g is the gravitational acceleration, β is the coefficient of thermal expansion, D_o is the outer diameter of the tube and v is the specific heat of air.

The total pressure drop in the heat exchanger consists of frictional, acceleration and gravitational components. Assuming the flow is fully developed, the gravitational component becomes negligible. Moreover, the acceleration effects are usually significantly smaller than the frictional effects, thus only the frictional pressure drop can be considered in the analysis. The pressure drop correlations (in kPas) with respect to the heat exchangers are given below (Lee and Yoo, 2000):

$$\Delta P_{evap} = 6 \times 10^{-6}\, Re_g^{1.6387} \;\; for\; 4000 < Re_g < 12000 \tag{5.17a}$$

$$\Delta P_{cond} = 6 \times 10^{-8}\, Re_g^{2} + 0.0009\, Re_g - 6.049 \;\; for\; 3000 < Re_g < 3 \times 10^{4} \tag{5.17b}$$

where Re_g is the Reynolds number given below:

$$Re = \frac{\dot{m}_{ref}\, D_{in}}{\mu\, A_i} \tag{5.18}$$

Here, $\dot{m}_{ref}$ is the mass flow rate of the refrigerant, D_{in} is the inner diameter of the tubes, μ is the dynamic viscosity and A_i is the tube cross-sectional area of the heat exchanger.

The air temperatures at the refrigerant evaporating/condensing exit states as well as superheating, desuperheating and subcooling states can be determined based on the heating and cooling loads, associated mass flow rates and the respective temperature differences. The temperature differences can be calculated based on the log mean temperature difference method (LMTD) for heat exchangers as given below:

$$LMTD = \frac{(T_{H,i} - T_{L,o}) - (T_{H,o} - T_{L,i})}{\ln\left(\dfrac{T_{H,i} - T_{L,o}}{T_{H,o} - T_{L,i}}\right)} \tag{5.19}$$

where the subscripts H and L represent high and low temperature sides and i and o refer to "in" and "out", respectively. The thermal performance of heat exchangers can be calculated with respect to the effectiveness-NTU method, so the effectiveness is defined as:

$$\varepsilon = \frac{\dot{Q}}{C_{min}(T_{hi} - T_{ci})} = 1 - \exp\left[\frac{1}{C^*}NTU^{0.22}(\exp(-C^* \times NTU^{0.78}) - 1\right] \quad (5.20a)$$

$$where\ C = \dot{m}C_p, \quad C^* = \frac{C_{min}}{C_{max}} \quad and \quad NTU = \frac{UA}{C_{min}} \quad (5.20b)$$

Here, C is the heat capacity rate, which is the mass flow rate times the specific heat, and C_{min} and C_{max} represents the smaller and larger heat capacity rates among the hot and cold sides, respectively. The heat transfer rates in the heat exchangers ($\dot{Q}$) are determined from the energy balance equations.

$$\varepsilon = \varepsilon_{max} = 1 - \exp(-NTU) \quad (5.21)$$

In the phase change regions of the evaporator and condenser, the effectiveness becomes maximum (ε_{max}) since C_{max} becomes considerably large. From the above equations, it can also be inferred that higher values of ε would require higher values of NTU which inherently means larger heat transfer coefficients on the hot and cold sides of the utilized heat exchangers.

5.3.3 Thermal Expansion Valve (TXV)

The thermal expansion valve controls the refrigerant flow into the evaporator by means of a capillary tube that has a thermal bulb. This bulb controls the width of the valve by balancing the thermal bulb and refrigerant internal pressures. In this system, TXV is modelled as an orifice where the liquid expands from the condensing to evaporating pressure. This process is usually considered to be isenthalpic as the sum of kinetic and potential energies between the entrance and the exit points as well as the heat transfer in between are relatively insignificant. The corresponding flow rate can be shown as:

$$\dot{m}_{ref} = C_{txv}A_{txv}\sqrt{(P_{txv,i} - P_{txv,o})/v_{txv,i}} \quad (5.22)$$

where $P_{txv,i}$ and $P_{txv,o}$ are the inlet and outlet valve pressures respectively. Also, A_{txv} is the flow area (in m^2) which can be calculated by the following equation:

$$A_{txv} = \frac{k_A}{\sqrt{2}C_{txv}} \quad (5.23)$$

Here, k_A is the thermal expansion valve characteristic parameter that relates to the evaporating temperature by the following correlation (Jabardo *et al.*, 2002):

$$k_A = 5.637 \times 10^{-5} + 1.358 \times 10^{-7}T_{evaporating} \quad (5.24)$$

Moreover, C_{txv} is the valve flow coefficient that is correlated from experimental studies in the literature as follows (Tian and Li, 2005):

$$C_{txv} = 0.187 + 4.84 \times 10^{-7}P_{txv,i} - 0.579x_{txv,o} \quad (5.25)$$

where $x_{txv,o}$ is the refrigerant quality at the valve exit.

5.3.4 Electric Battery

The electric battery plays a significant role on the overall vehicle performance and its efficiency is inherently linked to reducing the discrepancy between the optimum and operating conditions of the selected batteries, as regulated by the vehicle TMS. In this model, it is assumed that a given portion of the generated heat in the battery is absorbed by the battery coolant. Chapter 4 can be referred for the information and the procedure regarding modeling the battery.

5.3.5 System Parameters

In real-life thermal management systems and applications, each input varies within certain ranges based on the operating conditions, constraints and environmental impact. These ranges are set based on the common standards in the literature along with physical and economical limitations. In the refrigeration cycle, usually up to 10°C of superheating and subcooling is utilized in order to improve the system efficiency. The evaporator and condenser air mass flow rates vary with respect to the vehicle speed and fan power. They are taken to be between 0.1 and 0.5 kg/s in most BTMSs. Moreover, relatively high ambient temperatures are used when observing the effects of cooling the electric battery under high temperatures, since hot weather conditions are a more significant concern than cold weather conditions due to the permanent effects of high temperatures on the battery performance as well as associated potential safety concerns. In addition, the heating is provided through heater(s) placed in the vehicle such as the battery and the cabin core which have high efficiencies and very little room for improvement. Furthermore, a cooling capacity needs to be selected to provide adequate cooling to the vehicle cabin under these ambient temperatures. The list of the major BTMSs parameters and their general ranges are given in Table 5.2.

Moreover, since the use of R134a will be terminated in the near future by the European Community (due to the requirement of using refrigerants with GWP less than 150), the use of alternative refrigerants in the BTMS can also be considered in the analysis (European Union, 2006). One of the possible solutions to avoid R134a is the use of natural refrigerants, such as hydrocarbons, which attracted renewed interest during the past few decades due to being environmentally benign with negligible GWP and

Table 5.2 Range of parameters commonly used in BTMSs.

Parameter	Range of variation
Compressor speed (rev/min)	1500–5000
Compression Ratio	1–5
Evaporating Air Temperature (°C)	0–15
Superheating Temperature (°C)	0–12
Evaporator Air Mass Flow Rate (kg/s)	0.1–0.5
Cooling Capacity (kW)	1–5
Condensing Air Temperature (°C)	40–55
Condenser Air Mass Flow Rate (kg/s)	0.1–0.5
Subcooling Temperature (°C)	0–12

zero ODP. They also have various additional advantages such as availability, low cost, high miscibility with conventional mineral oil and compatibility with existing refrigerating systems. On the other hand, their main drawback is potential flammability and safety hazards. The characteristics of these refrigerants along with R134a can be seen in Table 5.3. Currently, hydrocarbons are already utilized in a few established applications around the world such as household refrigerators and small heat pump applications. It should be noted that R-744 (CO_2) is not considered among the prospective hydrocarbons even though it offers a number of desirable properties such as ready availability, low toxicity, low GWP and cost, due to the need for implementing a transcritical cycle and additional safety standards that require significant modifications to the baseline system, based on its different thermophysical properties relative to R134a.

In addition, there are also certain other refrigerants that could be utilized in EV BTMSs, such as R1234yf and dimethyl ether (DME), and therefore is also considered. Among the fluorinated propene isomers, R1234yf is one of the major candidates as a replacement for R134a in automotive applications due to its ability to be used with compatible materials and oils as well as having a low GWP (about 4) and low normal boiling temperatures with respect to R134a. Moreover, several studies have shown that the environmental impact of R1234yf is significantly lower than R134a in most cases (Koban, 2009). However, it also has certain drawbacks such as additional costs, relative flammability, miscibility with oil as well as stability problems in the presence of small amounts of water and air in the BTMS. DME is another good candidate due to being non-toxic during normal usage, widely available, environmentally safe, excellent material compatibility and better heat transfer properties as well as lower costs than R134a. The main drawback is its high flammability, which is about twice as high as the other hydrocarbons.

5.4 Energy and Exergy Analyses

In thermal management of batteries, another significant topic is energy analysis, which comes from the first-law of thermodynamics, and exergy analysis, which comes from the second-law of thermodynamics. Due the fact that the first-law is knows as conservation law (referring to the fact that energy is neither created nor destroyed), the role of energy analysis is quite limited due to the conservation law and cannot cover irreversibilities and/or destructions. This requires that we should go one step ahead and include exergy analysis as a potential tool. When it comes to system analysis and assessment, both energy and exergy analyses are utilized for practical systems and applications. This clearly indicates the importance of this section in regards to batter management system design, analysis and assessment.

5.4.1 Conventional Analysis

In the first step of the exergy analysis, the mass, energy, entropy and exergy balances are needed in order to determine the heat input, rate of entropy generation and exergy destruction as well as the energy and exergy efficiencies. In general, a balance equation for a quantity in a system may be written as follows:

$$Input + Generation - Output - Consumption = Accumulation \tag{5.26}$$

Table 5.3 Characteristics of R134a and various alternative refrigerants [a].

Code	Chemical formula/ Common name	Mol. mass	NBP[b] (°C)	T_{crit} (°C)	P_{crit} (bar)	Latent heat (kJ/kg)	Lower flam. limit (vol. %)[c]	ODP	GWP
R134a	CH_2FCF_3	44.1	−42.1	96.7	42.5	216.8	Non-flammable	0	1300
R290	C_3H_8/Propane	44.1	−42.1	96.7	42.5	423.3	2.3−7.3	0	20
R600	C_4H_{10}/Butane	58.1	−0.5	152.0	38.0	385.7	1.6−6.5	0	20
R600a	C_4H_{10}/Isobutane	58.1	−11.7	134.7	36.4	364.2	1.8−8.4	0	20
R1234yf	$CF_3CF{=}CH_2$/Tetrafluorpropene	N/A	−29.0	95	33.8	175	6.2−13.3	0	4
RE170 (DME)	CH_3OCH_3/Dimethylether	46.0	−24.7	126.9	53.7	410.2	3.4−17	0	<3−5

a) Data taken from Wongwises *et al.*, 2006
b) Normal Boling Point (NBP) is at 101.325 kPa (°C)
c) Explosive limits in air % by volume

where input and output terms refer to quantities entering and exiting through the system, respectively, whereas generation and consumption terms refer to quantities produced or consumed within the system, and the accumulation term refers to potential build-up of the quantity within the system.

In steady-state conditions, however, all properties are unchanging with time and therefore, all the transient accumulation terms become zero. Thus, under the steady-state assumption, the balance equations for mass, energy, entropy and exergy can be written as follows:

$$\dot{m}_{in} = \dot{m}_{out} \tag{5.27a}$$

$$\dot{E}_{\text{in}} = \dot{E}_{out} \tag{5.27b}$$

$$\dot{S}_{\text{in}} + \dot{S}_{gen} = \dot{S}_{out} \tag{5.27c}$$

$$\dot{E}x_{\text{in}} = \dot{E}x_{out} + \dot{E}x_D \tag{5.27d}$$

where

$$\dot{S}_{gen} = \dot{m}\Delta s \tag{5.27e}$$

$$\dot{E}x_D = T_0\,\dot{S}_{gen} \tag{5.27f}$$

In the first two equations, $\dot{m}$ and $\dot{E}$ are associated with the mass flow rate and energy transfer rate and show that the respective total rates in / out across the boundary are conserved (neglecting reactions). In the third equation, $\dot{S}$ is the entropy flow or generation rate. The amount transferred out of the boundary must exceed the rate in which entropy enters, the difference being the rate of entropy generation within the boundary due to associated irreversibilities. Similarly, in the equation (5.27d), $\dot{E}x$ is the exergy flow rate and it shows that exergy transferred out of the boundary must be less than the rate inwhich exergy enters, the difference being the rate of exergy destruction (or lost work) within the boundary due to associated irreversibilities which can be calculated by the dead-state temperature (T_0) multiplied by the entropy generation rate as given in equation (5.27f) (based on the Gouy-Stodola theorem). Minimum exergy destruction, or minimum entropy generation, design characterizes a system with minimum destruction of available work, which in the case of refrigeration plants, is equivalent to the design with a maximum refrigeration load, or minimum mechanical power input (Bejan, 1997). In cooling systems, T_0 usually equals to the temperature of the high-temperature medium T_H.

In addition, the specific flow exergy associated with the coolant medium is given below:

$$ex_{coolant} = (h - h_0) + \frac{1}{2}V^2 + gZ - T_0(s - s_0) \tag{5.28}$$

Considering a system at rest relative to the environment, kinetic and potential terms can be ignored,

$$ex_{coolant} = (h - h_0) - T_0(s - s_0) \tag{5.29}$$

The exergy rate is determined as

$$\dot{E}x = \dot{m} * ex \tag{5.30}$$

In this regard, the TMS configuration and relevant parameters are described and fundamental principles of the exergy are introduced, the TMS can be studied with respect to energy and exergy analyses based on the aforementioned system model. Ideally, in the thermal management system, the refrigerant travels through the condenser at constant pressure by heat absorption and exits the condenser as a saturated liquid. Moreover, the refrigerant is compressed isentropically in the compressor before entering the condenser and expanded isenthalpically in the thermal expansion valve before entering the evaporator. The refrigerant also flows through the evaporator at constant pressure by heat rejection and exits the evaporator as a saturated vapor. However, practical applications deviate from ideal conditions due to pressure and temperature drops associated with the refrigerant flow and heat transfer to/from the surroundings. During the compression process, entropy changes due to the irreversibilities and heat transfer to/from the surroundings. There is also some pressure drop as the refrigerant flows through the condenser and evaporator as modeled in the previous section. Furthermore, the refrigerant is subcooled as it is leaves the condenser (and may drop further before reaching the expansion valve) and slightly superheated (due to the pressure losses caused by friction) as it leaves the evaporator (and enters the compressor). The temperature of the refrigerant further increases as it flows to the compressor, increasing its specific volume, which increases the work of the compressor. On the coolant side, the coolant is pumped to the battery, where the pressure increases significantly with a slight increase on its temperature. The coolant then exchanges heat with the battery module without any phase change in the medium. Subsequently, the coolant enters the chiller in order to transfer the heat to the refrigerant cycle and enters the pump again to make up for the lost pressure before re-entering the battery.

For the compressor:

M.B.E

$$\dot{m}_1 = \dot{m}_2 = \dot{m}_r \tag{5.31a}$$

E.B.E

$$\dot{m}_1 h_1 + \dot{W}_{comp} = \dot{m}_2 h_2 \tag{5.31b}$$

En.B.E

$$\dot{m}_1 s_1 + \dot{S}_{gen,comp} = \dot{m}_2 s_2 \tag{5.31c}$$

Ex.B.E

$$\dot{m}_1 ex_1 + \dot{W}_{comp} = \dot{m}_2 ex_2 + \dot{E}x_{D,comp} \tag{5.31d}$$

$$\dot{E}x_{D,comp} = T_0\, \dot{S}_{gen,comp} = \dot{m} T_0 (s_2 - s_1) \tag{5.31e}$$

Exergy efficiency:

$$\eta_{ex,comp} = \frac{(\dot{E}x_{2,act} - \dot{E}x_1)}{\dot{W}_{comp}} = 1 - \frac{\dot{E}x_{D,comp}}{\dot{W}_{comp}} \tag{5.31f}$$

where $\dot{W}_{comp}$ is the compressor power input in kW. Moreover, the isentropic efficiency of the adiabatic compressor is defined as

$$\eta_{comp} = \frac{\dot{W}_s}{\dot{W}} = \frac{h_{2,s} - h_1}{h_2 - h_1} \tag{5.32}$$

Here, $\dot{W}_s$ is the isentropic power and $h_{2,s}$ is the isentropic (i.e, reversible and adiabatic) enthalpy of the refrigerant leaving the compressor.

For the condenser:

M.B.E $$\dot{m}_2 = \dot{m}_3 = \dot{m}_r \quad (5.33a)$$

E.B.E $$\dot{m}h_2 = \dot{m}h_3 + \dot{Q}_{cond} \quad (5.33b)$$

En.B.E $$\dot{m}_2 s_2 + \dot{S}_{gen,cond} = \dot{m}_3 s_3 \quad (5.33c)$$

Ex.B.E $$\dot{m}_2 ex_2 = \dot{m}_3 ex_3 + Ex_{Q_H} + \dot{E}x_{D,cond} \quad (5.33d)$$

$$\dot{E}x_{D,cond} = T_0\, \dot{S}_{gen,cond} = \dot{m}T_0\left(s_3 - s_2 + \frac{q_H}{T_H}\right) \quad (5.33e)$$

Exergy efficiency

$$\eta_{ex,cond} = \frac{\dot{E}x_{\dot{Q}_H}}{\dot{E}x_2 - \dot{E}x_3} = 1 - \frac{\dot{E}x_{D,cond}}{\dot{E}x_2 - \dot{E}x_3} \quad (5.33f)$$

with

$$\dot{E}x_{\dot{Q}_H} = \dot{Q}_H\left(1 - \frac{T_0}{T_H}\right) \quad (5.33g)$$

where $\dot{Q}_H$ is the heat rejection from the condenser to the high-temperature environment.

For the thermal expansion valve before the evaporator (the expansion process is considered isenthalpic):

M.B.E $$\dot{m}_3 = \dot{m}_4 = \dot{m}_r \quad (5.34a)$$

E.B.E $$h_3 = h_4 \quad (5.34b)$$

En.B.E $$\dot{m}_3 s_3 + \dot{S}_{gen,TXV} = \dot{m}_4 s_4 \quad (5.34c)$$

Ex.B.E $$\dot{m}_3 ex_3 = \dot{m}_4 ex_4 + \dot{E}x_{D,TXV} \quad (5.34d)$$

$$\dot{E}x_{D,TXV} = T_0\, \dot{S}_{gen,TXV} = \dot{m}T_0(s_4 - s_3) \quad (5.34e)$$

Exergy efficiency:

$$\eta_{ex,TXV} = \frac{\dot{E}x_4}{\dot{E}x_3} \quad (5.34f)$$

For the evaporator:

M.B.E $$\dot{m}_4 = \dot{m}_1 = \dot{m}_r \quad (5.35a)$$

E.B.E $$\dot{m}h_4 + \dot{Q}_L = \dot{m}h_1 \quad (5.35b)$$

En.B.E $$\dot{m}_4 s_4 + \dot{S}_{gen,evap} = \dot{m}_1 s_1 \quad (5.35c)$$

Ex.B.E $$\dot{m}_4 ex_4 + Ex_{Q_L} = \dot{m}_1 ex_1 + \dot{E}x_{D,evap} \quad (5.35d)$$

$$\dot{E}x_{D,evap} = T_0\, \dot{S}_{gen,evap} = \dot{m}T_0\left(s_1 - s_4 + \frac{q_L}{T_L}\right) \quad (5.35e)$$

Exergy efficiency

$$\eta_{ex,evap} = \frac{\dot{E}x_{\dot{Q}_L}}{\dot{E}x_4 - \dot{E}x_1} = 1 - \frac{\dot{E}x_{D,evap}}{\dot{E}x_4 - \dot{E}x_1} \quad (5.35f)$$

with

$$\dot{E}x_{\dot{Q}_L} = -\dot{Q}_L\left(1 - \frac{T_0}{T_L}\right) \quad (5.35g)$$

where $\dot{Q}_L$ is the heat taken from the low-temperature environment to the evaporator.

For the chiller:

M.B.E $$\dot{m}_5 = \dot{m}_1 = \dot{m}_{ref,b}\ \dot{m}_6 = \dot{m}_7 = \dot{m}_{cool} \quad (5.36a)$$

E.B.E $$\dot{m}_{ref,b}h_5 + \dot{Q}_{ch} = \dot{m}_{ref,b}h_{1''} \quad (5.36b)$$

En.B.E $$\dot{m}_{ref,b}s_5 + \dot{m}_{cool}s_6 + \dot{S}_{gen} = \dot{m}_{ref,b}s_{1''} + \dot{m}_{cool}s_7 \quad (5.36c)$$

Ex.B.E $$\dot{m}_{ref,b}ex_5 + \dot{E}x_{\dot{Q}_{ch}}\ \dot{m}_{ref,b}ex_{1''} + \dot{E}x_{D,ch} \quad (5.36d)$$

Exergy efficiency:

$$\eta_{ex,ch} = \frac{\dot{E}x_{\dot{Q}_{ch}}}{\dot{E}x_5 - \dot{E}x_{1''}} \quad (5.36e)$$

with

$$\dot{Q}_{ch} = \dot{m}_{cool}(h_7 - h_6)\ and\ \dot{E}x_{\dot{Q}_{ch}} = -\dot{Q}_{ch}\left(1 - \frac{T_0}{T_7}\right) \quad (5.36f)$$

The enthalpy and entropy changes in the water/glycol mixture of 50/50 by weight are calculated by assuming the specific heat remains constant as follows:

$$h_7 - h_6 = C_{wg}(T_7 - T_6)\ and\ s_7 - s_6 = C_{wg}\ln\left(\frac{T_7}{T_6}\right) \quad (5.37a)$$

For the pump:

M.B.E $$\dot{m}_7 = \dot{m}_8 = \dot{m}_{cool} \quad (5.38a)$$

E.B.E $$\dot{m}_{cool}h_7 + \dot{W}_{pump} = \dot{m}_{cool}h_8 \quad (5.38b)$$

En.B.E $$\dot{m}_{cool}s_7 + \dot{S}_{gen} = \dot{m}_{cool}s_8 \quad (5.38c)$$

Ex.B.E $$\dot{m}_{cool}ex_7 + \dot{W}_{pump} = \dot{m}_{cool}ex_8 + \dot{E}x_{D,pump} \quad (5.38d)$$

Exergy efficiency:

$$\eta_{ex,pump} = \frac{(\dot{E}x_{8,act} - \dot{E}x_7)}{\dot{W}_{pump}} \quad (5.38e)$$

For the battery:

M.B.E $$\dot{m}_8 = \dot{m}_6 = \dot{m}_{cool} \quad (5.39a)$$

E.B.E $$\dot{m}_{cool}h_8 + \dot{Q}_{bat} = \dot{m}_{cool}h_6 \quad (5.39b)$$

En.B.E $$\dot{m}_{cool}s_8 + \dot{S}_{gen} = \dot{m}_{cool}s_6 \quad (5.39c)$$

Ex.B.E $$\dot{m}_{cool}ex_8 + \dot{E}x_{\dot{Q}_{bat}} = \dot{m}_{cool}ex_6 + \dot{E}x_{D,bat} \quad (5.39d)$$

For the entire cooling system, the energetic coefficient of performance (COP) becomes

$$COP_{en,system} = \frac{\dot{Q}_{evap} + \dot{Q}_{ch}}{\dot{W}_{comp} + \dot{W}_{pump}} \quad (5.40)$$

Note that actual cooling systems are less efficient than the ideal energy models due to irreversibilities in the actual systems. As given in the previous equations, a smaller temperature difference between the heat sink and heat source provides higher cooling system efficiency. Thus, the aim of the exergy analysis is to determine the system irreversibilities by calculating the exergy destruction in each component and to calculate

Table 5.4 Exergy destruction rates for each component in the TMS.

Component	Exergy destruction rates
Compressor	$\dot{E}x_{D,comp} = T_0\dot{m}_r(s_2 - s_1)$
Condenser	$\dot{E}x_{D,cond} = T_0[\dot{m}_c(s_{c2} - s_{c1}) - \dot{m}_r(s_2 - s_3)]$
Evaporator TXV	$\dot{E}x_{D,evap,TXV} = T_0\dot{m}_{r1}(s_4 - s_3)$
Chiller TXV	$\dot{E}x_{D,ch,TXV} = T_0\dot{m}_{r2}(s_5 - s_3)$
Evaporator	$\dot{E}x_{D,evap} = T_0[\dot{m}_e(s_{e2} - s_{e1}) - \dot{m}_{r1}(s_4 - s_1)]$
Chiller	$\dot{E}x_{D,ch} = T_0[\dot{m}_{cool}(C_{wg}\ln(T_6/T_7)) - \dot{m}_{r2}(s_5 - s_1)]$
Pump	$\dot{E}x_{D,pump} = T_0\dot{m}_{cool}(C_{wg}\ln(T_6/T_0))$
Battery	$\dot{E}x_{D,bat} = T_0\dot{m}_{cool}(C_{wg}\ln(T_6/T_8))$

the associated exergy efficiencies. This methodology helps to focus on the parts where the greatest impact can be achieved on the system since the components with larger exergy destruction also have more potential for improvements. The exergy destruction calculations and results for each component can be observed in Table 5.4. For the overall system, the total exergy destruction of the cycle can be calculated by adding the exergy destruction associated with each component that was previously calculated.

For the overall system, the total exergy destruction of the system can be calculated by adding the exergy destruction associated with each component that was previously calculated.

$$\begin{aligned}\dot{E}x_{D,system} &= \dot{E}x_{D,comp} + \dot{E}x_{D,cond} + \dot{E}x_{D,evap,TXV} + \dot{E}x_{D,ch,TXV} + \dot{E}x_{D,evap} \\ &\quad + \dot{E}x_{D,ch} + \dot{E}x_{D,pump} + \dot{E}x_{D,battery}\end{aligned} \tag{5.41}$$

Finally, for the thermodynamic analysis, using the aforementioned exergy equations, the exergetic COP of the system can be calculated as

$$COP_{ex,system} = \frac{\dot{E}x_{\dot{Q}_{evap}} + \dot{E}x_{\dot{Q}_{ch}}}{\dot{W}_{comp} + \dot{W}_{pump}} \tag{5.42}$$

5.4.2 Enhanced Exergy Analysis

In the previous section, the exergy analysis is conducted in order to determine the exergy destructions associated with each component as well as the overall system. However, the conventional exergy analysis does not evaluate the mutual interdependencies among the system components. For this reason, the irreversibilities within the components are divided into two categories; the irreversibilities related to the specific entropy generation within the component ($s_{gen,k}$) and the ones related to the system structure and inefficiencies of the other components in the system (mainly with respect to the changes in the mass flow rate) by conducting so called enhanced exergy analysis. For this reason, *endogenous* and *exogenous* exergy destruction concepts are introduced as given below:

$$\dot{E}_{D,k} = \dot{E}_{D,k}^{EN} - \dot{E}_{D,k}^{EX} \tag{5.43}$$

The endogenous exergy destruction for a given component is associated only with the component itself and would still exist even if all the other components in the system would operate in an ideal way. On the other hand, exogenous exergy destruction is the remaining part of the entire exergy destruction within the component where it depends both on the inefficiencies associated with the component itself and the remaining components in the system. This distinction plays a key role in improving component design since the efforts spent on decreasing the endogenous exergy destruction in a component can often promote a decrease in exogenous part of the exergy destruction in other components. For the given BTMS, the boundaries of all exergy balances are taken at the ambient temperature where the exergy loss is zero (for the individual components) and therefore all the thermodynamic inefficiencies are solely due to exergy destructions in the components.

In addition, the exergy destruction associated with a component may not be able to be reduced based on technological limitations (such as the availability and cost of materials and manufacturing processes). This portion of the exergy destruction in a component is called unavoidable part of exergy destruction ($\dot{E}_{D,k}^{UN}$), whereas the remaining part is called avoidable part of the exergy destruction ($\dot{E}_{D,k}^{AV}$). This distinction between the parts of exergy destructions can be useful in providing a realistic measure of the potential for improving the thermodynamic efficiency of a component (Morosuk and Tsatsaronis, 2008).

The combination of endogenous/exogenous and avoidable/unavoidable exergy destructions can be very helpful in determining the components needed to be focused on in order to reduce the exergy destruction of the overall system and the portion of this exergy destruction that can possibly be reduced. In this regard, endogenous avoidable part of the exergy destruction can be reduced through improving the efficiency of the component, whereas the exogenous avoidable exergy destruction can be reduced improving the efficiency of the remaining components as well as the efficiency of the analyzed component. Moreover, the endogenous unavoidable exergy destruction cannot be reduced due to technical limitations on the components, whereas exogenous unavoidable exergy destruction cannot be reduced due to the technical limitations of the remaining components.

The battery thermal management systems are considered is a closed loop system, where compressor and pump works are used as the primary inputs and evaporator and chiller cooling loads are the primary outputs in a system where the output of one component is used as the input of the next component. In addition, the product for all the remaining components is the fuel of the component that follows them. Thus, the rates of exergy destruction should be calculated very carefully, since a part of the exergy destruction of each component is caused by inefficiencies of the remaining components. The exergy destruction in each component depends on the efficiency of the individual components along with the temperature and mass flow rates of the main and secondary working fluids.

In the studied system, the analysis can be conducted using either the total product ($\dot{E}_{P,TOT}$) or the total fuel ($\dot{E}_{F,TOT}$) to be constant in the system. This distinction does not affect any results in avoidable/unavoidable exergy destruction calculations since the components are considered in isolation during the analysis. However, it affects the endogenous/exogenous exergy destruction calculations since the mass flow rates of the working fluid changes based on the parameter that is considered to be constant (product or fuel). Since the analysis is trying to minimize the fuel consumption for a given system (with a fixed output), constant product assumption is used in the analysis.

In this regard, the exergy destruction of the evaporator is completely endogenous since $\dot{E}_{D,evap}$ is a function of the component's exergy destruction exclusively. On the other hand, $\dot{E}_{D,TXV}$ depends on the exergy efficiencies of the evaporator TXV and the evaporator. Similarly, $\dot{E}_{D,cond}$ depends on the exergy efficiency of condenser, evaporator TXV and evaporator, and $\dot{E}_{D,comp}$ depends on the exergy efficiency components compressor, condenser, evaporator TXV and evaporator in the refrigerant loop.

Moreover, in order to be able to split the exergy destruction into parts, a thermodynamic based approach (so called "Cycle method") is applied (Morosuk and Tsatsaronis, 2006). The exergy destruction in each component depends on the substance as well as the temperature/pressure and mass flow rates associated with the working fluids along with the efficiencies of each component. These fluids are composed of R134a for the refrigeration cycle, 50/50 water-glycol mix for the coolant cycle and air as the secondary fluids in the condenser and evaporator. The calculation of the exergy destructions associated with the TMS components are provided in Section 5.3.

In order to have a better understanding of the exergy destruction in each component and the associated interdependencies among them, the exergy destructions are split it into endogenous and exogenous components by analyzing the TMS under theoretical cycles. This is achieved with respect to assuming minimum exergy destruction associated with each component (zero if possible). Based on this theoretical cycle, the compression process is considered isentropic ($\dot{E}^{th}_{D,comp} = 0$). On the other hand, since the throttling process is always irreversible, it is replaced with an ideal expansion process for the theoretical cycle. Moreover, temperature difference of 0°C between the primary and secondary working fluids is used in the heat exchangers. However, the temperatures and mass flow rate associated with the evaporator and chiller secondary working fluids are kept the same in order to keep the cooling loads constant and therefore the associated exergy destructions within these components are considered only to be endogenous ($\dot{E}_{D,evap} = \dot{E}^{EN}_{D,evap}$, $\dot{E}_{D,chil} = \dot{E}^{EN}_{D,chil}$) (for further details, see Kelly *at al.*, 2009).

Additionally, only a part of the total thermodynamic inefficiencies can be avoided in each component while other parts cannot. The improvement efforts should be concentrated in the avoidable part of the irreversibilities, thus it becomes imperative to separate the avoidable and unavoidable parts of the exergy destruction for each component. In order to split the exergy destruction into avoidable and unavoidable parts, an additional cycle is developed where only unavoidable exergy destructions occur within each component. These unavoidable exergy destructions occur in the cycle as a result of unavoidable temperature differences in the heat exchangers, efficiencies in the compressor and pump and by the throttling processes. These occur due to technological limitations (availability and cost of material and manufacturing) that prevent exceeding a certain upper limit of the component exergetic efficiency regardless of the amount of investment (Tsatsaronis and Park, 2002).

It should be noted that in order to assess the endogenous/exogenous and available/unavailable exergy destructions, simultaneous computations of the parallel cycles need to be calculated. This requires providing the system components with multiple inputs at once and obtaining multiple results where the differences among them could be evaluated. For this reason, EES is determined to be the most compatible software for the conducted analysis based on the ease of running alternative scenarios simultaneously in the software, and is used to develop the conventional and enhanced exergy analyses.

5.5 Illustrative Example: Liquid Battery Thermal Management Systems

Based on the aforementioned methods and parameters, a liquid battery thermal management system is analyzed using the thermodynamic modelling introduced in this chapter. The liquid systems is based on the TMS shown in Figure 5.3 and includes three cooling media – an R134a refrigerant is used in the refrigerant cycle, water/glycol mixture of 50/50 by weight is used in the battery coolant cycle, and ambient air is utilized in the evaporator and condensers in the system. In the baseline model, ambient air conditions of 35 °C and 1 ATM are used to study the effects of the TMS on the battery. The refrigerant mass flow rates are determined from thermal expansion valve correlations and the cooling capacity is calculated accordingly. For the baseline model, the temperature of the passenger cabin is set at 20 °C. Temperatures of 5 °C and 55 °C are used for evaporating and condensing temperatures along with 5 °C superheating and subcooling in the evaporator and condenser, respectively. The refrigerant mass flow rate in the chiller is determined with respect to the amount of battery heat transferred from the water/glycol mix in the coolant circuit to the refrigerant circuit via the chiller. In the refrigerant cycle, the refrigerant flow in the evaporator and chiller is combined in the system before it is compressed to the condenser. For the coolant circuit, the battery coolant temperature is assumed to be 19 °C (since it operates in a temperature range of 19 °C to 25 °C) before entering the battery, and the heat generated by the battery is considered to be 0.35 kW, where the mass flow rate of the battery coolant is determined accordingly. Finally, the parameters to calculate the unavoidable cycle are selected as 0.5°C for the heat exchangers and 0.95 for the compressor, regardless of the technological improvements in the system.

In this system, the exergy efficiencies and exergy destruction rates associated with each component are provided in Figure 5.6. Throughout the exergy analysis; the exergy efficiencies and exergy destruction rates are calculated for each component in the thermal management system. Among these components, the heat exchangers have the lowest exergy efficiencies with respect to the high temperature differences and phase change which results in more entropy generation between the refrigerant and coolants.

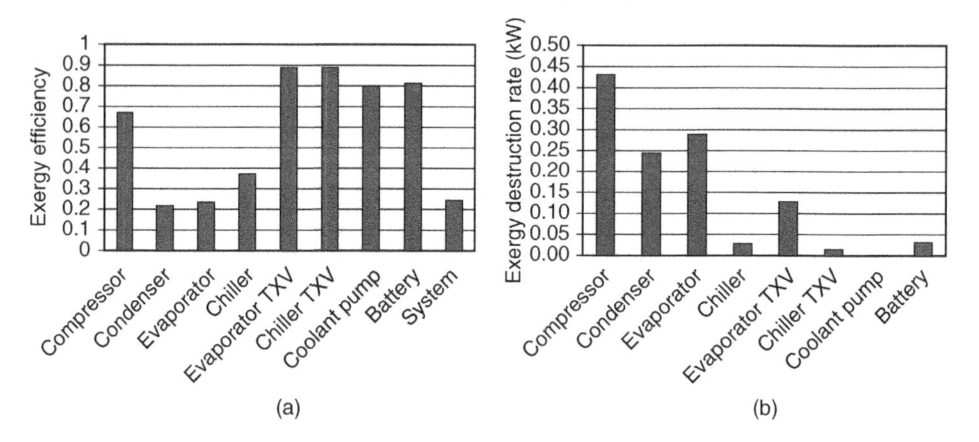

Figure 5.6 Baseline model (a) exergy efficiency and (b) exergy destruction rate of each component in the refrigerant and coolant cycles.

In the evaporator, the exergy losses are relatively high since (aside from the frictional losses) only part of the heat rejection occurs during the phase change process with large temperature differences between the working fluid in the evaporator and the vehicle cabin. Thus, reducing the mean temperature difference would reduce the exergy losses. One way of reducing the mean temperature difference is to increase the evaporator surface area, however, it should be weighed against the increase in the cost of installation (which will be analyzed in Chapter 6).

The condenser is calculated to have a lower exergy efficiency than the evaporator and the chiller, mainly due to the relatively higher temperature difference between the condenser exit and ambient air (taken at 35 °C), when compared to the differences between the evaporator exit and vehicle cabin temperature as well as the refrigerant and coolant temperatures.

Among the remaining components, the compressor has high compression pressure ratio and change in temperature of the refrigerant passing through the compressor, which contributes to an increase in exergy destruction. The exergy loss in the compressor can be reduced by using a compressor with higher isentropic efficiencies. Moreover, since the compressor power is highly dependent of the inlet and outlet pressures, proper sealing inside the compressor, heat exchanger improvements (such has reducing ΔT) and the implementation of multistage compression would reduce the exergy losses, thus reducing the compressor power. Furthermore, since a part of the irreversibilities occurs with respect to the frictional losses inside the compressor, utilizing appropriate lubricating oil that is miscible with the refrigerant (such as polyolester oil for R-134a) would reduce the respective exergy losses.

There is also significant research being conducted on the effects of using additives with a high conductivity (certain lubricant based nanofluids) in the refrigerant in order to improve the heat transfer rate, thus reducing the difference in the operating temperatures, which also reduced the exergy losses. However, proper care must be taken in the utilization of the lubricant in order to prevent the deposition of the lubricant in the evaporator wall. The interaction between the cooling and battery coolant cycles also helps in reducing the compressor requirements significantly. The transfer of excess heat from the battery coolant to the cooling cycle via the chiller helps allocate the thermal energy appropriately, since otherwise, the cooling cycle would need to supply the additional energy which uses a compressor. Therefore, further utilizing this interaction would also be beneficial. Moreover, irreversibilities in the system occur due to high temperature differences in heat exchangers, and therefore reducing these differences would reduce the associated irreversibilities.

The exergy efficiencies for the evaporator TXV and chiller TXV are higher (over 80%) since the processes are isenthalpic and have little or no heat loss. Therefore the exergy losses occur mainly due to a pressure drop in the expansion valve. The exergy losses in these TXVs can be reduced by lowering (or sub-cooling) the temperature of the refrigerant exiting the condenser, which can be feasible by utilizing the refrigerant vapor exiting the evaporator (Kumar *et al.*, 1989; Arora, 2008). The coolant pump also has a relatively higher efficiency (81%) since there is no significant heat loss from the pump.

It should be noted that the battery is modeled as a system in this chapter and thus the internal efficiencies for the battery are not considered. In this regard, the battery has high efficiencies within the target operating temperature range (up to 50 °C). However,

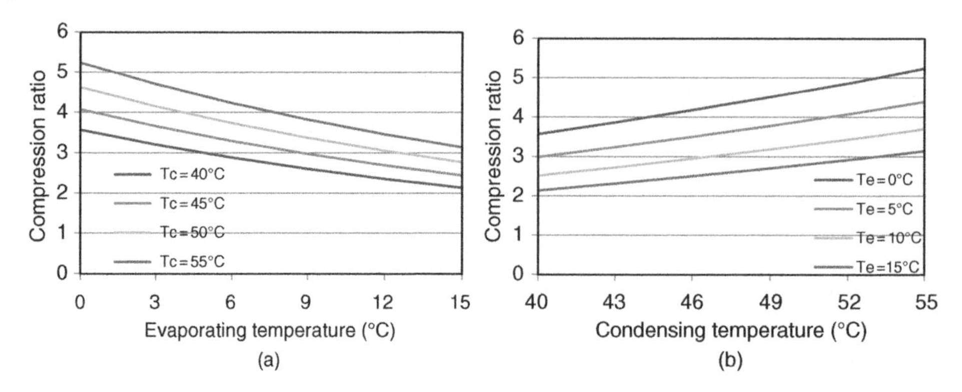

Figure 5.7 Compression ratio with respect to (a) evaporating and (b) condensing temperatures.

the associated efficiency would decrease significantly as the battery is heated up beyond this range. The thermodynamic analysis on a battery cell level can be seen in Chapter 4.

In order to understand the effects of different conditions and constraints, the TMS model is also analyzed based on the effects of condensing and evaporating temperatures, as well as subcooling and superheating temperatures, compressor speed, heat exchanger pressure drop and battery heat generation rates. Baseline values are used for all non-varied parameters in the parametric studies. Moreover, the use of various alternative refrigerants such as R290 (propane), R600 (butane), R600a (isobutane), R1234yf (tetrafluorpropene) and dimethyl ether (DME) are also investigated under various conditions. Condensing and evaporating temperatures affect the compression ratio, cooling load, COP and volumetric heat capacity. Compression ratio is a useful parameter on which to predict the volumetric performance of the compressor since lower compression reduces the likelihood of the high-pressure vapor to leak back to the low pressure side, which reduces the compressor volumetric efficiency. Figure 5.7 shows that the compression ratio is reduced by decreasing the condensing temperature or increasing the evaporator temperature.

In addition, as the evaporator temperature increases, the temperature of the refrigerant vapor before entering the compressor also increases. The refrigerant vapor specific volume reduction increases the associated refrigerant mass flow rate, and therefore increases the system cooling output. On the other hand, an increase in the condensing temperature leads to an increase in the temperature of the refrigerant discharged from the compressor along with the compression ratio. However, the compression capacity of the compressor will be reduced. Moreover, the refrigerant circulated per unit of time will be lower, which reduces the cooling load as shown in Figure 5.8.

Moreover, since energy consumption of the compressor is also proportional to the pressure ratio, this reduction in the condensing temperature or increase in the evaporator temperature increases the COP of the system by reducing the compression ratio. This shows that the required compressor power to a certain cooling capacity drops as the condensing temperature decreases or the evaporating temperature increases. Moreover, the throttling losses also decrease with decreasing temperature change, hence leading to an increase in the COP as shown in Figure 5.9.

Furthermore, the exergetic COP of the system also increases since reducing the condensing temperatures reduces the mean temperature difference between the refrigerant and the ambient air. Increasing the evaporating temperatures reduces the mean

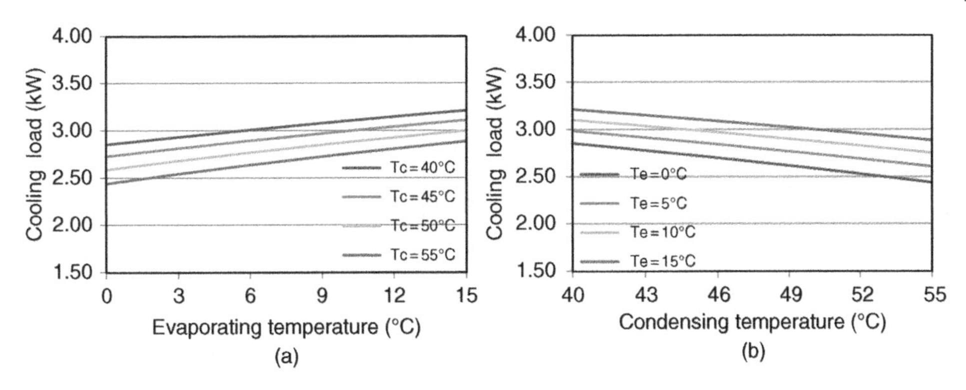

Figure 5.8 Cooling load with respect to (a) Evaporating and (b) Condensing temperatures.

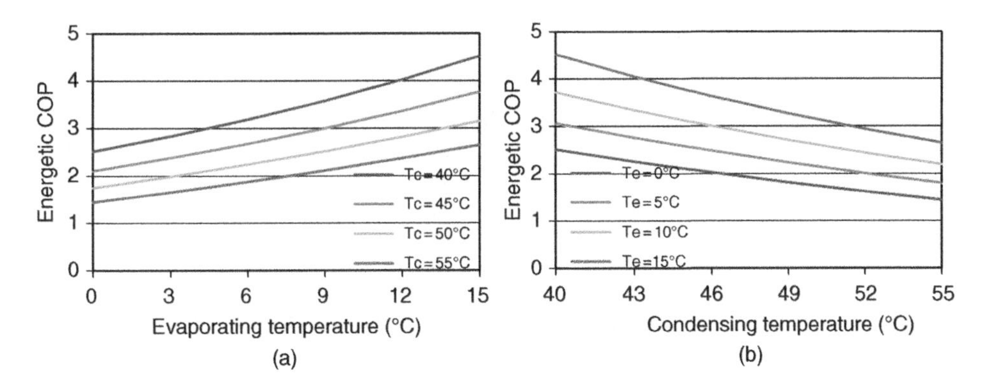

Figure 5.9 Energetic COP with respect to (a) Evaporating and (b) Condensing temperatures.

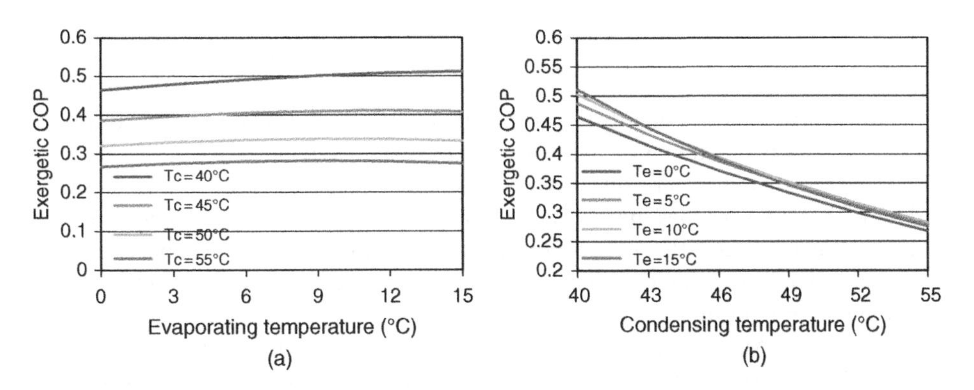

Figure 5.10 Exergetic COP with respect to (a) Evaporating and (b) Condensing temperatures.

temperature difference between the refrigerant and the cabin air, both reducing the associated exergy destructions as shown in Figures 5.10 and 5.11.

It is determined that increasing the degree of superheating can also lead to an increase in the refrigerant enthalpy, which results in extracting additional heat and increasing the refrigeration effect per unit mass of the evaporator. As a result of the larger refrigerating effect per unit mass of the superheated cycle, the associated mass flow rate of the

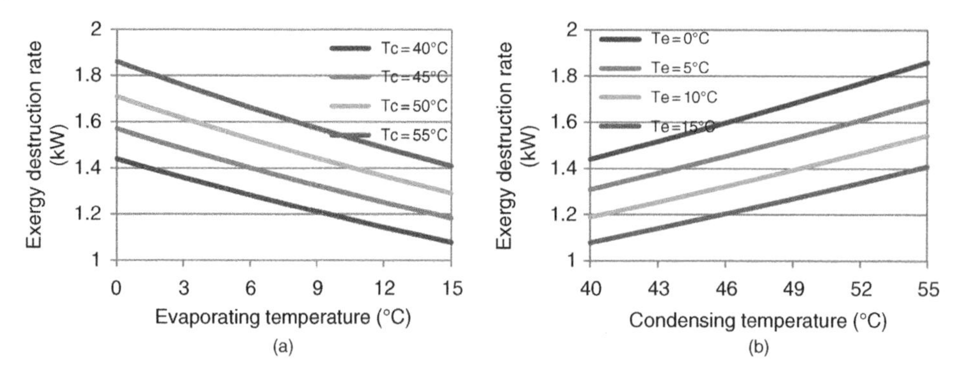

Figure 5.11 Exergy destruction rate with respect to (a) Evaporating and (b) Condensing temperatures.

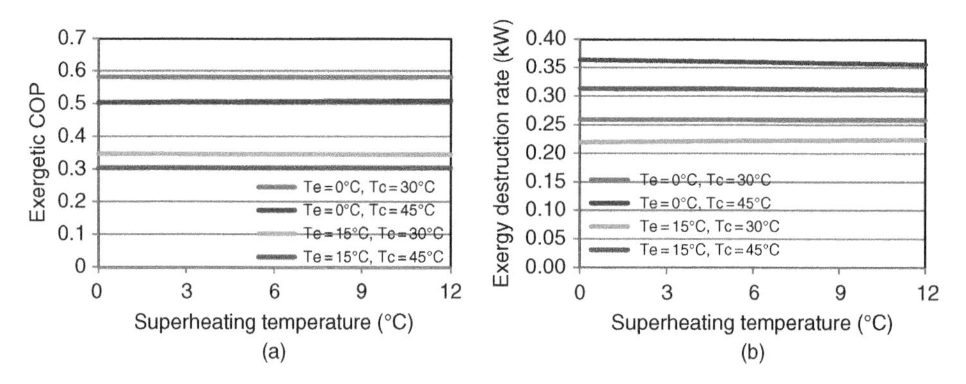

Figure 5.12 (a) Exergetic COP and (b) Exergy destruction rate with respect to superheating temperatures.

refrigerant per unit capacity decreases. In addition, the specific volume of suction vapor as well as the work of compression per unit mass also increases. However, the increase in the refrigerating effect is slightly larger than that of the work of compression, thus the exergetic COP of the system increase is negligible as shown in Figure 5.12.

The refrigerating effect per unit mass can also be increased by subcooling the saturated liquid before it reaches the TXV, due to a lower mass flow rate of refrigerant per unit capacity compared to that of the saturated cycle. The volume of vapor that the compressor must handle per unit capacity decreases since the refrigerant vapor entering the suction line inlet (and thus the specific volume of the vapor entering the compressor) remains the same. Moreover, since the heat of compression per unit mass also remains the same, the increase in refrigerating effect per unit mass increases the heat absorbed in the refrigerated space without increasing the quantity of the energy input to the compressor, and thus increases the exergetic COP of the system as shown in Figure 5.13.

The pressure drop in the heat exchangers also has a certain effect in the system parameters. The increase in pressure drop decreases the cooling capacity due to the reduction in the specific refrigerating effect. In addition, the associated pressure ratio across the compressor increases, leading to an increase in the corresponding compressor work. Both of these effects assist in reducing the exergetic COP of the system while increasing the exergy destruction. The effects of the air mass flow rates on the pressure drops as

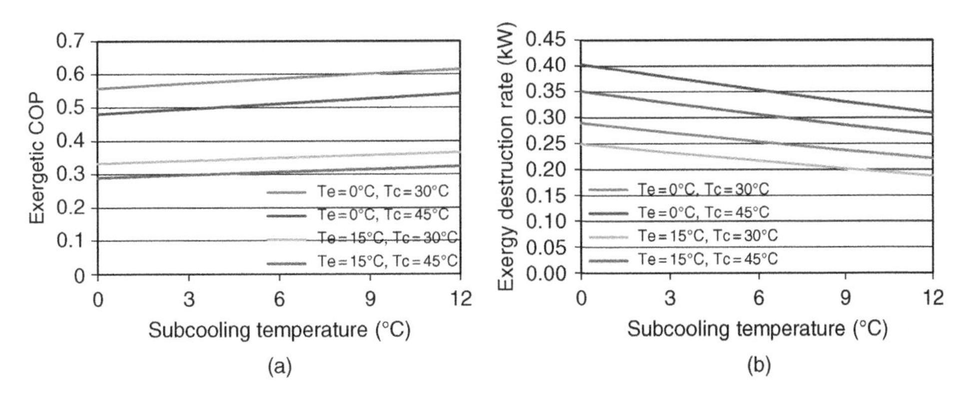

Figure 5.13 (a) Exergetic COP and (b) Exergy destruction rate with respect to subcooling temperatures.

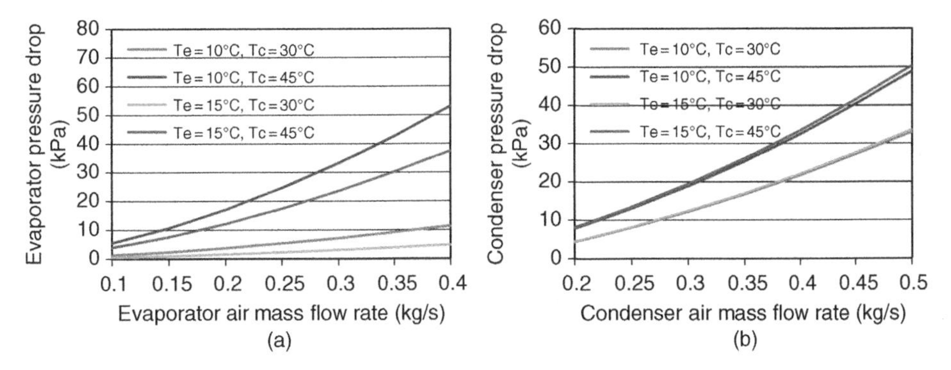

Figure 5.14 Pressure drop with respect to (a) Evaporator and (b) Condenser air mass flow rates.

well as the pressure drop on the exergetic COP and exergy destruction rate are shown in Figures 5.14 and 5.15.

The compressor ratio is another important parameter since it has a significant impact on compressor work, cooling capacity and energetic and exergetic COPs of the system. As the compressor speed increases, the average compressor work also increases, resulting in higher refrigerant mass flow rates, discharge pressure, compression ratio and lower suction pressure and volumetric efficiency. It is also found that the increase in the compressor ratio leads to an increase in the cooling capacity while decreasing the corresponding energetic COP of the system. The exergetic COP of the system also decreases since the associated pressure difference across the compressor and expansion valve increases the overall exergy destruction of the system. The effects of the resulting compression ratio on the system exergetic COP and exergy destruction rate are shown in Figure 5.16.

Moreover, parametric studies are conducted with respect to various refrigerants using EES and REFPROP software packages. In order to have a consistent comparison between these different refrigerants, the same cooling capacity (3 kW), condensing and evaporating temperatures (55 and 5°C, respectively), along with superheating and subcooling temperatures (5°C), are used in each model. The parameters for the model with different refrigerants are given in Table 5.5

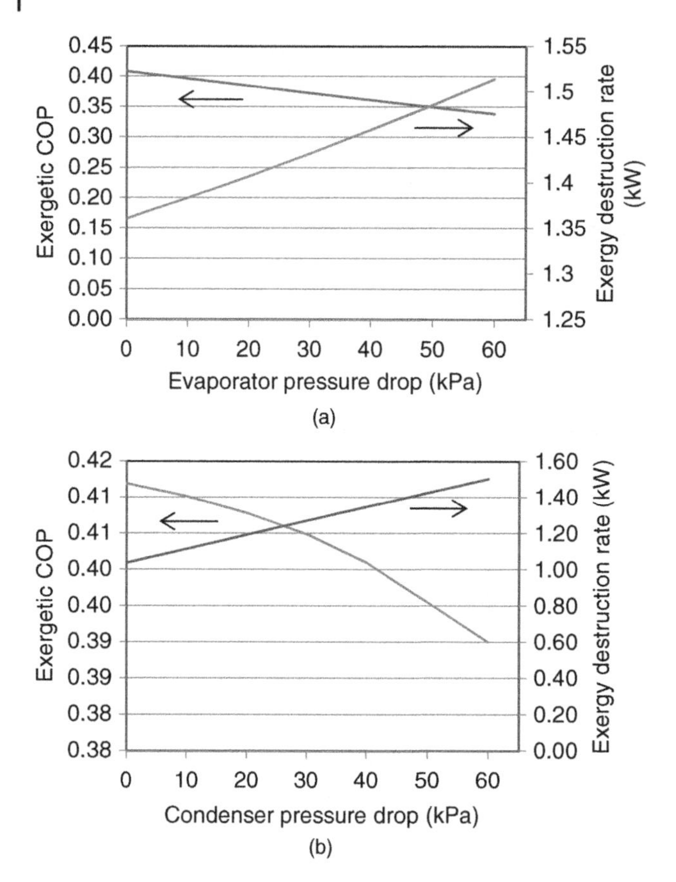

Figure 5.15 (a) Exergetic COP and (b) Exergy destruction rate with respect to evaporator pressure drop.

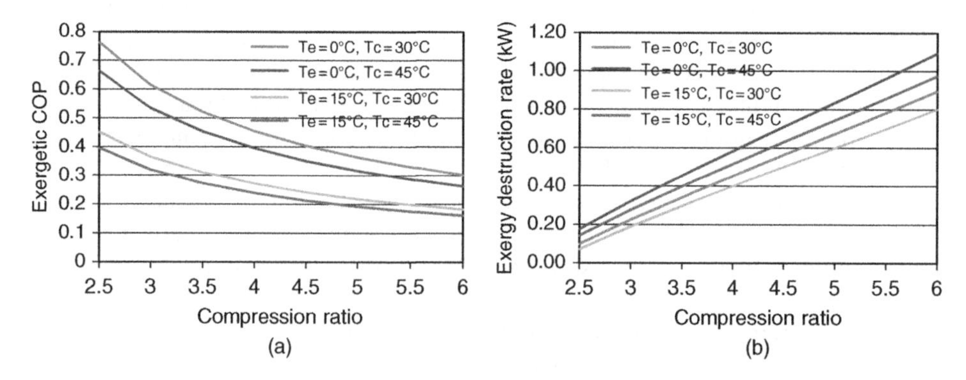

Figure 5.16 (a) Exergetic COP and (b) Exergy destruction rate with respect to compression ratio.

Table 5.5 Operational parameters of a standard EV TMS for various refrigerants at baseline conditions.

Refrigerant	$\dot{m}_{ref,a}$ (kg/s) × 10^2	$\dot{m}_{ref.b}$ (kg/s) × 10^2	$\dot{W}_{comp}$ (kW)	x_{evap}	$T_{sat,dis}$ (°C)	P_{dis} (bar)	ΔP_{cond} (kPa)	ΔP_{evap} (kPa)
R134a	2.21	0.26	1.30	0.31	81.92	14.92	25.11	29.41
R290	1.18	0.14	1.27	0.32	77.93	19.07	24.60	32.97
R600	1.08	0.13	1.26	0.28	73.62	5.64	8.64	14.00
R600a	1.25	0.15	1.26	0.32	68.71	7.64	10.68	17.64
R1234yf	2.98	0.35	1.37	0.39	65.60	14.64	53.61	64.51
(DME)	0.92	0.11	1.21	0.25	95.33	12.97	6.28	13.25

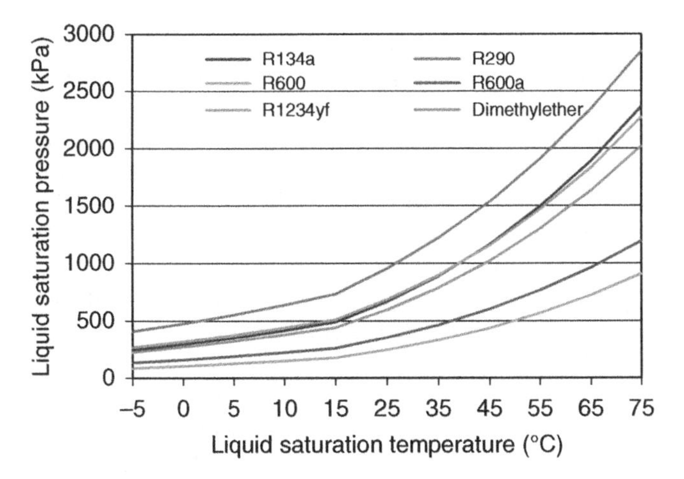

Figure 5.17 Liquid saturation temperature versus pressure for various refrigerants.

In order for a refrigerant to be a suitable replacement for R134a, its compressor capacity should be similar to avoid a different size compressor in the cycle to accommodate the difference in capacity. For this reason, the vaporization temperature of the liquid in the evaporator (which is the suction or evaporating temperature) becomes one of the critical properties in considering a drop-in replacement refrigerant for the thermal management system, since refrigerants with similar vapor pressure evaporates and condenses at the same pressures. Thus, a refrigeration cycle designed with a particularly high and low side pressure would perform comparably for two refrigerants with comparable vapor pressures. This would prevent a different size compressor in the cycle, since the compressor size decreases for fluids with higher vapor pressure and increases for ones with higher vapor pressure in order to provide the same cooling load. Moreover, since the expected capacities are proportional to the vapor pressure, the saturation pressure and temperature of the refrigerant alone would be good indicators of the compressor displacement volume. Thus, a convenient way to compare vapor pressure for multiple refrigerants is a saturation temperature - pressure plot as shown in Figure 5.17. As shown in Figure 5.17, R290, R1234yf and Dimethylether have more

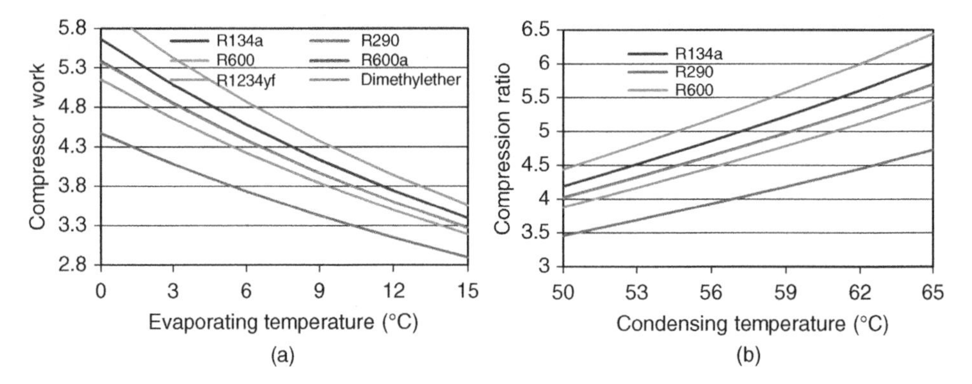

Figure 5.18 Compression ratio of the TMS with respect to (a) Evaporating and (b) Condensing temperatures using various refrigerants.

compatible drop-in replacements (with the least changes in compressor physical dimensions) based on their compressor capacities, compared to R600 and R600a.

The compression ratio is also a useful parameter on which to predict the volumetric performance of the compressor, since lower compression ratios can reduce the amount of potential leakage, and therefore can be used to compare the performance of the TMS using various refrigerants. Figure 5.18 shows that TMSs using R-600, R600a and R134a have higher compression ratios compared to the other systems. Systems utilizing R1234yf and Dimethylether exhibit the closest behavior to that of R134a with the compression ratio slightly lower than R134a system. Furthermore, the lowest compression ratio is achieved by TMS using R290, where it outperforms the system using R-134a up to 18% depending on the condensing and evaporating temperatures.

Moreover, the compressor work is also compared for the TMS using different refrigerants based on various evaporator and condenser temperatures, since it has a significant impact on the overall efficiency of the cycle. It can be seen that even though the TMS using R1234yf has a very low compression ratio among the refrigerants, it has the highest compressor work under baseline conditions due to its highest mass flow rate, as shown in Figures 5.18a and 5.18b. On the other hand, the TMS using dimethylether has the lowest compressor work due to having the lowest mass flow rate as well as a relatively low compression ratio under baseline conditions. The systems using the rest of the studied refrigerants are calculated to have similar but slightly less compressor work, compared to R134a, due to lower compression ratios and significantly lower mass flow rates in the system as shown in Figures 5.18 through 5.20.

Moreover, since the energy consumption of the compressor is also proportional to the pressure ratio and refrigerant mass flow rate, the COP of the system also varies for the same cooling loads and different refrigerants. Among the TMS studied, all of the systems, except for the using R1234yf, have lower exergy destruction rates and higher energetic and exergetic COPs compared to the baseline R134a system for the range of evaporating and condensing temperatures. TMS using dimethylether has the highest energetic and exergetic COPs with 7.3% and 7.7% higher than the baseline R134a system, respectively. However, due to the high flammability of this substance, in order to reduce the associated safety concerns, a secondary loop should be implemented to the thermal management system, where the conventional evaporator is replaced by a secondary fluid

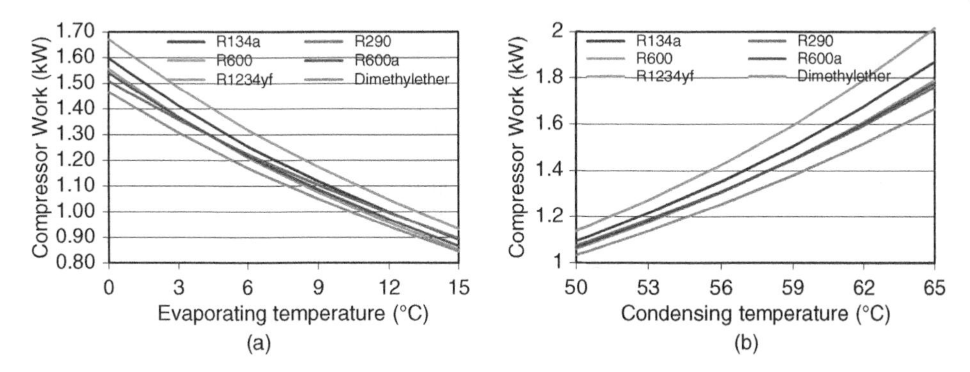

Figure 5.19 Compressor work of the TMS with respect to (a) Evaporating and (b) Condensing temperatures using various refrigerants.

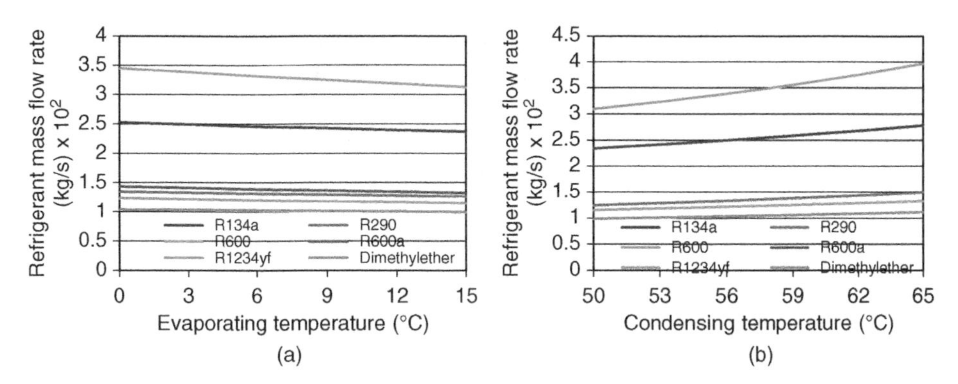

Figure 5.20 Refrigerant mass flow rate with respect to (a) Evaporating and (b) Condensing temperatures using various refrigerants.

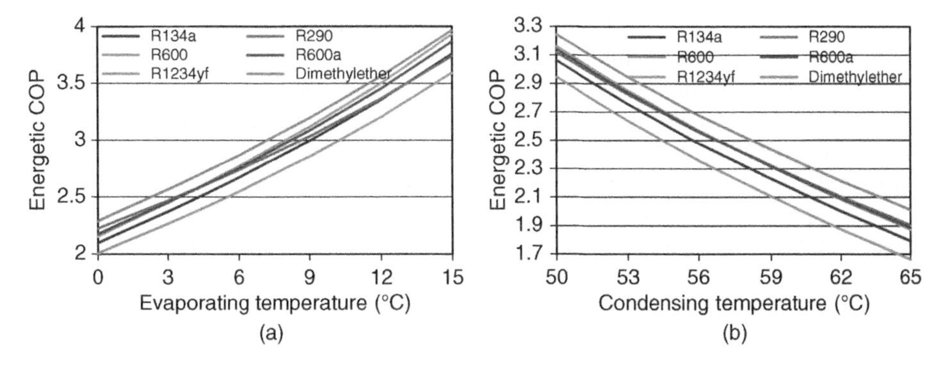

Figure 5.21 Energetic COP of the TMS with respect to (a) Evaporating and (b) Condensing temperatures using various refrigerants.

heat exchanger, which transfers heat between the primary and secondary loops. Thus, the overall efficiency of the system using this refrigerant may decrease for more practical applications. The energetic and exergetic COPs and exergy destruction of TMS with respect to evaporating and condensing temperatures using various refrigerants can be observed from Figures 5.21 through 5.23, respectively.

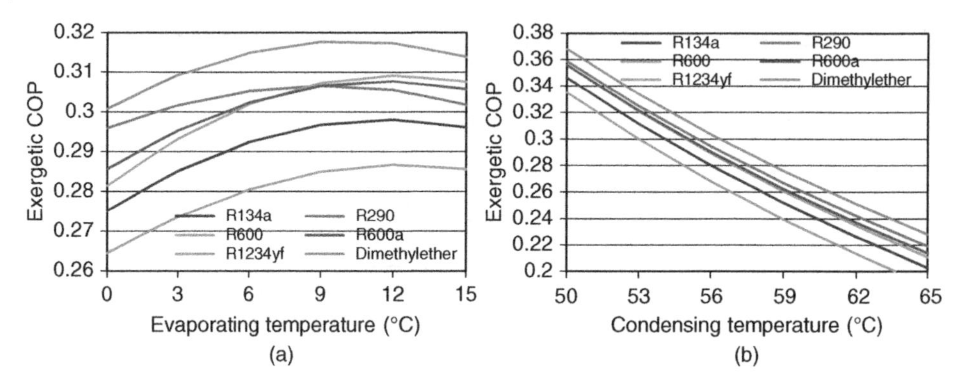

Figure 5.22 Exergetic COP of the TMS with respect to (a) Evaporating and (b) Condensing temperatures using various refrigerants.

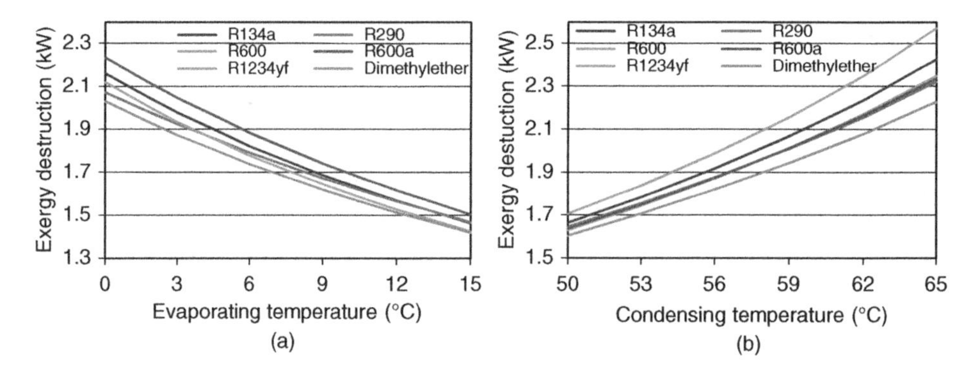

Figure 5.23 Exergy destruction of the TMS with respect to (a) Evaporating and (b) Condensing temperatures using various refrigerants.

Once the TMS COPs are calculated for various refrigerants, the corresponding indirect emissions and the sustainability indices are determined with respect to the system parameters of the baseline model. The sustainability index is a good indicator of how efficiently the resources are utilized in the TMS. Thus, it is therefore directly related to the exergetic COP and exergy destruction rates associated with each TMS. Moreover, the indirect GHG emissions are produced from electricity generation associated with the compressor and pump for the TMS. Figure 5.24 shows the GHG emissions and sustainability index with respect to the exergetic COP for the baseline TMS using R134a. In the figure, as the efficiency of the baseline TMS increases, the power input required for the TMS decreases under the same cooling loads. Hence, the corresponding emissions decrease and the sustainability index increases. It should be noted that the emissions in Figure 5.24a are determined based on the U.S. average energy generation mix composed of 49% coal, 20% natural gas, 20% nuclear, 7% hydro and 4% other renewables (Yang and Maccarthy, 2009) and therefore the associated indirect emissions will be different under other energy generation cases with different carbon intensities. Figure 5.24b shows that the emissions produced from electricity generation almost double under a high-carbon scenario, where the electricity is primarily generated using coal. This reduces significantly under a low-carbon scenario, where electricity is produced through a natural gas combined cycle.

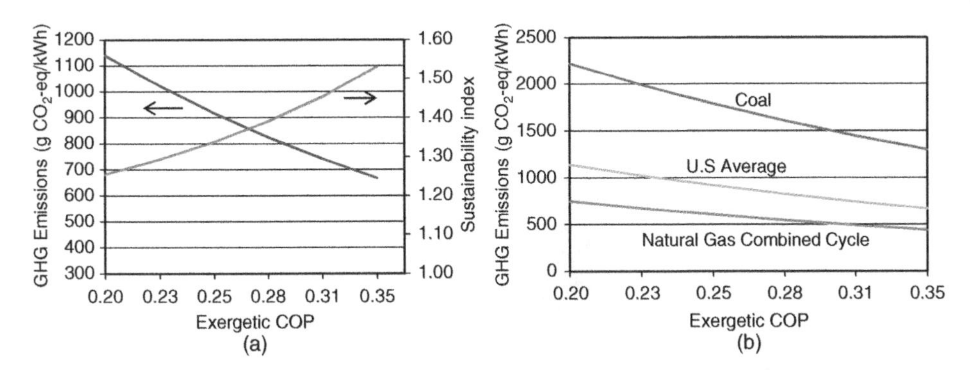

Figure 5.24 (a) GHG Emissions and sustainability index with respect to baseline TMS exergetic COPs (b) Under various carbon intensity of electricity generation.

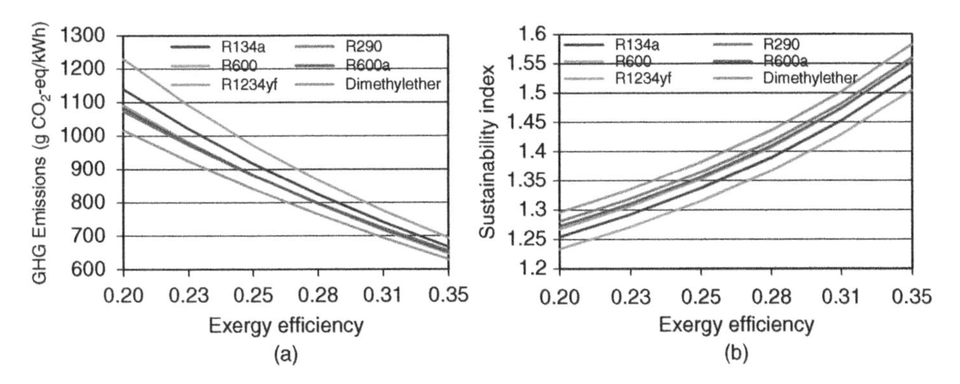

Figure 5.25 (a) GHG emissions and (b) Sustainability index with respect to exergetic COPs of the TMSs using various refrigerants.

In addition, the calculated baseline TMS GHG emissions and sustainability indices are compared against TMSs using various refrigerants. Figures 5.25a and 5.25b show that the TMS using R1234yf generates the highest indirect emissions and lowest sustainability index (6% and −1.6% over the baseline TMS, respectively) due to having the lowest system efficiency. The case using dimethylether generates the lowest indirect emissions and highest sustainability index (−8.3% and 3.3% over the baseline TMS, respectively), among the studied TMSs based on high system efficiency.

Moreover, TMS is analyzed with respect to theoretical thermodynamic and "unavoidable thermodynamic" cycles in order to split the exergy destructions associated with each component into endogenous/exogenous and avoidable/unavoidable parts. The normalized exergy destruction values for the major components can be seen in Figures 5.26 through 5.32. It should be noted that even though the total exergy destruction in a component cannot be negative, the exogenous portions can be negative, which would indicate a negative correlation between endogenous exergy destruction within a component and the exogenous exergy destruction within the remaining components.

Based on the Figures 5.26 through 5.32, it can be seen that the exogenous exergy destruction is small but significant portion of the total exergy destruction in each component, which shows that there is a moderate level of interdependencies among

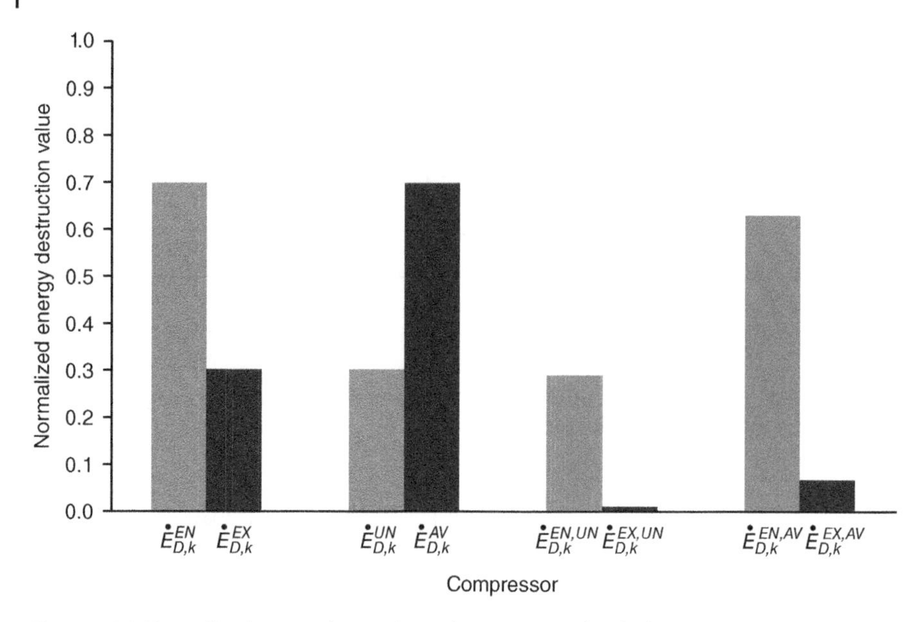

Figure 5.26 Normalized exergy destruction values associated with the compressor based on the conducted enhanced exergy analysis.

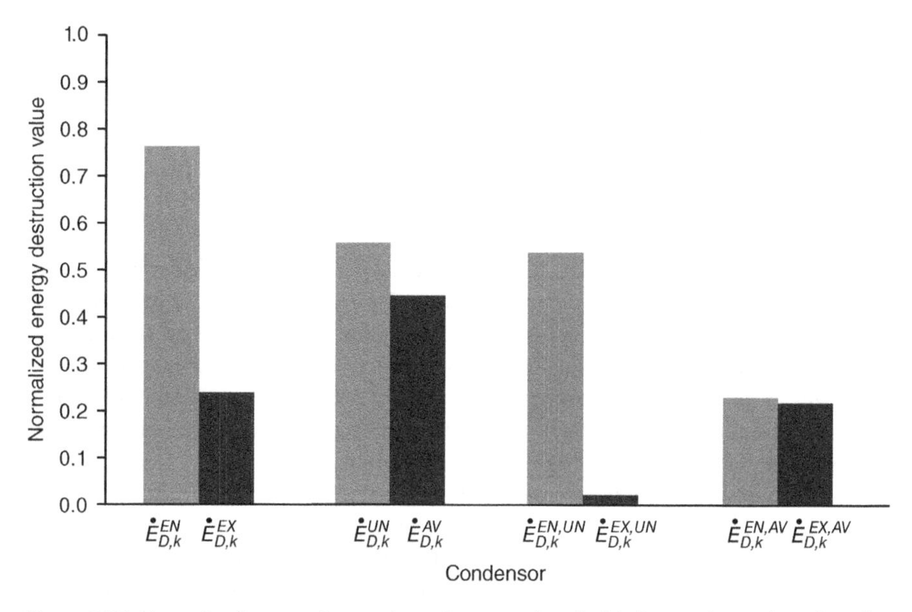

Figure 5.27 Normalized exergy destruction values associated with the condenser based on the conducted enhanced exergy analysis.

the components. Furthermore the exogenous exergy destruction is lower than the total exergy destruction for each component ($\dot{E}^{EX}_{D,k}$ is positive), which indicates that a reduction in the endogenous exergy destruction within a component will yield a reduction in the exogenous exergy destruction within the remaining components.

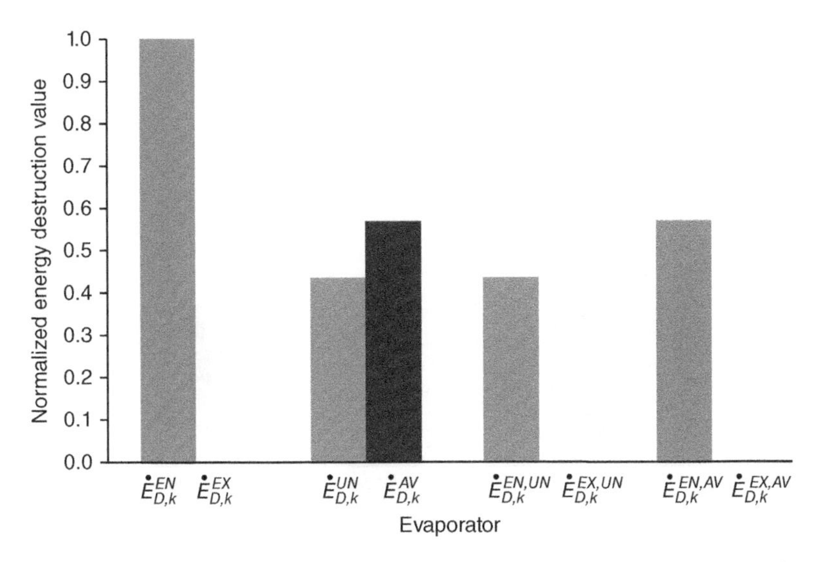

Figure 5.28 Normalized exergy destruction values associated with the evaporator based on the conducted enhanced exergy analysis.

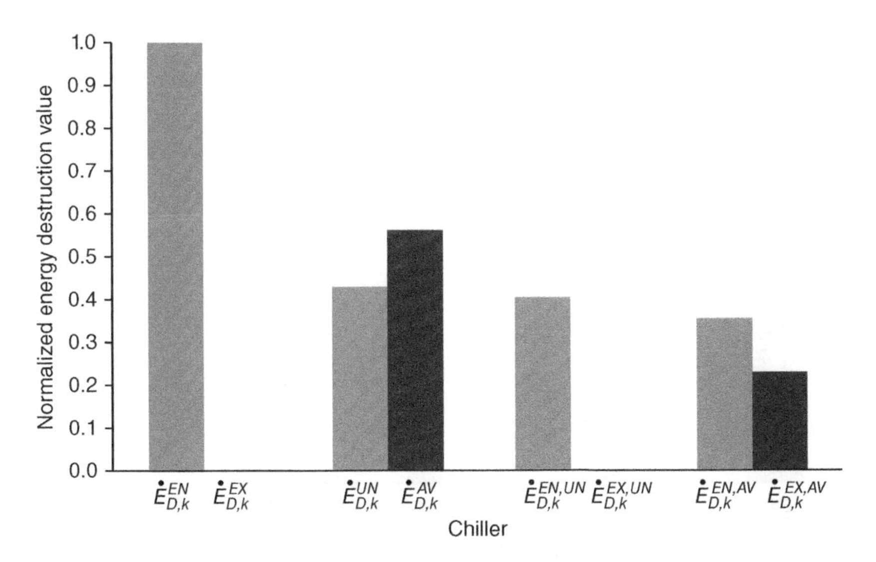

Figure 5.29 Normalized exergy destruction values associated with the chiller based on the conducted enhanced exergy analysis.

5.6 Case Study: Transcritical CO_2-Based Electric Vehicle BTMS

In this case study, the use of transcritical CO_2-based electric vehicle battery thermal management system (EV BTMS) is investigated in a single-stage vapour compression cycle along with its various modifications that include internal heat exchanger, two-stage vapour compression and an intercooler. The system is evaluated against conventional cycles with various commercial refrigerants including R152a, R134a, R290, R600, R600a,

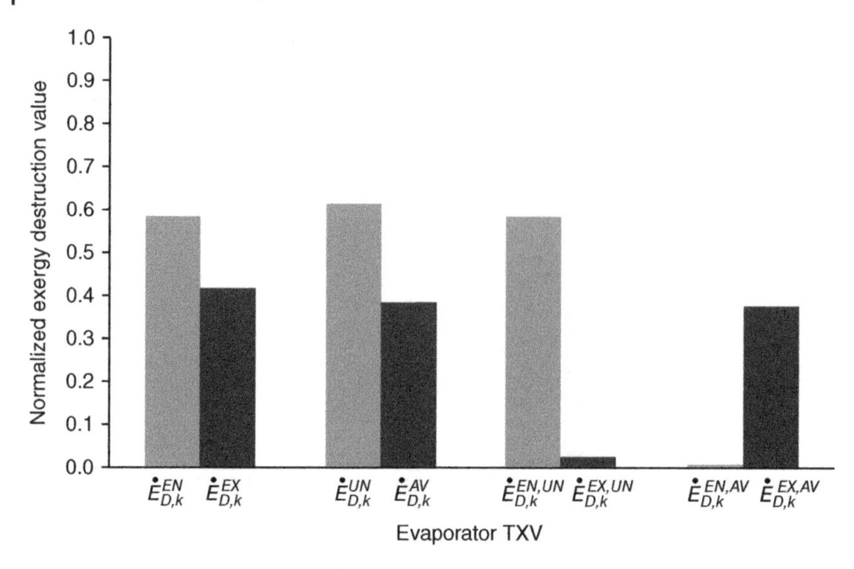

Figure 5.30 Normalized exergy destruction values associated with the evaporator TXVs Based on the conducted enhanced exergy analysis.

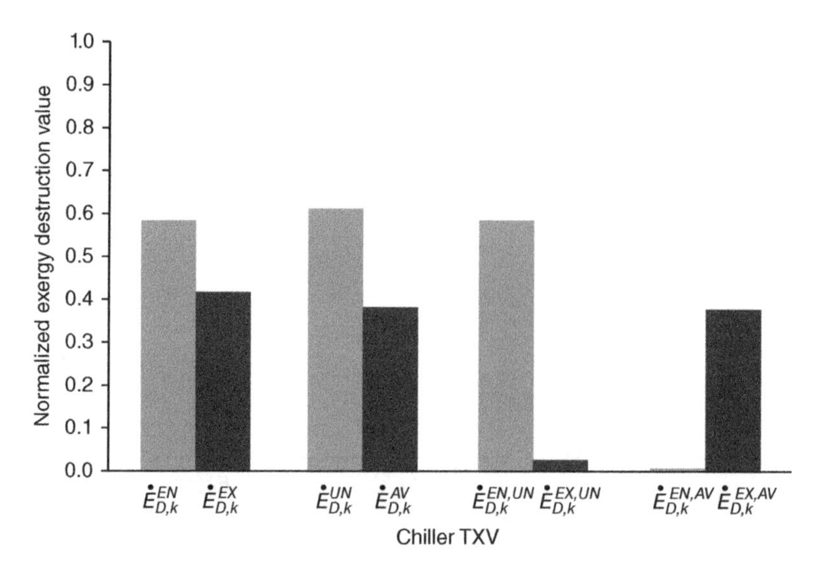

Figure 5.31 Normalized exergy destruction values associated with the chiller TXVs based on the conducted enhanced exergy analysis.

R1234yf and dimethylether and evaluated with respect to exergy analysis under various conditions and input parameters.

5.6.1 Introduction

Halogenated refrigerants have dominated the vapour compression-based systems due to their thermodynamic and thermo-physical properties. In the past decades however, the

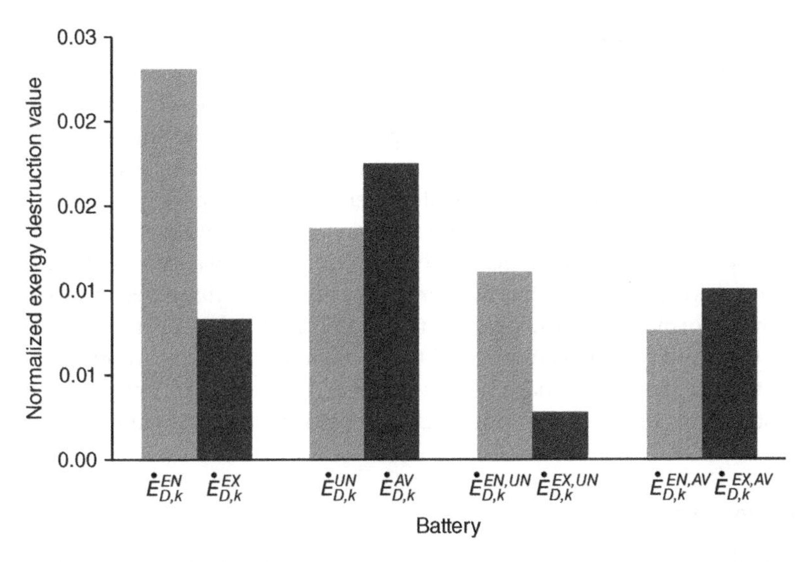

Figure 5.32 Normalized exergy destruction values associated with the battery based on the conducted enhanced exergy analysis.

refrigeration industry has changed dramatically, especially with the restrictions made in the Montreal and Kyoto protocols, as the chlorofluorocarbons (CFCs) were replaced by hydrochlorocarbons (HCFCs) and later by hydrofluorocarbons (HFCs) due to their negative impact on the ozone layer. However, even though most HFCs have virtually zero ozone depletion potential (ODP), they still have high global warming potential (GWP) and therefore still take part in the GHG emissions. This progression has also impacted the automotive sector significantly, where it went through rapid modifications for the refrigerants used in the vehicle air conditions due to corresponding environmental concerns and associated changes in the regulations. In 1990s, R12 was replaced with R134a, but this did not provide a long-term solution since the GWP of R134a is 1300 times that of carbon dioxide. Thus, significant research is still being conducted for fluids with a low GWP to be potentially used in vehicle TMS applications, especially in electric vehicles. Among these, natural refrigerants are started to be considered as one of the best potential solutions since they have zero ODP, low GWP and various other advantages when compared to the aforementioned alternatives.

Among these refrigerants, carbon dioxide (R744) has acquired significant attention due to its favorable heat transfer properties, having no toxicity, no inflammability, high volumetric capacity, lower pressure ratio, high compatibility with normal lubricants, wide availability, lower cost and recyclability (Yang *et al.*, 2005). In addition, the use of carbon dioxide in the vapour compression cycle prevents the large pressure lift between the evaporating and condensing temperatures, which increases the compression ratio in the system. Moreover, it also provides solution to the high leakage rate of R-12 and R134a from the flexible nylon, butyl rubber and the compressor shaft seal. Furthermore, since the CO_2 systems operate at high pressures, the specific volumes are reduced (compared to traditional refrigerants), which makes it possible to design highly compact heat exchangers (Mathur, 2000). On the other hand, the critical temperature of CO_2 is lower

(31.1°C) than typical values of other refrigerants in air conditioning systems, which does not makes it possible to transfer heat to the ambient above this critical temperature by condensation, causing the refrigerant to operate in transcritical mode when exposed to typical ambient temperatures. Thus, in CO_2 refrigeration systems, the conventional condenser is typically replaced by a gas cooler and various modifications are usually applied to the system in order to improve the system performance, capacity and adaptation to heat rejection temperatures for a given system and component size (Kim *et al.*, 2004). However, CO_2 can also have high power consumption and capacity loss at high ambient temperatures, which makes it hard to compete with conventional refrigerants in terms of system performance. Thus, significant research is still needed to be conducted to analyze the underlying reasons behind the inefficiencies in the carbon dioxide based electric vehicle thermal management systems and find ways to improve the corresponding system with respect to thermodynamic, cost and environmental impact criteria.

5.6.2 System Development

A simplified thermal management system of an electric vehicle battery with carbon dioxide vapour compression and liquid battery cooling is considered in Figure 5.33. The system is composed of two cycles, namely a refrigerant and battery coolant cycle. The refrigerant cycle enables air conditioning of the vehicle cabin, while the coolant cycle keeps the electric battery operating within its ideal temperature range. The air conditioning portion consists of a vapour compression cycle that includes a compressor, condenser/gas cooler, expansion device and an evaporator.

Even though a condenser is used for most conventional refrigerants, due to the low critical temperature of carbon dioxide (which makes the cycles operate in super-critical state at high temperatures and sub-critical state at low ambient temperatures), no saturation condition exists in the supercritical state, the condenser is replaced with a

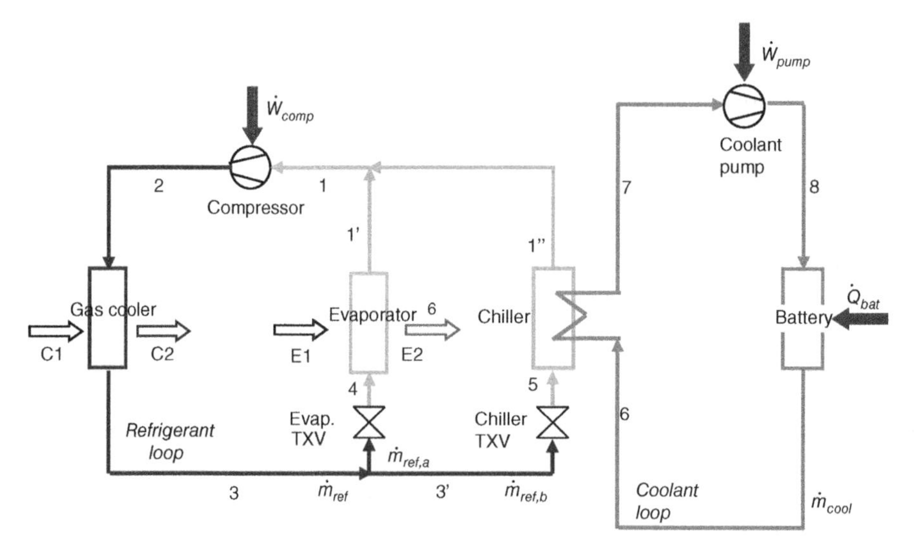

Figure 5.33 Representation of the studied basic EV BTMS system using carbon dioxide and 50–50 water/glycol mix.

gas-cooler in the cycle. The working fluid from the suction line enters the compressor (state 1), the high-pressure valve leaves the compressor (state 2) and enter the condenser/gas cooler. The cooled refrigerant splits into two and the first stream enters the evaporator expansion device (state 3) and expands to the evaporator (state 4) providing conditioning to the vehicle cabin. In conventional subcritical cycles, the specific enthalpy of state 3 is mainly a function of temperature, however pressure also influences the enthalpy at the supercritical high-side conditions. The second stream enters to the chiller expansion device (state 3) and expands to the chiller (state 5) providing heat exchange between the refrigerant and coolant loops. After the refrigerant exchanges heat in the evaporator and chiller, the two streams combine to re-enters the compressor (state 1).

In the battery coolant loop, the 50/50 water-glycol coolant is pumped to the battery (state 8) to absorb the heat dissipated by this component and then enters to the chiller (state 6) to transfer the heat to the refrigerant loop. Once the temperature of the coolant is reduced (state 7), it is pumped back into the battery. As described above, the chiller is used to exchange heat among the loops to provide supercooling to the battery cooling as it passes through the chiller unit. This increases the efficiency of the system significantly since cooling via refrigeration circuit would consume more energy than operating the battery coolant circuit due to the need of air compressor in the first case.

Subsequently, a modified version of the system is also studied, where a liquid/suction line internal heat exchanger (IHX) is used, which provides sub-cooling to the refrigerant liquid and evaporates the liquid before the compressor inlet, thus protecting the compressor from harmful liquid slugging as well as improving the efficiency of the system at high heat rejection temperatures. Even though internal heat exchangers may not usually be helpful to improve the system efficiency in refrigerants previously and currently used in automotive applications (such as R-22 and R134a), its benefits for carbon dioxide based systems can be considerable as the COP-optimizing discharge pressure becomes lower when an internal heat exchanger is added to the system. In addition, since adding IHX brings capacity and efficiency maximizing discharge temperatures together, it can also enable using simpler control systems and strategies in the system. In the modified system, enthalpy-based heat transfer effectiveness was used to determine the performance of the cycle with the internal heat exchanger. The battery coolant cycle is kept the same in all the modified systems in order to provide a fair comparison among the evaluated systems.

In order to further improve the efficiency of the system, a two-stage compressor cycle is also used to mitigate the performance deterioration of the basic single compression systems, and the refrigerant is compressed in two stages with intercooler in between. Finally, in order to find the best configuration, these two modified systems are combined to develop a carbon dioxide system with both an internal heat exchanger and multi-stage compression to evaluate the corresponding impact on the system parameters as shown in Figure 5.34.

For the analysis, processes in each component are assumed to be at steady state with negligible changes in potential and kinetic energy. Moreover, the heat loses to surroundings are assumed to be negligible and the expansion process is taken to be isenthalpic. Moreover, the transcritical carbon dioxide vapour compression cycle is compared with cycles using various other commercially available refrigerants. The

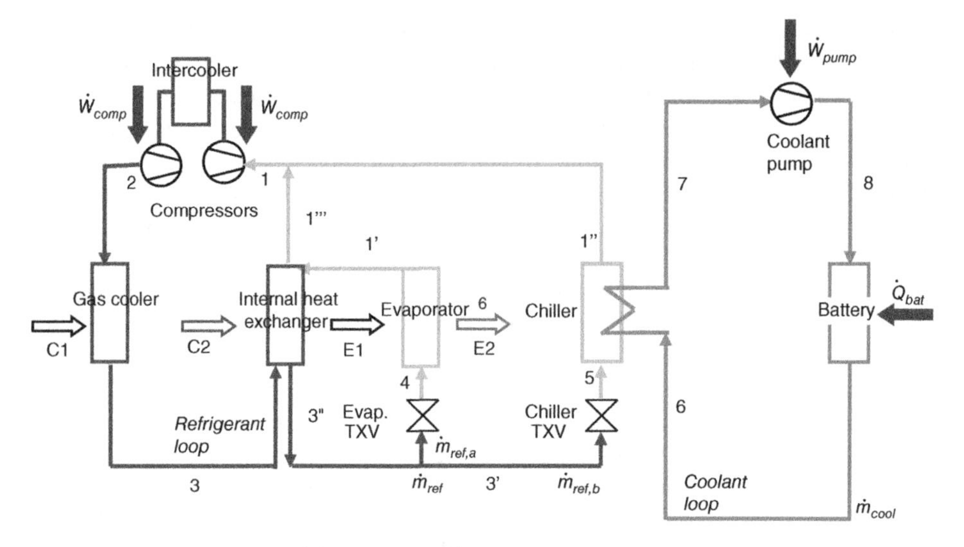

Figure 5.34 Representation of the EV BTMS with two-stage compression, intercooler and internal heat exchanger, using carbon dioxide and 50–50 water/glycol mix.

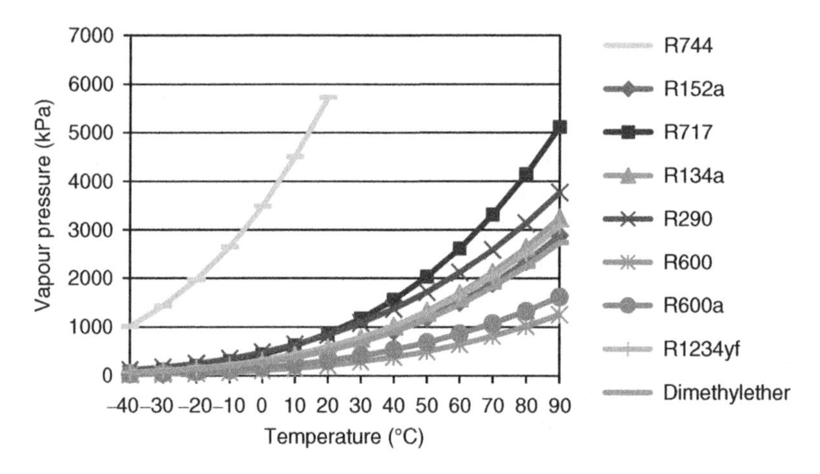

Figure 5.35 Vapour pressure for R744 and the compared refrigerants.

vapour pressure and slope of the saturation temperature curves for these refrigerants are provided in Figures 5.35 and 5.36.

It can be seen from Figure 5.35 that the vapour pressure of carbon dioxide is much higher than the compared refrigerants with much steeper slope near the critical temperature, causing smaller temperature change for a given pressure change, which plays an important role in the heat exchangers in the system and has significant impact on the overall efficiency and the system design. In addition, the operating parameters of the studied EV BTMS with these refrigerants are also provided in Table 5.6. EV BTMSs should be designed to have high performance and operate as efficiently as possible for

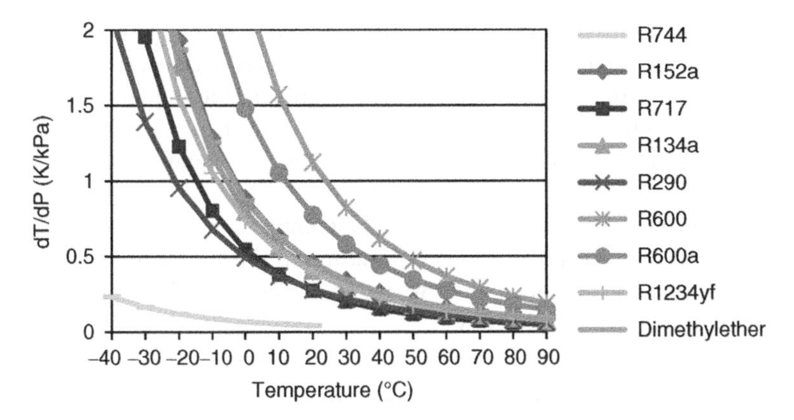

Figure 5.36 Slope of saturation pressure curve for R744 and the compared refrigerants.

Table 5.6 Operating parameters of studied EV BTMS with various refrigerants at baseline conditions.

Refrigerant	$\dot{m}_{ref,a}$ (kg/s) × 10^2	$\dot{m}_{ref,b}$ (kg/s) × 10^2	$\dot{W}_{comp}$ (kW)	$T_{sat,dis}$ (°C)	P_{dis} (kPa)	ΔP_{cond} (kPa)	ΔP_{evap} (kPa)
R744	1.62	0.19	1.51	126.20	13000	78	92
R152a	1.32	0.15	1.24	95.7	1333	12.8	17.4
R134a	2.20	0.26	1.30	81.9	1492	24.9	29.2
R290	1.17	0.14	1.26	77.9	1907	24.4	32.6
R600	1.07	0.13	1.26	73.6	564	8.5	13.9
R600a	1.24	0.15	1.26	68.7	764	10.6	17.5
R1234yf	2.88	0.34	1.33	67.8	1465	50.9	61.3
Dimethylether	0.91	0.11	1.21	95.3	1297	6.2	13.1

given restrictions under a wide range of operating conditions. Thus, exergy-based analysis is conducted for the aforementioned systems in the following section.

5.6.3 Thermodynamic Analysis

Thermodynamic analysis through energy and exergy approaches is critical for better design, analysis, assessment and improvement of thermal management systems of batteries. Energy analysis comes from the first law of thermodynamics (so called: conservation law), and exergy analysis comes from the second law of thermodynamics (so-called: non-conservation law due to inefficiencies, losses, irreversibilities, destructions, etc.). The performance assessments of such thermal management systems are performed through energy and exergy efficiencies, as well as some other energy and exergy based performance criteria (such as energetic and exergetic coefficient of performance values).

Based on the aforementioned model and parameter ranges, a steady-state thermodynamic model was created with Engineering Equation Solver (EES) and REFPROP software. In order to create a consistent comparison between different refrigerants, the same operating conditions are used in each model. Once the properties are determined at each point in the system, the mass, energy, entropy and exergy balance equations are used to calculate the exergy efficiency and exergy destruction rate of each component as well as the overall system with respect to the information and the equations used in this chapter.

Based on the work input to the system (in terms of the compressor and pump) and the associated cooling load (with regards to the evaporator and the chiller) under the defined boundary conditions, the energetic and exergetic COPs of the BTMS are also determined based on equations provided below.

$$COP_{en,sys} = \frac{\dot{m}(h_3 - h_2) + \dot{m}_{cool}(h_7 - h_6)}{\dot{m}_2[(h_2 - h_1) + (h_2 - h_1)] + \dot{m}_{cool}(h_8 - h_7)} \tag{5.44}$$

$$COP_{ex,sys} = \frac{\dot{m}(h_3 - h_2)\left(1 - \frac{T_0}{T_H}\right) + \dot{m}_{cool}(h_7 - h_6)\left(1 - \frac{T_0}{T_7}\right)}{\dot{m}_2[(h_2 - h_1) + (h_2 - h_1)] + \dot{m}_{cool}(h_8 - h_7)} \tag{5.45}$$

5.6.4 Results and Discussion

Based on the developed software code in EES with respect to the balance equations and system parameters provided in the previous sections, the exergy efficiencies and exergy destruction rates associated with the components in the vapour compression cycle of each system are provided in Table 5.7.

Moreover, the compressor work, energetic and exergetic COP and exergy destruction rates associated with each BTMS configuration is provided in Figures 5.37 through 5.40. Even though adding an IHX increases the temperature difference over the gas cooling process, which increases the exergy destruction rate of the gas cooler, and in turn creates a negative impact on the system efficiency; the exergy destruction rates through the expansion process is also decreased through subcooling, which increases enthalpy difference across the evaporator. Thus, in total, the overall energetic and exergetic COP of the system increases.

Table 5.7 Exergetic COP and exergy destruction rate associated with the refrigerant cycle of different transcritical carbon dioxide-based EV BTMSs.

	Single		*Single with IHX*		*Two-stage with IC*		*Two-stage with IHX and IC*	
Component	Ex_{eff}	$Ex_D(kW)$	Ex_{eff}	$Ex_D(kW)$	Ex_{eff}	$Ex_D(kW)$	Ex_{eff}	$Ex_D(kW)$
Compressor	0.73	0.60	0.75	0.58	0.77	0.28	0.78	0.26
Gas Cooler	0.14	0.53	0.16	0.50	0.18	0.31	0.19	0.29
Evaporator	0.10	0.41	0.13	0.39	0.16	0.34	0.18	0.32
Internal Heat Exchanger	–	–	0.48	0.10	–	–	0.49	0.09
Second Compressor	–	–	–	–	0.77	0.24	0.79	0.25
Intercooler	–	–	–	–	0.69	0.27	0.70	0.26
Evaporator TXV	0.89	0.38	0.89	0.38	0.90	0.36	0.93	0.35

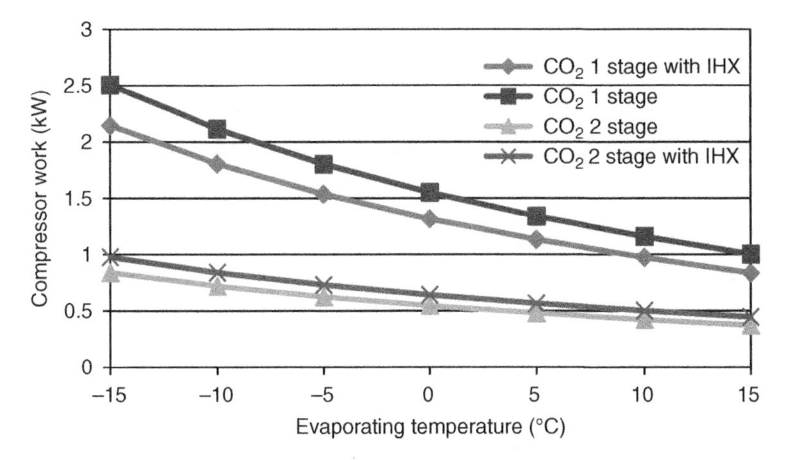

Figure 5.37 Compressor work of different EV BTMSs using carbon dioxide as refrigerant under varying evaporating temperatures.

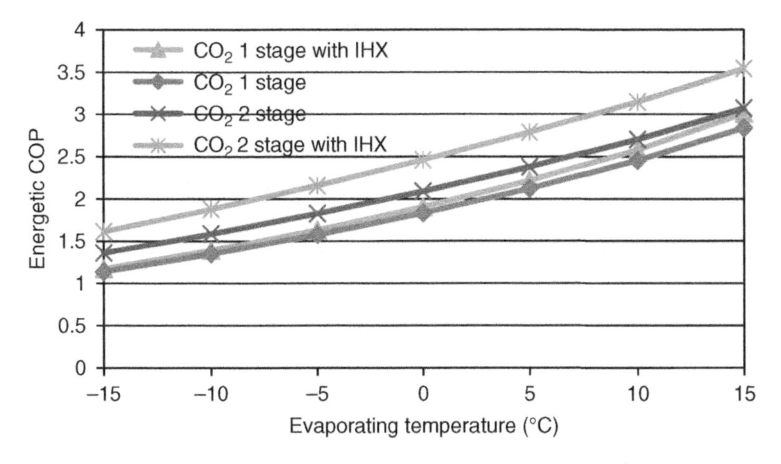

Figure 5.38 Energetic COP of different EV BTMSs using carbon dioxide as refrigerant under varying evaporating temperatures.

Furthermore, by developing a multi-stage compression in the vapour compression cycle of the TMS, an increase in system exergy efficiency up to 8% is achieved. By integrating IHX to the multi-compression system, exergy efficiency improvements over 22% can be achieved under the evaluated parameters in the study.

In addition, the effects of evaporating temperatures on these configurations are also provided in these figures as it has significant impact on the compression ratio (and hence, compressor work), cooling load, volumetric heat capacity and energetic and exergetic COP of the systems. Compression ratio is reduced by increasing the evaporating temperature and since energy consumption of the compressor is also proportional to the pressure ratio, this increase in the evaporator temperature increases the COP of all systems. This indicates that the required compressor power to a certain cooling capacity drops as the evaporating temperature increases. Moreover, the throttling losses also decrease with decreasing temperature change, which also leads to an increase in the

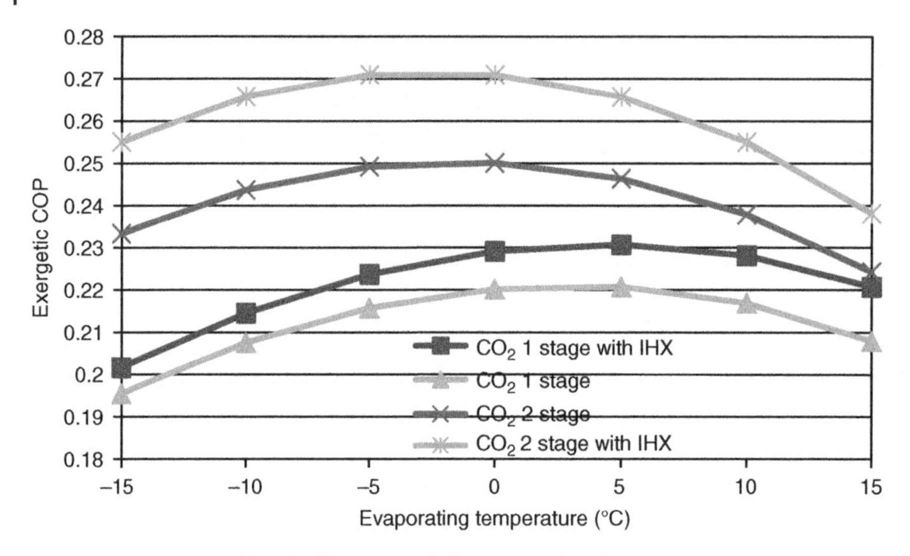

Figure 5.39 Exergetic COP of different EV BTMSs using carbon dioxide as refrigerant under varying evaporating temperatures.

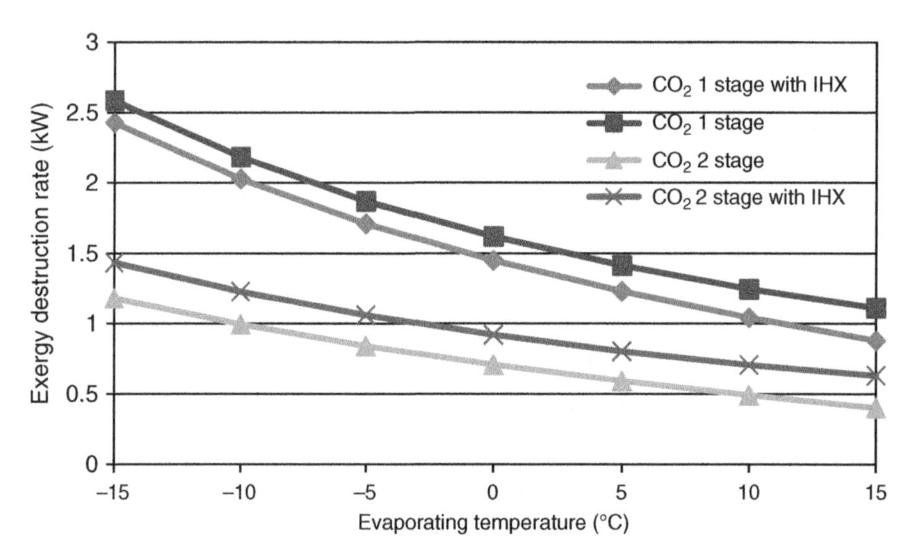

Figure 5.40 Exergy destruction of different EV BTMSs using carbon dioxide as refrigerant under varying evaporating temperatures.

COP. Furthermore, the COP of these systems also increase since increasing the evaporating temperatures reduce the mean temperature difference between the refrigerant and the cabin air, both reducing the associated exergy destruction. However, as the cooling capacity and compressor ratio decreases, the average air temperature at the evaporator increases, reducing the gained exergy as the evaporating temperature increases.

Since the efficiencies associated with the system are calculated to be highest in the system incorporating a multi-compression cycle with an intercooler and an internal heat exchanger, the further analyses and system comparisons (with BTMSs incorporating

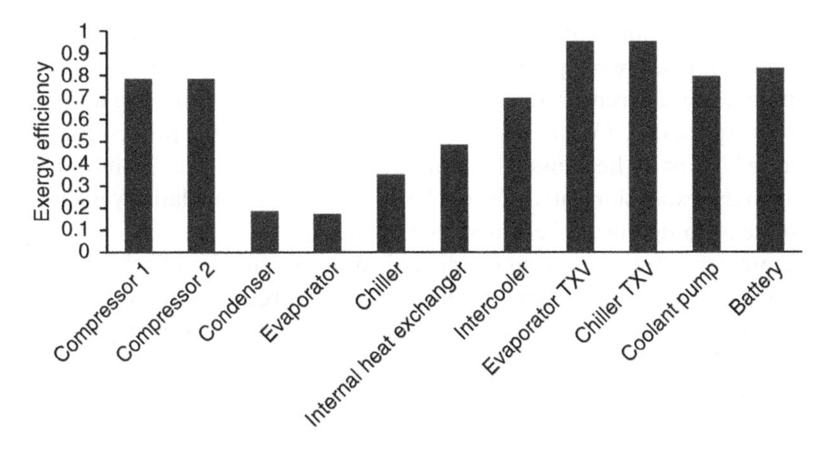

Figure 5.41 Exergy efficiency of each component in the two-stage compression system with intercooler and an internal heat exchanger.

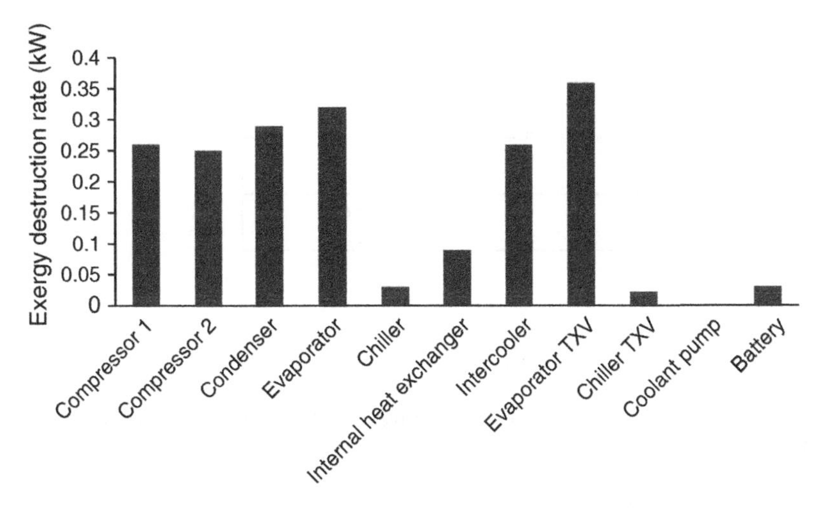

Figure 5.42 Exergy destruction rates of each component in the two-stage compression system with intercooler and an internal heat exchanger.

various other refrigerants) are conducted using this carbon dioxide based system for the rest of the study. The overall exergy efficiencies and exergy destruction rates associated with the complete system are provided in Figures 5.41 and 5.42.

In the evaluated system using multi-stage compression with intercooler and internal heat exchanger, the evaporator and gas cooler have the lowest exergy efficiencies with respect to phase change and temperature differences respectively, which results in more entropy generation between the refrigerant and the cooling air. In the evaporator, the exergy losses are relatively high since only part of the heat rejection occurs during the phase change process with large temperature differences between the working fluid in the evaporator and the vehicle cabin. Moreover, the exergy destruction in the compressor occurs due to the compression pressure ratio and the change in temperature of the refrigerant flowing through the compressor. However, it is significantly lower than the previous setups of the vapour compression cycle using carbon dioxide. The exergy efficiency of the internal heat exchanger is higher than the evaporator due to the relatively smaller temperature differences across the component. Furthermore, the exergy efficiencies for the evaporator TXV and chiller TXV are higher since the processes are isenthalpic and have little or no heat loss. Therefore the exergy losses occur mainly due to a pressure drop in the expansion valve. The coolant pump also has a relatively higher efficiency since there is no significant heat loss from the pump.

In addition, parametric studies are also conducted to determine the effects of key variables on the overall system efficiencies. Heat rejection pressures are varied in Figures 5.43 and 5.44, and it is determined that as the heat rejection pressure increases, the heat absorbed by the evaporator increases along with the work required to drive the compressor. In this region, since the rate of increase for the work required to drive the compressor is found to be greater than the rate of increase at which the heat

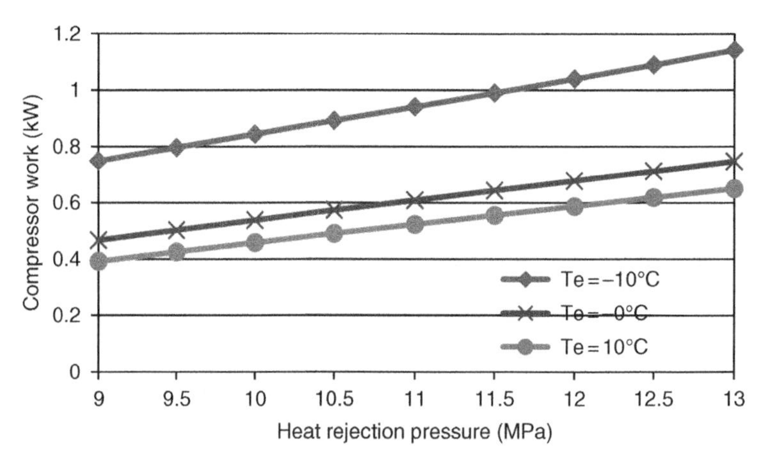

Figure 5.43 Compressor work associated with the system using carbon dioxide as refrigerant under varying heat rejection pressures.

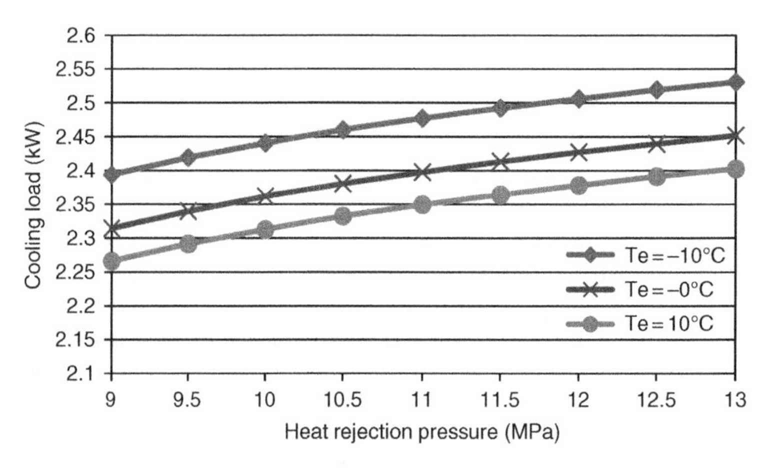

Figure 5.44 Cooling load associated with the system using carbon dioxide as refrigerant under varying heat rejection pressures.

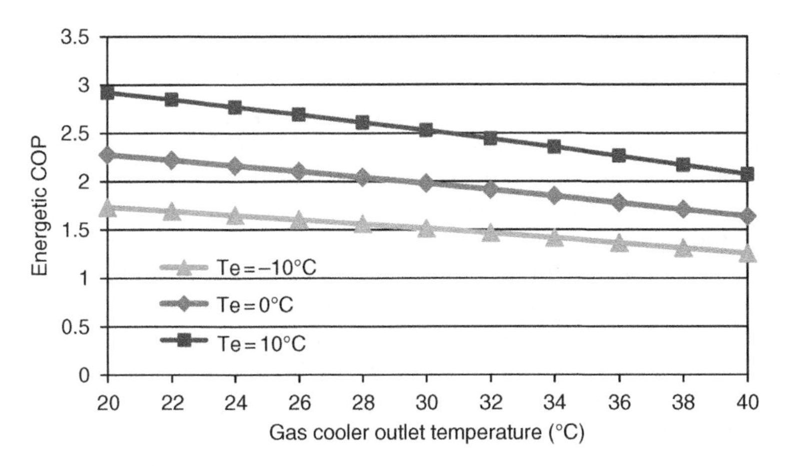

Figure 5.45 Energetic COP of the system using carbon dioxide as refrigerant under varying gas cooler outlet temperatures.

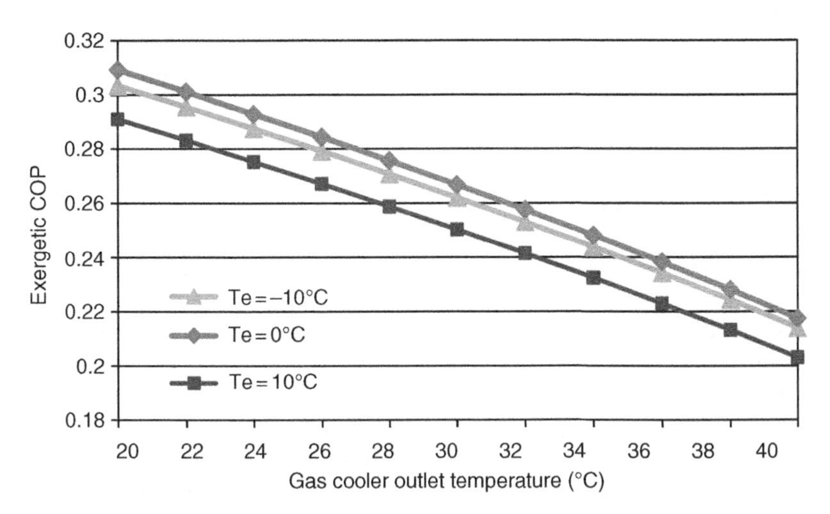

Figure 5.46 Exergetic COP of the system using carbon dioxide as refrigerant under varying gas cooler outlet temperatures.

is absorbed by the evaporator, the total energetic and exergetic COP of the system decreases.

Moreover, outlet temperature of the gas cooler is also varied since it is also one of the important parameters in the system. As the gas cooler outlet temperature increases, the exergy destruction associated with the compressor, evaporator, evaporator and chiller TXVs increases, which dominates the exergy destruction reduction in the gas cooler and therefore reduces the energetic and exergetic COP of the overall system as shown in Figures 5.45 and 5.46 for different evaporating temperatures.

5.6.5 Case Study Conclusions

In this case study, performance analysis of an electric vehicle BTMSs with transcritical carbon dioxode vapour compression cycle is investigated using single-stage vapour compression and two-stage vapour compression with/without an internal heat exchanger. Based on the analysis, the following main results were obtained.

- The energetic and exergetic COP of the system is increased by up to 18% and 22% respectively, by adding two-stage compression with intercooler and an internal heat exchanger. Even though the efficiency of the system is increased through the applied configuration, it is still calculated to be lower than the systems with the compared refrigerants in the study.
- Increasing heat rejection pressure in the studied region increases the compressor work and cooling capacity and therefore decreases the energetic and exergetic COP of the system, meanwhile increasing the outlet temperature of the gas cooler in the studied region reduces the energetic and exergetic COP of the system.
- The cooling load, volumetric heat capacity, and energetic and exergetic COPs of the system increase with increasing evaporating temperatures while compressor work, compression ratio, discharge temperature and pressure as well as the total exergy destruction increase with decreasing evaporating temperatures of the system.

Based on the analyses, it is determined that the exergy efficiency of the transcritical CO_2-based system can be enhanced by up to 22% with two-stage compression and internal heat exchanger. However, the efficiency is still lower compared to systems using the analyzed refrigerants.

5.7 Concluding Remarks

In this chapter, a walkthrough of developing the thermodynamic model of BTMS is presented to provide the readers with the necessary background, up-to-date information and tools to construct and/or improve their systems from a thermodynamic perspective. The content includes the comparison among the most utilized types of thermal management systems and the modelling of their major components. The focus is placed on the liquid thermal management system as it is considered to be the state of the art approach for BTMS for many applications. In order to provide an illustrative example, a given liquid thermal management system is analysed based on the mention principles and techniques and the results are provided to provide a reference point for the readers. Furthermore, a case study on a transcritical CO_2 based TMS is also included to introduce alternative systems and their performance with respect to the most commonly used TMSs today. This chapter contains some of the necessary knowledge for engineers to model their TMS and to improve the efficiency of their system. In the next chapter, economic and environmental considerations will also be included in developed models to approach the faced problems from a much wider perspective and provide the necessary steps to solve them in a multi-faceted methodology.

Nomenclature

A area (m^2)
A_{txv} flow area (m^2)
C_p specific heat capacity (J/kg K)

C	heat capacity rate (kW/K)
C_{txv}	valve flow coefficient
D	diameter (m)
$\dot{E}n$	energy rate (kW)
$\dot{E}x$	exergy rate (kW)
h	specific enthalpy (kJ/kg)
$\overline{h}$	heat transfer coefficient (W/m^2 K)
k	thermal conductivity (W/m C)
$\dot{m}$	mass flow rate (kg/s or L/min)
P	pressure ($kg/m\,s^2$)
Pr	Prandtl number
$\dot{Q}$	heat transfer rate (kW)
Ra	Rayleigh Number
Re	Reynolds number
s	specific entropy (kJ/kg K)
T	temperature (K or °C)
T_0	ambient temperature (K or °C)
U	overall heat transfer coefficient (W/m^2 K)
$\dot{W}$	work rate or power (kW)
V	velocity (m/s)
v	specific volume (m^3/kg)
Z	height (m)

Greek Letters

β	coefficient of thermal expansion
Δ	change in variable
δ	thickness (m)
η	efficiency
θ	pressure ratio
μ	dynamic viscosity (kg/ms)
ν	kinematic viscosity (m^2/s)

Subscripts

act	actual
bat	battery
cool	coolant
ch	chiller
comp	compressor
c,cond	condenser
D	destruction
en	energy
ex	exergy
e,evap	evaporator
f	fin
g	gas

H	high
i	inlet or inside
L	low
o	outlet or outside
ref	refrigerant
s	isentopic
txv	thermal expansion valve
w	wall
wg	water/glycol mix

Acronyms

COP	coefficients of performance
EV	electric vehicle
GHG	greenhouse gas
GWP	global warming potential
LMTD	logarithmic mean temperature difference
NTU	number of transfer units
TMS	thermal management system
TXV	thermal expansion valve

Study Questions/Problems

5.1 How would you define exergy and how does its use differentiate from the conventional energy analysis?

5.2 How is the concept of exergy tied to sustainability and environment? Please provide some conceptual examples.

5.3 What are the main considerations when developing a capable battery thermal management system?

5.4 Please select a component for battery TMS and evaluate it with respect to its exergy efficiency.

5.5 What are the effects of condensing/evaporating temperatures, compressor speed, heat exchanger pressure drop and battery heat generation rates on the battery TMS? Please use Ts and Pv diagrams to illustrate the changes from a thermodynamic perspective.

5.6 What are endogenous/exogenous and avoidable/unavoidable exergy destruction concepts? How can they be used to improve thermal systems?

5.7 Please re-iterate the analysis conducted in Section 5.5, using evaporating and condensing temperatures of 10 °C and 60 °C to keep the battery operating at 25 °C for a battery heat dissipation rate of 0.5 kW and cooling load of 2 kW.

5.8 In problem 5.7, what would be the change in your system efficiency with respect to the increasing the subcooling/superheating temperatures?

5.9 In problem 5.7, please select a component and comment on how much of its exergy destruction is based on technological limitations of the component with respect to the principles explained in enhanced exergy analysis.

References

Al-Hallaj S, Selman JR. (2002). Thermal modeling of secondary lithium batteries for electric vehicle/hybrid electric vehicle applications. *Journal of Power Sources* **110**:341 – 348.

Arcaklioglu E, Çavuşoglu A, Erisen A. (2005). An algorithmic approach towards finding better refrigerant substitutes of CFCs in terms of the second law of thermodynamics. *Energy Conversion and Management* **46**:1595 – 1611.

Arora A. and Kaushik S. C. (2008). Theoretical analysis of a vapour compression refrigerant system with R502, R404a and R507a. *International Journal of Refrigeration* **31**:998 – 1005.

Behr GmbH & Co. KG. (n.d.). Press Official Website, Technical Press Day. Available at: http://www.behrgroup.com/Internet/behrcms.eng.nsf [Accessed February 2011].

Bejan A. (1996). *Entropy Generation Minimization: The Method of Thermodynamic Optimization of Finite-Size Systems and Finite-Time Processes.* CRC Press LLC, Florida, U.S.

Bejan A. (1997). Thermodynamic optimization of heat transfer and fluid flow processes. *Developments in the design of thermal systems.* Cambridge University Press, Cambridge U.K.

Bhatti MS. (1999). *Enhancement of R134a Automotive Air Conditioning System.* International Congress and Exposition, Detroit, Michigan.

Bornakke C, Sonntag RE. (2009). *Fundemantals of Thermodynamics*, 7th ed. Wiley, Hoboken, NJ.

Brown JS, Yana-Motta, SF, Domanski PA. (2002). Comparative analysis of an automotive air conditioning systems operating with CO_2 and R134a. *International Journal of Refrigeration* **25**:19 – 32.

Churchill SW, Chu, HHS. (1975). Correlating Equations for Laminar and Turbulent Free Convection from a Vertical Plate. *International Journal of Heat Mass Transfer* **18**:1323 – 1329.

Dincer I, Rosen MA. (2013). *Exergy: Energy*, Environment and Sustainable Development, Oxford, England, UK: Elsevier.

Dittus SJ, Boelter LMK. (1930). *University of California Publications in Engineering* **2**, 443.

European Union. (2006). Directive 2006/40/EC of the European parliament and of the Council. *Official Journal of the European Union* **161**(12).

Jabardo JMS, Mamani WG, Ianekka MR. (2002). Modeling and experimental evaluation of an automotive air conditioning system with a variable capacity compressor. *International Journal of Refrigeration* **25**:1157 – 1172.

Kelly S, Tsatsaronis G, Morosuk R. (2009). Advanced exergetic analysis: Approaches for splitting the exergy destruction into endogenous and exogenous parts. *Energy* **24**:384–391.

Kim GH, Pesaran A. (2006). *Battery Thermal Management System Design and Modeling, 22nd International Battery, Hybrid and Fuel Cell Electric Vehicle Conference and Exhibition*, Yokohama, Japan.

Kim MH, Pettersen L, Bullard CW. (2004). Fundamental process and system design issues in CO2 vapor compression systems. *Progress in Energy Combustion Science* **30**:119–174.

Koban M. (2009). HFO-1234yf low GWP refrigerant LCCP analysis. In: Proceedings of SAE World Congress. Detroit, MI, USA.

Kumar S., Prevost M. and Bugarel R. (1989). Exergy Analysis of a Compression Refrigeration System. *Heat Recovery Systems & CHP* **9**:151–157.

Kuper C, Hoh M, Houchin-Miller G, Fuhr J. (2009). *Thermal Management of Hybrid Vehicle Battery Systems, 24th International Battery, Hybrid and Fuel Cell Electric Vehicle Conference and Exhibition*, Stavanger, Norway.

Lee GH, Yoo JY. (2000). Performance analysis and simulation of automobile air conditioning system. *International Journal of Refrigerant* **23**:243–254.

Mathur GD. (2000). Carbon dioxide as an alternative refrigerant for automotive air conditioning systems, Energy Conversion Engineering Conference and Exhibit (IECEC), 35th Intersociety. *Las Vegas, NV*, **1**:371–379.

Morosuk T, Tsatsaronis G. (2008). A new approach to the exergy analysis of absorption refrigeration machines, *Energy* **33**:890–907.

Morosuk T, Tsatsaronis G. (2006). The "Cycle Method"used in the exergy analysis of refrigeration machines: from education to research." In: Proceedings of the 19th international conference on efficiency, cost, optimization, simulation and environmental impact of energy systems 1:12–14.

Tian C, Li X. (2005). Numerical simulation on performance bad of automotive air conditioning system with a variable displacement compressor, *Energy Conversion and Management* **46**:2718–2738.

Tsatsaronis G, Park M. (2002). On avoidable and unavoidable exergy destructions and investment costs in thermal systems. *Energy Conversion Management* **43**:1259–1270.

Yang JL, Ma YT, Li MX, Guan HQ. (2005). Exergy analysis of transcritical carbon dioxide refrigeration cycle with an expander. *Energy* **30**:1162–1175.

Yang C, Maccarthy R. (2009). *Electricity Grid: Impacts of Plug-In Electric Vehicle Charging*, Recent Work, Institute of Transportation Studies (UC Davis).

Yumrutas R, Kunduz M, Kanoglu M. (2011). Exergy analysis of vapor compression refrigeration systems. *Exergy International Journal* **2**:266–272.

Wang SW, Gu J, Dickson T, Dexter T, McGregor I. (2005). Vapor quality and performance of an automotive air conditioning system. *Experimental Thermal Fluid Science* **30**:59–66.

Wongwises S, Kamboon A, Orachon B. (2006). Experimental investigation of hydrocarbon mixtures to replace HFC-134a in an automotive air conditioning system. *Energy Conversion Management* **47**:1644–1659.

6

Cost, Environmental Impact and Multi-Objective Optimization of Battery TMSs

6.1 Introduction

In a world with finite natural resources and increasing energy demand and prices, developing systems that are efficient, cost-effective and environmentally benign is one of the most prominent challenges that many engineers face today. In the past decades, the energy prices have been increasing while the legislations that aim to mitigate environmental problems (such as ozone layer depletion and global warming) have become more stringent. In this regard, exergy analysis has been used to improve thermal management system components and designs by determining the locations, types and true magnitude of inefficiencies in systems. However, exergy analysis does not provide any information on the financial and environmental aspect of the improvements. Hence, an integrated procedure that combines all these concerns needs to be developed to find a viable solution. For this reason, a multi-objective optimization that couples the second law of thermodynamics with economics and environmental impact becomes crucial to develop a powerful tool for the systematic study of the TMSs. Thus, in this chapter, engineers involved in electric vehicle battery development are expected to find solutions to the following corresponding issues:

- What are the largest contributors to the cost in BTMS systems and how can the operating parameters and associated cost flows be changed to reduce the investment and variable costs?
- How much is the environmental footprint does the BTMSs have through its lifetime and what are the major environmental impact potentials through different stages of utilization?
- What are the most applicable optimization methodologies to be used in developing BTMSs in a multi-faceted fashion and what are the most critical steps to ensure the accuracy and the reliability of the multi-objective optimization outcomes?

In order to answer these questions, detailed steps of conducting exergoeconomic and exergoenvironmental analysis and multi-objective optimization is provided in this chapter along with real life scenarios and case studies for given TMSs. Exergoeconomic analysis combines exergy analysis and economic principles, such as costs associated with purchase of equipment, input energy resources and maintenance and incorporates the associated costs of the thermodynamic inefficiencies in the total product cost of the system. These costs can be used to find the most and least cost-effective components and improving the overall system design. Meanwhile, exergoenvironmental analysis

Thermal Management of Electric Vehicle Battery Systems, First Edition.
İbrahim Dinçer, Halil S. Hamut and Nader Javani.

combines exergy analysis and the environmental impact, associated with construction, operation and maintenance and disposal stages, and allocates the corresponding impacts to the exergy streams, in order to point out the components causing the highest environmental impact and suggesting possibilities and trends for improvement, based on the calculated exergoenvironmental variables. Subsequently, multi-objective optimization with respect to these aforementioned analyses is described in order to compensate shortcomings of traditional single objective approaches (namely single objective exergy, exergoeconomic and exergoenvironmental optimizations) by allowing a larger perspective and determining a more complete spectrum of solutions that optimize the design according to more than one objective at a time. In most practical decision making problems, the objectives are conflicting in nature and a unique optimal solution cannot be identified. Thus, Pareto optimality is introduced to determine whether a solution is really one of the best possible trade-offs. By following these steps the readers are able to further the BTMS analysis that was started in the previous chapter by incorporating two most important constraints that most engineers have to face in their product development (namely product cost and environmental impact) and learn to optimize their systems accordingly. A simplified version of the code for the aforementioned analyses is also provided in the Engineering Equation Solver (EES) in the appendices section.

6.2 Exergoeconomic Analysis

Even though the thermodynamic analyses (especially exergy analysis) can be used to improve the efficiencies of the components and corresponding systems, the feasibility of applying these improvements is generally constrained by the limitation of financial resources. Moreover, in many cases, the approaches taken by purely scientific motivation may not always be cost effective. Thus, in order to achieve the optimum design for energy systems, techniques combining scientific disciplines (mainly thermodynamics) with economic disciplines (mainly cost accounting) should be utilized.

Design of various thermal management systems is normally performed by conventional methods based on scientific analyses, experimental data and practical experience. Most of these systems are often operating outside of their optimum parameters which results in inefficient use of resources, increasing production costs and adverse environmental impact. The objective of exergoeconomic analysis is to determine the inefficiencies in the system and calculate the associated costs. In this section, a walkthrough of exergy costing method (SPECO method) is provided for TMSs (e.g., Tsatsaronis and Lin, 1990; Lazzaretto and Tsatsaronis, 2006).

6.2.1 Cost Balance Equations

In order to conduct an exergoeconomic analysis, the cost flow rate, $\dot{C}$ ($\$/h$), is defined for each flow in a system, and a cost balance needs to be written for each component to provide exergy costing as follows:

$$\dot{C}_{q,k} + \sum_{i} \dot{C}_{i,k} + \dot{Z}_k = \sum_{e} \dot{C}_{e,k} + \dot{C}_{w,k} \tag{6.1}$$

where

$$\dot{C}_j = c_j \dot{E}x_j \tag{6.2}$$

Exergy transfer by entering and exiting streams as well as by power and heat transfer rates are written respectively as follows:

$$\dot{C}_i = c_i \dot{E}x_i = c_i \, \dot{m}_i ex_i \tag{6.3a}$$

$$\dot{C}_e = c_e \dot{E}x_e = c_e \, \dot{m}_e ex_e \tag{6.3b}$$

$$\dot{C}_w = c_w \dot{W} \tag{6.3c}$$

$$\dot{C}_q = c_q \dot{E}x_q \tag{6.3d}$$

However, before an exergoeconomic analysis can be started, the fuel and product exergies need to be defined for each component. The product exergy is defined according to the purpose of owning and operating a component under consideration, while the fuel represents the resources consumed in generating the product, where both are expressed in terms of exergy. The fuel and products for each component can be seen in Table 6.1.

By combining exergy and exergoeconomic balance equations, the following equation can be obtained as

$$\dot{E}x_{F,k} = \dot{E}x_{P,k} + \dot{E}x_{D,k} \tag{6.4}$$

The cost rate of exergy destruction is defined as follows:

$$\dot{C}_{D,k} = c_{F,k} \dot{E}x_{D,k} \tag{6.5}$$

Here, the component exergy destruction costs are determined by evaluating the exergy destruction rates associated with each component ($\dot{E}x_{D,k}$) with respect to the previously given exergy balance equations. Moreover, from Equation 6.1, the steady state form of the control volume cost balance can be written as given in Equation 6.6. The cost balances are generally written, so that all terms are positive:

$$\sum_e (c_e \dot{E}x_e)_k + c_{w,k} \dot{W}_k = c_{q,k} \dot{E}x_{q,k} + \sum_i (c_i \dot{E}x_i)_k + \dot{Z}_k \tag{6.6}$$

Table 6.1 Fuel and product definitions with respect to the system.

Component	*Fuel*	*Product*
Compressor	$\dot{W}_{comp}$	$\dot{E}x_2 - \dot{E}x_1$
Condenser	$\dot{E}x_2 - \dot{E}x_3$	$\dot{E}x_{10} - \dot{E}x_0$
Evaporator TXV	$\dot{E}x_{3a}$	$\dot{E}x_4$
Chiller TXV	$\dot{E}x_{3b}$	$\dot{E}x_5$
Evaporator	$\dot{E}x_4 - \dot{E}x_{1a}$	$\dot{E}x_9$
Chiller	$\dot{E}x_{1b} - \dot{E}x_5$	$\dot{E}x_7 - \dot{E}x_6$
Pump	$\dot{W}_{pump}$	$\dot{E}x_8 - \dot{E}x_7$
Battery	$\dot{W}_{bat}$	$\dot{E}x_6 - \dot{E}x_8$

which states that the total cost of the exiting exergy streams equals the total expenditure to obtain them, namely the cost of the entering exergy streams plus the capital and other costs. In general, there are "n_e" exergy streams exiting the component, "n_e" unknowns and only one equation, the cost balance. Thus, "$n_e - 1$" auxiliary equations need to be formulated using F and P rules.

The F rule (fuel rule) refers to the removal of exergy from an exergy stream within the considered component when exergy differences between the inlet and outlet are considered in the fuel definition for this stream. Thus, this rule states that the specific cost (cost per exergy unit) associated with this fuel stream exergy removal must be equal to the average specific cost at which the removed exergy was supplied to the same stream in upstream components. This provides an auxiliary equation for each removal of exergy, which equals the number of exiting exergy streams and "$n_{e,F}$" that are associated with the definition of the fuel for each component. The P rule (product rule) refers to the supply of exergy to an exergy stream within the component and states that each exergy unit is applied to any stream associated with the product at the same average cost. Since this corresponds to an exiting stream, the number of auxiliary equations provided by this rule always equals $n_{e,P} - 1$, where $n_{e,P}$ is the number of exiting exergy streams that are included in the product definition. Thus, since each exiting stream is defined as either fuel or product, the total number of exiting streams is equal to "$n_{e,F} + n_{e,P}$", which provides "$n_e - 1$" auxiliary equations (Lazzaretto and Tsatsaronis, 2006).

6.2.2 Purchase Equipment Cost Correlations

On the economic side, the capital investment rate can be calculated with respect to the purchase cost of equipment and capital recovery as well as maintenance factor over the number of operation hours per year as given below:

$$\dot{Z}_k = \frac{Z_k \cdot CRF \cdot \varphi}{N} \tag{6.7}$$

where N is the annual number of operation hours for the unit and φ is the maintenance factor, generally taken as 1.06 (Bejan *et al.*, 1996). *CRF* is the capital recovery factor which depends on the interest rate (i) and equipment life-time in years (n) as

$$CRF = \frac{i \times (1+i)^n}{(1+i)^n - 1} \tag{6.8}$$

Here, Z_k is the purchase equipment cost of the thermal management system components that should be written in terms of design parameters. The correlations for each component are given below (Valero, 1994):

$$Z_{comp} = \left(\frac{573\dot{m}_{ref}}{0.8996 - \eta_s}\right)\left(\frac{P_{cond}}{P_{evap}}\right)\ln\left(\frac{P_{cond}}{P_{evap}}\right) \tag{6.9}$$

where

$$\eta_s = 0.85 - 0.046667\left(\frac{P_{cond}}{P_{evap}}\right) \tag{6.10}$$

Here, $\dot{m}_{ref}$ is the refrigerant mass flow rate (kg/s) and η_s is the isentropic efficiency of a scroll compressor. For the heat exchangers the cost correlations developed by Selbas *et al.* (2006) are used. The fixed cost associated with the heat exchangers is neglected due to being insignificant relative to the variable costs as well as a lack of reliable data.

$$Z_{cond} = 516.621A_{cond} \tag{6.11}$$

$$Z_{evap} = 309.143A_{evap} \tag{6.12}$$

$$Z_{chil} = 309.143A_{chil} \tag{6.13}$$

where A_{cond}, A_{evap} and A_{chil} are the heat transfer areas associated with the condenser and evaporator respectively.

$$Z_{pump} = 308.9\dot{W}_{pump}^{C_{pump}} \tag{6.14a}$$

$$C_{pump} = 0.25 \text{ for } 0.02 \text{ kW} < \dot{W}_{pump} < 0.3 \text{ kW} \tag{6.14b}$$

$$C_{pump} = 0.45 \text{ for } 0.3 \text{ kW} < \dot{W}_{pump} < 20 \text{ kW} \tag{6.14c}$$

$$C_{pump} = 0.84 \text{ for } 20 \text{ kW} < \dot{W}_{pump} < 200 \text{ kW} \tag{6.14d}$$

Here, $\dot{W}_{pump}$ is the pumping power in kW and C_{pump} is the pump coefficient with respect to the corresponding pumping power ranges, provided above (Sanaye and Niroomand, 2009):

$$Z_{evap,\ txv} = k_{txv}\dot{m}_{ref,a} \tag{6.15}$$

$$Z_{chil,txv} = k_{txv}\dot{m}_{ref,b} \tag{6.16}$$

where k_{txv} is the cost per mass flow rate of refrigerant.

$$Z_{bat} = C_{bat}K_{bat} \tag{6.17}$$

Here, C_{bat} is the typical lithium-ion battery pack costs per kilowatt-hour, and K_{bat} is the battery pack energy that is associated with powering the thermal management system.

6.2.3 Cost Accounting

Cost balances for each component need to be solved in order to estimate the cost rate of exergy destruction in each component. In the cost balance equations with more than one inlet or outlet flow, the number of unknown cost parameters exceeds the number of cost balances for that component. Thus, auxiliary exergoeconomic equations developed by F and P rules are needed to equate the number of unknowns with the number of equations. Implementing Equation 6.6 (for most commonly used BTMS components) together with the auxiliary equations form a system of linear equations as follows:

$$[\dot{E}x_k] \times [c_k] = [\dot{Z}_k] \tag{6.18}$$

where the equation entails matrices of exergy rate (from exergy analysis), exergetic cost vector (to be evaluated) and the vector of $\dot{Z}_k$ factors (from economic analysis) respectively (Ahmadi *et al.*, 2011). The matrix form of Equation 6.18 is given below for a BTMS with most commonly used components:

$$
\begin{bmatrix}
1 & -1 & 0 & 0 & 0 & 0 & 0 & 0 & 0 & 0 & -\dot{W}_{comp} & 0 & 0 & 0 \\
0 & 1 & -1 & 0 & 0 & 0 & 0 & 0 & 0 & 0 & 0 & 0 & 1 & -1 \\
0 & \dot{E}x_3 & -\dot{E}x_2 & 0 & 0 & 0 & 0 & 0 & 0 & 0 & 0 & 0 & 0 & 0 \\
0 & 0 & 1 & -1 & 0 & 0 & 0 & 0 & 0 & 0 & 0 & 0 & 0 & 0 \\
-\dot{E}x_4 & 0 & 0 & \dot{E}x_1 & 0 & 0 & 0 & 0 & -1 & 0 & 0 & 0 & 0 & 0 \\
1 & 0 & 0 & -1 & 0 & 0 & 0 & 0 & 0 & 0 & 0 & 0 & 0 & 0 \\
0 & 0 & 1 & 0 & -1 & 0 & 0 & 0 & 0 & 0 & 0 & 0 & 0 & 0 \\
-1 & 0 & 0 & 0 & 1 & 1 & -1 & 0 & 0 & 0 & 0 & 0 & 0 & 0 \\
0 & 0 & 0 & 0 & 0 & 1 & -1 & 0 & 0 & 0 & 0 & 0 & 0 & 0 \\
0 & 0 & 0 & 0 & 0 & 0 & 1 & -1 & 0 & 0 & 0 & \dot{W}_{pump} & 0 & 0 \\
0 & 0 & 0 & 0 & 0 & 0 & 0 & 1 & 0 & -\dot{W}_{bat} & 0 & 0 & 0 & 0 \\
0 & 0 & 0 & 0 & 0 & 0 & 0 & 0 & 0 & 0 & 1 & 0 & 0 & 0 \\
0 & 0 & 0 & 0 & 0 & 0 & 0 & 0 & 0 & 0 & 0 & \dot{W}_{pump} & -\dot{W}_{comp} & 0 \\
0 & 0 & 0 & 0 & 0 & 0 & 0 & 0 & 0 & 0 & 0 & 0 & 1 & 0
\end{bmatrix}
\times
\begin{bmatrix}
\dot{C}_1 \\ \dot{C}_2 \\ \dot{C}_3 \\ \dot{C}_4 \\ \dot{C}_5 \\ \dot{C}_6 \\ \dot{C}_7 \\ \dot{C}_8 \\ \dot{C}_9 \\ \dot{C}_{10} \\ \dot{C}_{11} \\ \dot{C}_{12} \\ \dot{C}_{13} \\ \dot{C}_{14}
\end{bmatrix}
=
\begin{bmatrix}
-\dot{Z}_{comp} \\ -\dot{Z}_{cond} \\ 0 \\ -\dot{Z}_{etxv} \\ -\dot{Z}_{evap} \\ 0 \\ -\dot{Z}_{ctxv} \\ -\dot{Z}_{chil} \\ 0 \\ -\dot{Z}_{pump} \\ -\dot{Z}_{bat} \\ -c_{elect} \\ 0 \\ 0
\end{bmatrix}
$$

The matrix is obtained based on the cost balance equations as given below:

$$\dot{C}_1 + \dot{Z}_{comp} + c_{elect}\dot{W}_{comp} = \dot{C}_2$$

$$\dot{C}_2 + \dot{C}_{13} + \dot{Z}_{cond} = \dot{C}_3 + \dot{C}_{14}$$

$$\dot{C}_2\dot{E}x_3 = \dot{C}_3\dot{E}x_2$$

$$\dot{C}_3 + \dot{Z}_{etxv} = \dot{C}_4$$

$$\dot{C}_4 + \dot{Z}_{evap} = \dot{C}_9 + \dot{C}_1$$

$$\dot{C}_4\dot{E}x_1 = \dot{C}_1\dot{E}x_4$$

$$\dot{C}_3 + \dot{Z}_{ctxv} = \dot{C}_5$$

$$\dot{C}_5 + \dot{C}_6 + \dot{Z}_{chil} = \dot{C}_1 + \dot{C}_7$$

$$\dot{C}_6\dot{E}x_7 = \dot{C}_7\dot{E}x_6$$

$$\dot{C}_7 + \dot{Z}_{pump} + c_{elect}\dot{W}_{pump} = \dot{C}_8$$

$$\dot{C}_8 + \dot{Z}_{bat} = \dot{C}_6 + \dot{W}_{bat}$$

$$\dot{C}_{11} = c_{elect}\dot{W}_{comp}$$

$$\dot{C}_{11}\dot{W}_{pump} = \dot{C}_{12}\dot{W}_{comp}$$

$$\dot{C}_{13} = 0$$

Here, c_{elect} is the unit cost of electricity. By solving these equations, the cost rate of each flow can be calculated, which can be used to determine the cost rate of exergy destruction in each system component.

6.2.4 Exergoeconomic Evaluation

Moreover, certain additional variables can also provide useful exergoeconomic evaluation. Among these variables, the total cost rate provides the component with the highest priority in terms of exergoeconomic viewpoint and is the combination of the cost rate with respect to the exergy destruction and investment cost rates as given below:

$$\dot{C}_{TOT,k} = \dot{C}_{D,k} + \dot{Z}_k \tag{6.19}$$

The exergoeconomic relevance of a given component with respect to total cost rate is determined by the sum of the cost of exergy destruction $\dot{C}_{D,k}$ and the component-related cost $\dot{Z}_{D,k}$. Furthermore, an exergoeconomic factor is also used to determine the contribution of non-exergy related costs to the total cost of a component. It is defined as

$$f_k = \frac{\dot{Z}_k}{\dot{Z}_k + c_{f,k}\dot{E}_{D,k}} \tag{6.20}$$

where $c_{f,k}$ is the unit exergy cost of the fuel of any k component and $\dot{E}_{D,k}$ is the associated exergy destruction and the denominator forms the total cost rate. When a component has a low exergoeconomic factor value, cost savings in the entire system might be achieved by improving the component efficiency even if the capital investment for that component will increase. On the other hand, a high value might suggest a decrease in the investment costs at the expense of its exergetic efficiency.

In addition, relative cost difference can also be used as a useful thermoeconomic evaluation, where it shows the relative increase in the average cost per exergy unit is between the fuel and product of the component, and is defined as

$$r_k = \frac{c_{p,k} - c_{f,k}}{c_{f,k}} \tag{6.21}$$

6.2.5 Enhanced Exergoeconomic Analysis

Exergoeconomic analysis is useful in understanding the relative cost importance of each system and the options for improving the overall system effectiveness. Since cost of EVs and HEVs are one of the biggest road blocks for widespread commercialization of these technologies, where the thermal management is a significant portion of the total cost, it is worthwhile to further analyze the cost formation of the system, break it down into avoidable and unavoidable costs and determine the cost interactions among the components. Thus, the investment cost rates are also is split into endogenous/exogenous and avoidable/unavoidable parts (so called advanced exergoeconomic analysis).

The unavoidable investment cost ($\dot{Z}_k^{UN}$) for a component can be calculated by assuming a vastly inefficient version of this component which would not be used in real life applications due to very high fuel costs associated with it. This cost is determined based on arbitrary selection of a set of thermodynamic parameters for the components that would result in so inefficient solutions that would be economically unpractical. These are composed of very low values for isentropic efficiencies of the compressor and small heat transfer area for the heat exchangers. The cost rates associated with unavoidable and avoidable exergy destruction along with unavoidable and avoidable investment costs are can be calculated as:

$$\dot{C}_{D,k}^{UN} = c_{F,k}\,\dot{E}_{D,k}^{UN} \tag{6.22}$$

$$\dot{C}_{D,k}^{AV} = c_{F,k}\,\dot{E}_{D,k}^{AV} \tag{6.23}$$

where $c_{F,k}$ is the cost of fuel and $\dot{E}_{D,k}^{UN}$ and $\dot{E}_{D,k}^{AV}$ are the unavoidable and avoidable cost of exergy destruction associated with the individual components calculated in Chapter 5. $\dot{Z}_k^{UN}$ and $\dot{Z}_k^{AV}$ are the unavoidable and avoidable investment costs that are calculated by the aforementioned thermodynamic parameters as provided below:

$$\dot{Z}_k^{UN} = \dot{E}_{P,k}\left(\frac{\dot{Z}}{\dot{E}_P}\right)_k^{UN} \tag{6.24}$$

$$\dot{Z}_k^{AV} = \dot{E}_{P,k}\left(\frac{\dot{Z}}{\dot{E}_P}\right)_k^{AV} \tag{6.25}$$

Finally, the exergoeconomic factor can also be modified with respect to avoidable costs as follows:

$$f_k^* = \frac{\dot{Z}_k^{AV}}{\dot{Z}_k^{AV} + \dot{C}_{D,\,k}^{AV}} \tag{6.26}$$

where f_k^* shows the contribution of the avoidable investment cost on the total avoidable cost associated with each specific component. The use of avoidable exergy destruction and avoidable cost provides the readers with a more accurate representation with respect to the potential reductions that can be done in the irreversibilities and cost of the components compared to the conventional exergoeconomic variables.

6.2.6 Enviroeconomic (Environmental Cost) Analysis

Most hybrid electric vehicles (HEVs) use electricity from the grid to power the TMS (thermal management system). The TMS has a significant role in reducing the associated GHG emissions compared to conventional vehicles. Even though these vehicles produce virtually zero GHG emissions through the tailpipe in all-electric mode during operation, there may still be indirect emissions associated with the generation of electricity. These emissions, especially under a high carbon derived electricity generation mix, can be significantly high (possibly even higher than conventional vehicles) and therefore the associated CO_2 GHG emissions and corresponding environmental costs should be calculated.

For the developed model, various electricity generation mixes are considered from one that mainly utilizes a natural gas combined cycle to less environmentally friendly options that primarily use coal and steam. The associated environmental assessment based on the corresponding CO_2 emissions can be calculated as given below:

$$x_{CO_2} = \frac{y_{CO_2} \times \dot{W}_{total} \times t_{total}}{10^6} \tag{6.27}$$

where x_{CO2} is the associated CO_2 emissions released in a year (tCO_2/year) and y_{CO_2} is the corresponding CO_2 emissions for a coal fired electricity generator, $\dot{W}_{total}$ is the total power consumption of the TMS and t_{total} is the total working hours of the system in a year.

In order to conduct an enviroeconomic analysis, a carbon price (or CO_2 emissions price) needs to be established along with calculating the quantity of the carbon released. The carbon price is an approach imposing a cost on the emission of greenhouse gases which cause global warming. The international carbon price is typically between 13 and 16 $/t$CO_2$ based on different carbon scenarios. The enviroeconomic parameter in terms of CO_2 emissions price in a year ($/year) can be calculated as given below:

$$C_{CO_2} = (c_{CO2})(x_{CO2}) \tag{6.28}$$

where c_{CO_2} is the CO_2 emissions price per tCO_2.

6.3 Exergoenvironmental Analysis

As mentioned in the previous sections, using exergy analysis to determine the exergy efficiencies and exergy destruction associated with each component can be used to make significant improvements on the system. However, improving the efficiencies of a system may often imply modifications in component design, which a lot of times lead to increasing a parameter (commonly an area, thickness or temperature) that results in an increase in the materials and energy needed for manufacturing the component. This in turn, may increase the consumption of natural resources to produce the component, pollutants generated during its operation or emissions associated with its disposal. Thus, the system should be evaluated with respect to the environmental impact associated with each component in addition to their thermodynamic efficiencies (Meyer *et al.*, 2009).

Exergoenvironmental analysis reveals the environmental impact associated with each system component and the real sources of the impact by combining exergy analysis with a comprehensive environmental assessment method, such as life cycle assessment (LCA), which is an internationally standardized method that considers the entire useful life cycle of the components or overall systems with respect to their impact to the environment determined by the environmental models.

In the environmental analysis, LCA is carried out in order to obtain the environmental impact of each relevant system components and input streams. It consists of goal definition, inventory analysis and interpretation of results, which incorporates the supply of the input streams (especially fuel) and full life cycle of components. The quantification of environmental impact with respect to depletion and emissions of a natural resource can be conducted using different methodologies. For the LCA analysis, various damage categories need to be covered, weighted and expressed in terms of Eco-indicator points (mPts).

6.3.1 Environmental Impact Balance Equations

Exergoenvironmental analysis is considered to be one of the most promising tools to evaluate energy conversion process from environmental point of view (Boyano *et al.*, 2012). In order to be able to perform the analysis, the allocation of environmental analysis results to exergy streams is performed analogous to the allocation of exergy stream costs in exergoeconomics. Initially, an environmental impact rate $\dot{B}_j$ is expressed in terms of Eco-indicator 99 points and is converted into hourly rates (mPts/hour). Subsequently, these values along with the previously conducted exergy analysis are used to calculated specific environment impact b_j for the streams in the system.

$$b_j = \frac{\dot{B}_j}{\dot{E}x_j} \tag{6.29}$$

The environmental impact rates associated with heat and work transfers can be calculated as follows:

$$\dot{B}_w = b_w \dot{W} \tag{6.30}$$

$$\dot{B}_q = b_q \dot{E}x_q \tag{6.31}$$

where

$$\dot{E}x_q = \left(1 - \frac{T_0}{T_j}\right)\dot{Q} \tag{6.32}$$

The values for internal and output streams can only be obtained by considering the functional relations among system components, which are done through formulating environmental impact balances and auxiliary equations. The basis for formulating impact balances is that all environmental impacts entering a component have to exit the component with its output streams. In addition, there is also the component-related environmental impact that is associated with the life cycle of each component.

In order to conduct an exergoenvironmental analysis, an environmental impact balance is written for each component to provide environmental impact formation as follows:

$$\dot{B}_{q,k} + \sum_i \dot{B}_{i,k} + \dot{Y}_k = \sum_e \dot{B}_{e,k} + \dot{B}_{w,k} \tag{6.33}$$

In the above equation, $\dot{Y}_k$ is the component related environmental impact associated with the life cycle of the component, which is an indicator of the reduction potential of environmental impact of the component. The environmental balance equation states that the sum of all environmental impacts associated with all input streams plus the component-related environmental impact is equal to the sum of environmental impacts associated with all output streams.

6.3.2 Environmental Impact Correlations

In order to be able to solve the environmental balance equations, the environmental impacts associated with each component need to be determined. Usually Eco-indicator 99 points are used, as it is an indicator especially developed to support decision making in design for the environment, since it enables a fair comparison among different components. These impact points are approximated with respect to a combination of correlations developed from numerous studies conducted in literature as well as available data as shown in Table 6.2 in order to enable the readers to improve their system component models without the need of cumbersome experimental relationships.

For the heat exchangers, the eco-indicator points are estimations based on the area, and are calculated from scaling down various case studies performed in the literature. The component-related heat exchanger environmental impacts associated

Table 6.2 Environmental impact correlations (eco-indicator 99) developed based on the compiled data.

Component	$\dot{Y}(mPts/h)$	Criteria
Compressor	0.89	m_{comp}
Condenser	0.27	A_{cond}
Evaporator	0.22	A_{evap}
Chiller	0.15	A_{chil}
Evaporator TXV	0.04	$\dot{m}_{ref,a}$
Chiller TXV	<0.01	$\dot{m}_{ref,b}$
Pump	0.13	m_{pump}

with the non-heat exchanging areas are neglected due to their relatively small size and unavailability of the data. The environmental impact for the compressor and pump is determined with respect to the weight of the components whereas for thermal expansion valves, it is based on the mass flow rate of the refrigerant. Finally, the thermal expansion valve component-related eco-indicator points determined per mass flow rate of the R134a refrigerant.

6.3.3 LCA of the Electric Battery

Even though EVs and HEVs can form part of the solution to environmental concerns such as urban air pollution and global warming compared to the conventional vehicles with ICEs, when the EVs and HEVs are evaluated, there are still environmental concerns associated with the electric battery itself. Thus, determining the battery environmental impact plays a significant role in accurately assessing the overall environmental impact of the system.

In the environmental analysis, LCA needs to be carried out in order to obtain the environmental impact of the battery assembly. It is a cradle to grave approach to study the environmental aspects throughout a product's life from raw material acquisition through production, use and disposal and provides a quantitative data to identify the potential environmental impacts of the material and/or production on the environment (ISO 14040, 1997).

It consists of goal definition, inventory analysis and interpretation of results, which incorporates the supply of the full life cycle of the battery. The quantification of environmental impact with respect to depletion and emissions of a natural resource can be conducted using different methodologies. For the LCA analysis, numerous damage categories need to be covered and the results need to be weighted and expressed in terms of a common measurement system.

The goal and scope of the analysis is to calculate the environmental impact associated with the lithium-ion battery in the TMS used and determine the parts/processes that have the largest contribution to the overall impact. In these analyses, the final environmental impact value is usually calculated as a single Eco-indicator 99 point based on a unit kg of the battery chemistry for a given electric generation mix and weighting set.

In Table 6.3, an example of a weighting set that belongs to the hierarchist perspective (H/H) is shown.

In order to conduct the necessary analysis for the battery, three stages of life cycle inventory (LCI) need to be used incorporated. In stage 1, the initial mining of raw materials and metal production stages during which raw materials are extracted and transported are considered. The raw materials depend on the battery chemistry and packaging type, but usually are the copper and aluminum used for the cathodes and anodes. Stage 2 included the conversion of materials to battery parts and associated machining processes. The main components are usually the electrodes, pastes, separators and electrolytes used in the battery cell along with the battery management system, module packaging, and the overall casing that are used to contain and protect the cells. The list of the major components considered in most analyses along with their corresponding weights per 1 kg of Li-ion battery is provided in Table 6.4.

Table 6.3 Normalization used for eco-indicator 99 H/H.

Normalization	Value
Human Healt[a)]	114.1
Ecosystem Quality[b)]	$1.75x10^{-4}$
Resources[c)]	$1.33x10^{-4}$

a) Unit: Disability adjusted life years
b) Unit: Potential disappeared fraction of plant species
c) Unit: MJ surplus energy

Table 6.4 Major battery components used in the LCA analysis and their corresponding weights.

Component	Weight (kg)
Electrode paste (+)	0.199
Electrode paste (−)	0.028
Cathode	0.034
Anode	0.083
Electrolyte	0.120
Separator	0.033
Casing	0.201
Module Packaging	0.170
BMS	0.029

Source: Majeau-Bettez *et al.* (2011)

Finally, stage 3 is composed of final assembly of the components to the battery. In this stage the analysis is usually conducted with respect to 1 kg of battery and is later scaled up to the full size of the considered battery assembly to reduce the complexity of the analysis. It should be noted that the determined eco-indicator points in these stages need to be converted into hourly rates (mPts/hour) to be consistent with the rest of the analyses.

6.3.4 Environmental Impact Accounting

Environmental impact balances for each component need to be solved in order to estimate the environmental impact rate of exergy destruction in each component. For the balance equations with multiple inlet and outlet flows, auxiliary exergoenvironmental equations (analogous to exergoeconomic equations) need to be developed to match the unknown impact parameters with the number of environmental impact balance equations. Implementing equation 6.33 for each component together with the auxiliary equations form a system of linear equations as follows:

$$[\dot{E}x_k] \times [b_k] = [\dot{Y}_k] \tag{6.34}$$

where the equation entails matrixes of exergy rate (from exergy analysis), environmental impact vector (to be evaluated) and the vector of $\dot{Y}_k$ factors (from environmental analysis) respectively. The matrix form of the equation 6.34 can be seen below:

$$\begin{bmatrix}
1 & -1 & 0 & 0 & 0 & 0 & 0 & 0 & 0 & 0 & -\dot{W}_{comp} & 0 & 0 & 0 \\
0 & 1 & -1 & 0 & 0 & 0 & 0 & 0 & 0 & 0 & 0 & 0 & 1 & -1 \\
0 & \dot{E}x_3 & -\dot{E}x_2 & 0 & 0 & 0 & 0 & 0 & 0 & 0 & 0 & 0 & 0 & 0 \\
0 & 0 & 1 & -1 & 0 & 0 & 0 & 0 & 0 & 0 & 0 & 0 & 0 & 0 \\
-\dot{E}x_4 & 0 & 0 & \dot{E}x_1 & 0 & 0 & 0 & 0 & -1 & 0 & 0 & 0 & 0 & 0 \\
1 & 0 & 0 & -1 & 0 & 0 & 0 & 0 & 0 & 0 & 0 & 0 & 0 & 0 \\
0 & 0 & 1 & 0 & -1 & 0 & 0 & 0 & 0 & 0 & 0 & 0 & 0 & 0 \\
-1 & 0 & 0 & 0 & 1 & 1 & -1 & 0 & 0 & 0 & 0 & 0 & 0 & 0 \\
0 & 0 & 0 & 0 & 0 & 1 & -1 & 0 & 0 & 0 & 0 & 0 & 0 & 0 \\
0 & 0 & 0 & 0 & 0 & 0 & 1 & -1 & 0 & 0 & 0 & \dot{W}_{pump} & 0 & 0 \\
0 & 0 & 0 & 0 & 0 & 0 & 0 & 1 & 0 & -\dot{W}_{bat} & 0 & 0 & 0 & 0 \\
0 & 0 & 0 & 0 & 0 & 0 & 0 & 0 & 0 & 0 & 1 & 0 & 0 & 0 \\
0 & 0 & 0 & 0 & 0 & 0 & 0 & 0 & 0 & 0 & 0 & \dot{W}_{pump} & -\dot{W}_{comp} & 0 \\
0 & 0 & 0 & 0 & 0 & 0 & 0 & 0 & 0 & 0 & 0 & 0 & 1 & 0
\end{bmatrix}
\times
\begin{bmatrix}
\dot{B}_1 \\ \dot{B}_2 \\ \dot{B}_3 \\ \dot{B}_4 \\ \dot{B}_5 \\ \dot{B}_6 \\ \dot{B}_7 \\ \dot{B}_8 \\ \dot{B}_9 \\ \dot{B}_{10} \\ \dot{B}_{11} \\ \dot{B}_{12} \\ \dot{B}_{13} \\ \dot{B}_{14}
\end{bmatrix}
=
\begin{bmatrix}
-\dot{Y}_{comp} \\ -\dot{Y}_{cond} \\ 0 \\ -\dot{Y}_{etxv} \\ -\dot{Y}_{evap} \\ 0 \\ -\dot{Y}_{ctxv} \\ -\dot{Y}_{chil} \\ 0 \\ -\dot{Y}_{pump} \\ -\dot{Y}_{bat} \\ -b_{elect} \\ 0 \\ 0
\end{bmatrix}$$

The matrix is obtained based on the environmental impact balance equations as given below:

$$\dot{B}_1 + \dot{Y}_{comp} + b_{elect}\dot{W}_{comp} = \dot{B}_2$$
$$\dot{B}_2 + \dot{B}_{13} + \dot{Y}_{cond} = \dot{B}_3 + \dot{B}_{14}$$
$$\dot{B}_2\dot{E}x_3 = \dot{B}_3\dot{E}x_2$$
$$\dot{B}_3 + \dot{Y}_{etxv} = \dot{B}_4$$
$$\dot{B}_4 + \dot{Y}_{evap} = \dot{B}_9 + \dot{B}_1$$
$$\dot{B}_4\dot{E}x_1 = \dot{B}_1\dot{E}x_4$$
$$\dot{B}_3 + \dot{Y}_{ctxv} = \dot{B}_5$$
$$\dot{B}_5 + \dot{B}_6 + \dot{Y}_{chil} = \dot{B}_1 + \dot{B}_7$$
$$\dot{B}_6\dot{E}x_7 = \dot{B}_7\dot{E}x_6$$
$$\dot{B}_7 + \dot{Y}_{pump} + b_{elect}\dot{W}_{pump} = \dot{B}_8$$
$$\dot{B}_8 + \dot{Y}_{bat} = \dot{B}_6 + \dot{W}_{bat}$$
$$\dot{B}_{11} = b_{elect}\dot{W}_{comp}$$
$$\dot{B}_{11}\dot{W}_{pump} = \dot{B}_{12}\dot{W}_{comp}$$
$$\dot{B}_{13} = 0$$

Here, b_{elect} is the unit environmental impact associated with the electricity generation mix used (from Eco-indicator 99). By solving these equations, the environmental impact rate of each flow can be calculated, which can be used to determine the environmental impact rate of exergy destruction in each system component.

6.3.5 Exergoenvironmental Evaluation

In order to evaluate the environmental performance of the BTMS components and provide suggestions and recommendations, exergoenvironmental variables need to be defined for the system and are analogous to most exergoeconomic variables. The environmental impact rate associated with the exergy destruction of a component is defined as follows:

$$\dot{B}_{D,k} = b_{F,k}\dot{E}x_{D,k} \quad (\text{if } \dot{E}x_{P,k} \text{ is constant}) \tag{6.35}$$

Here, the environmental impact of component exergy destructions are determined by evaluating the exergy destruction rates associated with each component ($\dot{E}x_{D,k}$) with respect to the exergy balance equations along with the exergy based specific environmental impact calculated by the aforementioned environmental impact matrix. The total environmental impact of a component is calculated by adding the sum of the environmental impact of exergy destruction to the previously calculated component-related environmental impact.

$$\dot{B}_{TOT,k} = \dot{B}_{D,k} + \dot{Y}_k \tag{6.36}$$

The exergoenvironmental approach identifies the relevance from the environment point of view of a given component with respect to total environmental impact that is determined by the sum of the environmental impact of exergy destruction $\dot{B}_{D,k}$ and the component-related environmental impact $\dot{Y}_{D,k}$. Moreover, the relative difference of specific environmental impacts $r_{b,k}$ is defined by

$$r_{b,k} = \frac{r_{P,k} - r_{F,k}}{r_{F,k}} \tag{6.37}$$

which is an indicator of reduction potential of the environmental impact associated with a component. In general, the higher the value of relative difference of specific environmental impacts of a component in a system, the smaller the effort it would be needed to reduce the environmental impact of that component. This variable represents the environmental quality, independently of the absolute value of environmental impact.

Furthermore, the sources for the formation of environmental impact in a component are compared using the exergoenvironmental factor $f_{b,k}$, which expresses the relative contribution of the component-related environmental impact $\dot{Y}_k$ to the total environmental impact for the component.

$$f_{b,k} = \frac{\dot{Y}_k}{\dot{Y}_k + b_{f,k}\dot{E}_{D,k}} \tag{6.38}$$

In general, it is considered that $f_{b,k}$ higher than approximately 0.7 signifies that the component-related environmental impact $\dot{Y}_k$ is dominant, whereas $f_{b,k}$ lower than approximately 0.3 signifies that exergy destruction is the dominant source. Thus, the higher the exergoenvironmental factor, the higher the influence of the component-related environment impact to the overall performance of the system from the environmental perspective.

Based on the exergoenvironmental variables above, an evaluation of the system can be conducted by examining the components with high total environmental impacts (indicated by $\dot{B}_{TOT,k}$) and selecting the ones with highest improvement potentials (indicated by $r_{b,k}$) and identifying the main source of the environmental impact (identified by $f_{b,k}$) associated with those components. Finally suggestions for reducing the overall environmental impact can be developed based on the results of the LCA and impact correlations if the component-related impact dominates the overall impact, or with the help of the exergy analysis, if the thermodynamics inefficiencies are the dominant source of the environmental impact being considered.

6.4 Optimization Methodology

An optimization of any process, system and application is critical in improving it, such as by increasing the efficiency and quality, reducing the cost and environmental impact, and hence increasing the sustainability. In an optimization is then important to define the objective function(s) and constraints accordingly. This section thoroughly discusses these aspects for optimization of battery TMSs.

6.4.1 Objective Functions

A multi-objective optimization problem requires the simultaneous satisfaction of a number of different and usually conflicting objectives characterized by distinct measure of performance. It should be noted that multi-objective optimization problems generally show a possible uncountable set of solutions which represents the best possible trade-offs in the objective function space and that no combination of decision variable values can minimize/maximize all the components of functions simultaneously (Sayyaadi and Babaelahi, 2011). For BTMSs, these objective functions are usually the combinations of exergy efficiency (to be maximized), the total cost rate of product (to be minimized) and environmental impact (to be minimized). In this regard, the corresponding objective functions can be expressed through equations 6.39–6.41. Even though each objective function varies in terms of the objective it is optimizing, they all have the same underlying parameters which are affected by the changes in the selected decision variables. It should be noted that all the objectives in the multi-objective optimization at this point are assumed to be equally important, and therefore no additional weighting criteria are assigned to the objectives in order to minimize subjectivity in the analysis. Instead, the LINMAP (linear programming technique for multidimensional analysis of preference) method is used where the point on the Pareto optimal frontier closest to an ideal unreachable point (where all selected objectives are optimized) is selected as the single best optimization point (Lazzaretto and Toffolo, 2004).

Exergy efficiency:

$$\psi_{system} = \frac{\dot{E}x_{\dot{Q}_{evap}} + \dot{E}x_{\dot{Q}_{ch}}}{\dot{W}_{comp} + \dot{W}_{pump}} \tag{6.39}$$

where the inputs are the work of the compressor and the pump and the outputs are the exergy of heat with respect to the evaporator and the chiller (refer to Chapter 5).

Total cost rate:

$$\dot{C}_{system} = \dot{Z}_k + \dot{C}_{D,k} \tag{6.40}$$

where the total cost rates of the system consists of the total investment cost and cost of exergy destruction respectively (refer to Section 6.2).

Environmental impact:

$$\dot{B}_{system} = \dot{B}_k + \dot{B}_{D,k} \tag{6.41}$$

where the total environmental impact of the system consists of the component-related environmental impact and the impact associated with exergy destruction respectively. The environmental impact points can be determined using a LCA software along with various correlations developed from experimentations.

6.4.2 Decision Variables and Constraints

In main decision variable that are most commonly utilized in BTMSs are provided below:

- the condenser saturation temperature (T_{cond}),
- the evaporator saturation temperature (T_{evap}),
- the magnitude of superheating in the evaporator (ΔT_{sh}),
- the magnitude of subcooling in the condenser (ΔT_{sc}),

- the evaporator air mass flow rate ($\dot{m}_e$),
- the compressor efficiency (η_{comp}).

In engineering application of the optimization problems, there are usually constraints on the trade-off decision variables that arise from appropriate feasibility, commercial availability and engineering constraints. The limitations on the minimum and maximum ranges of decision variables are given in Table 6.5. As can be seen from Table 6.5, the lower bound for the evaporator temperature is taken to be higher than 0°C since lower temperatures would cause icing on the surface of the evaporator due to the formation of the water droplets. This reduces the volume of air flowing through the evaporator and in turn reduces the efficiency of the system. On the other hand, the upper bound of the evaporator is limited by the cabin cooling temperatures. For condenser, the lower temperature bound is based on the ambient temperature, whereas the upper bound is constraint with respect to the compression ratio of the compressor, since very high compression ratios increase the probability of the high pressure vapor to leaking back to the low pressure side and even cause compressor failure. Moreover, constraints are provided between the evaporator and condenser temperatures and the incoming air temperatures in order to have feasible and adequate heat transfer in the heat exchangers. Furthermore, the compressor and pump efficiencies are limited to 0.95 due to previously mentioned technological limitations, whereas the air mass flow rates are limited with respect to the vehicle speed and fan power.

6.4.3 Genetic Algorithm

Currently, there are many search techniques that are used to deal with multi-objective optimization problems. These include, but are not limited to, generic algorithm, simulated annealing, tabu and scatter search, ant system, particle swarm and fuzzy programing. Among these, there is no technique that provides the optimum results for all problems and thus the best method should be selected with respect to the current system. In this research, a generic algorithm is used since it requires no initial conditions,

Table 6.5 Common constraints associated with the decision variables selected for the BTMS.

Constraints
$0°C \leq T_{evap} \leq T_{cabin} - \Delta T_{evap.min}$
$T_0 + \Delta T_{cond.min} \leq T_{cond} \leq 65°C$
$T_{evap} < T_{evap,air,in} - \Delta T_{evap.min} - T_{sh}$
$T_{cond} > T_{cond,air,in} + \Delta T_{air.min} + \Delta T_{cond.min} + \Delta T_{sc}$
$0°C \leq \Delta T_{sh} \leq 10°C$
$0°C \leq \Delta T_{sc} \leq 10°C$
$\eta_{comp} \leq 0.95$
$\eta_{pump} \leq 0.95$
$\dot{m}_{air} \leq 0.35\ kg/s$

works with multiple design variables, finds global optima (as opposed to local optima), utilizes populations (as opposed to individuals) and uses objective function formation (as opposed to derivatives).

In the last decades, genetic algorithms (GAs) have been extensively used as search and optimization tools in various problem domains due to their broad applicability, ease of use and global perspective. The concept of GAs was first conceived by Holland in 1970s in order to simulate growth and decay of living organisms in a natural environment and various improvements were conducted ever since. GAs today apply an iterative and stochastic search strategy to drive its search towards an optimal solution through mimicking nature's evolutionary principles and have received increasing attention by the research community as well as the industry to be used in optimization procedures. Based on the inspired evolutionary process, the weak and unfit species are faced with extinction while the strong ones have greater opportunity to pass their genes to future generation via reproduction. Throughout this process, given long enough time line, the species carrying the suitable combination in their genes become the dominant population.

In these analyses, the GA terminology adopted by Konak *et al.* (2006) is used. Based on this terminology, a solution vector is called an individual or a *chromosome*, which consists of discrete units called *genes*. Each gene controls one or more features of the chromosome, which corresponds to a unique solution in the solution space. Moreover, the collection of these chromosomes are called a *population*, which are initialized randomly at first and includes solutions with increasing fitness as the search evolves until converging to a single solution. Furthermore, operators called *crossover* and *mutation* are used to generate new solutions from existing ones. Crossover is one of the key operators where two chromosomes, called *parents*, are combined together to form new chromosomes called *offspring*. Due to the having preference towards fitness, these offsprings will inherit good genes from the parents and through the iterative process, and therefore the good genes are expected to appear more frequently in the population, where they eventually converge to an overall good solution.

The mutation operator on the other hand introduces random changes into the characteristics of the chromosomes at the gene level. Usually the mutation rate (probability of changing properties of a gene) is very small and therefore the new chromosome produced will not be very different than the original one. The key here is that, while the crossover leads the population to converge (by making the chromosome in the population alike), the mutation reintroduces genetic diversity and assists to the escape from local optima.

Reproduction involves selection of chromosomes for the next generation, where the fitness of an individual usually determined the probability of its survival. The selection procedures can vary depending on how the fitness values are used (such as proportional selection, ranking and tournament). The basic schematic for the evolutionary algorithm is given in Figure 6.1. The GA has major advantages since constraints of any type can be easily implemented and that they can find more than one near-optimal point in the optimization space, which enables users to pick the most applicable solution for the specific optimization problem and therefore are widely used for various multi-objective optimization approaches.

Even though maximizing/minimizing a criterion would be beneficial, many real-world problems involve multiple measures of performance, or objectives, which should be

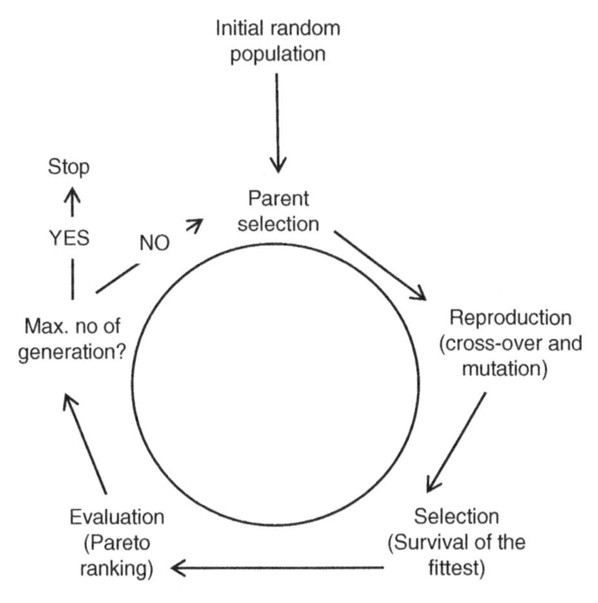

Figure 6.1 Sample schematic for the evolutionary algorithm used (adapted from Ghaffarizadeh, 2006).

optimized simultaneously. Objectives that are optimized individually can provide optimal results with respect to their own criteria while providing very low performance in other objective functions. Thus, a trade-off is needed among the different dimensions in order to obtain a family of optimal "acceptable" solutions for the problem (Fonseca and Fleming, 1995). This ability along with not requiring the user to prioritize, scale or weigh objectives makes them unique in solving multi-objective optimization problems.

The first real application of EAs for finding multiple trade-off solutions in one single simulation run was suggested and used by David Schaffer in 1984. He used vector-evaluated genetic algorithm (VEGA) to capture multiple trade-off solutions for a small number of iterations. This is followed by David Goldberg who suggested using 10-line sketch of a plausible multi-objective evolutionary algorithm optimization using the concept of domination. Consequently, many different implementations of MOEAs have been developed such as weight-based GA, non-dominating sorting GA, Pareto-GA (NPGA), fast non-dominating sorting generic algorithm (NSGA-II) and multi-objective evolutionary algorithm along with different ways of using EAs to solve multi-objective optimization problems such as diploidy, weight-based and distance based approaches.

One of the most prominent differences of classical search and optimization algorithms is that EAs use population of solutions in each iteration (instead of single solutions), which produces a final outcome of a population of multiple non-dominated solutions (that are in parallel) by taking advantage of similarities in the family of possible solutions. Since usually, EAs usually do not converge in a single solution (due to conflicting criteria), EA captures multiple optimum solutions in its final population. These solutions are called "Pareto optimal", where no other feasible solution can reduce some objective function without causing a simultaneous increase in at least no other objection function. The objective function values corresponding to these feasible non-dominating solutions

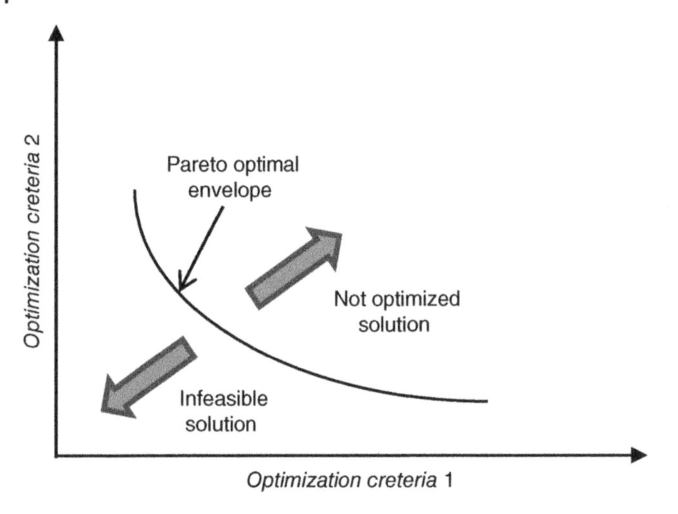

Figure 6.2 A general pareto optimal curve.

are called "Pareto optimal frontier". The general concept of Pareto optimal frontier is illustrated in Figure 6.2.

6.5 Illustrative Example: Liquid Battery Thermal Management Systems

In order to illustrate an application of the aforementioned steps and procedures to the readers, a real life example is provided in this section. Thus, the BTMS from Chapter 5 is continued with additional information to further analyze the system with respect to cost, environmental impact and multi-objective optimization. The additional parameters effecting the cost and environmental impact are provided below.

The battery pack used in the analysis weights 200 kgs with a typical lithium-ion battery pack cost per kilowatt-hour of \$500. The analyzed battery has an energy storage capacity of 16 kWh, where only 12.9 kWh can be utilized for charging and driving in order to extend the life of the battery. Among this, only 9.6 kWh is used to propel the car and the accessories. The cost per mass flow rate (kg/s) of refrigerant is \$5,000. The unit cost of electricity is 0.075 $\$/kWh$. Moreover, the unit environmental impact associated with the electricity generation mix used (from Eco-indicator 99) is 22 mPts/kWh. For the analysis, the parameters to calculate the unavoidable investment costs are selected as temperature differences of 29°C and 18°C for the condenser and evaporator respectively, and an efficiency of 0.6 for the compressor. The total working hours of the system is taken as 1,460 hours based on 4 hours of daily driving.

Furthermore, the final environmental impact value is calculated as a single Eco-indicator 99 point based on 1 kg lithium-ion battery with European electric generation mix and weighting set belonging to the hierarchist perspective (H/H). An eco-invent lithium-ion battery model is modified and improved in order to calculate the environmental impacts of the components and processes associated with the production of the battery.

6.5.1 Conventional Exergoeconomic Analysis Results

In the illustrative example of Chapter 5, exergy analysis for a given BTMS is provided in order to gain a further understanding of the true efficiencies of each component and corresponding irreversibilities. However, this does not provide any information regarding the economic constraints on improving the efficiency of the components or the associated costs. Thus, an exergoeconomic analysis is also conducted where the cost formation can be determined for the thermal management system as provided in Table 6.6. In the table, state 13 has 0 values, since it is available ambient air entering the condenser.

Based on the calculated costs, the exergy destruction costs are also determined for each component with respect to the selected baseline parameters. In Figure 6.3, it can be seen that the evaporator has the highest cost rate of exergy destruction, followed by the condenser, battery and compressor. The high exergy destruction cost of the battery is mostly associated with the high fuel cost, while the majority of the exergy destruction cost of the compressor, condenser and evaporator is associated with relatively high exergy destruction rates for these components.

However, before any remarks can be made regarding design or investment changes, the components should be analyzed with respect to the cost distribution, their exergoeconomic significance and the impact of improving the component efficiency on the total capital investment costs. The cost distribution among investment and exergy destruction rates can be seen in Figure 6.4.

From an exergoeconomic viewpoint, the components that have the highest priority are the ones that have the highest sum of total capital investment and exergy destruction cost rate ($\dot{Z} + \dot{C}_D$). Among these components, the relationship between the exergy efficiency investment costs of the components is investigated with the help of the exergoeconomic factor. These values for each component are provided in Table 6.7.

Table 6.6 Exergy flow rates, cost flow rates and the unit exergy cost associated with each state of BTMS.

State	$\dot{E}x$ (kW)	$\dot{C}$ (\$/h)	c (\$/kW)
1	0.71	0.13	0.18
2	1.58	0.25	0.16
3	1.27	0.20	0.16
4	1.01	0.18	0.18
5	0.12	0.02	0.18
6	0.02	0.02	0.83
7	0.04	0.03	0.83
8	0.04	0.04	0.91
9	0.36	0.31	0.85
10	0.01	0.05	0.01
11	1.30	0.10	0.08
12	<0.01	<0.01	0.08
13	0.00	0.00	0.00
14	0.07	0.06	0.95

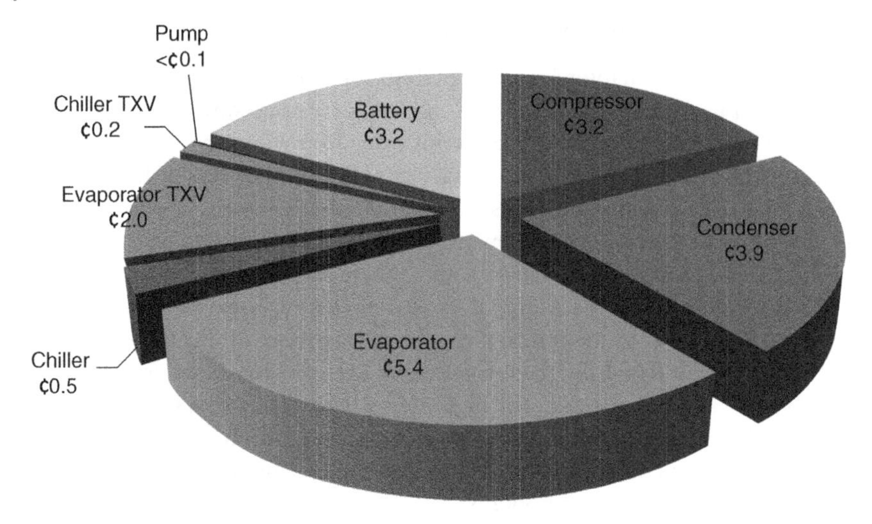

Figure 6.3 Cost rate of exergy destruction for thermal management system components.

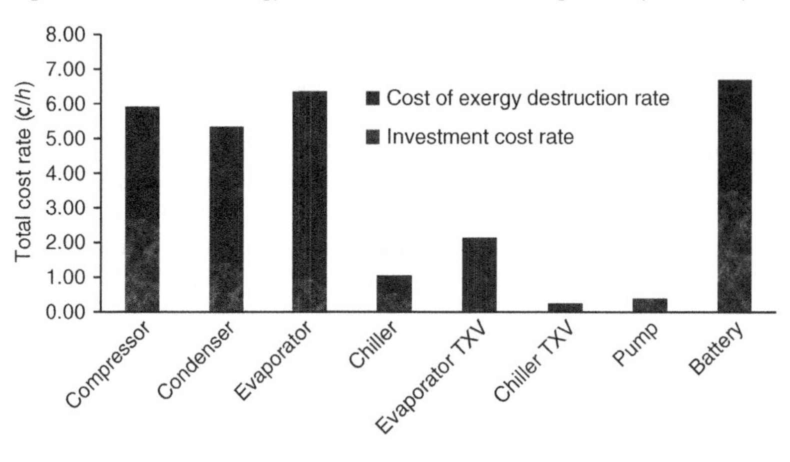

Figure 6.4 Cost distribution among investment and exergy destruction rates for the BTMS components.

Table 6.7 Investment cost rate, cost rate of exergy destruction, total cost rate, exergoeconomic factor and relative cost difference associated with the BTMS components.

Component	$\dot{Z}_k$ (¢/h)	$\dot{C}_{D,k}$ (¢/h)	$\dot{Z}_k + \dot{C}_{D,k}$ (¢/h)	f_k (%)	r_k (−)
Compressor	2.7	3.2	5.9	45.4	0.9
Condenser	1.4	3.9	5.2	26.6	4.9
Evaporator	0.9	5.4	5.7	14.9	3.7
Chiller	0.6	0.5	1.2	52.4	3.6
Evaporator TXV	0.1	2.0	0.8	4.7	0.1
Chiller TXV	0.0	0.2	0.2	4.6	0.1
Pump	0.4	<0.1	0.4	99.6	25.1
Battery	3.5	3.2	6.4	52.4	2.7

When the components are analyzed with respect to $\dot{Z} + \dot{C}_D$, an electric battery has the highest total cost rate compared to the rest of the components, mainly due to having significantly larger investment costs, as provided in Table 6.7. After the battery, the highest sum of total capital investment rate and cost rate of exergy destruction are determined to be the compressor followed by the evaporator and condenser. These components are followed by the pump and thermal expansion valves which have relatively insignificant cost rates compared to the rest of the system components. In the battery, compressor and chiller, the non-exergy related costs and total cost of a component are divided rather equally, thus the current investment cost for this components are found to be reasonable. Condenser and evaporator are determined to have low exergoeconomic factors, where reducing the investment cost on this component should be investigated at the expense of their exergetic efficiencies to improve the effectiveness of the system. On the other hand, for the pump improving the component efficiency would be more cost effective even if the capital investment for that component will increase.

6.5.2 Enhanced Exergoeconomic Analysis Results

In order to improve the accuracy and validity of the analysis, an enhanced exergoeconomic study is also conducted for the thermal management system. Initially, the investment cost is split into avoidable and unavoidable parts in order to determine how much of the total investment can be actually eliminated as seen in Table 6.8.

The total cost rates and exergoeconomic factor is also included in order to provide a comparison between the conventional and enhanced exergoeconomic analysis. The ratio of available to total cost rates indicates that up to 81% of the total cost rates could be theoretically avoided in the system. From the exergoeconomic factor based on the avoidable costs, it can be seen that the dominant factor in the total cost rate for the condenser, evaporator and thermal expansion valves are the cost of exergy destructions and therefore the exergy efficiency of these components should be increased, even at the expense of increased investment costs. On the other hand, the most prominent factor in the total cost rate for the chiller, pump and the electric battery is determined to be the investment costs and therefore the investment cost needs to be reduced for these components to improve the cost effectiveness of the system. Even though the similar trends are achieved using available cost rates (compared to the total cost rates), the use

Table 6.8 Comparison of total and avoidable cost rates of the respective exergoeconomic factors associated with the components of the BTMS.

Component	$\dot{Z}_k + \dot{C}_{D,k}$ (¢/h)	$\dot{Z}_k^{AV} + \dot{C}_{D,k}^{AV}$ (¢/h)	$\dot{Z}_k^{AV} + \dot{C}_{D,k}^{AV}/\dot{Z}_k + \dot{C}_{D,k}$ (%)	f_k (%)	f_k^* (%)
Compressor	5.9	3.3	55.4	45.4	54.1
Condenser	5.2	2.7	52.0	26.6	35.5
Evaporator	5.7	3.1	53.7	14.9	23.7
Chiller	1.2	0.7	81.0	52.4	58.5
Evaporator TXV	0.8	0.7	36.7	4.7	9.2
Chiller TXV	0.2	0.1	36.7	4.6	9.2
Pump	0.4	<0.1	9.6	99.6	84.6
Battery	6.4	2.6	41.0	52.4	65.4

of avoidable costs revealed how far the components really are to the ideal parameters to optimize the cost distribution and provided a much realistic measure on what approach should be taken (and how much) to improve the effectiveness of each component and enabled comparison of dissimilar components with each other.

Moreover, the relationship between investment cost and exergy destruction for the compressor, condenser and evaporator is further examined in order to provide a more detailed information on their correlation since these components are the major contributor to the total cost and exergy destruction of the system and can be optimized accordingly. For the compressor, the compression ratio associated with the system is varied, which in turn changes the isentropic efficiency of the compressor and therefore effects the exergy destruction and investment cost associated with the compressor as seen in Figure 6.5. In the figures, the asymptotes in the X-axis and Y-axis provide the unavoidable cost and exergy destruction rates respectively.

Furthermore, in order to further evaluate the heat exchangers, different evaporating and condensing temperatures are used which in turn varied the heat exchanging area associated with the system and thus altered the exergy destruction and investment costs for the condenser and evaporator as shown in Figures 6.6 and 6.7.

In addition, a sensitivity analysis is also conducted in order to determine the effects of the interest rates used in the analysis. Thus, the investment and exergy destruction rates with respect to various interest rates are shown in Figure 6.8–6.10. In order to provide a comparison among different components, exergy destruction rates and investment cost rates are provided in terms of "per product unit exergy".

Subsequently, parametric studies are also conducted based on different compressor efficiencies and condensing and evaporating temperatures in order to see their corresponding effects on investment and exergy destruction related costs which are shown in Figures 6.11–6.13.

In order to provide the emissions associated with the system in terms of cost, the indirect amount of CO_2 emissions released to the environment as a result of the electricity consumed from the grid are also calculated in terms of a cost input. The emissions

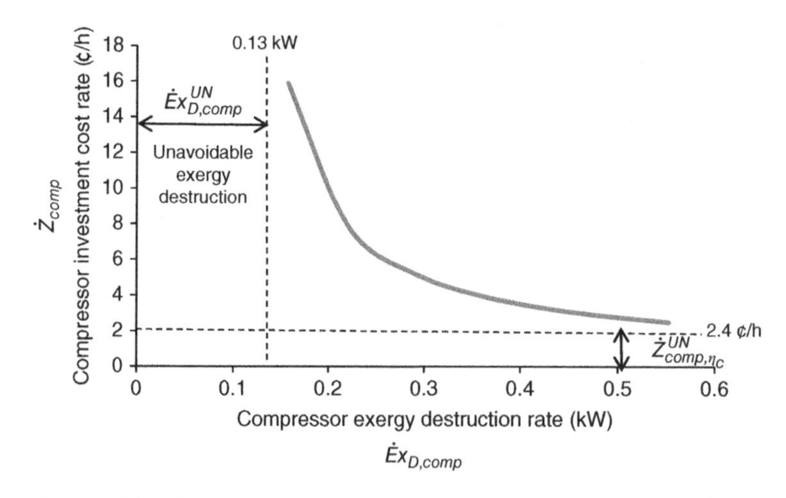

Figure 6.5 Relationship between compressor exergy destruction rate and investment cost rate.

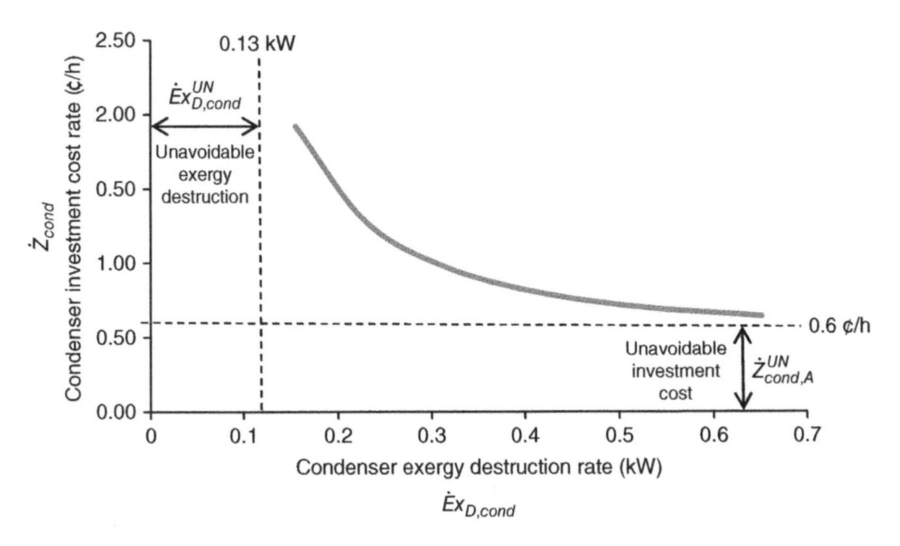

Figure 6.6 Relationship between condenser exergy destruction rate and investment cost rate.

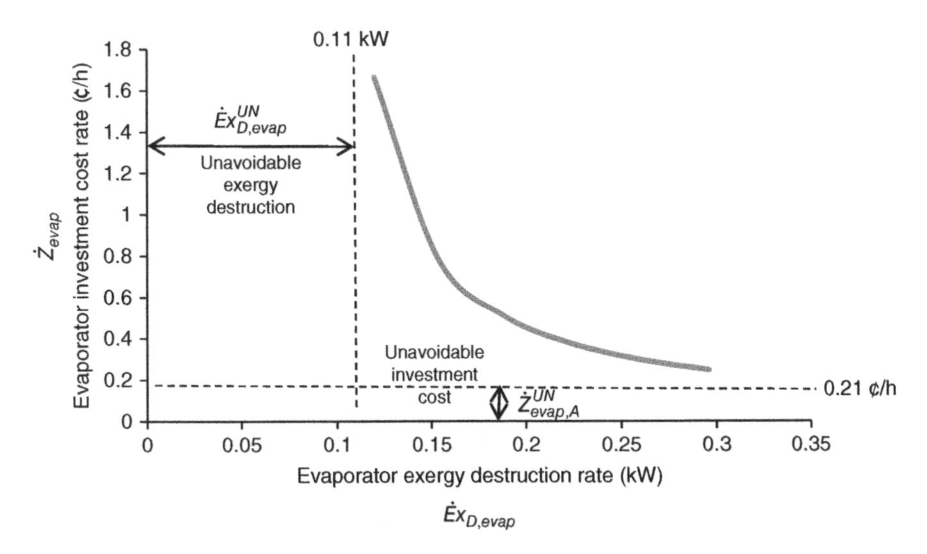

Figure 6.7 Relationship between evaporator exergy destruction rate and investment cost rate.

are calculated with respect to various electricity generation mixes, including one that utilizes a natural gas combined cycle to one that uses primarily coal/steam, with a range of 400 to 1,118 gCO_2eq/kWh including life cycle estimates for electricity production. The associated emissions with respect to various electricity generation mixes can be seen in Figure 6.14a. Subsequently, a carbon price is established and the associated cost of corresponding CO_2 emissions are determined accordingly under various carbon price ranges as shown in Figure 6.14b.

If the steps taken in this section are summarized; the exergoeconomic and enviroeconomic (environmental cost) analyses of hybrid electric vehicle thermal management

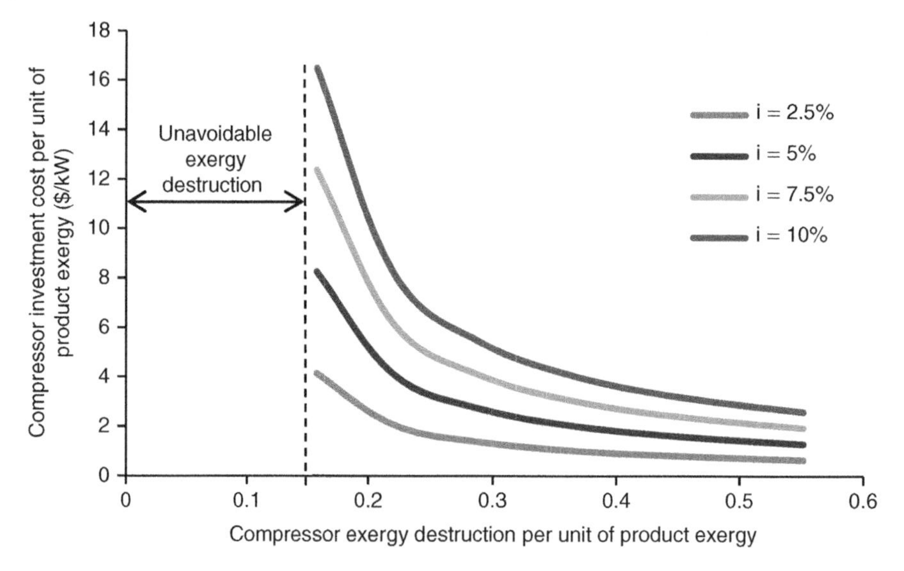

Figure 6.8 Relationship between compressor exergy destruction rate and investment cost rate per unit product exergy under different interest rates.

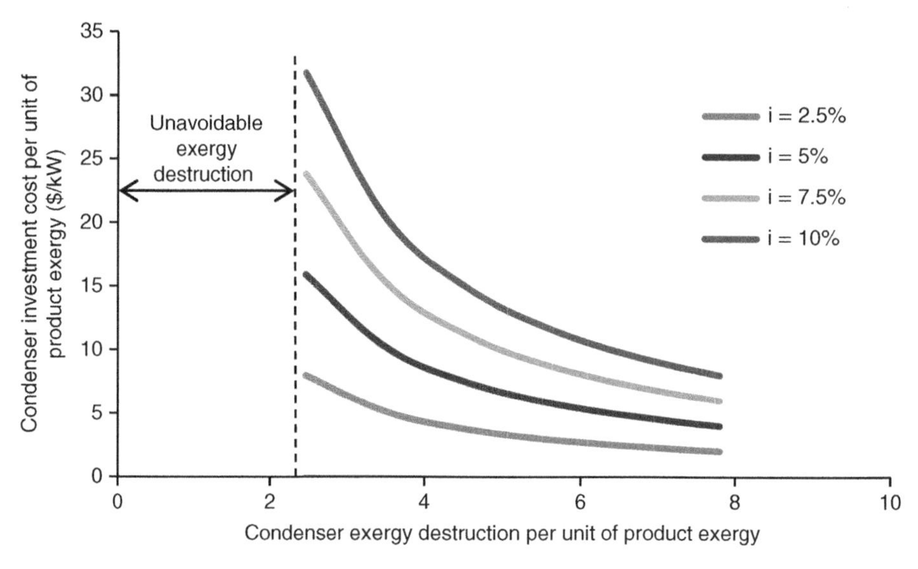

Figure 6.9 Relationship between condenser exergy destruction rate and investment cost rate per unit product exergy under different interest rates.

systems are conducted with respect to various system parameters as well as operating conditions. In the analysis, the investment cost rates are calculated with respect to equipment costs, which are determined by cost correlations for each system component, and capital recovery factors. Thus, by combining it with previously conducted exergy analysis, an exergoeconomic model is developed whereby the exergy streams are identified, fuel and products are defined and cost equations are allocated for each component.

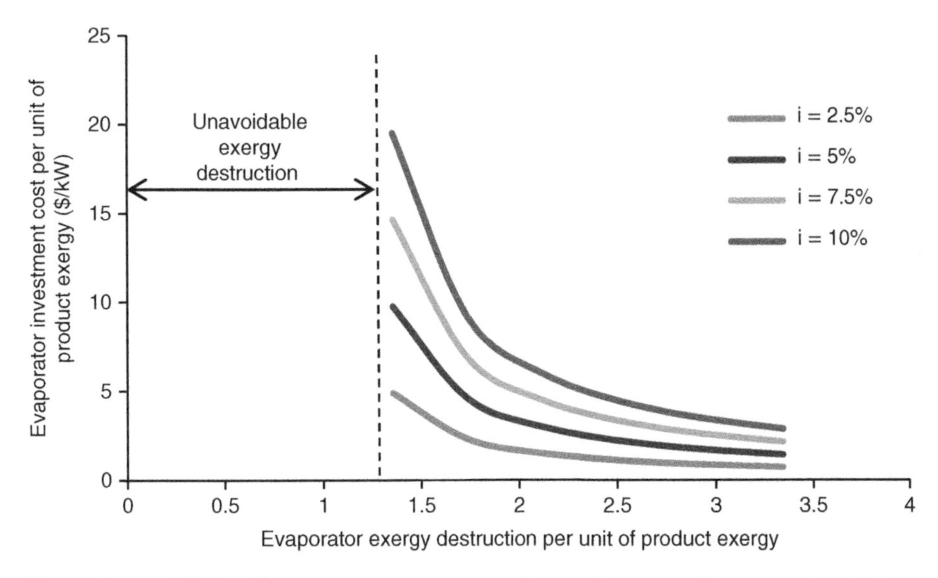

Figure 6.10 Relationship between evaporator exergy destruction rate and investment cost rate per unit product exergy under different interest rates.

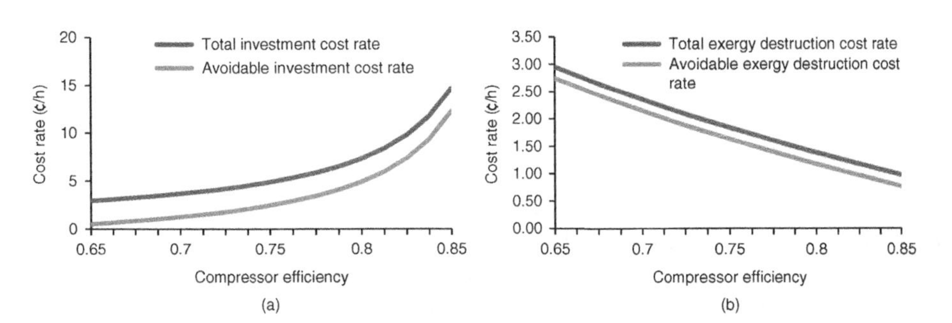

Figure 6.11 Total and avoidable cost rates with respect to (a) investment and (b) exergy destruction for the compressor based on various compressor efficiencies.

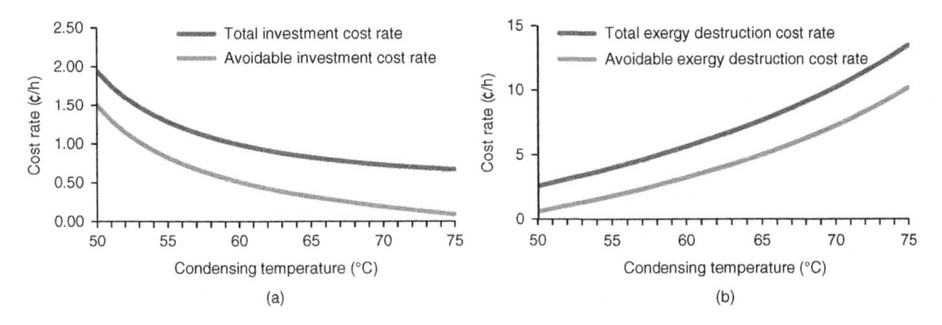

Figure 6.12 Total and avoidable cost rates with respect to (a) investment and (b) exergy destruction for the condenser based on various condensing temperatures.

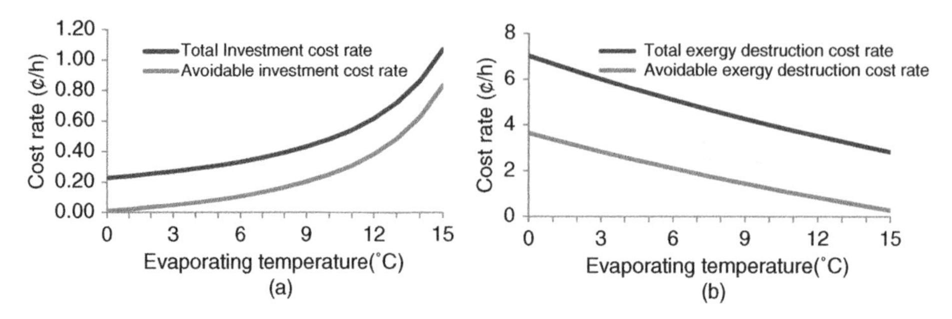

Figure 6.13 Total and avoidable cost rates with respect to (a) investment and (b) exergy destruction for the evaporator based on various evaporating temperatures.

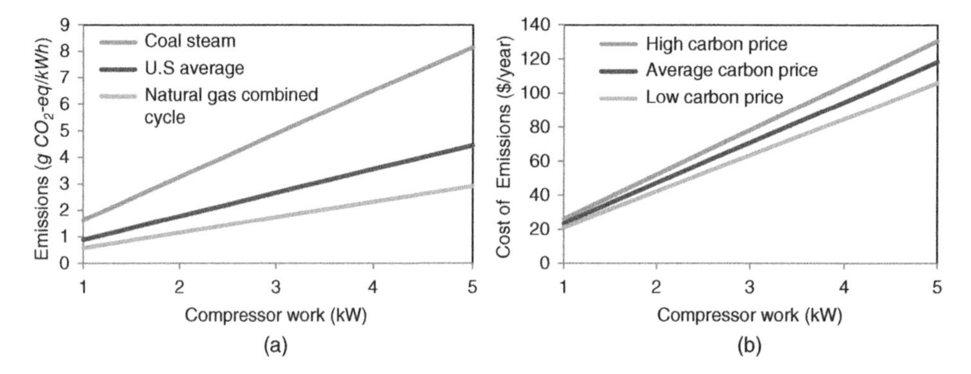

Figure 6.14 (a) Amount of emissions released and (b) associated imposed cost with respect to varying compressor work under different electricity generation mixes.

The costs from the economic analysis are used to determine the unit cost of exergy, cost rate of exergy destruction as well as other useful exergoeconomic variables for each component. Moreover, an enviroeconomical (environment cost) analysis is also conducted based on the established carbon price associated with the released CO_2 to the environment, corresponding to the indirect emissions from the electricity used in the TMS under varying carbon prices and electricity generation mixes.

6.5.3 Battery Environmental Impact Assessment

Before the exergoenvironmental analysis can be conducted for the BTMS, the environmental impact associated with the battery needs to be determined. Since the battery has one of the highest environmental impacts, it is worth further analyzing it using a LCA tool, namely SimaPro, which is used to calculate the environmental impact for each battery sub-component as well as the production procedures.

SimaPro is a life cycle assessment software package that has the capability of collecting, analyzing and monitoring the environmental performance of products and services and can model and evaluate complex life cycles in a systematic and transparent way following the ISO 13030 series recommendations. The software is integrated with an ecoinvent database (Ecoinvent, 2012) that is used for a variety of applications including carbon footprint calculations, product design/eco-design as well as assessing the environmental impact with respect to various parameters. The software can define non-linear relationships in the model, conduct analysis of complex waste treatment and recycling scenarios and allocate multiple output processes. Thus, it provides significant value in conducting LCA for the system components.

Initially the LCA of the battery is conducted in order to acquire environmental impact potential associated with the electric battery in terms of eco-indicator points. The battery assembly components and their respective environmental impacts are illustration in Figure 6.15.

From the figure, it can be seen that anode has the highest impact in the battery due to the high amount of copper used, followed by the battery thermal system with respect to the gold used in the integrated circuits which accounts for over 40% and 26% of the total impact score respectively. Electrode paste on the other hand, has a relatively small contribution to the total environmental impact even though it encompasses a significant portion of the battery weight. In addition, other auxiliary components such as the module packaging and the battery case also add to the battery impact along with the electricity, heat and natural gas used to produce the battery. The impacts associated with the battery are also provided in Figure 6.16.

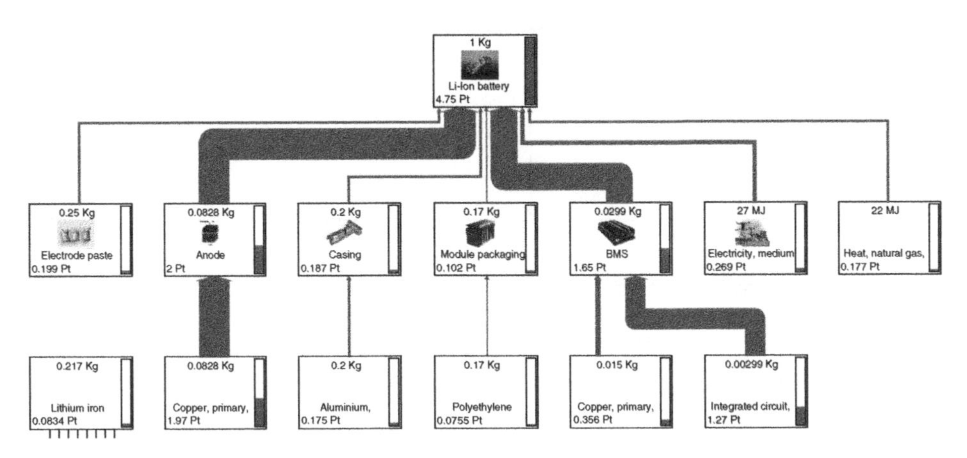

Figure 6.15 Illustration of the lithium-ion battery using SimaPro 7.

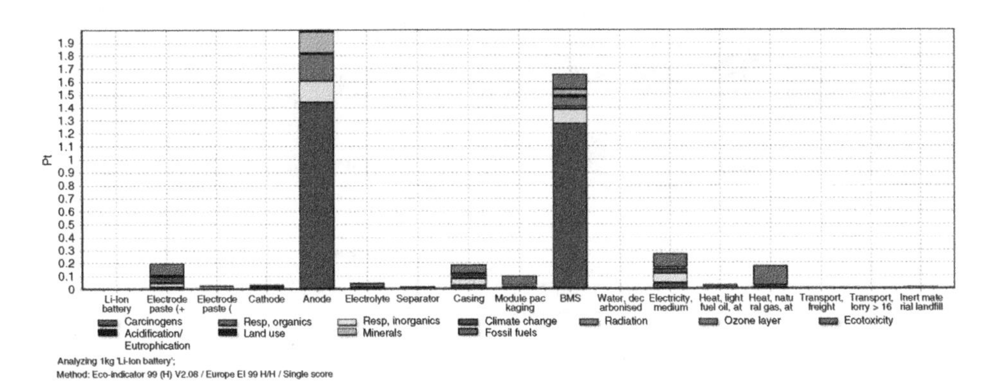

Figure 6.16 Various environmental impact potentials associated with each battery sub-component.

Moreover, Figure 6.16 is divided into production, energy and transportation categories in order to provide their contributions to the total environmental impacts as shown in Figure 6.17. From the analysis, it can be seen that up to 90% of the total emissions come from the direct production of the battery while majority of the remaining impact corresponds to that of the energy used during the production. In addition, the battery components are also investigated in terms of various environmental impact potentials as shown in Figure 6.18.

6.5.4 Exergoenvironmental Analysis Results

Once the environmental impact associated with the battery is calculated and combined with that of the remaining BTMS components, it is used to determine the environmental impact formation of the system. In Table 6.9, it can be seen that the highest environmental impact rate is achieved at state 10, which is the exit state of the electric battery and the lowest environmental impact rate is associated with the pump input. In the table, state 13 has a value of zero, since it is available ambient air entering the condenser.

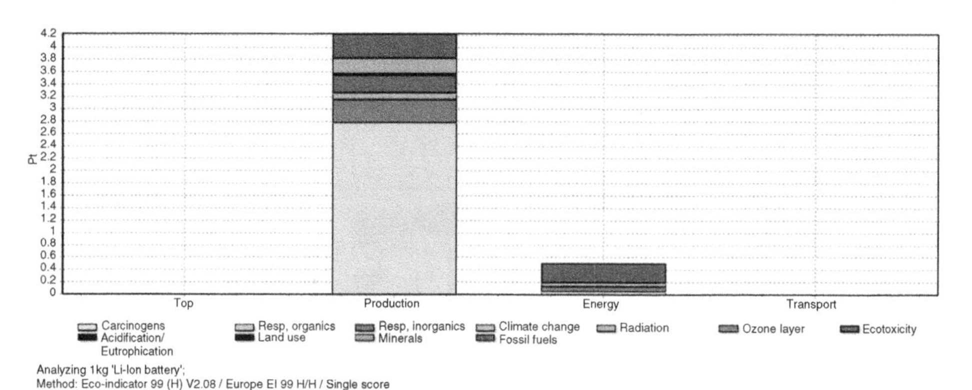

Figure 6.17 Eco-indicator points associated with production, energy usage and transport of the battery.

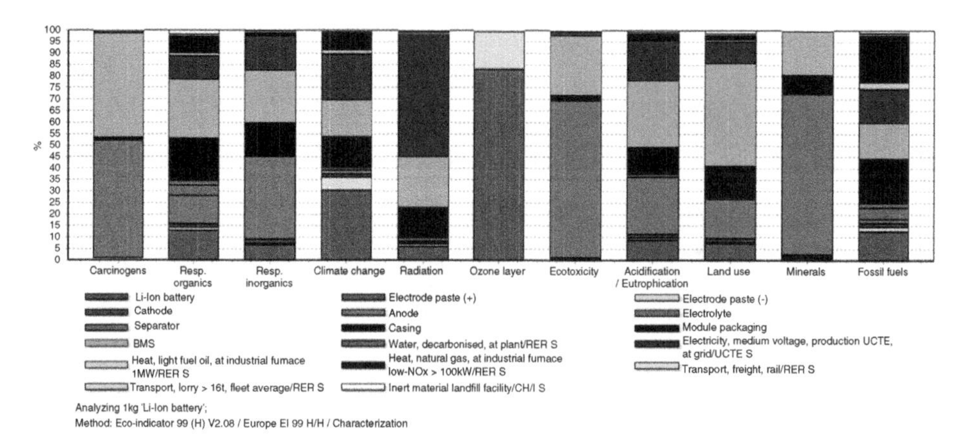

Figure 6.18 Percentage contribution of each component to the environmental impact with respect to eco-indicator 99 points.

Table 6.9 Exergy flow rates, environmental impact due flow rates and the unit environmental impact cost associated with each state of BTMS.

State	$\dot{E}x$ (kW)	$\dot{B}$ (mPts/h)	b (mPts/kJ)
1	0.71	30.21	42.67
2	1.58	59.84	37.83
3	1.27	48.15	37.88
4	1.01	43.17	42.62
5	0.12	5.01	42.69
6	0.02	2.74	123.34
7	0.04	4.77	123.32
8	0.04	4.92	124.92
9	0.36	64.15	178.27
10	0.01	35.98	7.20
11	1.30	28.76	22.05
12	<0.01	<0.10	22.04
13	0.00	0.00	0.00
14	0.07	12.01	177.91

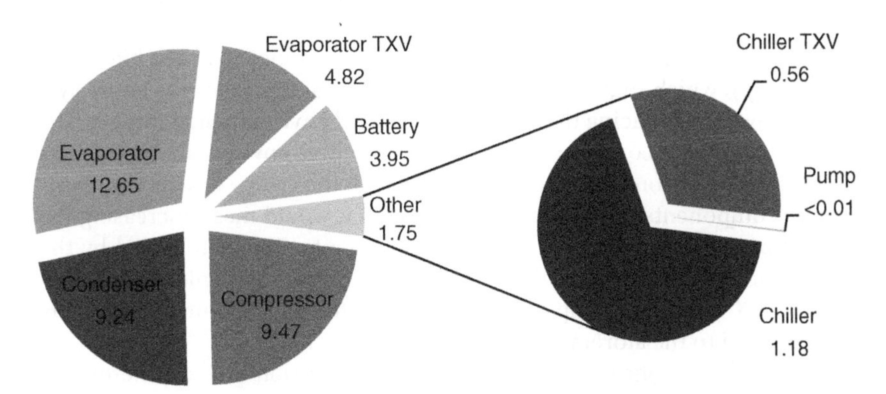

Figure 6.19 Environmental impact eco-indicator 99 points associated with exergy destruction for thermal management system components.

Moreover, based on the environmental impact associated with component flow and the exergy destruction rates, the environmental impact due to exergy destruction rates are also determined for each component. In Figure 6.19, it can be seen that the evaporator has the highest environmental impact due to exergy destruction rates, followed by the condenser, compressor and the battery. The environmental impact of the battery is determined to be mostly component-related, while the environmental impact of the compressor, condenser and evaporator is associated with relatively high exergy destruction rates for these components.

When evaluated from an exergoenvironmental point of view, the most important component would be the one with the highest sum of component-related environmental

Table 6.10 Total environmental impact, exergoenvironmental factor and relative difference of exergy-related environmental impacts associated with the BTMS components.

Component	$\dot{Y}_k(mPts/h)$	$\dot{B}_{D,k}(mPts/h)$	$\dot{Y}_k + \dot{B}_{D,k}(mPts/h)$	$f_b(\%)$	$r_b(-)$
Compressor	0.90	9.47	10.37	8.64	0.55
Condenser	0.28	9.24	9.514	2.91	6.54
Evaporator	0.22	12.65	12.87	1.72	4.05
Chiller	0.13	1.18	1.31	10.13	4.10
Evaporator TXV	0.02	4.82	4.83	0.36	0.14
Chiller TXV	<0.01	0.56	0.56	0.36	0.13
Pump	0.13	<0.01	0.13	96.94	8.19
Battery	33.78	3.95	37.72	89.54	2.02

impact and environmental impact due to exergy destruction rate ($\dot{Y} + \dot{B}_D$). Moreover, exergoenvironmental factor and relative difference of exergy-related environmental impacts are also calculated to provide the relationship between these two factors. The capital environmental impact, exergy destruction impact rate and exergoenvironmental factor for each component is provided in Table 6.10.

When the components are analyzed with respect to $\dot{Y} + \dot{B}_D$, the electric battery by far has the highest environmental impact mainly due to the high copper mass used in the lithium-ion battery anodes. Moreover, the battery is also determined to have a high exergoeconomic factor (f), which suggests that the environmental impact of the entire system could be improved by reducing the component-related environmental impact. The evaporator, compressor and condenser have the next highest environmental impacts respectively, where the environmental impact related with the exergy destruction associated with these components should be reduced even if it would mean increasing the environmental impact during production of the components. This is followed by the evaporator TXV and the chiller where the environmental impact is significantly lower. Finally, the chiller TXV and the pump impacts are found to be exergoenvironmentally insignificant compared to the aforementioned components.

Moreover, due to the significance of the electricity generation mix on the overall environmental impact, a sensitivity analysis is also conducted where the environmental impact related to the exergy destruction rate of the system is determined with respect to the electricity generation mixes for various countries as shown in Table 6.11.

If the steps taken in this section are summarized; exergoenvironmental analysis for the BTMS is conducted with respect to the environmental impact from LCA along with the data compiled from the literature in order to obtain the impact of each relevant system components and input streams in terms of Eco-indicator 99 points, which are assigned to the corresponding product exergy streams. Subsequently, exergoenvironmental variables (such as environmental impact of product, fuel and components, environmental impact rate of exergy destruction as well as relative difference of specific environmental impacts and exergoenvironmental factor) are calculated and exergoenvironmental evaluation is performed in order to identify the environmentally most relevant system components and provide information about possibilities and trends for design improvements.

Table 6.11 Environmental impact related to the exergy destruction rate for BTMS components using electricity generation mixes for various countries.

	U.S	Europe Average	Switzerland	Italy
Component		($mPts/h$)		
Compressor	9.47	11.19	3.62	20.60
Condenser	9.24	10.86	3.70	19.76
Evaporator	12.65	14.88	5.07	27.05
Chiller	1.18	1.39	0.47	2.53
Evaporator TXV	4.82	5.66	1.93	10.30
Chiller TXV	0.56	0.66	0.23	1.21
Pump	<0.01	0.01	<0.01	0.01
Battery	3.95	4.55	1.87	7.91

6.5.5 Multi-Objective Optimization Results

Once the model is developed in the light of each selected criterion and evaluated individually, a multi-objective optimization with aforementioned objective functions (Equations 6.39–6.41), constraints (Table 6.5) and six decision variables are performed for this example with the help of genetic algorithms. In the analysis, five optimization scenarios with the objective functions of exergy efficiency (single-objective), total cost rate (single-objective), environmental impact rate (single-objective), along with exergoeconomic (multi-objective) and exergoenvironmental (multi-objective) optimizations are performed. The corresponding optimization scenarios can be seen in Figures 6.20–6.24.

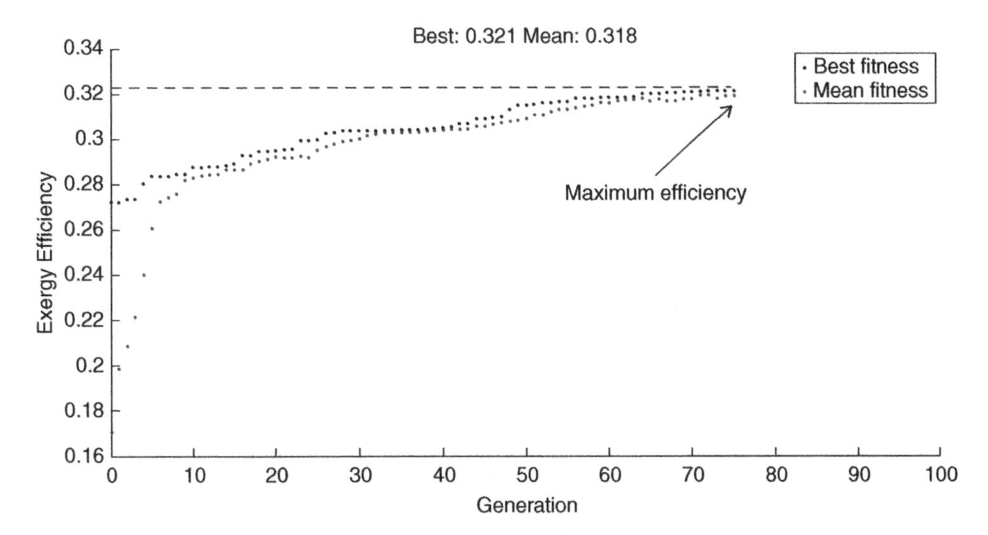

Figure 6.20 Single objective optimization of BTMS over generations with respect to exergy efficiency.

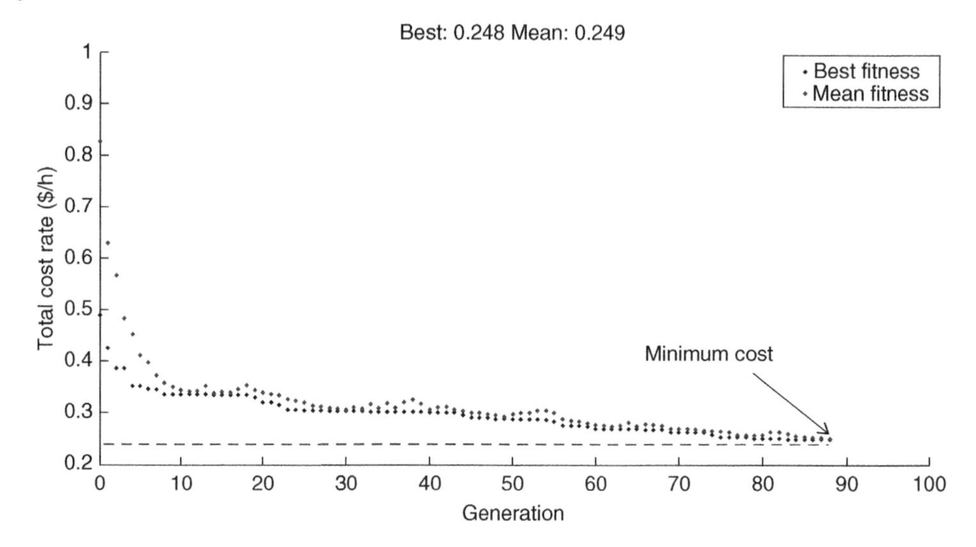

Figure 6.21 Single objective optimization of BTMS over generations with respect to product cost rate.

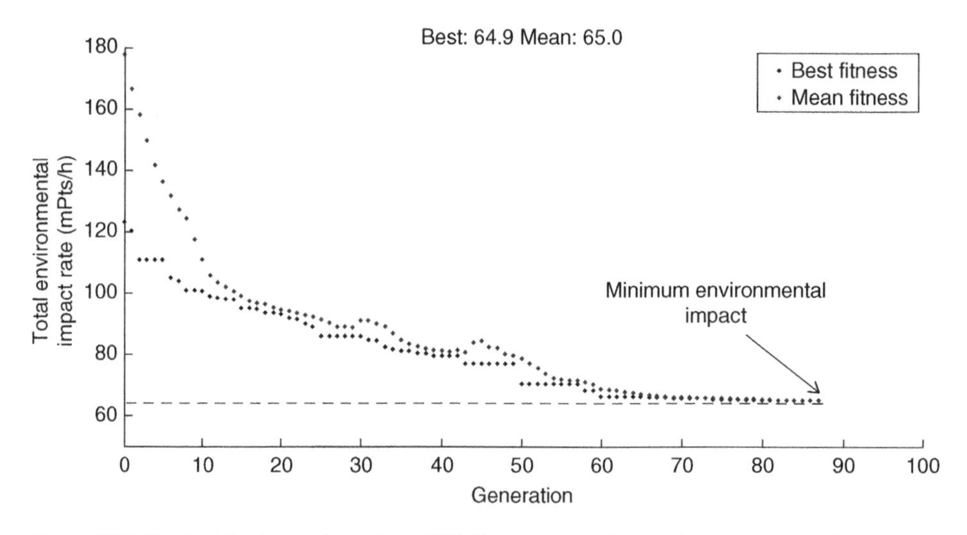

Figure 6.22 Single objective optimization of BTMS over generations with respect to product cost rate.

As previously mentioned, all the points on the Pareto optimum frontier are potentially an optimum solution for the analysis and therefore a weighting factor needs to be assigned for each objective and/or decision needs to be made (often based on experience or importance of each objective) in order to select a single final solution among them. In this selection process, a traditional method called LINMAP decision-making is used to select a desirable final solution as shown in Figures 6.23 and 6.24. This method creates a hypothetical ideal point in which all objectives have their corresponding optimum values independent of each other and would stay below the Pareto optimum frontier. Even though this point would be impossible in reality, it would serve a useful purpose

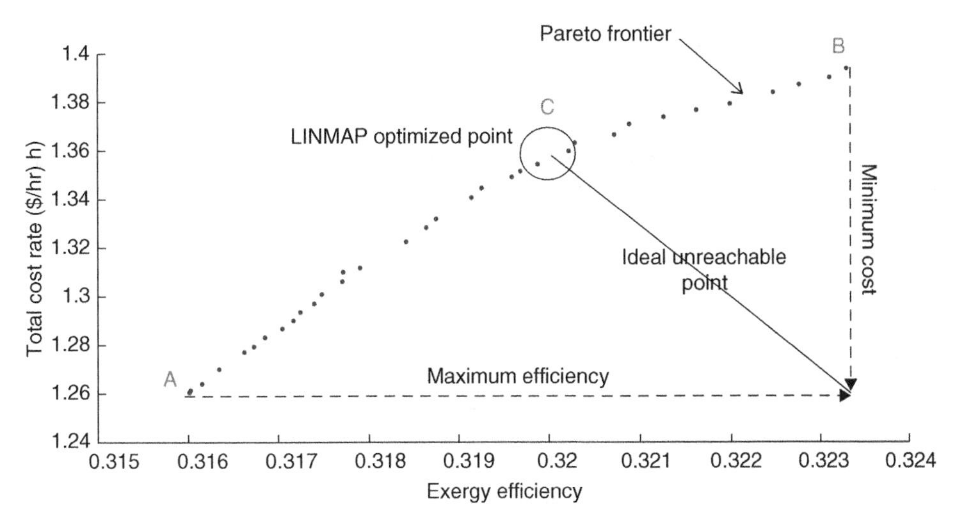

Figure 6.23 Multi-objective optimization of BTMS with respect to exergy efficiency and total cost rate.

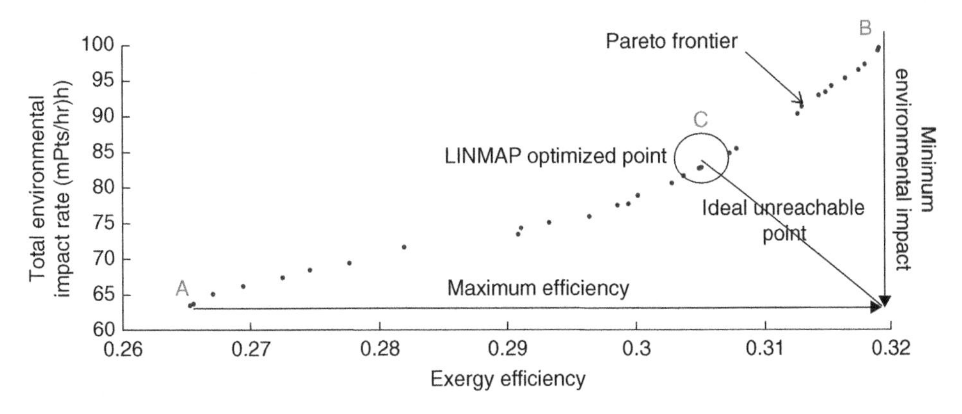

Figure 6.24 Multi-objective optimization of BTMS with respect to exergy efficiency and total environmental impact rate.

by assisting the decision makers to select the point on the Pareto optimum frontier that has the closest distance to this ideal point as the desirable final solution.

Table 6.12 shows the values for the decision variables in the base case design along with the four different optimization criteria. In addition, the results of exergy, economic and environmental analyses for each optimization criteria are shown in Tables 6.12–6.15. It should be noted that the values for the decision variables are considered to be continuous over the determined constraints for the multi-objective optimization problem. However, usually parameters associated with some of these variables (especially size and efficiency) are only available in discrete units. Therefore, in a case where the determined parameter values are not available, the closest available values should be utilized in the system for most optimal results.

In the above tables, it can be seen that each single objective optimization approach pays attention only to its own criterion without taken others into consideration.

Table 6.12 Decision variables for the base case design under various optimization criteria.

Decision Variable	Base Case Design	Single-Obj. Exergetic	Single-Obj. Economic	Single-Obj. Environmental	Multi-Obj. Exergo-economic	Multi-Obj. Exergo-Environmental
T_{cond} (°C)	55	55.23	54.20	55.18	56.01	55.25
T_{evap}(°C)	5	0.40	8.94	8.92	8.93	8.82
ΔT_{sh}(°C)	5	4.75	3.86	2.40	9.69	0.96
ΔT_{sc}(°C)	5	9.94	9.68	4.90	9.99	1.71
$\dot{m}_e(kg/s)$	0.17	0.35	0.19	0.18	0.21	0.25
η_{comp}	0.63	0.80	0.66	0.79	0.72	0.79

Table 6.13 Exergetic analysis results for the base case design under various optimization criteria.

Decision Variable	Base Case Design	Single-Obj. Exergetic	Single-Obj. Economic	Single-Obj. Environmental	Multi-Obj. Exergo-economic	Multi-Obj. Exergo-Environmental
$\eta_{ex,comp}$	0.67	0.82	0.69	0.81	0.75	0.81
$\eta_{ex,cond}$	0.22	0.24	0.26	0.21	0.23	0.20
$\eta_{ex,evap}$	0.24	0.21	0.27	0.27	0.26	0.28
$\eta_{ex,chil}$	0.37	0.32	0.43	0.42	0.43	0.42
$\eta_{ex,etxv}$	0.89	0.88	0.92	0.91	0.92	0.88
$\eta_{ex,ctxv}$	0.89	0.89	0.92	0.91	0.92	0.89
$\eta_{ex,pump}$	0.80	0.83	0.84	0.80	0.78	0.77
$\eta_{ex,bat}$	0.81	0.83	0.85	0.83	0.84	0.83

Table 6.14 Economic analysis results for the base case design under various optimization criteria.

Decision Variable ($/h)	Base Case Design	Single-Obj. Exergetic	Single-Obj. Economic	Single-Obj. Environmental	Multi-Obj. Exergo-economic	Multi-Obj. Exergo-Environmental
$\dot{Z}_{comp}$	5.90	20.38	5.10	6.87	5.57	9.80
$\dot{Z}_{cond}$	5.32	10.96	5.08	5.31	5.46	7.27
$\dot{Z}_{evap}$	6.32	18.39	5.75	6.44	6.39	9.23
$\dot{Z}_{chil}$	1.06	1.38	0.92	1.01	0.91	1.04
$\dot{Z}_{etxv}$	2.13	5.56	1.47	2.15	1.53	3.68
$\dot{Z}_{ctxv}$	0.25	0.33	0.16	0.25	0.15	0.30
$\dot{Z}_{pump}$	0.38	0.38	0.38	0.38	0.38	0.38
$\dot{Z}_{bat}$	6.67	7.45	6.38	6.66	6.37	6.74

Table 6.15 Environmental analysis results for the base case design under various optimization criteria.

Decision Variable ($mPts/h$)	Base Case Design	Single-Obj. Exergetic	Single-Obj. Economic	Single-Obj. Environmental	Multi-Obj. Exergo-economic	Multi-Obj. Exergo-Environmental
$\dot{B}_{comp}$	9.34	9.13	8.58	4.93	9.11	6.61
$\dot{B}_{cond}$	8.48	12.62	8.77	7.10	9.61	8.64
$\dot{B}_{evap}$	12.83	24.05	10.49	9.09	14.06	12.54
$\dot{B}_{chil}$	1.31	1.29	0.98	0.92	0.90	0.95
$\dot{B}_{etxv}$	4.82	7.56	3.22	3.55	5.12	6.01
$\dot{B}_{ctxv}$	0.57	0.44	0.35	0.41	0.31	0.49
$\dot{B}_{pump}$	0.14	0.14	0.14	0.14	0.14	0.14
$\dot{B}_{bat}$	37.73	37.42	37.01	36.79	36.77	35.67

Exergetic single-objective optimization scenario maximizes the exergy efficiencies for each component; however no attention is paid to economic or environmental objectives. Similarly, exergoeconomic single-objective optimization scenario has the lowest unit costs for each component at the expense of exergy efficiency and environmental impact. And finally, exergoenvironmental single-objective optimization has the lowest Eco-indicator 99 points for each component at the expense of exergy efficiencies and cost. In multi-objective optimization scenario however, these objectives are considered simultaneously, which provided optimized solutions with values in between the extremes yielded by the single-objective approaches as a result of the trade-offs made between the solutions of the two conflicting objectives. Normalized value of the objectives with respect to each optimization criteria is provided in Figure 6.25. Moreover, when the exergoeconomic and exergoenvironmental optimizations are compared against the ones using energy efficiencies, the selected values for the decision variables in the LINMAP optimization points are determined to have 4.8% lower cost

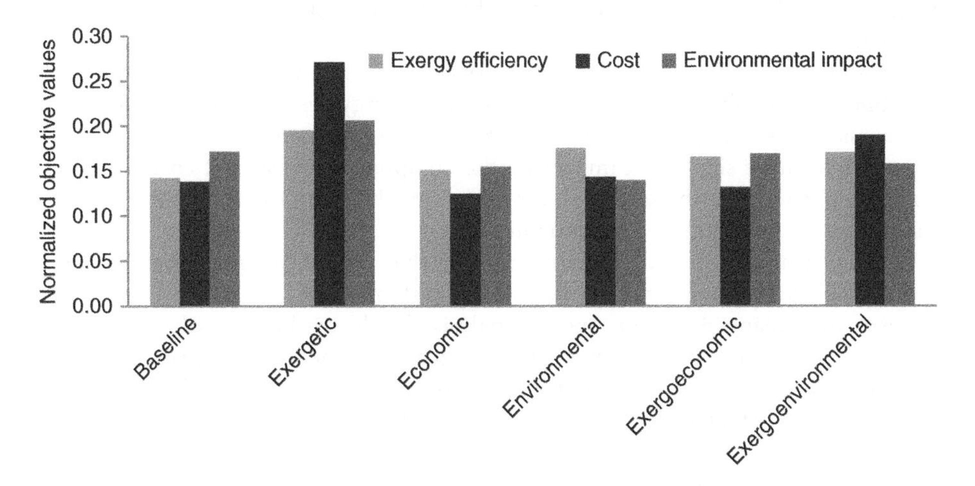

Figure 6.25 Normalized values of different objectives with respect to various optimization functions.

and 3.9% lower environmental impact rates than the one calculated by the energy approach, which yields total cost and environmental impact rates of 1.41 $/h and 87.27 mPts/h respectively.

If the steps taken in this section are summarized; the BTMS of a hybrid vehicle is optimized using a multi-objective evolutionary algorithm using exergoeconomic and exergoenvironmental objectives. The optimization is performed in order to maximize the exergy efficiency (based on exergetic efficiency), minimize the unit exergy cost (based on cost of unit exergy destruction and investment costs) and unit environmental impact (based on Eco-Indicator 99 impact points). Condensing and evaporating, superheating and subcooling temperatures, evaporator air mass flow rate and compressor efficiency are selected as the decision variables for the analyses and various constraints are applied based on appropriate feasibility and engineering constraints. The decision variables along with exergy efficiency, total cost and environmental impact (for each component) are compared under each optimization approach. In the multi-objective optimization, a Pareto frontier is obtained and a single desirable optimal solution is selected based on LINMAP decision-making process. The corresponding solutions are compared against each exergetic, exergoeconomic and exergoenvironmental single objective optimization results. Even though the single-objective approaches provided optimal solutions for their objectives, they have provided very poor solutions for the remaining objectives. Thus, the multi-objection optimization approach provided a solution set within the extremes of the single-objective results by evaluating two objectives simultaneously and providing trade-off between them to obtain desirable solution sets.

6.5.5.1 Case Study Conclusions

Based on the case study, the following specific concluding remarks were made:

- The exergy destruction associated with each component is split into various parts and determined that the exogenous exergy destruction is small but significant portion of the total exergy destruction in each component, which shows that there is a moderate level of interdependencies among the components of the BTMS.
- The exergy destruction is split into avoidable/unavoidable parts and determined that a big portion of the exergy destruction within the components could be potentially avoided.
- An inverse relationship is determined between the exergy destruction rates and investment costs, where the minimum achievable exergy destruction and cost values are calculated for the specified range over varying interest rates.
- Interest rates have significant impact on the component investment cost rates and would help determine the so-called "cut-off" points between the component investment rate and the exergy destruction rates.
- The ratio of available to total cost rates indicates that a significant portion of the total cost rates could be theoretically avoided in the system.
- Based on the enhanced exergoeconomic analysis, the electric battery is determined to have the highest total cost rate due to the significantly higher initial investment rates. In addition, the investment costs of the condenser and evaporator should be reduced to improve the cost effectiveness of the system while keeping the compressor and the chiller the same.
- The chiller TXV and the pump are calculated to be relatively insignificant compared to the rest of the components in terms of the cost calculated by conventional and

enhanced exergoeconomic analysis due to low mass flow rates in coolant loop under baseline battery heat dissipation rates.

Based on the conducted analyses, exergy efficiency, total cost rate and environmental impact for the baseline system is determined to be 0.29, ¢28/h and 77.3 mPts/h respectively. The exergy efficiency could be increased by up to 27% (by single objective exergy) and the cost and environmental impact can be reduced by 10% (by single objective cost) and 19% (by single objective environmental impact) respectively, at the expense of the non-optimized outputs. Moreoverö based on the exergoeconomic optimization, it is concluded that a 14% higher exergy efficiency and 5% lower cost can be achieved, compared to baseline parameters at an expense of 14% increase in the environmental impact. And finally, based on the exergoenvironmental optimization, a 13% higher exergy efficiency and 5% lower environmental impact can be achieved at the expense of 27% increase in the total cost.

6.6 Concluding Remarks

In this chapter, a walkthrough of developing exergoeconomic analysis is presented in order to give the readers the necessary tools to analyze the investment costs associated with their system components and assess the economic feasibility of the suggested improvements. In the economic portion of the analysis, the investment cost rates for the BTMSs are demonstrated with respect to equipment costs and capital recovery factors. Subsequently, by combining them with previously demonstrated exergy analysis (in Chapter 5), the steps to determine the exergoeconomic variables is provided and the associated methodology on understanding the precedence among the components from an exergoeconomic perspective is demonstrated. In addition, methods of conducting enhanced exergoeconomic analyses are also shown in order to improve the reader's understanding of the interdependencies among the BTMS components, enable comparison of dissimilar components and provide a more realistic measure on how to improve the effectiveness of each component by determining how much of the component costs can be avoided.

Furthermore, a run-through of developing exergoenvironmental analysis is also presented in order to determine the environmental impact associated with a BTMS. Procedures to determine the exergy streams are shown and a databank of environmental impact correlations are provided for the readers to help them model their system components with an extensive accuracy without the need of cumbersome experimental relationships. Moreover, life cycle assessment analysis (LCA) for the electric battery is also presented since it generally has significantly higher impact than the rest of the components. Moreover, impact points are assigned to exergy streams for each BTMS component to point out the components causing the highest environmental impact and suggesting possibilities and trends for improvement, based on the calculated exergoenvironmental variables. Finally, the vital steps for conducting a multi-objective optimization study is for BTMS is carried on where the results from exergy, exergoeconomic and exergoenvironmental analyses are used according to the developed objective functions and system constraints in order to illustrate the methods to optimize the system parameters under different operating conditions with respect to these criteria using Pareto Optimal optimization techniques.

Nomenclature

b	specific environmental impact (m Points/kJ)
$\dot{B}$	environmental impact rate (m Points/s)
c	cost per unit of exergy ($/kJ)
$\dot{C}$	cost rate associated with exergy ($/h)
$\dot{E}$	energy rate (kW)
$\dot{E}x$	exergy rate (kW)
h	specific enthalpy (kJ/kg)
$\dot{m}$	mass flow rate (kg/s or L/min)
P	pressure (kg/m s2)
$\dot{Q}$	heat transfer rate (kW)
s	specific entropy (kJ/kg K)
$\dot{S}$	entropy (kJ/K)
T	temperature (K or °C)
T_0	ambient temperature (K or °C)
V	velocity (m/s)
$\dot{W}$	work rate or power (kW)
$\dot{Y}$	component-related environmental impact rate associated with LCA (m/Points/s)
$\dot{Z}$	cost rate associated with the sum of capital investment ($/h)
Z	height (m)

Greek Symbols

Δ	change in variable
ψ	exergy

Subscripts

0	ambient
act	actual
bat	battery
cool	coolant
c, cond	condenser
ch	chiller
comp	compressor
D	destruction
e	exit
elect	electricity
en	energy
ex	exergy
evap	evaporator
F	fuel
g, gen	generation
i	in
k	component

P	product
q	heat
ref	refrigerant
s	isentopic
tot	total
txv	thermal expansion valve
w	work

Acronyms

A/C	air conditioning
BTMS	battery thermal management system
COP	coefficients of performance
EES	Engineering Equation Solver
EV	electric vehicle
HEV	hybrid electric vehicle
ICE	internal combustion engine
LCA	life cycle assessment
LINMAP	linear programming technique for multidimensional analysis of preference
TMS	thermal management system
TPIM	traction power inverter module
TXV	thermal expansion valve

Study Questions/Problems

Please answer the following questions with respect to the BTMS you have considered in Chapter 5. If you have not considered a system, you can use the transcritical CO_2 based electric vehicle BTMS provided in the case study of Chapter 5.

6.1 Please define F and P rules and determine the "fuel" and "product" for each component?

6.2 By following the steps provided in this chapter, please form a cost and environmental impact accounting matrix for the BTMS of the previous question.

6.3 How much does the results you have calculated depend on the battery pack, electricity and the equipment life-time?

6.4 Please calculate the exergoeconomic and exergoenvironmental factors for the components and elaborate on which ones need to be improved first and how.

6.5 What portion of the cost in your individual components is associated with exergy destruction rate and investment cost rate? What are the avoidable and unavoidable portions of these costs?

6.6 What are the main decision variables and the constraints in your system? Which constraint has the largest impact on your multi-objective optimization and how sensitive are your results to this constraint?

References

Ahmadi P, Dincer I, Rosen MA. (2011). Exergy, exergoeconomic and environmental analyses and evolutionary algorithm based multi-objective optimization of combined cycle power plants. *Energy* **36**:5886–5898.

Bejan A. (1996). "*Entropy Generation Minimization: The Method of Thermodynamic Optimization of Finite-Size Systems and Finite-Time Processes*", CRC Press LLC, Florida, U.S.A.

Boyano A, Morosuk T, Blanco-Marigota AM, Tsatsaronis G. (2012). Conventional and exergoenvironmental analysis of steam methane reforming reactor for hydrogen production. *Journal of Cleaner Production* **20**:152–160.

Ecoinvent. (n.d.). "Ecoinvent Centre Database". Available at: www.ecoinvent.org/database [Accessed October 2012].

Fonseca CM, Fleming PJ. (1995). An overview of evolutionary algorithms in multiobjective optimization, *Evolutionary Computation* **3**:1–15.

Ghaffarizadeh A. (2006). *Investigation on Evolutionary Algorithms Emphasizing Mass Extinction*. B.Sc. Thesis, Shiraz University of Technology.

ISO 14040. (2006). *Environmental Management e Life Cycle Assessment Principals and Framework*, International Organization for Standardization, Geneva, Switzerland.

Konak A, Coit DW, Smith AE. (2006). Multi-objective optimization using genetic algorithms: A tutorial. *Reliability Engineering & System Safety* **91**:992–1007.

Lazzaretto A, Toffolo A. (2004). Energy, economy and environment as objectives in multi-criterion optimization of thermal systems design. *Energy* **29**:1139–1157.

Lazzaretto A, Tsatsaronis G. (2006). SPECO: A systematic and general methodology for calculating efficiencies and costs in thermal systems. *Energy* **31**:1257–1289.

Majeau-Bettez G, Hawkins TR, Strømman AH. (2011). Life Cycle Environmental Assessment of Lithium-Ion and Nickel Metal Hydride Batteries for Plug-in Hybrid and Battery Electric Vehicles. *Environmental Science and Technology* **45**:4548–4554.

Meyer L, Tsatsaronis G, Buchgeister J, Schebek L. (2009). Exergoenvironmental analysis for evaluation of the environmental impact of energy conversion systems. *Energy* **34**:75–89.

Sanaye, S., Niroomand, B. (2009). Thermal-economic modeling and optimization of vertical ground-coupled heat pump. *Energy Conversion Management* **50**:1136–1147.

Sayyaadi H, Babaelahi M. (2011). Multi-objective optimization of a joule cycle for re-liquefaction of the Liquefied Natural Gas. *Applied Energy* **88**:3012–3021.

Selbaş R, Kizilkan Ö, Şencan A. (2006). Thermoeconomic optimization of subcooled and superheated vapor compression refrigeration cycle. *Energy* **31**:2108–2128.

Tsatsaronis G, Lin L. (1990). On Exergy Costing in Exergoeconomics, Computer-Aided Energy Systems Analysis. *American Society of Mechanical Engineers* **21**:1–11.

Valero A. (1994). CGAM problem: definition and conventional solution. *Energy* **19**:268–279.

7

Case Studies

7.1 Introduction

In this chapter, various case studies are presented to the readers that employ the tools, methodology and procedures presented throughout the book and conduct analyses on real-life applications to further illustrate their efficacy on the design, development and optimization of electric vehicle battery thermal management systems. In order to follow the progression provided in the book chapters, the case studies start from examples in evaluating different types of vehicles in terms of performance, cost and environmental impact (with the focus on electric vehicles). Next, the characteristics of electric vehicle Li-ion battery thermal problems are investigated and solutions are offered with respect to various thermal management systems. Subsequently, applications of current and novel phase change materials are also provided to illustrate methods of enhancing the efficiency and effectiveness of the electric vehicle battery thermal management systems.

7.2 Case Study 1: Economic and Environmental Comparison of Conventional, Hybrid, Electric and Hydrogen Fuel Cell Vehicles

7.2.1 Introduction

In this case study, published data from various sources in academic and industry are used to perform economic and environmental comparisons of four types of vehicles: conventional, hybrid, electric and hydrogen fuel cell. Within this context, the production and utilization stages of the vehicles are taken into consideration. The comparison is based on a mathematical procedure which includes normalization of economic indicators (prices of vehicles and fuels during the vehicle life and driving range) and environmental indicators (greenhouse gas and air pollution emissions), and evaluation of an optimal relationship between the types of vehicles in the fleet.

In assessing the vehicle systems, stages involved in the vehicle's life cycle are considered, where they range from the extraction of natural resources to produce fuels to the final transformation of fuel to mechanical energy in an engine. The efficiency and environmental impact related to the fuel use are defined by both engine quality and

Thermal Management of Electric Vehicle Battery Systems, First Edition.
İbrahim Dinçer, Halil S. Hamut and Nader Javani.

the efficiency and environmental impact associated with the all life cycle stages preceding fuel utilization. The overall environmental impact of vehicle use also includes the impacts associated with vehicle production and end-of-life utilization measures (Dhingra *et al.*, 1999). Thus, the case study evaluates economic and environmental indicators (based on actual data), for vehicle production and utilization stages, and uses them to perform a comparison of four kinds of vehicles: conventional, hybrid, electric and hydrogen fuel cell. The evaluation methodology and procedures are provided to inform and assist the readers in the decision making processes with regards to vehicles using different propulsion technology based on the aforementioned criteria (Granovskii *et al.*, 2006).

7.2.2 Analysis

7.2.2.1 Economic Criteria

The following criteria are taken to be key economic characteristics of vehicles: vehicle price (including the price for changing batteries for hybrid and electric vehicles), and fuel costs (which are related to vehicle lifetime), and driving range (which defines the number of refueling stations required). Four particular vehicles, with different release years, are taken as representative of each vehicle category: Toyota Corolla (conventional), Toyota Prius (hybrid), RAV4 EV (electric) and Honda FCX (hydrogen fuel cell). The characteristics of each vehicle were based on published specifications. The price of the Honda FCX fuel cell vehicle was listed as 2,000,000 US$, but is estimated to be reduced to 100,000 US$ in regular production. This reduced price is considered here to make the comparisons reasonable. The cost used for batteries is based on the average kWh cost for hybrid and electric cars at the time the study is conducted. A 40-liter tank is considered for conventional and hybrid vehicles in order to calculate driving range. Table 7.1 lists technical and economic vehicle parameters.

The average prices of gasoline, hydrogen and electricity for the selected years are used to calculate the prices of fuels (listed in column 3 of Table 7.1). Data are not available for the price of hydrogen, but according to an analysis which shows the price of gasoline is about two times that of crude oil, the price of hydrogen is about two times that of natural

Table 7.1 Economic characteristics for four vehicle technologies.

Vehicle Type	Fuel	Price (thousands US $)	Fuel consumption[a], MJ/100 km	Fuel price, US $/100 km	Driving range, km	Price of battery changes during life cycle[b] of vehicle, (thousands US $)
Conventional	Gasoline	15.3	236.8[c]	2.94	540	1×0.1
Hybrid	Gasoline	20.0	137.6	1.71	930	1×1.02
Electric	Electricity	42.0	67.2	0.901	164	2×15.4
Fuel cell	Hydrogen	100.0	129.5	1.69	355	1×0.1

a) Fuel consumption based on 45% highway and 55% city driving.
b) Life cycle of vehicle is taken as 10 years.
c) Heat content of conventional gasoline is assumed to be its lower heating value (LHV), fixed at 32 MJ/l.
Source: Granovskii *et al.*, 2006.

gas. The efficiencies of producing gasoline from crude oil and hydrogen from natural gas are similar (Granocskii *et al.*, 2005). As the prices of natural gas and gasoline have not varied greatly, the ratio of price to lower heating value (LHV) of hydrogen assumed to be equal to that of gasoline. But because the density of gaseous hydrogen is very low, in order to use it as a fuel in a vehicle, it must be compressed, liquefied or stored in a chemical or physical bonded form. In order to compress hydrogen from 20 atm (the typical pressure after natural gas reforming) to 350 atm (the pressure in the hydrogen tank of the Honda FCX), about 50 kJ of electricity is consumed per MJ of hydrogen on board the vehicle.

7.2.2.2 Environmental Impact Criteria

In this study, environmental impact is considered by examining air pollution (AP) and greenhouse gas (GHG) emissions. The main gases in GHG emissions are CO_2, CH_4, N_2O and SF_6 (sulfur hexafluoride), which have GHG impact weighting coefficients relative to CO_2 1, 21, 310 and 24,900, respectively (Houghton *et al.*, 1996). Sulfur hexafluoride is used as a cover gas in the process of magnesium casting. Impact weighting coefficients (relative to NO_x) for the airborne pollutants CO, NO_x and VOCs (volatile organic compounds) are based on those obtained by the Australian Environment Protection Authority (Australian Greenhouse Office, 2005) using cost-benefit analyses of health effects. The weighting coefficient of SO_x relative to NO_x is estimated using Ontario Air Quality Index data (Basrur, 2001). Thus, for considerations of air pollution, the airborne pollutants CO, NO_x, SO_x and VOCs are characterized by the following weighting coefficients: 0.017, 1, 1.3 and 0.64, respectively.

The environmental impact related to the vehicle production stage is associated with material extraction and processing, manufacturing, and end-of-life utilization steps. Data on the gaseous emissions accompanying a typical vehicle are taken from (Dhingra *et al.*, 1999) and presented in Table 7.2. The AP_m emissions per unit curb mass of a conventional car are obtained by applying weighting coefficients to the masses of air pollutants in accordance with formula:

$$AP_m = \sum_{1}^{4} m_i w_i \tag{7.1}$$

where i is the index denoting an air pollutant (CO, NO_x, SO_x, VOCs), m_i is the mass of air pollutant i, and w_i is the weighting coefficient of air pollutant i. The results of the environmental impact evaluation for the vehicle production stage for the vehicle types considered are presented in Table 7.3.

Table 7.2 Gaseous emissions per kilogram of curb mass of a typical vehicle.

Industrial stage	CO, kg	NO_x, kg	GHG emissions, kg
Extraction	0.0120	0.00506	1.930
Manufacturing	0.000188	0.00240	1.228
End-of-life	$1.77 \cdot 10^{-6}$	$3.58 \cdot 10^{-5}$	0.014
Total	0.0122	0.00750	3.172

Source: Dhingra *et al.*, 1999.

Table 7.3 Environmental impact associated with vehicle production stages.

Type of car	Curb mass, kg	GHG emissions, kg	AP emissions, kg	GHG emissions per 100 km of vehicle travel[a], kg/100 km	AP emissions per 100 km of vehicle travel, kg/100 km
Conventional	1134	3595.8	8.74	1.490	0.00362
Hybrid	1311	4156.7	10.10	1.722	0.00419
Electric	1588	4758.3	15.09	1.972	0.00625
Fuel cell	1678	9832.4	42.86	4.074	0.0178

a) During vehicle's life time (10 years), an average car drives 241,350 km.
Source: Pehnt, 2001.

It is assumed that GHG and AP emissions are proportional to the vehicle mass, but the environmental impact related to the production of special devices in hybrid, electric and fuel cell cars; the utilized batteries and fuel cell stacks, are evaluated separately. Accordingly, the AP and GHG emissions are calculated for conventional vehicles as

$$AP = m_{car} AP_m \tag{7.2a}$$

$$GHG = m_{car} GHG_m \tag{7.2b}$$

for hybrid and electric vehicles as

$$AP = (m_{car} - m_{bat})\, AP_m + m_{bat} AP_{bat} \tag{7.3a}$$

$$GHG = (m_{car} - m_{bat})\, GHG_m + m_{bat} GHG_{bat} \tag{7.3b}$$

and for fuel cell vehicles as

$$AP = (m_{car} - m_{fc})\, AP_m + m_{fc} AP_{fc} \tag{7.4a}$$

$$GHG = (m_{car} - m_{fc})\, GHG_m + m_{fc} GHG_{bat} \tag{7.4b}$$

where, m_{car}, m_{bat} and m_{fc} are respectively the masses of cars, batteries and the fuel cell stack, AP_m, AP_{bat} and AP_{fc} are air pollution emissions per kilogram of conventional vehicle, batteries and the fuel cell stack, GHG_m, GHG_{bat} and GHG_{fc} are greenhouse gas emissions per kilogram of conventional vehicle, batteries and fuel cell stack. The masses of batteries for hybrid and electric cars are 53 kg (1.8 kWh capacity) and 430 kg (27 kWh capacity), respectively. The mass of the fuel cell stack is about 78 kg (78 kW power capacity). According to Rantik (Rantik, 1999), the production of 1 kg of the used battery requires 1.96 MJ of electricity and 8.35 MJ of liquid petroleum gas.

The environmental impact of battery production is presented in Table 7.4, assuming that electricity is produced from natural gas with an average 40% efficiency (which is reasonable since the efficiency of electricity production from natural gas varies from 33% for gas turbine units to 55% for combined-cycle power plants, with about 7% of the electricity dissipated during transmission). The material inventory for a polymer exchange membrane fuel cell (PEMFC), from, is presented in Table 7.5.

The environmental impact of the fuel cell stack production stage, is used to express environmental impact in terms of AP and GHG emissions (Table 7.4, last line). Compared to the selected batteries, the data indicates that the PEMFC production stage accounts for relatively large GHG and AP emissions. Manufacturing of electrodes

Table 7.4 The environmental impact related to the production of nickel metal hydride (NiMH) batteries and polymer exchange membrane fuel cell (PEMFC) stacks.

Equipment	mass, kg	Number per life of vehicle	AP emissions per life of vehicle, kg	GHG emissions per life of vehicle, kg
Battery for hybrid car	53	2	0.507	89.37
Battery for electric car	430	3	6.167	1087.6
PEMFC stack for fuel cell car	78	1	30.52	4758.0

Source: Granovskii *et al.*, 2006.

Table 7.5 Material inventory of a polymer exchange membrane fuel cell stack.

Component	Material	Mass (kg)
Electrode	Platinum	0.06
	Ruthenium	0.01
	Carbon paper	4.37
Membrane	Nafion membrane	5.64
Bipolar plate	Polypropylene	16.14
	Carbon fibers	16.14
	Carbon powder	21.52
End-plate	Aluminum alloy	2.80
Current collectors	Aluminum alloy	1.14
Tie-rod	Steel	2.05
Total		69.87

Source: Granovskii *et al.*, 2006.

(including material extraction and processing) and bipolar plates constitute a major part of the emissions.

GHG and AP emissions also emanate from fuel production and utilization stages. The corresponding environmental impact has been evaluated in numerous life cycle assessments of fuel cycles. Various data from industry and academia are incorporated in the analysis.

Three scenarios for electricity production are considered here: (1) electricity is produced from renewable energy sources including nuclear energy; (2) 50% of the electricity is produced from renewable energy sources and 50% from natural gas with an efficiency of 40%; (3) all electricity is produced from natural gas with an efficiency of 40%. Nuclear/renewable weighted average GHG emissions were taken as 18.4 tonnes CO_2-equiv. per GWh of electricity. These emissions are embedded in material extraction, manufacturing and decommissioning for nuclear, hydro, biomass, wind, solar and geothermal power generation stations. AP emissions were calculated assuming that GHG emissions for plant manufacturing correspond entirely to natural gas combustion. GHG and AP emissions embedded in manufacturing a natural gas power generation plant are negligible compared to the direct emissions during its

Table 7.6 Greenhouse gas and air pollution emissions per MJ of electricity produced.

Scenario	GHG emission, g	AP emission, g
Scenario 1	5.11	0.0195
Scenario 2	77.5	0.296
Scenario 3	149.9	0.573

Source: Granovskii *et al.*, 2006.

Table 7.7 Greenhouse gas and air pollution emissions per MJ (LHV) of hydrogen and gasoline from combustion in fuel cell and internal combustion engine vehicles.

Fuels	GHG emission, g	AP emission, g
Hydrogen from natural gas scenarios		
Scenario 1	78.5	0.0994
Scenario 2	82.1	0.113
Scenario 3	85.7	0.127
Gasoline from crude oil	84.0	0.238

Source: Granovskii *et al.*, 2006.

utilization. Taking all these factors into account, GHG and AP emissions for the three scenarios for electricity generation are calculated and presented in Table 7.6.

As noted above, hydrogen use in a fuel cell vehicle requires its compression and, as a consequence, electricity to power a compressor. Table 7.7 lists GHG and AP emissions from gasoline and hydrogen utilization in vehicles depending on the electricity generation scenario.

Table 7.8 presents the environmental impact as a result of the fuel utilization stage, and the overall environmental impact, which includes the fuel utilization and car production stages.

7.2.2.3 Normalization and General Indicator

In order to allow different cars to be compare when different kind of indicators are available, a normalization procedure is performed. The value of a normalized indicator of 1 is chosen to correspond to the best economic and environmental performance among the cars considered. Therefore, normalized indicators for vehicle and fuel costs, and greenhouse gas and air pollution emissions, are proposed according to the following expression:

$$(NInd)_i = \frac{(1/Ind)_i}{(1/Ind)_{\max}} \tag{7.5}$$

where $(1/Ind)_i$ are reciprocal values of indicators like vehicle and fuel costs, greenhouse gas and air pollution emissions (see Tables 7.1 and 7.8), $(1/Ind)_{max}$ is the maximum of the reciprocal values of those indicators, $(NInd)_i$ is the normalized indicator, and the index i denotes the vehicle type (from the four kinds of vehicles considered here).

Table 7.8 Greenhouse gas/air pollution emissions with respect to fuel utilization stage and overall environmental impact for various propulsion systems.

Car Type		Fuel utilization stage		Total	
		GHG emissions per 100 km of vehicle travel, kg/100 km	AP emissions per 100 km of vehicle travel, kg/100 km	GHG emissions per 100 km of vehicle travel,[a] kg/100 km	AP emissions per 100 km of vehicle travel, kg/100 km
Conventional		19.9	0.0564	21.4	0.0600
Hybrid		11.6	0.0328	13.3	0.0370
1[b]	Electric	0.343	0.00131	2.31	0.00756
	Fuel cell	10.2	0.0129	14.2	0.0306
2	Electric	5.21	0.0199	7.18	0.0262
	Fuel cell	10.6	0.0147	14.7	0.0324
3	Electric	10.1	0.0385	12.0	0.0448
	Fuel cell	11.1	0.0165	15.2	0.0342

a) During vehicle life time (10 years), an average car drives 241,350 km.
b) Numbers in this column denote scenario for electricity production.
Source: Granovskii *et al.*, 2006.

But for driving range (distance on one full tank of fuel or on one full charge of batteries) indicators, the normalized indicators $(NInd)_i$ are expressible as

$$(NInd)_i = \frac{(Ind)_i}{(Ind)_{\max}} \tag{7.6}$$

where $(Ind)_i$ denotes the driving range indicator for four types of vehicles (implied by index i) considered here, and $(Ind)_{max}$ denotes the maximum value of the driving range indicator.

After normalization of the information, normalized economic and environmental indicators for four types of vehicles were been obtained for the all three scenarios of electricity generation (Table 7.9). The generalized indicator represents the product of the calculated normalized indicators (which is a simple geometrical aggregation of criteria with an absence of weighting coefficients). The "ideal car" is associated with a generalized indicator of 1, as such a vehicle possesses all the advantages of those considered. The calculated values of general indicators provide a measure of "how far" a given car is from the ideal one, for the factors considered.

7.2.3 Results and Discussion

In order to simplify the comparisons of the vehicles, the general indicator has been also normalized according to Equation 7.6. Figure 7.1 shows the dependence of the normalized general indicator *NGInd* (column 8 in Table 7.8) on the electricity-generation scenario. According to those results, hybrid and electric cars are competitive if nuclear and renewable energies account for about 50% of the energy to generate electricity. If fossil fuels (in this case natural gas) are used for more than 50% of the energy to generate electricity, the hybrid car has significant advantages over the other three.

Table 7.9 Normalized economic and environmental indicators for four types of cars.

	Car type	Normalized indicators					General indicator	Normalized general indicator
		Car cost	Range	Fuel cost	Greenhouse gas emissions	Air pollution emissions		
1[a]	Conventional	1	0.581	0.307	0.108	0.126	0.00243	0.0651
	Hybrid	0.733	1	0.528	0.174	0.205	0.0138	0.370
	Electric	0.212	0.177	1	1	1	0.0374	1
	Fuel cell	0.154	0.382	0.532	0.163	0.247	0.00126	0.0336
2	Conventional	1	0.581	0.307	0.336	0.436	0.0261	0.176
	Hybrid	0.733	1	0.528	0.541	0.708	0.148	1
	Electric	0.216	0.177	1	1	1	0.0374	0.252
	Fuel cell	0.154	0.382	0.532	0.488	0.807	0.0123	0.0832
3	Conventional	1	0.581	0.307	0.599	0.628	0.0670	0.197
	Hybrid	0.733	1	0.528	0.911	0.967	0.341	1
	Electric	0.212	0.177	1	1	0.824	0.0308	0.0903
	Fuel cell	0.154	0.382	0.532	0.794	1	0.0248	0.0728

a) Numbers in this column denote scenario for electricity generation.
Source: Granovskii *et al.*, 2006.

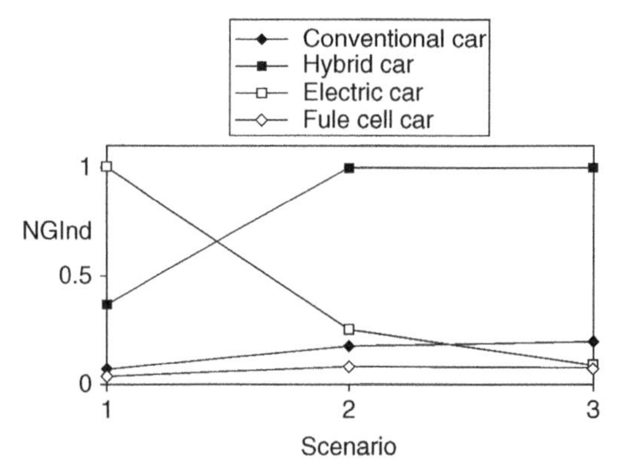

Figure 7.1 The dependence of the normalized general indicator, *NGInd*, on electricity-generation scenario for four types of cars (adapted from Granovskii *et al.*, 2006).

An optimization has been performed to obtain the optimal relationship between vehicles in a fleet. The optimal relationship is considered here to be the maximum value of the general indicator in accordance with following equations:

$$\sum_{i=1}^{4} \beta_i = 1 \tag{7.7}$$

$$\prod_{j=1}^{5} \sum_{i=1}^{4} \beta_i \cdot NInd_i^j = \max imum \tag{7.8}$$

where β_i is the fraction of a given type of car in the fleet, $NInd_i^j$ is the normalized economic or environmental indicator for a given type of car, the index i denotes the vehicle

Table 7.10 Optimal relationship in fleet between different types of cars.

Electricity Generation Scenario	Conventional vehicle (%)	Hybrid Vehicle (%)	Electric Vehicle (%)	Fuel cell Vehicle (%)	General Indicator
1	0	38	62	0	0.079
2	0	78	22	0	0.159
3	0	100	0	0	0.341

Source: Granovskii *et al.*, 2006.

type, and the index j denotes the five kinds of economic and environmental indicators from Table 7.9.

Table 7.10 presents the optimal relationship between different types of cars in the fleet, depending on the scenario for electricity generation. The best result occurs for a fleet of 20% of hybrid cars and 80% of electric cars for scenario 1 for electricity generation. If the nuclear and renewable energy fraction is reduced (scenarios 2 and 3), the electric car becomes uncompetitive with respect to the hybrid car. The hydrogen fuel cell car is not competitive for the all scenarios considered here, but it has the best air pollution emissions indicator for scenario 3.

As seen in Table 7.9 (scenario 3), the electric car is inferior to the hybrid one in terms of car price, range and air pollution emissions. The simplest technical solution to increase its range is to produce electricity on-board the vehicle. Since the efficiency of electricity generation by means of an internal combustion engine is lower than that of a gas turbine unit (typically the efficiency of a thermodynamic cycle with fuel combustion at constant pressure is higher than that one at constant volume), it could make sense on thermodynamic grounds to incorporate a gas turbine engine into electric car. The application of fuel cell systems (especially solid oxide fuel cell stacks) within gas turbine cycles allows their efficiency to be increased to 60%.

The pressure of the natural gas required to attain a range equal to the range of a hybrid car is more than two times less than the pressure of hydrogen in the tank of the fuel cell vehicle. So, corresponding to the efficiency of electricity generation from natural gas $\eta = 0.4–0.6$, the required pressure in the tank of a hypothetical electric car could reduce from 170 to 115 atm.

Assuming the cost and GHG and AP emissions corresponding to the hypothetical electric car production stage are equal to those for the electric prototype, the normalized indicators for the different on-board electricity-generation efficiencies can be calculated (see Table 7.11). An optimization is needed to obtain the optimal relationship between capacities of batteries and a gas turbine engine.

Figure 7.2 presents the optimal fractions of hybrid and hypothetical electric cars in a fleet to increase the general indicators in Table 7.11. From Table 7.11, it can be seen that if electricity is generated with an efficiency of about 50–60% by a gas turbine engine connected to a high-capacity battery and electric motor, the electric car becomes superior.

The gas turbine engine has many advantages over the conventional internal combustion engine: the opportunity to use various kinds of liquid and gaseous fuels, quick starts at low air temperatures, high traction qualities and simplicity of design. The main reason the implementation of gas turbine engines into light-duty vehicles in the 1960s failed was their poor ability to change fuel consumption with varying traffic conditions. Then,

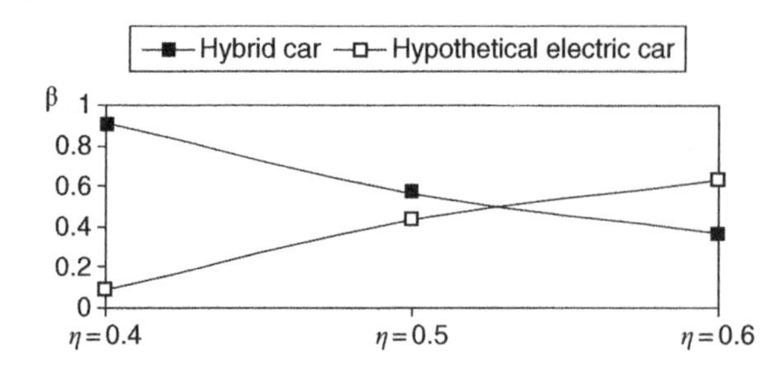

Figure 7.2 The optimal fraction (β) for hybrid and hypothetical electric cars in the fleet (adapted from Granovskii *et al.*, 2006).

Table 7.11 Normalized economic and environmental indicators for hybrid and hypothetical electric car with different efficiencies for on-board electricity generation.

	Normalized indicators					
Car type	Car cost	Range	Fuel cost	Greenhouse gas emissions	Air pollution emissions	General indicator
Hybrid	1	1	0.316	0.720	0.954	0.217
Electric, $\eta = 0.4$	0.289	1	0.663	0.725	0.718	0.0997
Electric, $\eta = 0.5$	0.289	1	0.831	0.867	0.863	0.180
Electric, $\eta = 0.6$	0.289	1	1	1	1	0.289

Source: Granovskii *et al.*, 2006.

the gas turbine engine was considered for use in directly converting fuel energy into mechanical work to drive an automobile. The application of a gas turbine unit only to generate electricity, permits this weakness to be overcome, when the gas turbine is integrated with a high-capacity battery and electric motor. The introduction of ion conductive membranes and fuel cells into a gas-turbine cycle can further increase the efficiency and decrease AP emissions.

7.2.4 Closing Remarks

Using actual data, an economic and environmental comparison is performed of four types of vehicles: conventional, hybrid, electric and hydrogen fuel cell. The analysis shows that the hybrid and electric cars can have significant advantages over its benchmarked technologies. The economics and environmental impact associated with use of an electric car depends substantially on the source of the electricity. If electricity comes from renewable energy sources, the electric car is advantageous to the hybrid vehicle. If the electricity comes from fossil fuels, the electric car remains competitive only if the electricity is generated on-board. If the electricity is generated with an efficiency of about 50–60% by a gas turbine engine connected to a high-capacity battery and electric motor, the electric car becomes superior in many respects. The implementation of fuel cells stacks and ion conductive membranes into gas turbine cycles could permit electricity generation efficiency to be further increased and air pollution emissions to be further decreased. It is concluded, therefore, that the electric car with capability for on-board electricity generation represents a beneficial option worthy of further investigation in the development of energy efficient and ecologically benign vehicles.

The main limitations of this study follow: (i) the use of data which may be controversial in some instances; (ii) subjective choice of indicators; (iii) the simplified procedure for building up the general indicator without using unique weighting coefficients; and (iv) the potential inaccuracy of some of data for today due to the latest advancements in the technology. In spite of these limitations, the study provides the necessary steps to conduct economic and environmental comparison for various types of vehicles for the readers to better understand the associated advantages/drawbacks of electric vehicles and their main components.

7.3 Case Study 2: Experimental and Theoretical Investigation of Temperature Distributions in a Prismatic Lithium-Ion Battery

7.3.1 Introduction

EVs and HEVs depend on the type of batteries and improving their life-time will reduce the runtime and the costs for the vehicle. These technologies have primarily shifted towards Lithium-ion batteries due to its energy storage capabilities over extended life-cycles (Damodaran *et al.*, 2011). Lithium-ion battery is widely used today in automobile industry because it possesses high specific energy and power densities, high nominal voltage and low self-discharge rate for EVs and HEVs. Apart from automobiles, laptops, cell phone or mobiles, toys, as well as many other consumer products use Lithium-ion batteries as the main or secondary power source (Abousleiman *et al.*, 2013). But the drawback of Lithium-ion battery is that precautions during charging and discharging must be taken. Exceeding voltage, current or power limits may be resulted in the battery cell damage. There is a possibility of thermal runaways. Also, Lithium-ion polymer batteries must be carefully monitored and managed (electrically and thermally) to avoid safety (inflammability) and performance issues.

Temperature is also one of the most important parameter for battery, which can affect both the time life and energy of the battery. The battery temperature should be within a temperature range which is considered optimum for the better performance and long life, for both use and storage. This temperature range differs between technologies and manufacturer. Therefore, thermal management of batteries is required. However, it is a challenging task and critical in achieving the desired performance in a low-temperature environment and the desired life in the high-temperature environment. The heat generated inside a battery must be dissipated to improve reliability and prevent failure. Lithium-ion batteries degrade rapidly at higher temperature ranges, while the power and energy output are reduced at cold temperature ranges, thereby limiting the driving range and/or performance capabilities.

This case study focuses on the batteries as it is the main component of the EV technology and deals with the surface temperature distributions on the principal surface of the battery at 1C and 3C discharge rates and different boundary conditions (cooling/operating/bath temperature) of 5 °C, 15 °C, 25 °C, and 35 °C. The air cooling and water cooling system is designed and developed based on a prismatic Lithium-ion that has 20 Ah capacity. In addition, the battery thermal model is developed which represents the main thermal phenomena in the battery cell in terms of temperature distribution.

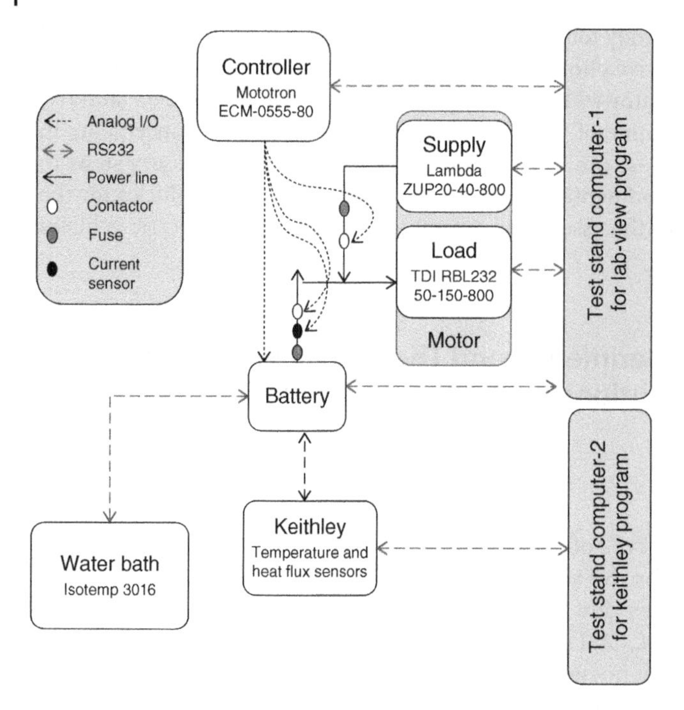

Figure 7.3 Schematic of experimental set up (adapted from Panchal *et al.*, 2016).

The developed model is validated with the experimental data collected including temperature and discharge voltage profile.

7.3.2 System Description

The schematic of the experimental set-up used for thermal characterization of the battery is shown in Figure 7.3. The components shown on the upper half of the schematic make up the battery cycling equipment. Whereas the components displayed on the lower half of the schematic comprise the thermal data collection, and battery cooling system. The compression apparatus is used to contain and insulate the battery while testing is underway with cold plates. The selected discharge rates were 1C (20Amp) and 3C (60Amp). These major components are described in detail in the following sections (Panchal *et al.*, 2016).

The test stand computer-1 manages a Lab-view program which record values at one second intervals and the test stand computer-2 manages the Keithley-2700 data acquisition system (for thermal measurements). The low voltage supply and the load were Lambda ZUP20-40-800 and TDI Dynaload RBL323-50-150. The MotoTron controller interfaces via RS232 communication to the test stand computer-1. The air cooling set up is shown in Figure 7.4a while the water cooling set up is illustrated in Figure 7.4b, and the computer-aided drawing of the battery cooling set-up is exhibited in Figure 7.4c which includes battery, cold plates and other components. The battery cooling set-up is designed and built to isolate the cell from the ambient thermal environment and maintain compression of the cooling plate/battery surface interface.

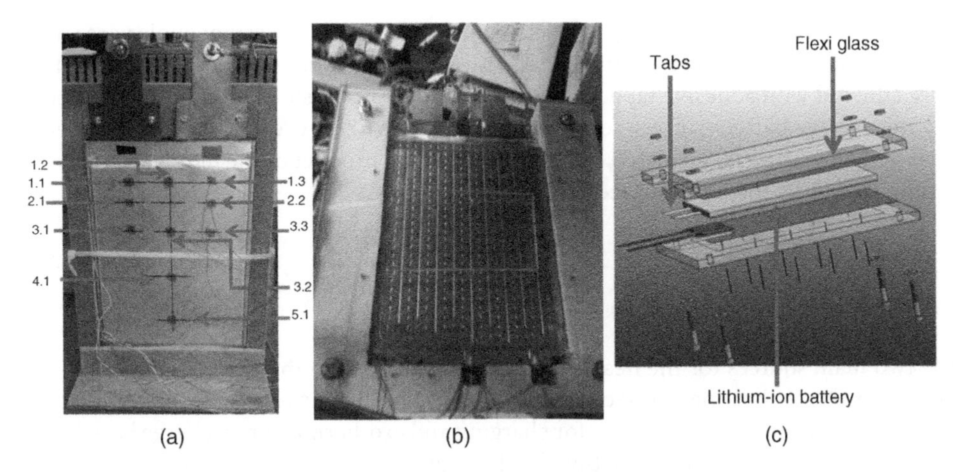

Figure 7.4 (a) Air cooling; (b) Water cooling; and (c) CAD of battery cooling set up (Panchal *et al.*, 2016).

For the water cooling set up, two cooling plates were used and out of two; one cold plate is placed at the top of the Lithium-ion battery and the other cold plate is placed at the bottom of the Lithium-ion battery. This type of a cold plate is characterized as having single flow channel with one inlet and one outlet. A Lithium-ion battery is insulated from three sides to prevent heat loss from the battery to the surrounding.

7.3.3 Analysis

7.3.3.1 Temperature Measurements

For the battery surface temperature measurements the thermocouples were installed on the principal surface of the battery. The positions of these thermocouples are listed in Table 7.12 and shown accordingly in Figure 7.4a.

Table 7.12 Positions of thermocouple locations.

	Thermocouple locations	X (mm)	Y (mm)
1	1,1	26.5	194
2	1,2	78.5	194
3	1,3	130.5	194
4	2,1	26.5	169
5	2,2	130.5	169
6	3,1	26.5	130
7	3,2	78.5	130
8	3,3	130.5	130
9	4,1	78.5	70
10	5,1	78.5	35

Source: Panchal *et al.*, 2016.

These thermocouples are T-type, special limits of error (SLE) thermocouple wire with uncertainty of 1°C (according to manufacturer's specifications). The thermocouples were adhered with adhesive tape to the Lithium-ion battery surface. In order to calculate the average battery surface temperature, Equation 7.9 is used. The average battery surface temperature is calculated by summation of the product of the temperature-area and dividing it by the total surface area.

$$T_{as} = \frac{\sum(T_{ij}A_{ij})}{A_{total}} \tag{7.9}$$

7.3.3.2 Heat Generation

The two main sources for the heat generation is evaluated in the analysis. First, Joule's heating or Ohmic heating and second, the entropy change due to electrochemical reactions. The heat can be endothermic for charging and exothermic for discharge based the electrode pair. The following equation for the heat generation in a battery is used:

$$\dot{Q} = I(V_{oc} - V_{ac}) - I\left[T\left(\frac{dV_{oc}}{dT}\right)\right] \tag{7.10}$$

Here, $I(V_{oc} - V_{act})$ is known as the Ohmic or Joule's heating and $I[T(dV_{oc}/dT)]$ is known as the heat generated or consumed due to the reversible entropy change which results from electrochemical reactions within the battery cell. Usually, the second term is small as compared to the first term, and therefore negligible for the EV and HEV current rates (Smith and Wang, 2006).

In a prismatic Lithium-ion battery, the current collectors create an additional heating (Joule or Ohmic heating) due to high current densities. In another work, Equation 7.11 was developed to include two added terms that account for heat generated in the current collector tabs (Gu and Wang, 2000)

$$\dot{Q} = I\left(V_{oc} - V_{ac} - T\left(\frac{dV_{oc}}{dT}\right)\right) + A_p R_p I_p^2 + A_n R_n I_n^2 \tag{7.11}$$

Thermal runaway can occur if this heat is not removed properly, as elevated temperatures trigger additional heat generating exothermic reactions. These reactions then increase battery temperature further; creating a positive feedback mechanism that causes battery temperature to climb sharply if heat is not dissipated well. As a result thermal runaway can occur that results into complete cell failure accompanied by fire or explosive gas release. Furthermore, even if thermal runaway does not occur, significant degradation of battery capacity can lead by consistently operating at elevated temperature (>50 °C) (Ramadass *et al.*, 2011). Therefore, the use of a BTMS becomes imperative for the performance.

7.3.4 Results and Discussion

7.3.4.1 Battery Discharge Voltage Profile

In order to evaluate the impact of the discharge rates, six experiments were performed on brand new battery with 1C charge rate and C/5, C/2, 1C, 2C, 3C, and 4C discharge rates at an ambient condition (22 °C). In all six tests the discharge was continued until the cut-off voltage of 2.0 was reached. Figure 7.5 shows the battery discharge voltage profile during above mentioned discharge rates versus the discharge capacity measured

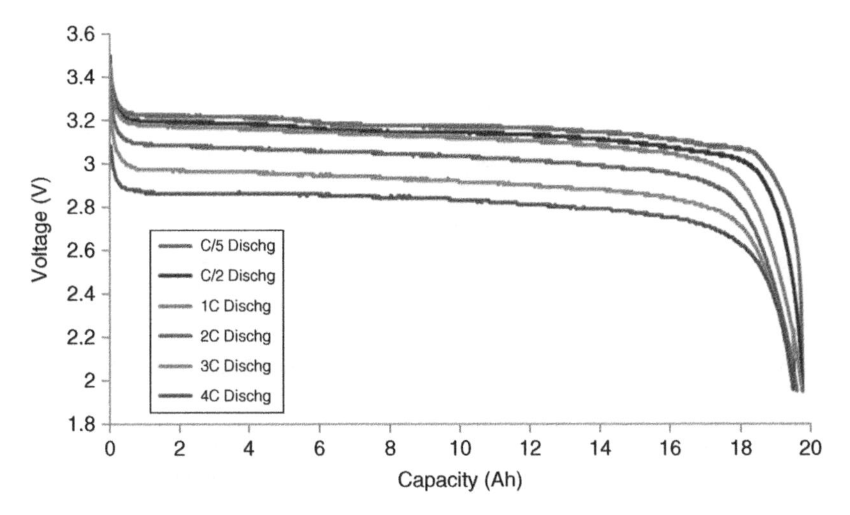

Figure 7.5 Voltage profile at C/5,C/2, 1C, 2C, 3C, 4C (adapted from Panchal *et al.*, 2016).

at an ambient condition. It was found that the battery capacity was observed to be same (20Ah) as reported by manufacturer's data sheet at the all discharge rates. For the low discharge rate, the discharge voltage is higher and it decreases when the increases, i.e. the voltage plateau of the battery cycles at higher discharge rates (3C and 4C) is shorter than the cycled at the lower discharge rates (C/2 and C/5).

7.3.4.2 Battery Internal Resistance Profile

Figure 7.6 shows the battery internal resistance profile during 1C, 2C, 3C, and 4C discharge rates as a function of capacity at an ambient conditions. A steep rise is seen in the first part of the discharge, the internal resistance is almost constant until 16Ah when a steady increase is observed. The increase appears to become steeper as discharge progresses and is highest just before the end of discharge. Furthermore, it can be seen

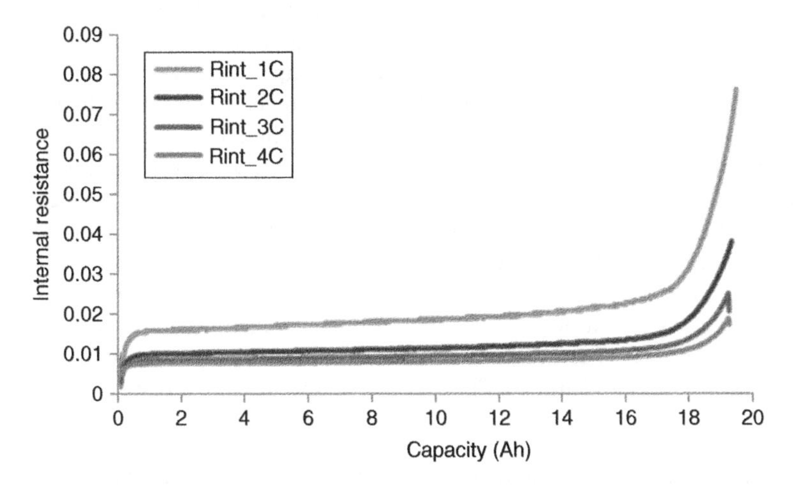

Figure 7.6 Internal resistance profile at 1C, 2C, 3C, 4C (adapted from Panchal *et al.*, 2016).

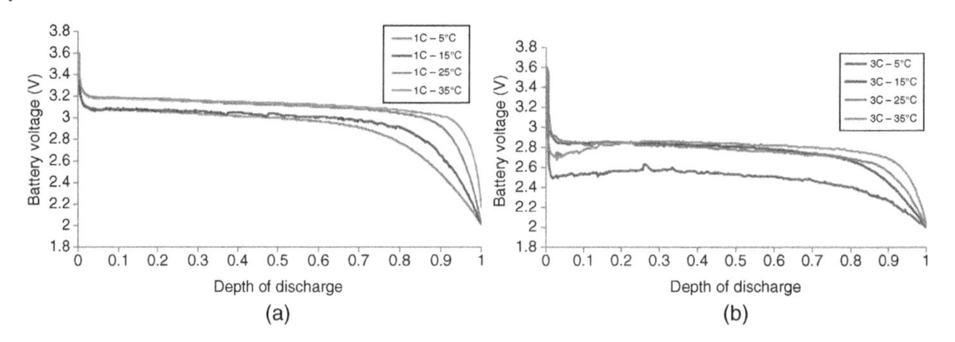

Figure 7.7 Discharge voltage profile at 1C and 3C and 5 °C, 15 °C, 25 °C, and 35 °C BC (adapted from Panchal *et al.*, 2016).

that the increase in the discharge rate and thus the discharge current causes consistent increase in the internal resistance from the beginning to the end of discharge.

7.3.4.3 Effect of Discharge Rates and Operating Temperature on Battery Performance

Figure 7.7 shows the battery discharge voltage profile corresponding to 1C and 3C discharge rates and 5°C,15°C, 25°C and 35°C boundary conditions. This illustrates the effect of discharge rate and operating temperature on the electrical performance of the batteries. This is of particular concern for EV and HEV development because the vehicle range is directly affected by the battery capacity. It was expected and can be seen that when operating temperature decreases from 35°C to 5°C, the discharge voltage profile also decreases.

Figure 7.8a-h shows the surface temperature distribution on the principal surface of the battery at 1C and 3C discharge rates and 5 °C, 15 °C, 25 °C, and 35 °C boundary conditions (BCs). Here, the battery surface temperature profile recorded by ten thermocouples is plotted as a function of time. It can be observed that the thermocouple location (1, 1), (1, 2), and (1, 3), that is, near electrodes has faster response. These thermocouples are placed nearest the negative and the positive electrodes of the battery and indicate the location of highest heat accumulation, and likely the rates of heat generation are highest near the electrodes (cathode and anode) due to higher discharge current.

By comparing the images, it can be seen that the higher temperature distribution is noted over entire surface of the battery for 3C at 35°C and the lower temperature distribution is noted for 1C at 5°C. Furthermore, it is observed that the increased discharge rates and increased operating conditions (or BCs) results in increased surface temperature distributions on the surface of the battery. The maximum average battery surface temperature at 1C and 3C discharge rates and different bath temperatures (or BCs) is also measured. It is found that at the average maximum surface temperature is 10.1°C and 15.7°C at 1C-5°C and 3C-5°C, while the average maximum surface temperature is 19.6°C and 25.4°C at 1C-15°C and 3C-15°C. Similarly, the average maximum surface temperature is 27.6°C and 32.6°C at 1C-25°C and 3C-25°C. Lastly, the average maximum surface temperature is 35.9°C and 40.3°C at 1C-35°C and 3C-35°C, respectively.

7.3.4.4 Model Development and Validation

A Simulink block diagram for battery model is shown in Figure 7.9. A battery thermal model is developed based upon data acquired from the experimental set-up using dual

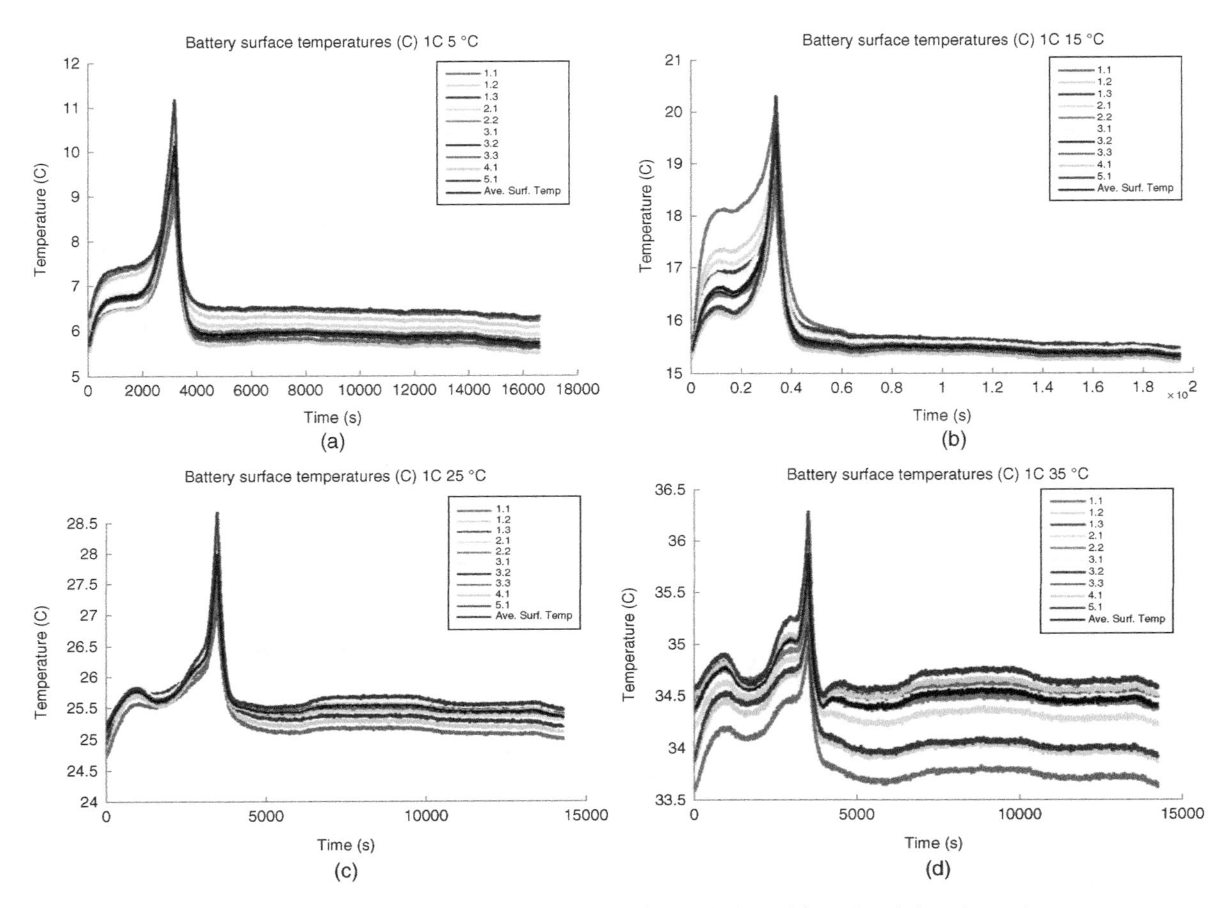

Figure 7.8 Surface temperature profile at 1C and 3C and 5 °C, 15 °C, 25 °C, and 35 °C BCs (adapted from Panchal *et al.*, 2016).

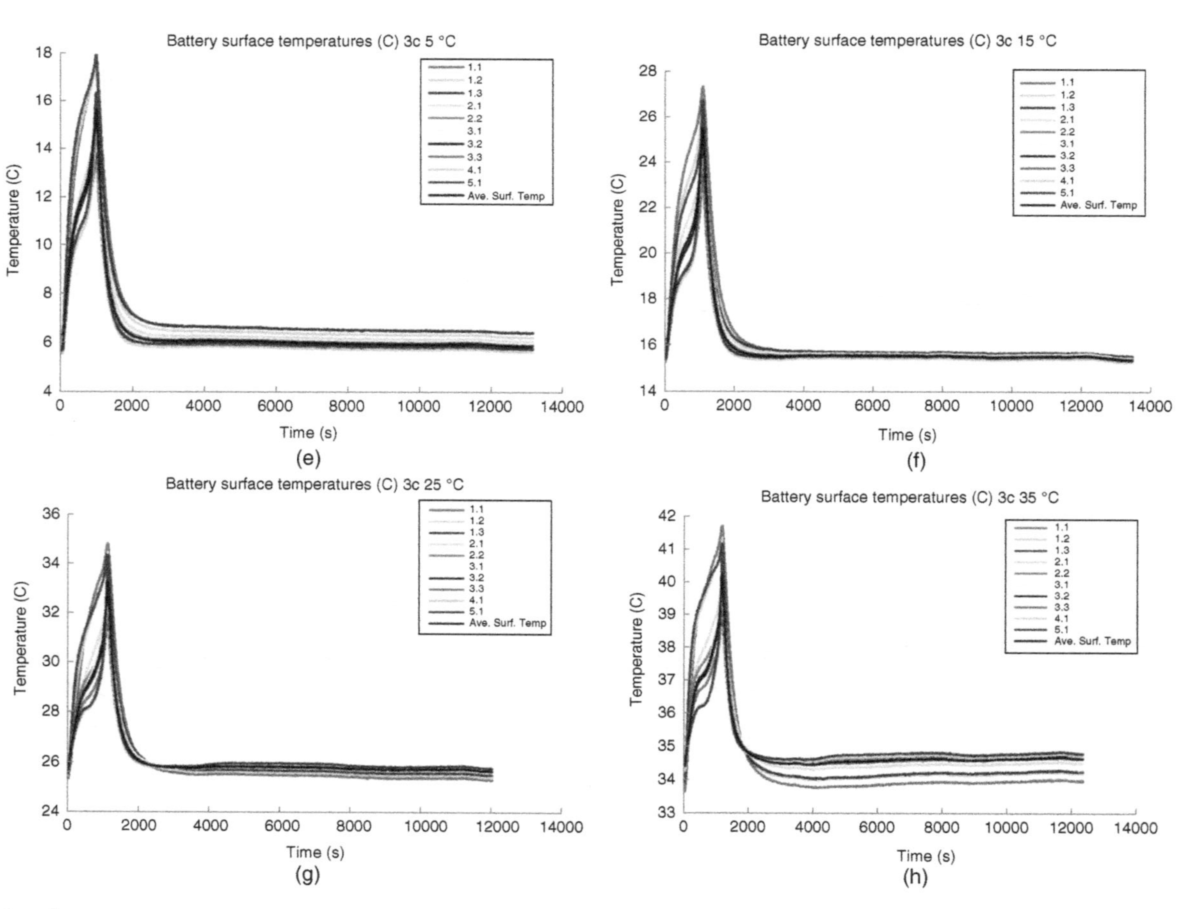

Figure 7.8 *(Continued)*

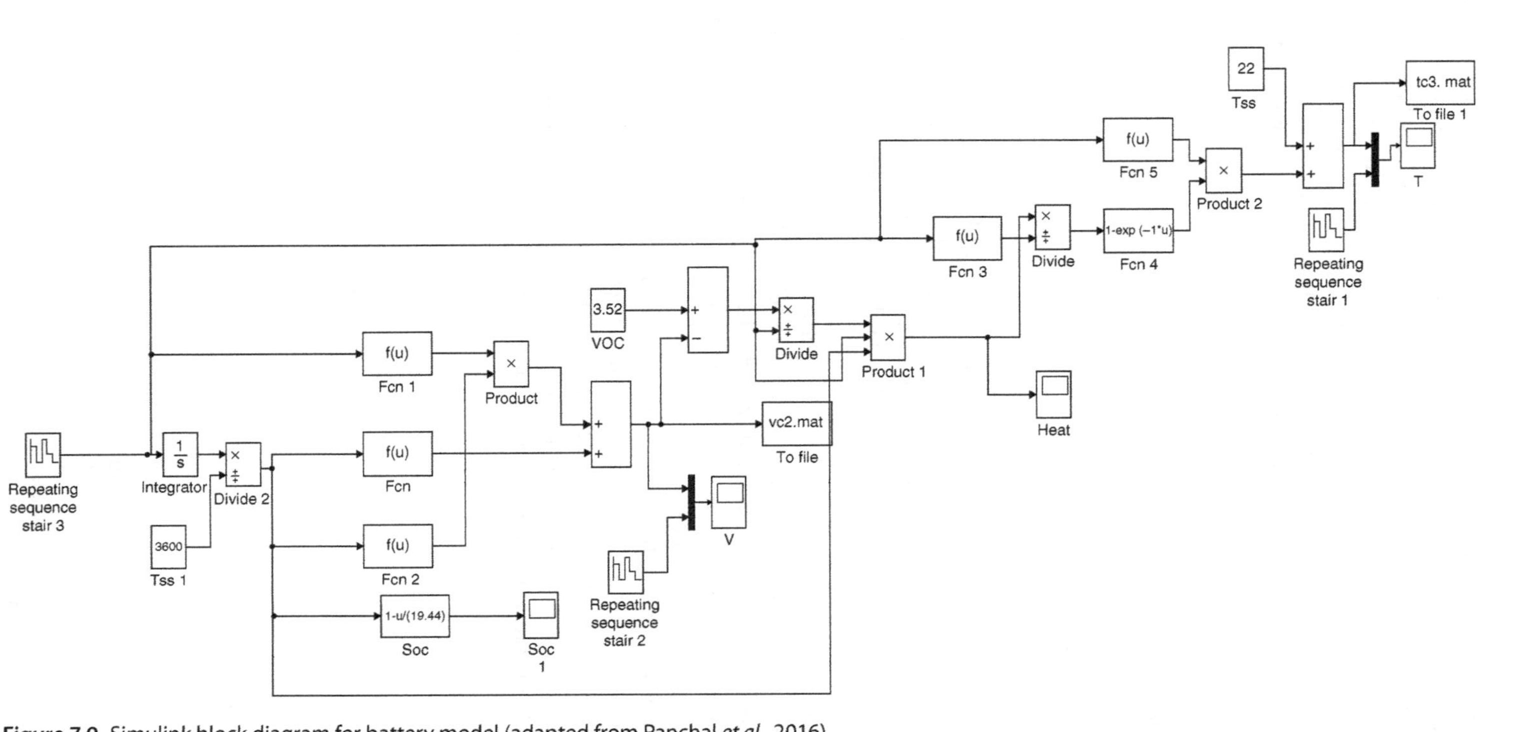

Figure 7.9 Simulink block diagram for battery model (adapted from Panchal *et al.*, 2016).

cold plates for a Lithium-ion battery. The acquired data is incorporated into a look up table along with simple algorithms written in MATLAB/Simulink to access the look-up table. The model works by using an input current draw to measure the battery voltage which together provides an estimate of the power required from the system.

The model is divided in to three sections called SOC, voltage and temperature. The instantaneous capacity is then divided by the maximum capacity to achieve the SOC estimate. The charge and discharge resistance is obtained from following equations:

$$R_d = \frac{\Delta V_d}{\Delta I_d} \tag{7.12}$$

$$R_d = \frac{\Delta V_c}{\Delta I_c} \tag{7.13}$$

The internal resistance is calculated by

$$R_{in} = (V_{oc} - V_{ac})/I_d \tag{7.14}$$

Figure 7.10 shows a comparison of the measured terminal voltage during charging cycle at 1C with the profile predicted by the battery thermal model at an ambient temperature. It shows the good agreement between measured and predicted values. At the extremes slight variations are noted. The simulated values are higher than the measured value.

Figure 7.11a-h show comparisons between experimental temperature profiles measured at 1C and 3C discharge rates and at 5 °C, 15 °C, 25 °C and 35 °C BCs and the profiles predicted by the battery thermal model. These experimental and theoretical comparisons show a good agreement. Here the predicted values follow expected trends but the slight discrepancies are observed. The model temperature response depends on the heat generated by losses in the battery cell, the thermal mass of the battery cell, and the heat transfer to the environment. Access to more definitive information on the battery cell properties would lead to better agreement between model and measurement.

As shown in Figure 7.12 (a to h), comparisons of the measured terminal voltage at 1C and 3C discharge rates and at 5 °C, 15 °C, 25 °C, and 35 °C BCs with the theoretical

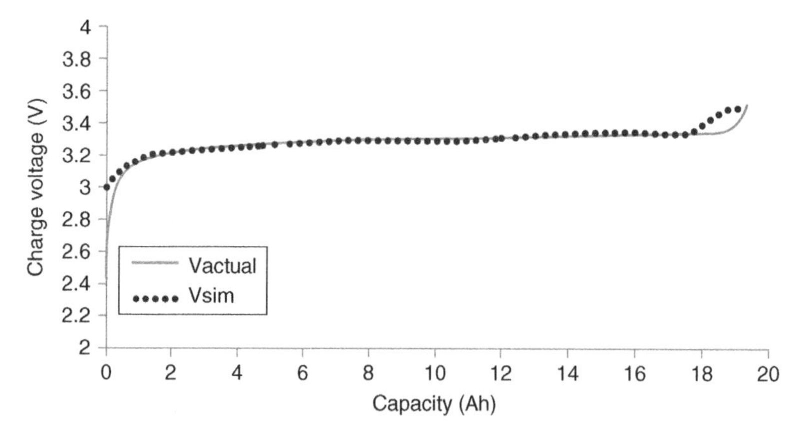

Figure 7.10 Comparison of actual and simulated charge voltage profile at 1C at an ambient condition (adapted from Panchal *et al.*, 2016).

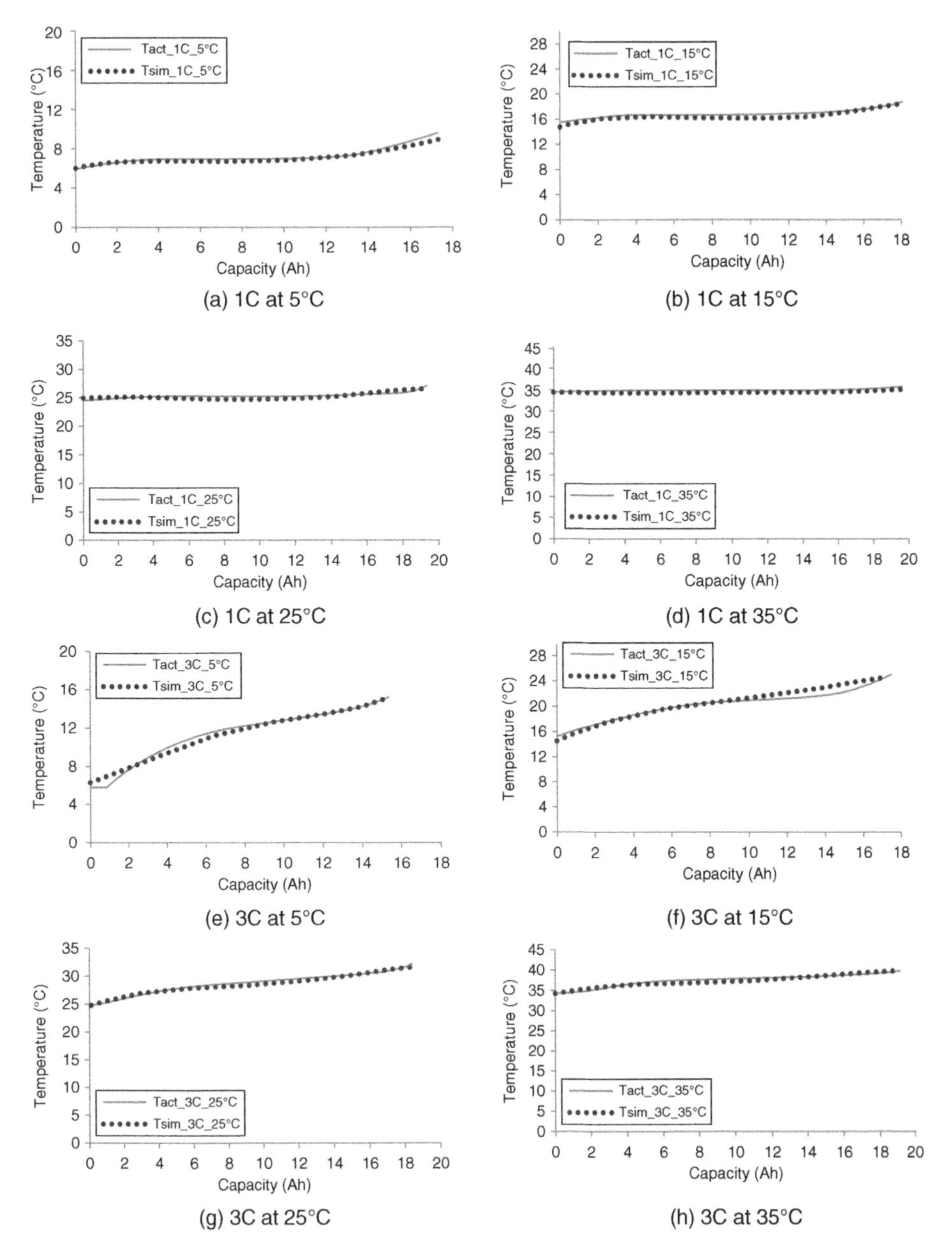

Figure 7.11 Comparison of actual and simulated discharge temperature profiles at 1C and 3C at 5 (adapted from Panchal *et al.*, 2016).

profiles are provided by the empirical battery thermal model. The voltage data is based on outputs from the model which utilized resistance value. Figure 7.12 shows a good agreement between data and model (curve fit) but slight discrepancies are observed. The battery model assumes linear behavior, and therefore the curve fit would be expected to provide good results near the point of intersection of a line and the curve, as is the case.

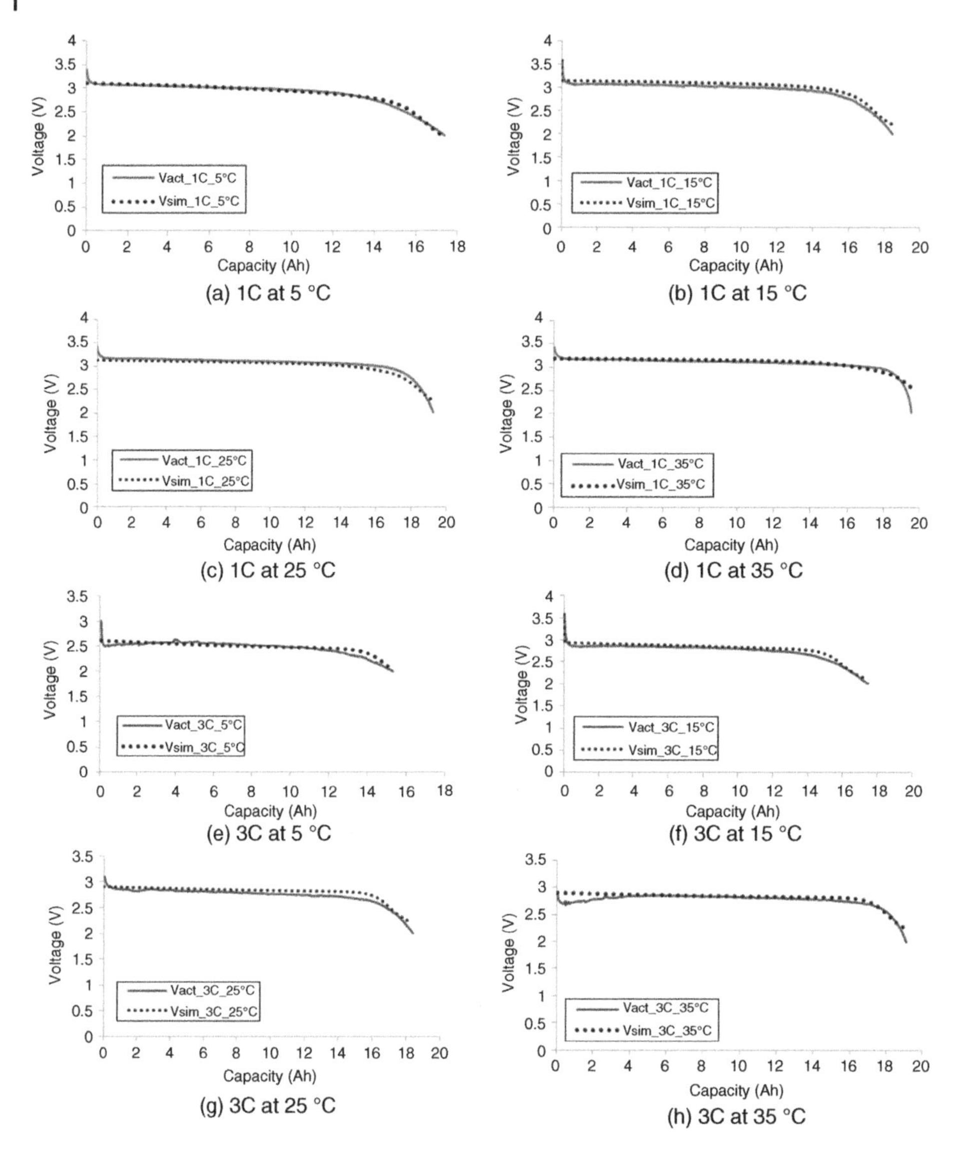

Figure 7.12 Comparison of actual and simulated discharge temperature and voltage profile at 1C and 3C at 5 °C, 15 °C, 25 °C, and 35 °C BCs (adapted from Panchal *et al.*, 2016).

7.3.5 Closing Remarks

An empirical battery thermal model was developed to accurately track the profiles observed in the experimental results. The model also showed good agreement between the measured and predicted profiles for charge voltage. In addition, the average battery surface temperature distribution was measured at 1C and 3C discharge rates with varying boundary conditions of 5°C, 15°C, 25°C, and 35°C. It was observed that the highest maximum average surface temperature (40°C) was for 3C at 35°C and the lowest maximum average surface temperature (10.1°C) was for 1C at 5°C.

7.4 Case Study 3: Thermal Management Solutions for Electric Vehicle Lithium-Ion Batteries based on Vehicle Charge and Discharge Cycles

7.4.1 Introduction

Due to environmental benefits and rising oil prices, major automobile manufacturers have spent millions of dollars on the research and development of hybrid and electric vehicles which are becoming more and more environmentally friendly by protecting the eco-system. Batteries, a major powertrain component of hybrid and electric vehicles, will undergo thousands of cycles during the life-time of a vehicle. Over this lifetime, a battery degrades potentially to the point of requiring replacement. Given the high cost of the battery and its importance in determining electric vehicle range, it is very desirable to postpone battery degradation as long as possible.

The performance, life-cycle cost, and safety of electric and hybrid electric vehicles (EVs and HEVs) depend strongly on the vehicle's energy storage system. Advanced batteries such as lithium-ion (Li-ion) batteries are quite viable options for storing energy in EVs and HEVs. Battery temperature impacts battery performance, SOH, and may even present a safety risk. Therefore, thermal management is essential for achieving the desired performance and life-cycle from a vehicle battery pack comprised of a particular battery cell or module.

This case study presents the thermal characteristics and associated thermal management options for a prismatic pouch battery composed of $LiFePO_4$ electrode material and modules. Characterization is performed via experiments that enable development of an empirical battery thermal model for vehicle simulations. As well electrical data is presented for the validation of electrochemistry based battery thermal models. In order to provide the readers a walkthrough of the necessary steps to arrange the needed experimental set up, procedures, evaluations and make the design and improvements, the case study is organized in the following structure:

An apparatus was designed to measure the surface temperature distribution, heat flux, and heat generation from a battery pouch cell undergoing various charge/discharge cycles. In this work, a prismatic lithium-ion pouch cell is cooled by two cold plates with numerous thermocouples and heat flux sensors applied to the battery at distributed locations. The total heat generation from a particular battery is obtained at various discharge rates (1C, 2C, 3C, and 4C) and different cooling bath temperature (5 °C, 15 °C, 25 °C, and 35 °C). Results show that the heat generation rate is greatly affected by the both discharge rate and boundary conditions. The developed experimental facility can be used for the measurement of heat generation from any prismatic battery, regardless of chemistry. Thermal images obtained at different discharge rates are presented within to enable visualization of the temperature distribution. An empirical battery thermal model is developed and validated with collected data from a test bench in terms of temperature, SOC and voltage profile.

7.4.2 System Description

The performance and life of Li-ion battery packs for PHEVs, HEVs, and EVs applications are greatly influenced by battery operating temperatures. In order to understand the thermal behaviour of batteries and its impact on battery performance and life, the first step experimentally is to study the battery temperature distributions and the heat generation profiles at different charge and discharge rates. And to make this study relevant to PHEVs, HEVs, and EVs, the charge and discharge rates must be typical of those seen and expected to be seen in vehicles. Operating lithium-ion batteries above 50 °C can accelerate the aging process and lead to significant degradation of battery capacity and electric range reduction (Ramadass *et al.*, 2008). Battery cell temperatures above 50 °C are very possible especially when cells are stacked into modules, and then packs, and if the ambient temperature is closer to 50 °C. The possibility of fire is also a major issue with high operating temperature where thermal runaway is a possibility. Thus, adequate battery cooling and thermal management are an integral part of the vehicle operation during electric mode operation. PHEVs, HEVs, and EVs require a robust battery thermal management system in order to ensure optimal (safe, good performance, and long battery life) vehicle operation.

Experimental data on the thermal characteristics of batteries is important not only to the battery pack designers and modellers, but also to those looking more fundamentally at electrochemical battery models. Battery modelling gives very important information on battery charging/discharging, SOC, SOH, and temperature. There are different methods for modeling batteries, for example, 1) electrochemical modeling, and 2) equivalent circuit modeling (Lo, 2013). Electrochemical modeling provides a deep understanding of the chemical and physical process inside the battery and is useful when building a cell or pouch cell, but high computational time makes this approach impractical for applications that involve multiple pouch cells, such as vehicle battery packs. On the other hand, in equivalent circuit modeling, battery losses are represented in terms of electrical circuit components, making this method more efficient in terms of computation.

Given the problems of 1) battery performance, 2) aging or degradation of batteries, and 3) fire issues, all due to high battery operating temperature as identified in the previous section, and given our limited knowledge of the thermal behaviour of vehicle batteries, it is quite important to do more research on performance of PHEV, HEV, and EV batteries undergoing realistic vehicle charge and discharge cycles. To date, significant amount of work has been conducted on battery modeling but limited published work exists experimentally with varying boundary conditions. Thus, one of the key objectives in this case study is to characterize the thermal behavior of a vehicle suitable pouch cell using dual cold plates to provide a large range of boundary conditions (Satyam, 2014).

7.4.3 Analysis

7.4.3.1 Design of Hybrid Test Stand For Thermal Management

In order to conduct the necessary analyses, a hybrid test stand was designed and built to perform component degradation studies and to enable scaled Component-In-The-Loop (CIL) testing . Due to cost and safety implications the system was designed at $1/50^{th}$ scale of the full powertrain to allow for the testing of smaller fuel cells and single cell batteries. The hybrid test stand was modified so that fuel cell components were not connected to the power bus. This enabled the test stand to behave in full EV mode, where only batteries are charged and discharged using the electrical distribution grid of the lab.

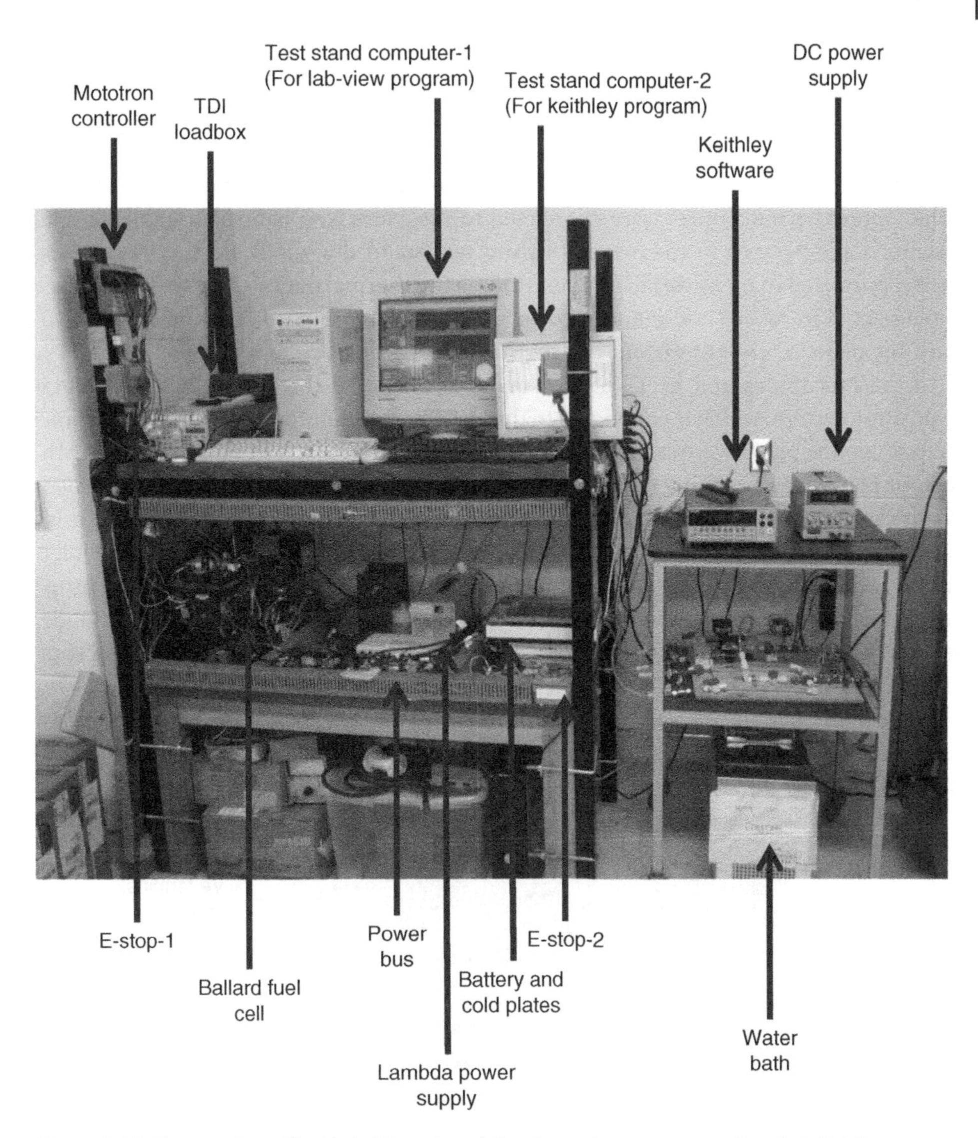

Figure 7.13 Picture of modified hybrid test bench for thermal management (Panchal, 2014).

In order to allow for significant flexibility, high and low voltage supplies and loads were integrated. The low-voltage supply and load were Lambda ZUP20-40-800 and TDI Dynaload RBL232 50-150-800 respectively. TDI Load box, Lambda power supply and all other necessary components are shown in Figure 7.13. The system was designed in order to allow for battery only, fuel cell only, and hybrid testing. From the main bus, DC/DC and fuel cell are removed during battery testing.

The voltage, current and temperature of the cell are monitored by the MotoTron controller (CTRLPCM00200). The MotoTron controller interfaces via RS232 communication to the test stand computer-1. The test stand computer-1 manages Lab-view program (measurements of battery voltage, current, charge-discharge cycle,

cycle number, and time) which record values at one second intervals and test stand computer-2 manages the Keithley-2700 (Data Acquisition) program (measurements of battery surface temperature, heat flux, and water inlet and outlet temperature for top and bottom cold plates). Similar test set up as the previous case study is gathered in order to provide compatible solutions to the associated battery temperature distributions. Three thermocouples were connected to the MotoTron controller allowing one thermocouple per cell in the original hybrid test stand design. These thermocouples were repurposed to measure lab temperature, as the thermal data collection system was implemented to record and monitor battery temperatures. Figure 7.14 and Figure 7.15 show the pictorial view of various cell thermal management system set ups.

All thermocouples and heat flux sensors are connected to the Keithley-2700 data acquisition system, which is connected to the test stand computer-2. Two different commercial cooling plates were selected from industry to remove heat from the battery. Both coolant plates were manufactured from two stamped aluminum plates that are joined in a nickel-brazing process. The first plate tested was the "zig-zag" plate as shown in Figure 7.16. This plate was characterized as having a single flow channel with the inlet placed at the top. The single flow channel ran down the length of the plate before turning back on itself, stepping one channel width across the plate with each turn. This flow pattern results in a thermal profile where coolant temperature gradient is largest across the width of the plate.

The second plate tested was the "U-turn" cold plate. This plate was characterized as having multiple flow channels with inlets and outlets (9-inlets and 9-outlets) placed on

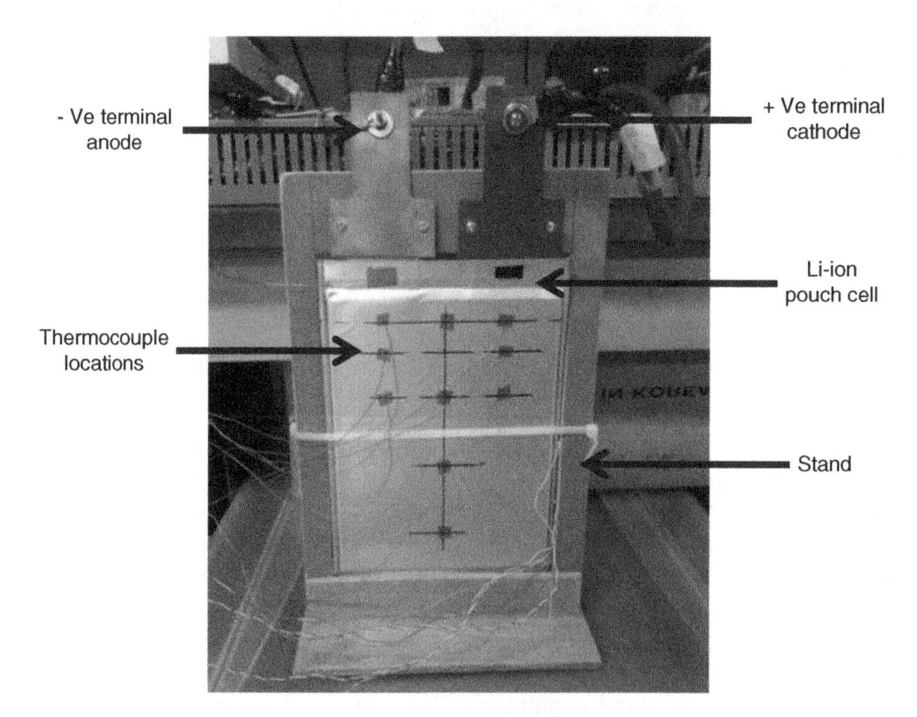

Figure 7.14 Passive air cooling set up (Panchal, 2014).

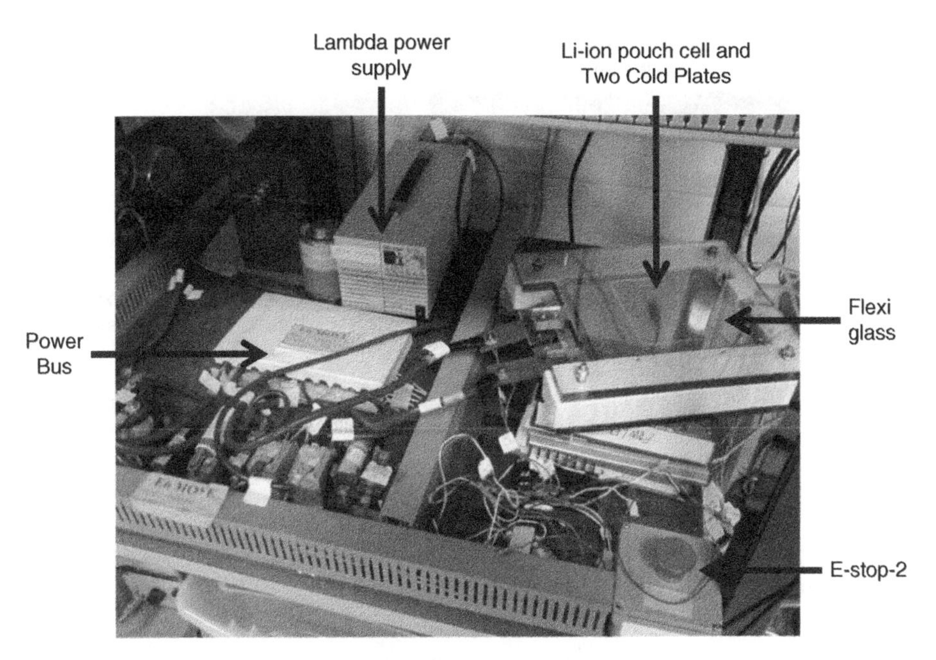

Figure 7.15 Pouch cell and two cold plates set up (Panchal, 2014).

Figure 7.16 Zig-Zag cold plate set up (active cooling) (Panchal, 2014).

the edges of the plate, near the bottom of the battery. The coolant flow paths are symmetrical down the center of the plate such that the flow channels were mirrored about the centerline. Figure 7.17 shows a picture of one of the two U-turn cold plates used in this test set-up. One cold plate is at the top of the pouch cell and the other cold plate at the bottom of the pouch cell. A Li-ion pouch cell is insulated from three sides (left, right, and bottom of the pouch cell along the height of the pouch cell) using polystyrene

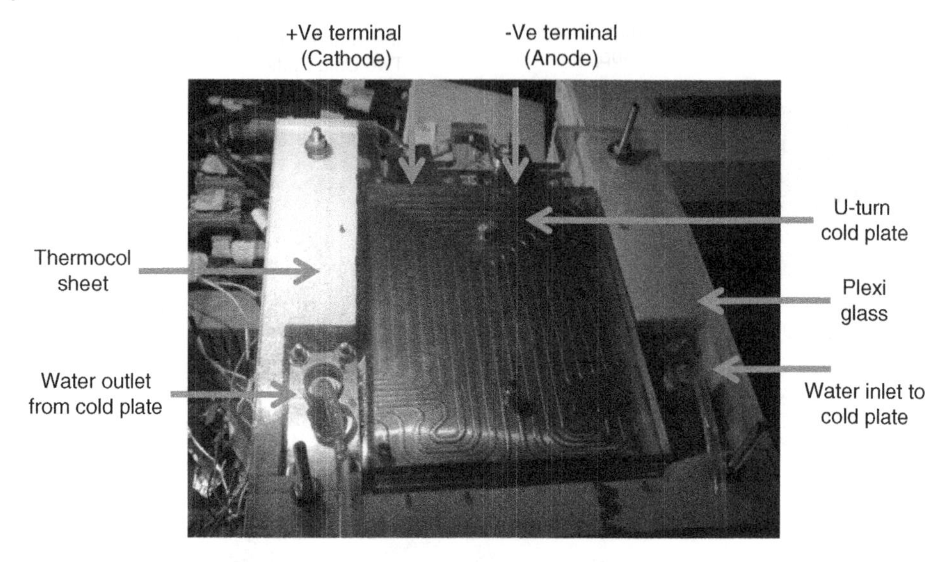

Figure 7.17 U-Turn cold plate set up (active cooling) (Panchal, 2014).

insulation to prevent heat loss from the pouch cell to the surrounding environment. The thermal systems used in the testing are provided in the next section.

7.4.3.2 Battery Cooling System

The battery cooling system consisted of a closed loop of tubing connecting two cooling plates (P1, P2) to a Fisher Scientific Isotemp 3016 fluid bath. Sensors were placed along the flow path to record properties of the fluid. A schematic of the system is shown in Figure 7.18. The cooling plates were placed within the compression rig directly against the principal surfaces of the battery such that heat generated within the battery was principally removed by conduction to the surfaces of the cooling plates.

7.4.3.3 Sensors and Flow Meter

Two types of sensors monitor the battery cooling system: thermocouples and flow meters. These sensors are required to determine the heat gained or lost by the fluid within the cooling plates. T-type insertion thermocouples are installed directly upstream and downstream of the inlet and outlet to each cooling plate. These thermocouples measure the bulk temperature of the fluid flowing within the coolant line. The Keithley 2700 records the output of these thermocouples.

A Microtherm FS1 30–300ml/min flow meter is installed directly upstream of the inlet to each cooling plate. An LCD display provides instantaneous measurement of volumetric flow. This volumetric flow value was manually recorded at the beginning and end of test cycles.

7.4.3.4 Compression Rig

For tests that include active cooling using the battery cooling system a special isolating rig was created. The rig was created to isolate the cell from the ambient thermal

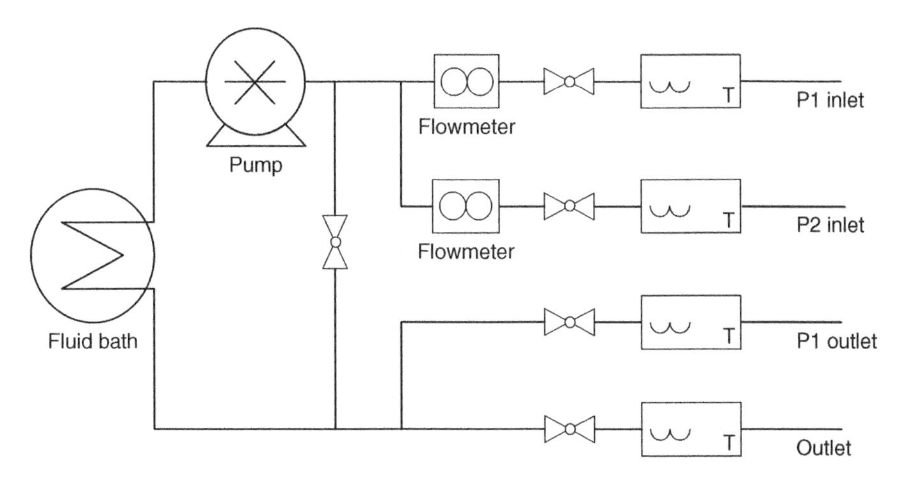

Figure 7.18 Schematic of cooling system flow from bath to upper and lower cold plates (adapted from Panchal, 2014).

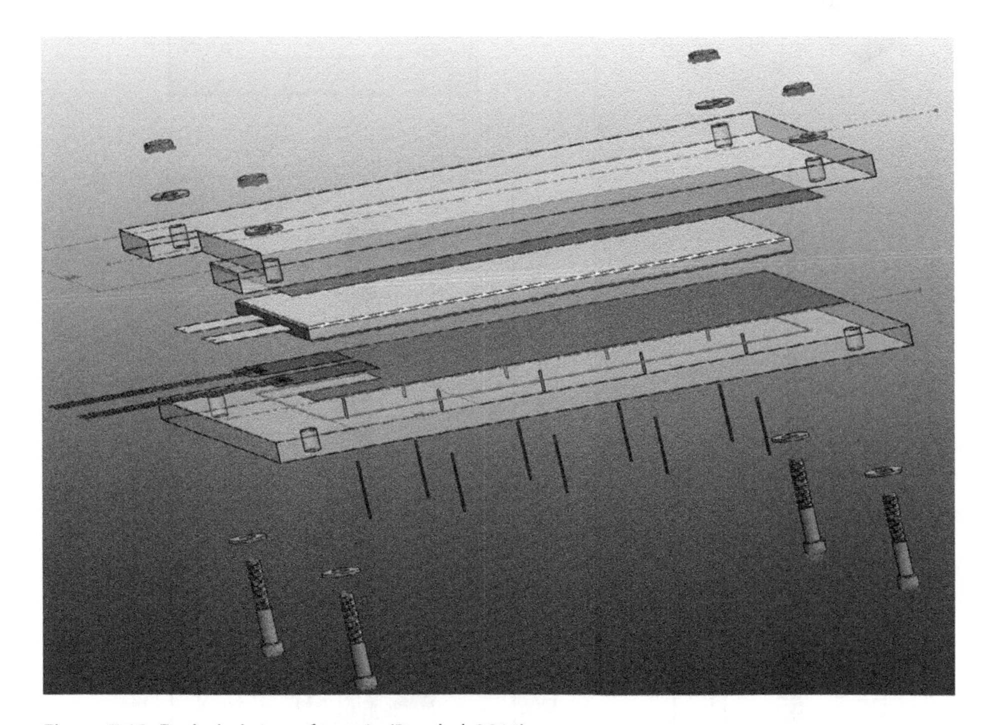

Figure 7.19 Exploded view of test rig (Panchal, 2014).

environment and maintain compression of the cooling plate/battery surface interface. An exploded assembly of the isolating rig without insulation is shown in Figure 7.19 and the CAD drawing of compression rig is shown in Figure 7.20.

The rig is composed of two transparent 12.7 mm thick polycarbonate sheets and in between the battery and cooling plates are sandwiched. A hole is present in each corner

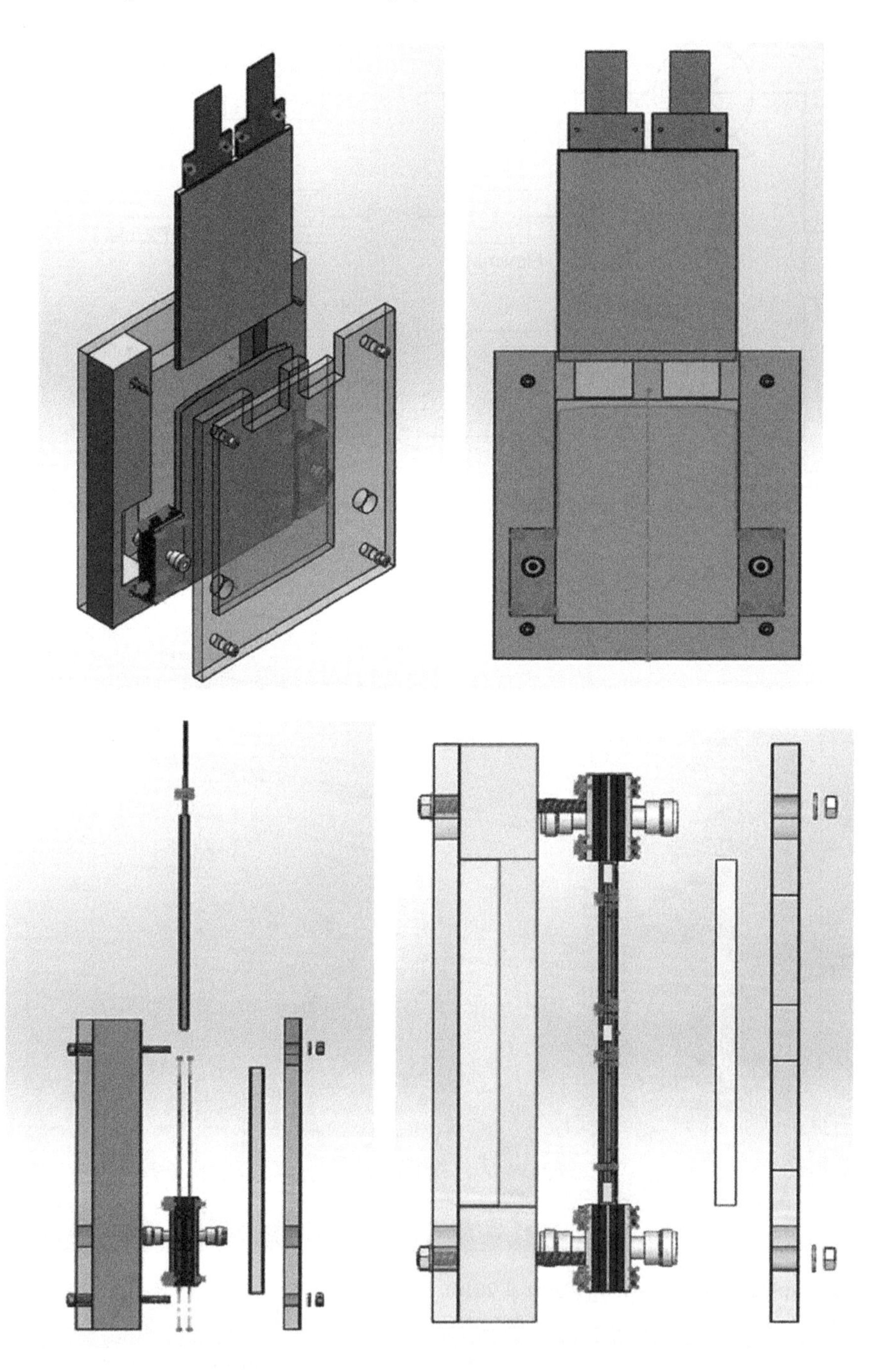

Figure 7.20 CAD drawings of compression rig (Panchal, 2014).

Figure 7.21 $LiFePO_4$- 20Ah lithium-ion prismatic pouch cell (Panchal, 2014).

to allow a threaded bolt to pass through. By tightening down the bolt fasteners, the battery is confined to the assembly by pressure. Spacers were constructed using 9.5 mm thick polycarbonate with dimensions equal to the battery dimensions. These spacers were required to change the distance between the cooling plate and the outer polycarbonate plate so that the inlet, outlet and tubing would fit.

7.4.3.5 Battery

In this work, 20Ah lithium-ion prismatic pouch cell, shown in Figure 7.21 is tested. The pouch cell specifications are given in Table 7.13. The two holes were made on tabs to connect to the power source and voltage sensing device.

7.4.3.6 Thermal Management System – Experimental Plan and Procedure

In this experiment, three different cooling types are tested: passive convection, active zig-zag cold plates, and active U-turn cold plates. For both cooling plates, four different coolant temperatures or boundary conditions are selected: 5 °C, 15 °C, 25 °C, and 35 °C. Four different discharge rates are selected: 1C, 2C, 3C, and 4C. The charge current is 1C. The experimental plan is shown in Table 7.14.

This procedure was followed to initiate battery cycling and the thermal data collection, and it does not directly describe the procedure for assembling the battery and cooling/instrumentation components within the compression rig. As such, this procedure assumes the cell and cooling components are correctly installed and fully connected to all other components as required.

1) The isothermal fluid bath and pump was turned on minimum 2 hours prior to beginning cycling to bring the battery, bath and compression rig to a steady state temperature. The valves leading to the cold plates were observed and set to open. The isothermal fluid bath was set to the desired cooling temperature or boundary conditions of 5 °C, 15 °C, 25 °C, and 35 °C for the test.

Table 7.13 $LiFePO_4$- 20Ah Lithium-ion prismatic pouch cell specifications.

Specification	Values
Cathode Material	$LiFePO_4$
Anode Material	Graphite
Electrolyte	Carbonate based
Nominal Capacity	20.0 Ah
Nominal Voltage	3.3 V
Nominal Energy	65 Wh
Energy Density	247 Wh/L
Mass	496 g
Discharge Power	1200
Dimensions	7.25 mm × 160 mm × 227 mm
Specific Power	2400 W/kg
Specific Energy	131 Wh/kg
Operating Temperature	−30°C to 55°C
Storage Temperature	−40°C to 60°C
Volume	0.263 L
Number of Cycles	Minimum 300, typical 2000
Max Discharge (SOC & Temperature dependent)	300 A
Max Charge (SOC & Temperature dependent)	300 A
Internal resistance	0.5 mΩ

Source: Panchal, 2014.

Table 7.14 Experimental plan.

Cooling Type	Ambient/Coolant/Bath Temperature [°C]	Discharge Rate
Passive (Ambient air only)	~ 22	1C, 2C, 3C, 4C
Active (Zig-zag turn cold plate)	5	1C, 2C, 3C, 4C
	15	1C, 2C, 3C, 4C
	25	1C, 2C, 3C, 4C
	35	1C, 2C, 3C, 4C
Active (U-turn cold plate)	5	1C, 2C, 3C, 4C
	15	1C, 2C, 3C, 4C
	25	1C, 2C, 3C, 4C
	35	1C, 2C, 3C, 4C

Source: Panchal, 2014.

2) The lab view code for the charge discharge stand was loaded and relevant test parameters were input to the program. Relevant test parameters include:
   a) Charge current
   b) Discharge current

   c) Number of cycles
   d) Drive cycle (charge and discharge cycle)
   e) Maximum voltage at end of charge
   f) Minimum voltage at end of discharge
   g) Measurement sample rate
3) The thermal data acquisition PC and Keithly 2700 were turned on and allowed to initialize. On the PC, the ExcelLink recording software was prepared for data acquisition. The following parameters were recorded within the recording software:
   a) Battery surface temperature (top and bottom surface of the pouch cell)
   b) Heat flux at top surface of the battery near electrodes
   c) Water inlet and outlet temperature at top and bottom cold plate
4) The internal clocks on both PCs were synchronized to the same time to allow combining data files.
5) The charge-discharge test stand and thermal data acquisition were then activated at the same time, such that charging/discharging and data acquisition begin at the same instant.
6) The test continued until the desired number of battery cycles was completed.
7) Two files were created for each test.
   a) Data from thermal data acquisition PC
   b) Electrical charge discharge data

7.4.3.7 Data Analysis Method

The data produced in the heat generation and cooling requirements experiment consists of electrical data (battery voltage, and current flow) as well as thermal data: surface temperature data (13 channels), heat flux meter output voltages (3 channels), inlet/outlet cooling system temperatures (4 channels). MATLAB software was used in order to manipulate the large spread sheets of data. The following energy balance equation was used to determine the actual heat output of the battery.

$$Q_{Total} = Q_{stored} + Q_{removed} + Q_{environment} \tag{7.15}$$

7.4.3.7.1 a) Sensible Heat The heat stored in the battery is termed sensible heat. It is evaluated based on the change in temperature of the battery in conjunction with the specific heat value. Equation 7.16 is used to evaluate the sensible heat energy stored in the battery when the battery temperature changes from some initial temperature to a final temperature.

$$\underset{t_1\ to\ t_2}{Q_{sensible}} = m_{battery} c_{p,battery} (T_{t_2} - T_{t_1}) \tag{7.16}$$

A standard method of determining the average temperature across the entire battery surface has been devised to enable sensible heat calculations. For each thermocouple, it is assumed that the measured temperature represents the average of an area extending around the sensor. The areas are determined by defining each area boundary by calculating the x and y midpoint distance between adjacent sensors. Equation 7.17 is used to evaluate the average battery surface temperature by summing the temperature-area products and dividing by the total area of the surface.

$$\underset{average}{T_{surface}} = \frac{\Sigma(T_{ij}A_{ij})}{A_{total}} \tag{7.17}$$

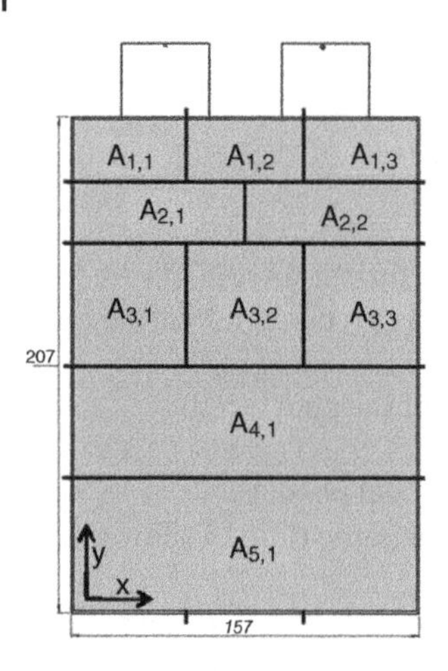

Figure 7.22 Distribution of areas used to determine average surface temperature (Panchal, 2014).

Table 7.15 X and Y component lengths of thermocouple areas.

Thermocouple	Area	X [mm]	Y [mm]	Area [m^2] (x10^{-3})
1,1	$A_{1,1}$	52.5	25.5	1.34
1,2	$A_{1,2}$	52	25.5	1.33
1,3	$A_{1,3}$	52.5	25.5	1.34
2,1	$A_{2,1}$	78.5	32	2.51
2,2	$A_{2,2}$	78.5	32	2.51
3,1	$A_{3,1}$	52.5	49.5	2.60
3,2	$A_{3,2}$	52	49.5	2.57
3,3	$A_{3,3}$	52.5	49.5	2.60
4,1	$A_{4,1}$	157	47.5	7.46
5,1	$A_{5,1}$	157	52.5	8.24

Source: Panchal, 2014.

The ten thermocouples measuring surface temperatures are each assigned the areas that correspond to their locations as shown in Figure 7.22. The physical size of the thermocouple areas is presented in Table 7.15. The rate of sensible heat accumulation is directly influenced by the battery heat generation rate and the heat transfer coefficient out of the system. The temperature of the battery increases as heat is generated due to the finite heat transfer coefficient to the surrounding. The rate of sensible heat accumulation is determined from the rate of change of Equation 7.18, where $\frac{dT}{dt}$ is the rate that the battery temperature changes.

$$\dot{Q}_{sensible} = m_{battery} c_{p,battery} \frac{dT}{dt} \tag{7.18}$$

The rate of temperature change is evaluated by measuring temperature at two times and calculating the rate using Equation 7.19;

$$\frac{dT}{dt} = \frac{(T_{t_2} - T_{t_1})}{t_2 - t_1} \tag{7.19}$$

The rate of sensible heat accumulation can then be determined via Equation 7.20.

$$\dot{Q}_{\substack{sensible \\ t_1\ to\ t_2}} = m_{battery}\, c_{p,battery} \frac{(T_{t_2} - T_{t_1})}{t_2 - t_1} \tag{7.20}$$

7.4.3.7.2 b) Heat Removed From Battery A fraction of the heat generated by the battery is removed by cooling processes. In one series of testing, the operating battery was placed in a vertical position and ambient air carried heat away by natural convection. In other tests, cooling plates were present that utilized forced coolant flow to remove heat.

1. Air Cooling For a fluid at known temperature contacting a surface of known area, heat is transferred via convection and observes Newton's law of cooling.

$$Q_{convection} = h_c A\,(T_{surface} - T_{\infty}) \tag{7.21}$$

If the heat transfer coefficient, h and surface temperature is known, the corresponding heat transfer can be calculated. Determining h represents the fundamental challenge of thermal convection. The coefficient is not itself a thermo-physical property of the fluid, but is dependent on numerous variables such that:

$$h_c = f\,(\rho,\, C_p,\, \mu,\, \beta, g,\, k,\, [T - T_f],\, L) \tag{7.22}$$

Nusselt numbers are proportional to h_c via the following relation:

$$Nu_L = \frac{h_c L_{surface}}{k_{fluid}} \tag{7.23}$$

Correlations for Nusselt numbers have been developed in experimental tests and are readily available in mainstream heat transfer texts. These correlations correspond to specific geometries and care must be taken when selecting the appropriate correlation to use. For buoyancy driven laminar fluid flow across a vertical plate, the following correlation is found (Incropera and Dewitt, 2007):

$$Nu_L = 0.68 + \frac{0.670(Ra_L)^{1/4}}{\left[1 + \left(\frac{0.492}{Pr}\right)^{9/16}\right]^{4/9}} \;\ldots\ldots\; Ra_L \leq 10^9 \tag{7.24}$$

The Prandtl number, Pr is tabulated for different fluid temperature values in heat transfer texts and is evaluated at the film temperature, which is the average temperature between the fluid and the surface to be cooled. The Rayleigh number is calculated by the standard relation:

$$Ra_L = \frac{g\beta}{\nu\alpha}(T_{surface} - T_{\infty})L^3 \tag{7.25}$$

2. Water Cooling (Cold Plates) Cooling plate heat removal rate is determined by the inlet and outlet thermocouple data, in conjunction with the recorded flow rates. The difference in inlet and outlet temperatures is due to heat conducted from the battery surface. The heat removed by a single cooling plate is calculated using Equation 7.26.

$$\dot{Q}_{Cooling\ Plate} = \dot{m}_w C_{p,w}(T_{w,o} - T_{w,i}) \tag{7.26}$$

The total amount of heat removed by the cooling plates for a time period Δt can be determined using Equation 7.27

$$Q_{Cooling\,Plate} = \dot{m}_w C_{p,w} (T_{w,out,average} - T_{w,in,\,average}) \Delta t \tag{7.27}$$

The term $T_{w,out,average}$ is the average measured outlet temperature during the period Δt, as in Equation 7.28 N_T represents the number of temperature readings in the summation.

$$T_{w,out,average} = \frac{\sum T_{w,o}}{N_T} \tag{7.28}$$

7.4.3.7.3 c) Heat from Environment The compression rig is not perfectly insulated and as such, cooling plates measurements incorporate a component of heat gain or loss from the environment. When the cooling is set to 5°C, a temperature difference of approximately 17°C is established between the inside surface of the compression rig and the ambient air. This results in heat transfer between the ambient environment and the cooling fluid. This additional heat is measured as an increased temperature difference between the inlets and outlets. For tests above ambient temperature, the opposite occurs. This additional heat affects the temperature difference between the inlets and outlets.

In order to evaluate this effect, the cooling system and thermal data acquisition was activated with the battery in place but no charging or discharging occurring. In this way, the temperature difference between the inlet and outlet of each cooling plate could be recorded. The average difference for each plate along with the respective flow rates used were used to quantify the heat removed or added by the environment using the method presented in above section of cold plates. The heat removed or added by the environment for different coolant temperature is shown in Table 7.16. The data is plotted in Figure 7.23.

7.4.4 Results and Discussion

In this section, the temperature distribution on the principle surface of battery, maximum surface temperature of battery, average and peak heat flux, the total heat generation and heat generation rate, effect of discharge rate and operating temperature on discharge capacity, and thermal images are presented and discussed in detailed with different discharge rates and various boundary conditions.

Table 7.16 Ambient heat flow to compression rig for different coolant temperature.

Bath Temperature [°C]	$\dot{Q}_{environment}$ [W]
5	−21.61
15	−11.24
25	4.12
35	18.10

Source: Panchal, 2014.

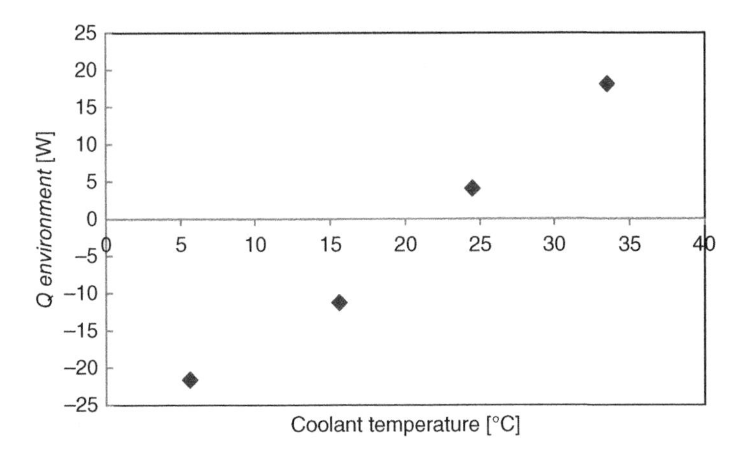

Figure 7.23 Ambient heat flow to compression rig for four coolant temperatures (adapted from Panchal, 2014).

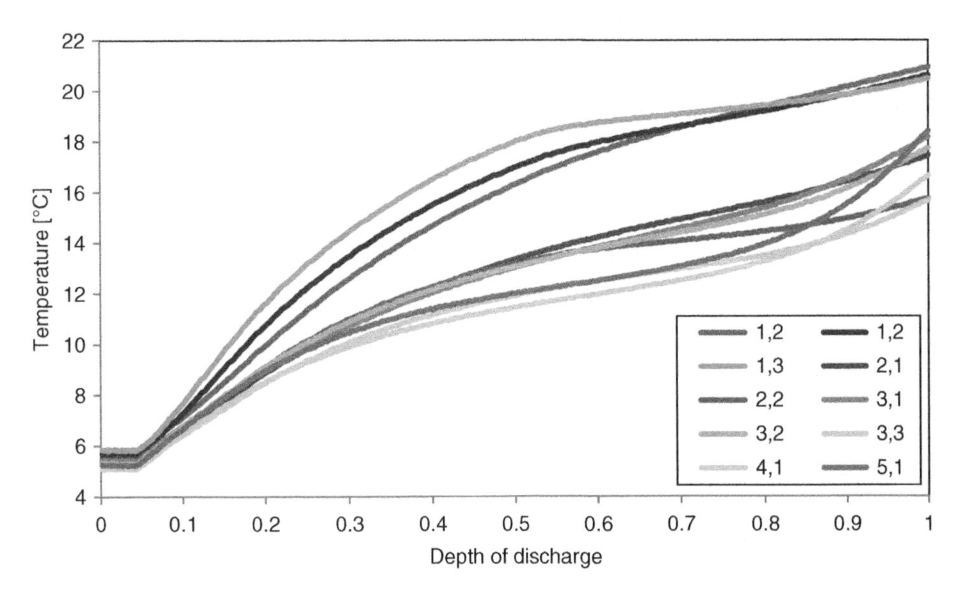

Figure 7.24 Surface temperature profile at 4C discharge and 5°C bath temperature (adapted from Panchal, 2014).

7.4.4.1 Battery Surface Temperature Profile

The battery surface temperature profile recorded by ten thermocouples at 4C discharge and 5 °C and 35 °C bath temperature are plotted as a function of depth of discharge (DOD) in Figure 7.24 and Figure 7.25. It can be observed that the response of the thermocouple at location (1, 1), (1, 2), and (1, 3) has the faster rate of increase over entire period of 0.0 and 1.0 depth of discharge (DOD). These thermocouples are nearest the negative and the positive electrodes of the battery and indicate the location of highest heat accumulation, and likely the rates of heat generation are highest near the electrodes.

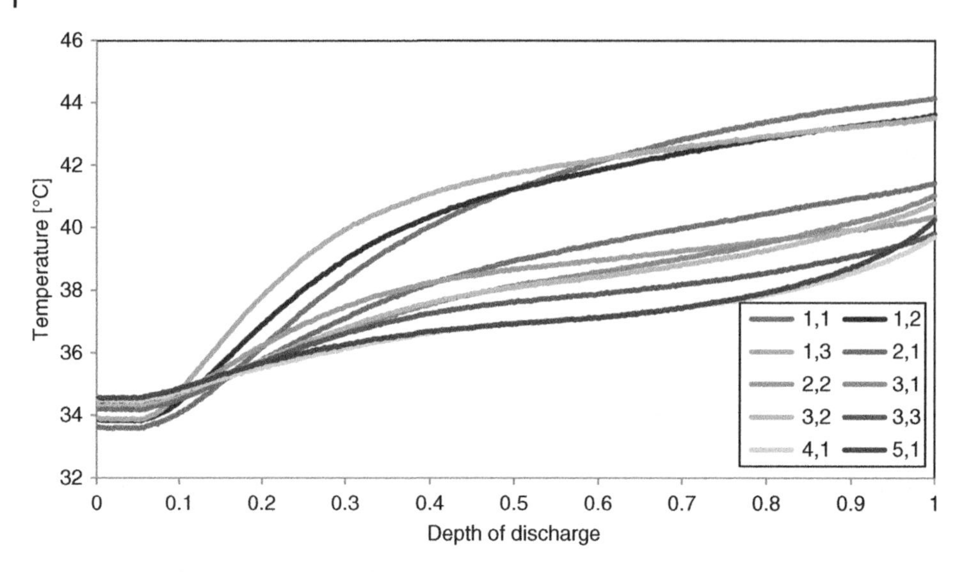

Figure 7.25 Surface temperature profile at 4C discharge and 35 °C bath temperature (adapted from Panchal, 2014).

By comparing the images, it can be seen that the higher temperature distribution is noted for 4C–35 °C as compared to 4C–5 °C.

7.4.4.2 Average Surface Temperature of Battery

In Figure 7.26 the maximum average surface temperatures of the battery for the four discharge rates and operating/cooling/bath temperatures has been plotted. The

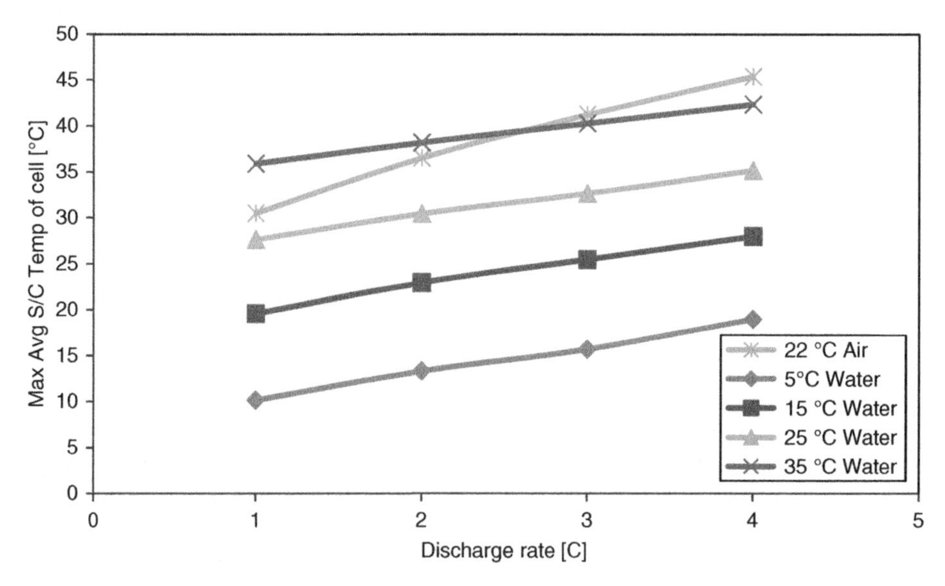

Figure 7.26 Maximum average surface temperature of battery for discharge rates and coolant/operating temperatures tested (adapted from Panchal, 2014).

Table 7.17 maximum average battery surface temperature for different discharge rates and bath temperatures.

Cooling Type	Ambient/Coolant/Bath Temperature [°C]	Maximum Temperature [°C]			
		1C	2C	3C	4C
Active (U-turn cold plates)	5 °C	10.1	13.3	15.7	18.9
	15 °C	19.6	22.9	25.4	28.0
	25 °C	27.6	30.4	32.6	35.2
	35 °C	35.9	38.2	40.3	42.3
Passive (Ambient air only)	~22 °C	30.5	36.5	41.2	45.4

Source: Panchal, 2014.

trend shows that regardless of operating temperature the maximum average surface temperature of the battery increases as the discharge rate is increased. In Table 7.17, the maximum average surface temperatures are summarized for the test conditions used.

Since the increase in coolant/operating/bath temperature can account for much of the differences in Figure 7.26, it is useful to consider the difference between the start of discharge average surface temperature and the end of discharge average surface temperature. This provides a measure of the buildup of heat within the battery. The differences are plotted in Figure 7.27. Table 7.18 shows the linear fit for four different bath temperatures.

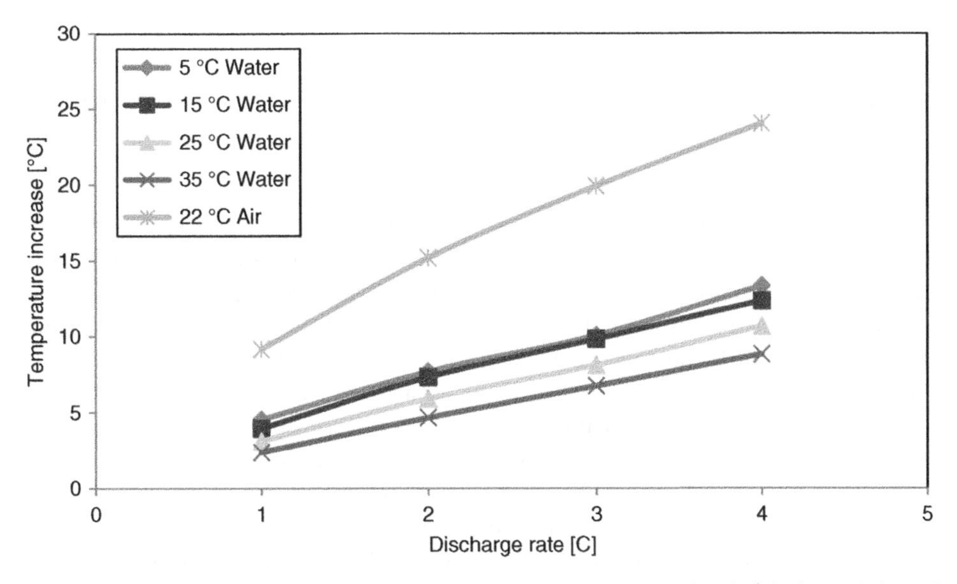

Figure 7.27 Difference in average surface temperature between start and end of discharge (adapted from Panchal, 2014).

Table 7.18 Linear fit for four different bath temperatures.

Bath Temperature	Linear Fit
5 °C	$2.879x + 1.711$
15 °C	$2.769x + 1.510$
25 °C	$2.483x + 0.764$
35 °C	$2.142x + 0.369$

Source: Panchal, 2014.

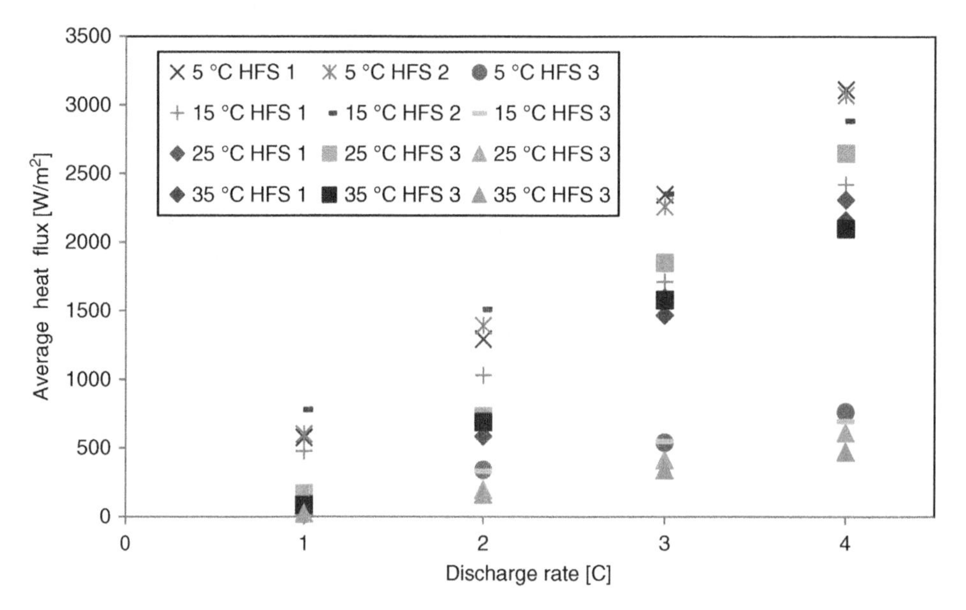

Figure 7.28 Average heat flux at 1C, 2C, 3C, 4C discharge rates and different boundary conditions (adapted from Panchal, 2014).

7.4.4.3 Average Heat Flux

In Figure 7.28 the average heat fluxes measured by the three heat flux sensors for all operating temperatures are plotted against discharge rate. To refresh the reader, HFS 1 is located near the positive electrode (cathode), HFS 2 is located near the negative electrode (anode), and HFS 3 is located at the middle of the cell along the height of the cell. It is observed that the highest average heat fluxes were measured at HFS 1 and HFS 2 for 4C discharge rates and 5 °C cooling. In general, for all tests the sensors nearest the electrodes (HFS 1 and 2) measured heat fluxes higher than the sensor located at the middle of the battery surface. The trend observed is that increased discharge rates (between 1C, 2C, 3C, and 4C) and decreased operating temperature (between 35 °C, 25 °C, 15 °C, and 5 °C) results in increased average heat fluxes at the three locations measured.

In Table 7.19, the values of Figure 7.28 are tabulated along with the average heat fluxes for the passive cooling natural convection cases. For the passive cases, the average heat flux of HFS 2, near the negative electrode is always highest. Liquid cooling cases do not

Table 7.19 Summary of average heat flux at 1C, 2C, 3C, 4C discharge rates and different boundary conditions.

Cooling Type	Ambient/Coolant/Bath Temperature [°C]	Average Heat Flux [W/m²]				
		Position	1C	2C	3C	4C
Passive (Ambient air only)	~22	1	42.3	130.3	238.4	343.6
		2	46.3	145.7	239.6	373.4
		3	24.6	71.1	108.6	145.3
Active (U-turn cold plates)	5	1	575.5	1294.5	2347.7	3112.2
		2	599.3	1390.8	2259.5	3072.8
		3	149.4	341.3	539.3	764.1
	15	1	475.7	1029.7	1711.8	2419.0
		2	781.4	1509.9	2351.6	2887.1
		3	157.9	331.7	548.4	697.3
	25	1	148.2	684.9	1597.3	2309.3
		2	168.9	733.2	1851.6	2648.2
		3	74.4	194.8	413.0	611.1
	35	1	47.6	585.6	1468.4	2160.2
		2	86.1	689.7	1579.9	2101.5
		3	25.2	163.3	340.6	471.8

Source: Panchal, 2014.

show a definitive pattern between HFS 1 and 2. This is likely due to the slightly uneven cooling gradient across the U-turn cold plate. The coolant temperature and thus plate temperature increases across the width of the battery surface as heat is absorbed. This is in contrast to the passive cooling case, where the vertical orientation of the battery provided a condition where cooling potential is approximately equal across the width of the surface. It could be inferred that the passive cooling cases are a better representation of the differences in heat generation between the three locations.

7.4.4.4 Peak Heat Flux

In Figure 7.29, the peak heat fluxes measured by the three heat flux sensors for all operating temperatures (5 °C, 15 °C, 25 °C, and 35 °C) are plotted against discharge rates of 1C, 2C, 3C, and 4C. It is observed that the highest peak heat fluxes were always measured at HFS 2. In general, for all tests the sensors nearest the electrodes (HFS 1 and 2) measured greater peak heat fluxes than the sensor located at the middle of the battery surface. The trend observed is that increased discharge rates and decreased operating temperature results in increased peak heat fluxes at the three locations measured.

In Table 7.20, the values of Figure 7.29 are tabulated along with the average heat fluxes for the passive cooling natural convection cases.

7.4.4.5 Heat Generation Rate

Figure 7.30 shows the measured heat generation rates of the battery for 2C discharge rates at 5°C , 15°C , 25°C ,and 35°C coolant/operating/bath temperature as a function

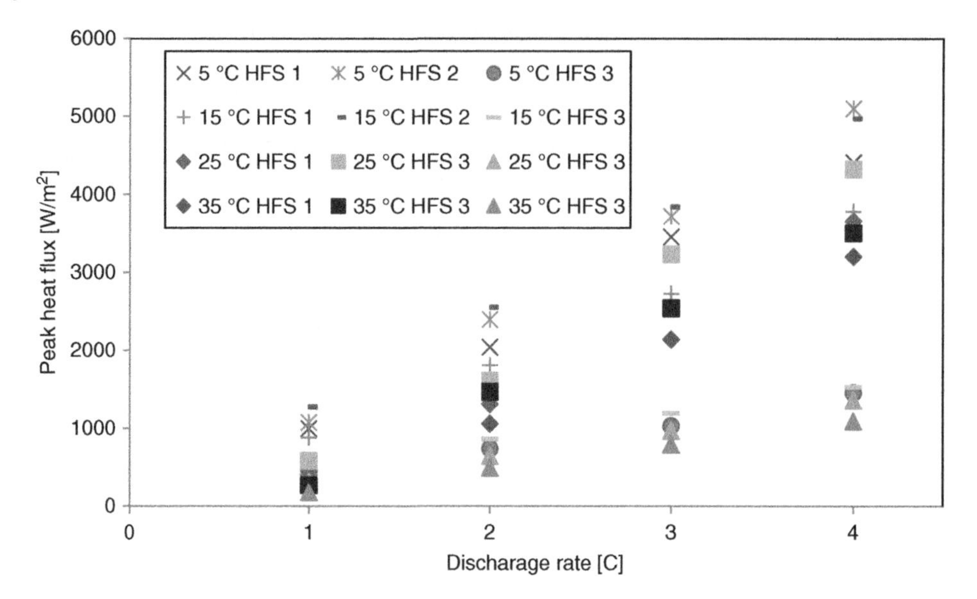

Figure 7.29 Peak heat flux at 1C, 2C, 3C, 4C discharge rates and different boundary conditions (adapted from Panchal, 2014).

Table 7.20 Summary of peak heat flux at 1C, 2C, 3C, 4C discharge rates and different boundary conditions.

Cooling Type	Ambient/Coolant/Bath Temperature [°C]	Peak Heat Flux [W/m^2]				
		Position	1C	2C	3C	4C
Passive (Ambient air only)	~22	1	134.4	315.7	446.6	630.0
		2	140.5	292.8	485.6	763.2
		3	105.9	195.9	246.6	340.8
Active (U-turn cold plate)	5	1	998.9	2043.4	3454.6	4410.9
		2	1076.7	2395.1	3718.5	5099.7
		3	421.3	744.4	1033.2	1461.3
	15	1	882.5	1808.5	2726.3	3783.9
		2	1275.8	2554.2	3842.7	4972.2
		3	443.2	864.4	1196.1	1532.3
	25	1	453.1	1311.7	2543.9	3660.9
		2	589.1	1613.7	3235.6	4327.5
		3	321.3	648.3	972.3	1363.8
	35	1	204.8	1059.9	2143.7	3202.4
		2	272.6	1469.4	2545.8	3508.4
		3	180.7	495.9	793.5	1094.5

Source: Panchal, 2014.

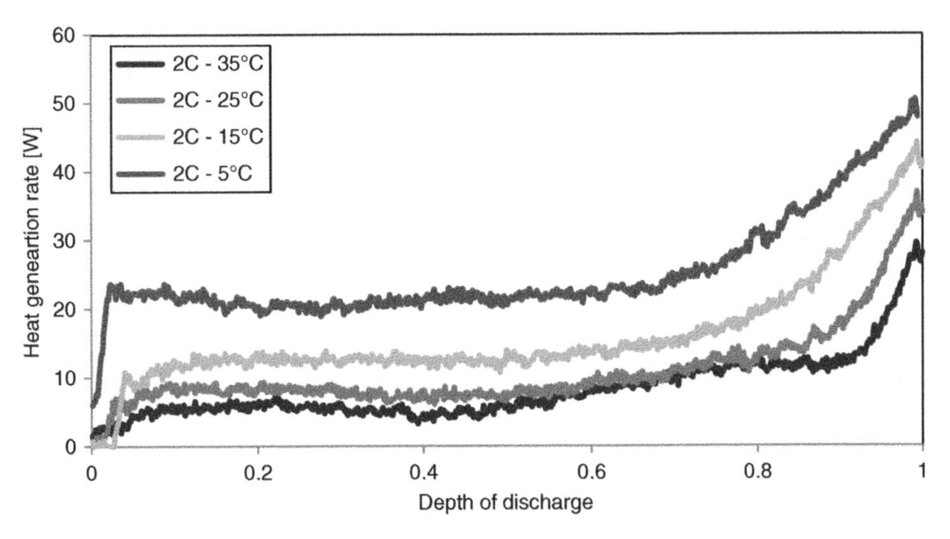

Figure 7.30 Heat generation rate at 2C discharge and different bath temperatures (adapted from Panchal, 2014).

of depth of discharge (DOD) varying from 0 to 1. A steep rise is seen in the first 0.04 of discharge, the heat generation rate becomes approximately constant until 0.5 DOD when a steady increase is observed. The increase appears to become steeper as the discharge progresses and is highest just before the end of discharge (>0.95 DOD). It was also noted that the highest heat generation rate was 90.7W measured at 4C discharge rate and 5°C coolant/bath temperature and minimum value (12.5W) was obtained at 1C and 35°C. Furthermore, it can be seen that the increase in discharge rate and thus discharge current causes consistent increase in the heat generation rate for equal depth of discharge points.

The increased heat generation can be accounted for by looking at Equation 7.21 and 7.22 . As the current is increased with the discharge rate, the irreversible Ohmic heating term becomes larger. From Equation 7.23, the current collector heat generation increases with the square of current. From this it follows that more heat will be generated at higher discharge rates. In Table 7.21, the maximum heat generation rates that the battery discharges with liquid cooling plates produced are tabulated.

Table 7.21 Summary of maximum heat generation rate at 1C, 2C, 3C, 4C discharge rates and different boundary conditions.

Cooling Type	Operating/Coolant Temperature [°C]	Maximum Heat Generation Rate [W]			
		1C	2C	3C	4C
Active (U-turn cold plate)	5	24.0	50.4	68.3	90.7
	15	22.5	44.2	73.8	88.7
	25	19.4	36.9	56.2	82.3
	35	12.5	29.5	38.7	48.5

Source: Panchal, 2014.

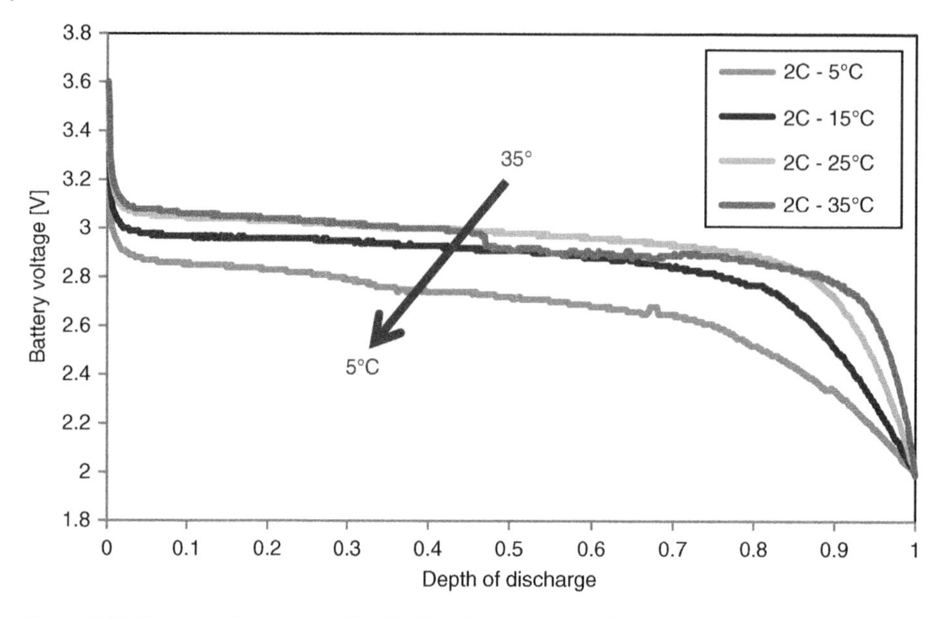

Figure 7.31 Battery voltage versus depth of discharge (adapted from Panchal, 2014).

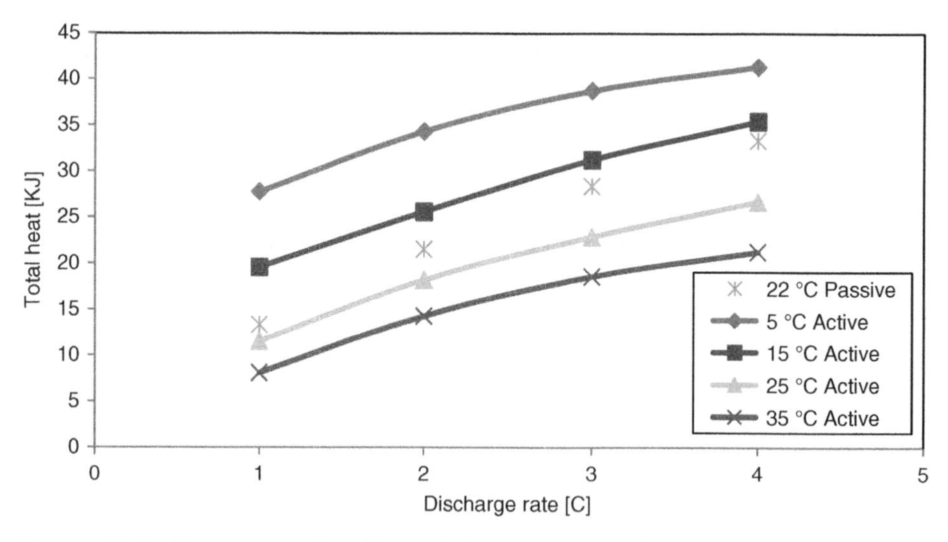

Figure 7.32 Total heat generated at different discharge rates and different boundary conditions (adapted from Panchal, 2014).

The battery discharge voltage profile corresponding to 2C discharge and four operating conditions shown in Figure 7.30 is described in Figure 7.31. It can be seen that when operating temperature (bath temperature) decreases from 35°C to 5°C, the discharge voltage profile also decreases.

Table 7.22 Summary of total heat generated at 1C, 2C, 3C, 4C discharge rates and different boundary conditions.

Cooling Type	Ambient/Coolant Temperature [°C]	Total Heat Generated [KJ]			
		1C	2C	3C	4C
Passive (Ambient air only)	~22	13.34	21.52	28.41	33.35
Active (U-turn cold plate)	5	27.76	34.31	38.74	41.34
	15	19.59	25.59	31.27	35.44
	25	11.54	18.20	22.88	26.70
	35	8.09	14.28	18.58	21.25

Source: Panchal, 2014.

7.4.4.6 Total Heat Generated

Figure 7.32 shows the total heat generated in KJ in the battery during four discharge rates (1C, 2C, 3C, and 4C) are determined using the energy balance equations. In general, the trend shows that increased discharge rate and decreased temperatures result in increased amount of heat produced during a discharge. In Table 7.22, the total heat generation values for the above mentioned discharge rates and operating temperatures is provided. The highest value of heat produced was found to be 41.34 kJ when the operating/coolant temperature was at 5 °C and the discharge rate 4C. This is almost twice the heat generated by the equivalent discharge at 35°C (an increase of approx. 30 °C).

7.4.4.7 Effect of Discharge Rate and Operating Temperature on Discharge Capacity

In Figure 7.33, the discharge capacity (Ah) of the battery is plotted against four discharge rates of 1C, 2C, 3C, and 4C for all five cooling conditions (Active cooling : 5°C, 15°C,

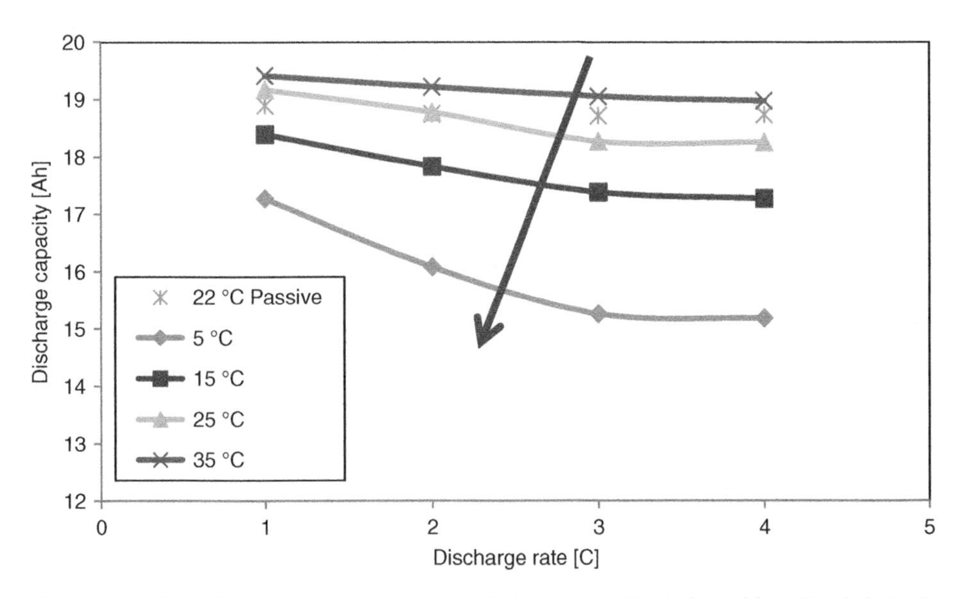

Figure 7.33 Effect of operating temperatures on discharge capacities (adapted from Panchal, 2014).

25°C, 35°C and passive cooling). This illustrates the effect of discharge rate and operating temperature on the electrical performance of the batteries. This is of particular concern for EV development as battery capacity directly impacts vehicle range. Due to the aggressive charging and discharging profiles used, the battery never discharged the nominal 20 Ah as specified by the manufacturer. Since the charging conditions were held constant between tests, the battery performance in each test is still comparable. In Table 7.23, the values displayed in Figure 7.33 are tabulated. In general, battery discharge capacity decreases as coolant/operating temperature decreases from 35 °C to 5 °C and battery discharge rate increases from 1C to 4C.

Another measure of discharge capacity is discharge time. The constant current discharges are within 0.1% of the nominal current draw at all times and thus increased discharge times indicate more energy discharged from the battery. The discharge times are summarized in Table 7.24.

7.4.5 Closing Remarks

In order to fulfill the objectives and experimental milestones aforementioned in the case study, an apparatus called "the thermal boundary condition test apparatus using dual cold plates for Li-ion pouch cell with indirect liquid cooling" was designed and built. The thermal behaviour in terms of the surface temperature distribution, heat flux, and

Table 7.23 Summary of discharge capacities at 1C, 2C, 3C, 4C discharge rates and different boundary conditions.

Cooling Type	Ambient/Coolant Temperature [°C]	Discharge Capacity [Ah]			
		1C	2C	3C	4C
Passive (Ambient Air Only)	~22	18.90	18.75	18.72	18.74
Active (U-turn cold Plates)	5	17.27	16.08	15.26	15.18
	15	18.40	17.83	17.38	17.27
	25	19.17	18.78	18.27	18.26
	35	19.41	19.22	19.05	18.98

Source: Panchal, 2014.

Table 7.24 Summary of discharge times at 1C, 2C, 3C, 4C discharge rates and different boundary conditions.

Cooling Type	Operating/Coolant Temperature [°C]	Discharge Time [s]			
		1C	2C	3C	4C
Passive (Ambient Air Only)	~22	3406	1689	1123	843
Active (U-turn cold Plates)	5	3100	1445	914	706
	15	3318	1604	1043	778
	25	3448	1689	1096	821
	35	3488	1729	1142	854

Source: Panchal, 2014.

heat generation from Li-ion pouch cell for PHEV, HEV and EV applications was studied in a lab and the impact of various passive and active thermal management systems on the battery cells are provided.

7.5 Case Study 4: Heat Transfer and Thermal Management of Electric Vehicle Batteries with Phase Change Materials

7.5.1 Introduction

A battery in an electric vehicle (EV) uses chemical energy stored in rechargeable battery packs. The vehicle's performance depends strongly on a number of factors including the battery pack performance. In general, battery temperature increases and non-uniformity are major concerns of the battery design. The battery temperature impacts vehicle performance, reliability and life cycle cost. Therefore, an effective thermal management system for battery packs is crucial if electric vehicles are to operate effectively over a range of weather conditions (Pesaran *et al.*, 1999) Thermal management using phase change material (PCM) is a promising passive technique to eliminate the need for additional cooling systems, and it improves the power use. Battery packs can be maintained at an optimum temperature with proper thermal management, by effectively integrating PCM into the battery design (Al-Hallaj and Selman, 2000). Hence the overall system volume could be reduced.

This case study examines a passive thermal management system for electric vehicle batteries, consisting of encapsulated phase change material (PCM) which melts during a process to absorb the heat generated by a battery. The study focuses on the issue of temperature rise in electric vehicle batteries in hot climates. A new configuration for the thermal management system, using double series PCM shells, is analyzed with finite volume simulations. A combination of computational fluid dynamics (CFD) and second law analysis is used to evaluate and compare the new system against the single PCM shells. Using a finite volume method, heat transfer in the battery pack is examined and the results are used to analyses the exergy losses. The simulations provide design guidelines for the thermal management system to minimize the size and cost of the system (Ramandi *et al.*, 2011).

7.5.2 System Description

In this section, the problem description, geometry, material and battery specifications will be outlined. Furthermore, the heat transfer and fluid flow equations, boundary conditions and initial conditions required to solve the problems will be described. In general, due to the heat generation inside the battery during the operation of an electric vehicle, battery temperature increases significantly, which can influence the battery life and operation. Figure 7.34 illustrates the geometry of the battery and proposed thermal management systems.

The thermal management system consists of the battery pack integrated with a single PCM shell (a) and double PCM shells (b) which contain PCM in the solid phase. The PCM shells are around the entire battery module. The PCM is integrated into the battery design and it serves as a heat sink for the heat generated within the battery. As heat is generated within the battery, it can be absorbed by the PCM. The solid PCM experiences a melting process and the battery temperature is maintained in the desired range. The

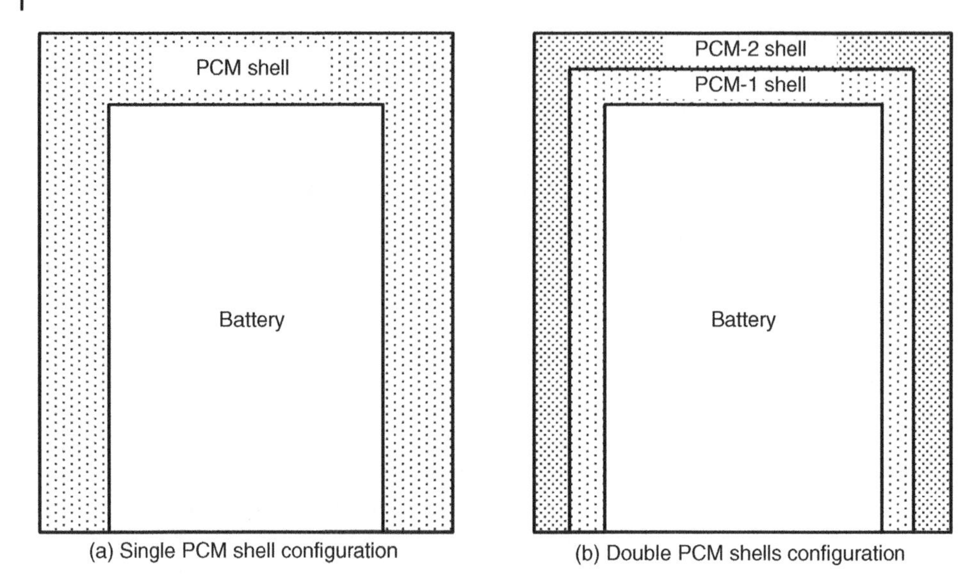

Figure 7.34 Schematic of the proposed thermal management systems (un-scaled) (adapted from Ramandi *et al.*, 2011).

Table 7.25 Battery specifications.

Parameter	Value
Number of cells	6
Nominal voltage	12 V
Capacity	20 Ah
Overall dimensions	$0.11 \times 0.16 \times 0.15$ m
Plastic case thickness	0.002 m
Weight	8 kg
Heat generation rate	5.83 W/cell
Cell core average thermal conductivity	6 W/m K
Plastic case (polypropylene) thermal conductivity	0.25 W/m K
Overall module heat capacity	900 J/Kg K

Source: Ramandi *et al.*, 2011.

specification of the selected battery pack is summarized in Table 7.25. The thickness of the shell in the single PCM system is 2 mm, and for the double PCM system, the thickness of each shell is 1 mm.

In the following analysis, a number of assumptions will be adopted. It is assumed that the discharge rate is constant so the heat generation is uniform inside the battery. The vehicle is assumed to stop after 2.5 h of use for the battery to be recharged. Radiation heat transfer to the PCM is negligible. Thermophysical properties of materials are assumed to be constant. All materials are assumed incompressible with a constant density. Based on these assumptions, the energy balance equation is obtained for the phase change material and battery.

Since phase change occurs during the operation, the energy equation should be solved separately for the solid and liquid phases. However, by using the enthalpy form of the energy equation, there is no need to consider liquid and solid phases separately. In this study, the enthalpy-porosity technique (Fluent Inc., 2006) has been used for modeling of the melting process. In this technique, the melt interface is not tracked explicitly. However, a quantity called the liquid fraction, which represents the fraction of the cell volume that is in a liquid form, is associated with each cell in the domain. The liquid fraction computed at each iteration is based on the enthalpy balance. Thus,

$$\frac{\partial}{\partial t}(\rho H) = \nabla \cdot (k\Delta T) + S_h \tag{7.29}$$

where

$$H = h + \Delta H \quad (enthalpy\ of\ PCM) \tag{7.30}$$

The total content of latent heat of phase change is

$$\Delta H = \beta H \tag{7.31}$$

and

$$h = h_{ref} + \int_{T_{ref}}^{T} c_p dT \tag{7.32}$$

Using the lever rule, the liquid fraction is

$$\beta = 0 \quad if \quad T < T_m \tag{7.33}$$

$$\beta = 1 \quad if \quad T > T_m \tag{7.34}$$

Furthermore, S_h is the heat generation rate which is assumed to be constant in battery and is zero in PCM shell.

Hundreds of PCM candidates have been reviewed in past literature. However, only a few of these materials have actually been used in actual PCM applications. The latent heat of fusion and melting temperature are two crucial parameters which need to be considered in PCM selection. However, when other properties such as thermal conductivity are considered, the number of the PCM candidates is considerably reduced to only a few potential PCMs. On the other hand, many PCMs with good thermal characteristics are hazardous or highly corrosive. Based on practical design criteria for electric vehicle battery packs, four PCMs with different properties have been identified in Table 7.26. The table summarizes the specifications of the PCMs that will be used in this research.

Table 7.26 Specification of PCMs utilized in thermal management systems.

PCM Name	Capric Acid	Eicosane	$Na_2(SO_4)\cdot10H_2O$	$Zn(NO_3)_2\cdot6H_2O$
PCM Number	1	2	3	4
Melting Temperature (K)	304.5	309.8	305.4	309.4
Latent Heat (J/Kg)	153,000	241,000	254,000	147,000
Density (kg/m^3)	884	778	1485	2065
Thermal Conductivity (W/m K)	2	0.27	0.544	0.31
Specific Heat (J/kg K)	2090	1900	1930	1340

Source: Alawadhi, 2001.

7.5.3 Analysis

7.5.3.1 Exergy Analysis

Exergy analysis is a valuable tool of thermodynamic analysis that uses the conservation of energy principle of the first law, together with the second law, for better design and analysis of energy systems. Thus, the exergy balance equation will be used to analyze the energy conversion aspects of the PCM melting processes in this section (Rosen *et al.*, 2011).

The exergy balance for the system can be written as

$$Ex_{in} - Ex_{out} - Ex_d = Ex_{sys} \tag{7.35}$$

The exergy change in the system is a result of the exergy changes associated with phase change material and battery:

$$\Delta Ex_{sys} = \Delta Ex_{PCM} + \Delta Ex_B \tag{7.36}$$

Since there is both a sensible and latent exergy change within the PCM, the total change in exergy will be the sum of the sensible exergy change, plus the exergy change due to melting of the PCM. For the latent exergy term, for any incompressible substance,

$$\Delta Ex_{sys} = \Delta E_{sys} - T_\infty \Delta S_{sys} \tag{7.37}$$

The change in entropy for the melting process is determined by the entropy of fusion as follows:

$$\Delta S_{PCM} = m_{PCM} \frac{L}{T_m} \tag{7.38}$$

By using the previous two equations and considering the liquid and solid fractions, the total exergy change can be calculated as follows:

$$\begin{aligned} \Delta Ex_{PCM} = {} & m_{PCM} C_s \left[T_m - \overline{T}_{ini,PCM} - T_\infty \ln\left(\frac{T_m}{\overline{T}_{ini,PCM}} \right) \right] \\ & + m_{PCM} x_l C_l \left[\overline{T}_{f,PCM} - T_m - T_\infty \ln\left(\frac{\overline{T}_{f,PCM}}{T_m} \right) \right] \\ & + m_{PCM} x_l L \left(1 - \left(\frac{T_\infty}{T_m} \right) \right) \end{aligned} \tag{7.39}$$

Also, the battery exergy change can be determined as follows:

$$\Delta Ex_B = m_B C_B \left[(T_{f,B} - T_{i,B}) - T_\infty \ln\left(\frac{T_{f,B}}{T_{i,B}} \right) \right] \tag{7.40}$$

The exergy destruction is given by

$$Ex_d = T_\infty S_{gen} \tag{7.41}$$

Since the exergy destruction depends on the entropy generation, an entropy balance must be performed on the system:

$$\Delta S_{sys} = S_{in} - S_{out} + S_{gen} \tag{7.42}$$

Assuming no heat loss to/from the system, the generated entropy occurs as a result of entropy generation from heat transfer to the shells:

$$S_{gen} = \Delta S_{sys} = \Delta S_B + \Delta S_{PCM} \tag{7.43}$$

The change of entropy of the battery and PCM can be obtained as follows:

$$\Delta S_B = m_B C_B \ln\left(\frac{T_{f,B}}{T_{i,B}}\right) \tag{7.44}$$

$$\Delta S_{PCM} = m_{PCM}\left[\frac{L}{T_m} + C_s \ln\left(\frac{T_{i,PCM}}{T_m}\right) + C_l \ln\left(\frac{T_m}{T_{f,PCM}}\right)\right] \tag{7.45}$$

Subsequently the exergy balance equation can be solved. Moreover, the exergy efficiency is determined by the ratio of the desired exergy output to the required exergy input. The exergy efficiency can then be calculated as the desired exergy output divided by the total exergy input, or

$$\psi = \left(1 - \frac{Ex_d}{Ex_{in}}\right) \times 100 \tag{7.46}$$

In the next section, the numerical formulation to perform this exergy analysis will be described.

7.5.3.2 Numerical Study

In order to analyze the heat transfer and exergy losses within the PCM and battery, a finite volume code (FLUENT 6.3.26) was used. The computational domain was divided into two sections: the battery module and the phase change material (PCM) and then was subdivided into a finite number of cells. The discretized governing equation was linearized implicitly and solved using a point implicit (Gauss-Seidel) linear equation solver. A second-order upwind discretization scheme for energy equations was chosen. The solution converged when the scaled residuals were less than 10^{-10}.

Once all initial and boundary conditions are known, the solver is able to begin solving the problem over time. The initial temperature of the battery module and PCMs is 303 K. Two types of boundary condition were considered for the case studies: insulated walls and convection heat transfer at the shell boundaries (using h = 35 W/ m^2 K).

7.5.4 Results and Discussion

7.5.4.1 CFD Analysis

A number of grid and time step independence tests were initially performed to verify that the results were sufficiently independent of grid spacing and time step size. The tests were performed for the following case: single PCM shell containing Eicosane, boundary condition of convection heat transfer, ambient temperature of 318 K, and initial temperature of 303 K. In Figure 7.35, the liquid fraction of PCM has been compared for a

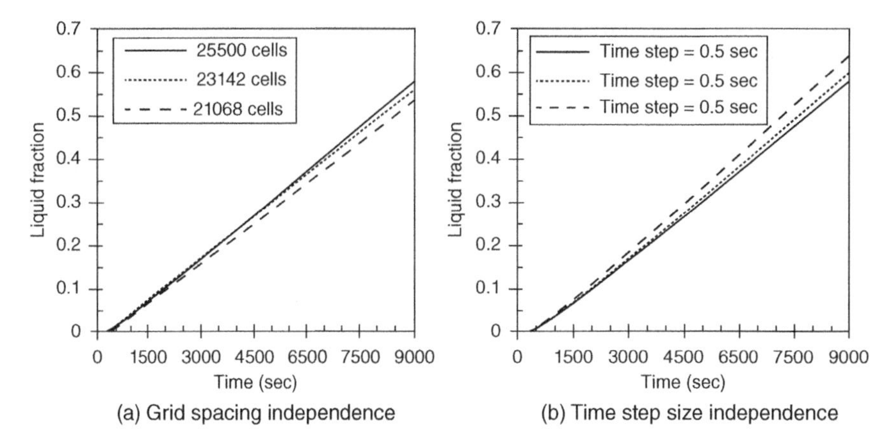

Figure 7.35 Independence tests for single PCM shell at ambient temperature of 318 K (adapted from Ramandi *et al.*, 2011).

different number of cells and time step sizes. The results converge closely to each other by about 25,500 cells and a time step size of 0.5 sec. As a result, these two values will be used in subsequent simulations.

Several simulations have been performed in order to investigate the effects of different single PCM and double PCM shell configurations in the presence of insulated walls and non-insulated walls of the thermal management system. The battery temperature and liquid fraction of PCM were then compared for different case studies. However, before moving to those results, a model validation was performed to ensure the model represents an accurate simulation of actual processes. The predictions are compared to a previous study (Al-Hallaj *et al.*, 2006), which showed good agreement and useful validation of the predictions (Figure 7.36). A CFD analysis for three additional cases was also investigated and the results (including PCM liquid fraction and battery temperature) will be discussed.

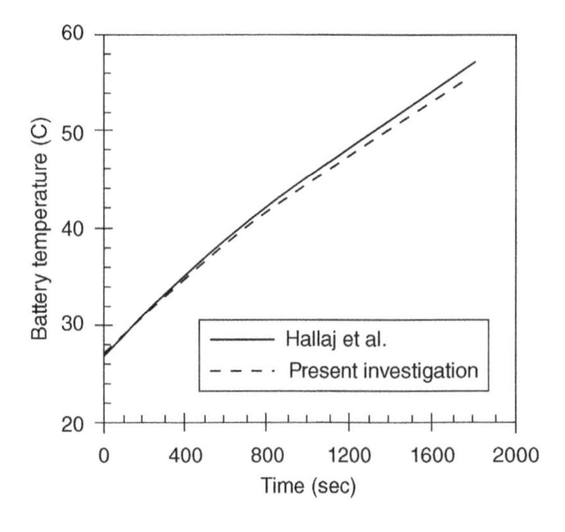

Figure 7.36 Comparison of predictions with past data of Al-Hallaj *et al.*, 2006 (Ramandi *et al.*, 2011).

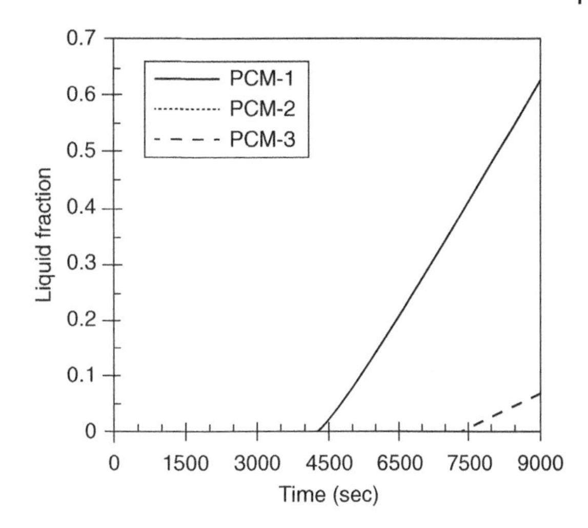

Figure 7.37 Liquid fraction of PCMs in shells for single PCM shell system (insulated walls) (adapted from Ramandi *et al.*, 2011).

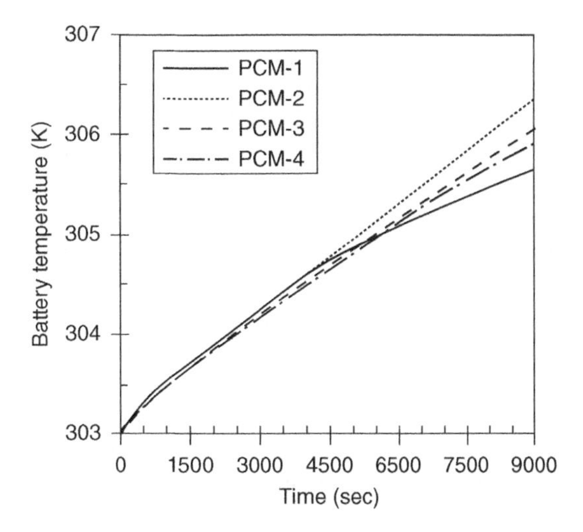

Figure 7.38 Battery temperature change during operation for single PCM shell (insulated walls) (adapted from Ramandi *et al.*, 2011).

7.5.4.1.1 a) Single PCM Shell System with Insulated Walls This case represents a single PCM shell system with insulated walls and the following parameters: PCM-1 CAPRIC ACID, PCM-2 EICOSANE, PCM-3 $Na_2SO_4.10H_2O$, PCM-4 $Zn(NO_3)_2.6H_20$, initial temperature of 303 K, and boundary condition of insulated walls.

The liquid fraction of the PCMs and battery temperature change during the operation for different PCMs are shown in Figures 7.37 and 7.38. During the operation, the heat generated in the battery is transferred to the PCM. In Figure 7.37, it is apparent that PCM-2 and PCM-4 have not melted within the real simulation time. Moreover PCM-1 has experienced the biggest effect. The melting process of PCM-1 has started earlier than the other PCM-3. Since there is no significant change in liquid fraction of the PCMs (except PCM-1), it seems that these PCMs (except PCM-1), are not able to control the battery temperature. But in Figure 7.38, there is no considerable difference between the battery final temperatures for different PCM systems. For all cases, it is below 307 K

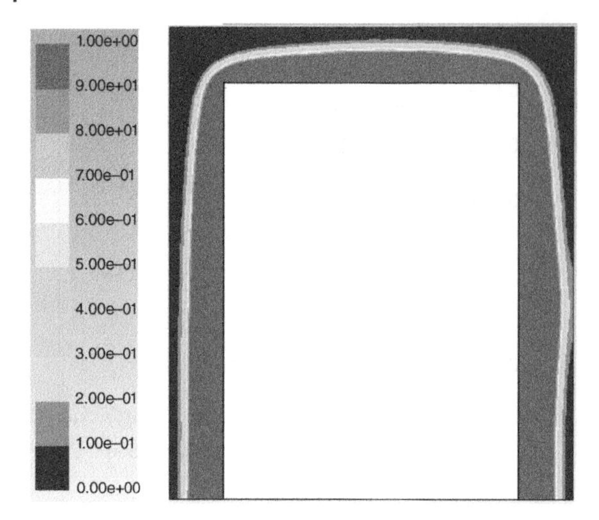

Figure 7.39 Liquid fraction of PCM-1 after 2.5 hours for a single PCM shell (insulated walls) (Ramandi *et al.*, 2011).

(34 °C), which demonstrates that the PCMs had the desired control over the battery temperature. The reason for different trends in liquid fraction is due to the different melting temperature. The lower melting temperature causes a faster melting process.

In this case, the sensible heat process is effective as with the latent heat effect for all PCMs except PCM-1. This result would help to optimize the thermal management system design, as a thinner shell which contains a smaller volume of PCM would be capable of removing the heat generated inside the battery. However, this trend may not apply for the higher heat generation rates which may not occur at higher discharge rates. In these circumstances, PCMs 2, 3 and 4 would be more efficient with the current system design while for PCM-1, the shell needs to be wider.

The contours shown in Figure 7.39 show the melting process at the end of a simulation (after 2.5 hours). The melting process has occurred in a region close to the battery walls, and the region near the shell walls has not been influenced (still is solid). Since the cost and weight of the PCM shells and consequently battery pack are important factors in thermal management system design, the width of PCM shells can be reduced.

7.5.4.1.2 b) Double PCM Shells System with Insulated Walls The next set of simulations involves double PCM shells with insulated walls. The simulation conditions for this case are as follows: PCM-1 + PCM-2: CAPRIC ACID-EICOSANE, PCM-1 + PCM-3: CAPRIC ACID-$Na_2SO_4.10H_2O$, PCM-1 + PCM-4: CAPRIC ACID-$Zn(NO_3)_2.6H_20$, initial temperature of 303 K, boundary condition of insulated walls.

The liquid fraction of the PCMs for both shells, and the battery temperature change during the operation for different PCMs, are shown in Figures 7.40 and 7.41. As shown in Figure 7.40a, the liquid fraction of the PCM in the first shell (close to the battery) increases dramatically for all cases and it reaches 1 before the end of the simulation time. Moreover, at any time, the liquid fraction of the PCM-1 in the first case (combination of PCM-1 and 2) is larger than the other two cases. The reason is that the second PCM (PCM in the outer shell) in the first case has the smallest thermal conductivity, so the heat cannot easily be transferred to the second PCM, and consequently the majority of heat generated in the battery is utilized to increase the temperature and as a result,

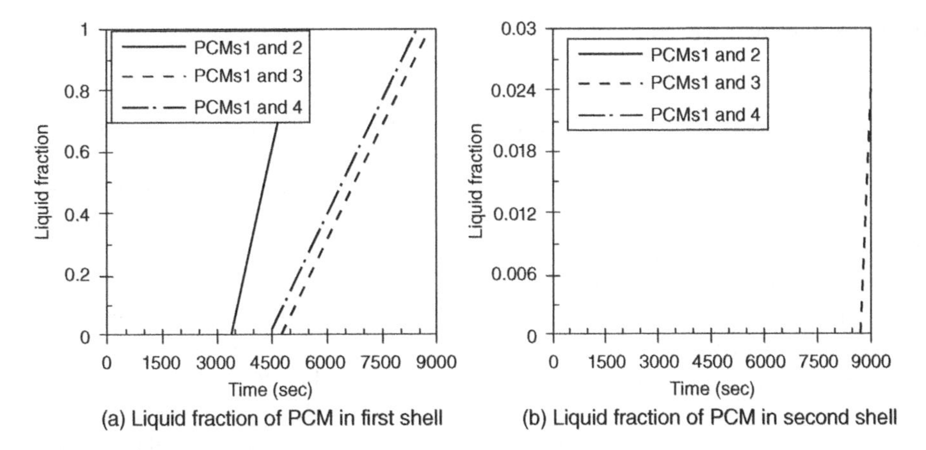

Figure 7.40 Liquid fraction of PCM in first and second shell for double systems (insulated walls) (adapted from Ramandi *et al.*, 2011).

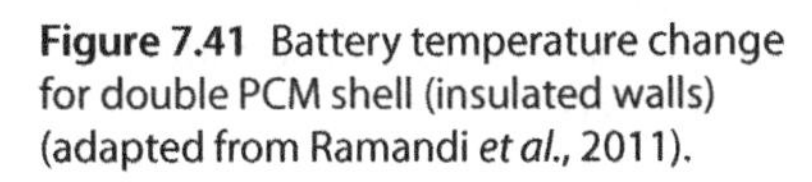

Figure 7.41 Battery temperature change for double PCM shell (insulated walls) (adapted from Ramandi *et al.*, 2011).

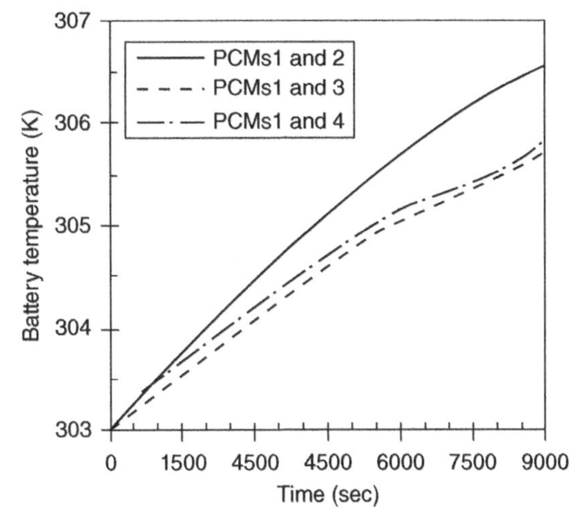

the liquid fraction of the PCM in the inner shell. The same argumentation occurs in the working order for the combination third case (combination of PCM-1 and 4).

In Figure 7.40b, it can be observed that the PCMs in the outer shell have not undergone the melting process, except in the second case (combination of PCM-1 and 3). PCM-1 has a good heat conductivity compared to the other PCMs and it transfers the generated heat to the outer shell. The absorbed heat by the PCM in the outer shell causes an increase in temperature. However, since PCM-3 has the lowest melting temperature, it melts before the other PCMs at the end of the process. The battery temperature (see Figure 7.41) illustrates that the temperature surges significantly for all cases, but the trend for the first case (combination of PCM-1 and 2) has a sharp rise. From Figure 7.40a, it can be seen that PCM-1 has melted over a period of 2000 seconds between 3500 and 5500 seconds. Afterwards, it is more difficult to control the temperature of the battery.

The previous cases indicate that they are able to perform the thermal management. The issue is then to find the best design for double shell systems. The thermal conductivity and melting temperature are the two most important parameters in the configuration of the shells.

*7.5.4.1.3 **c) Single PCM Shell System with Non-Insulated Walls*** The next cases consider a single PCM shell system with non-insulated walls. The simulation conditions for this case are shown as follows: PCM-1: CAPRIC ACID, PCM-2: EICOSANE, PCM-3: $Na_2SO_4.10H_2O$, PCM-4: $Zn(NO_3)_2.6H_20$, initial temperature of 303 K, and boundary condition of h=35 W/m^2 K, T_{am} = 318 K.

The liquid fraction of the PCMs and battery temperature change during the operation for different PCMs are shown in Figures 7.42 and 7.43. Results in Figure 7.42 indicate that for all cases, the melting process is initiated immediately and increases significantly until the end of the electric vehicle operation. Compared to other PCMs, the trend of liquid fraction of PCM-1 is sharper, as PCM-1 has the lowest melting point and the largest thermal conductivity. Compared to the single PCM shell system with insulated walls

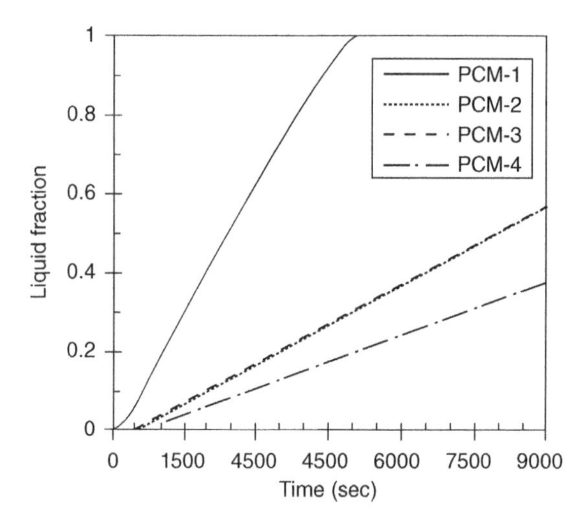

Figure 7.42 Liquid fraction of PCMs in shells for single PCM shell (non-insulated walls) (adapted from Ramandi *et al.*, 2011).

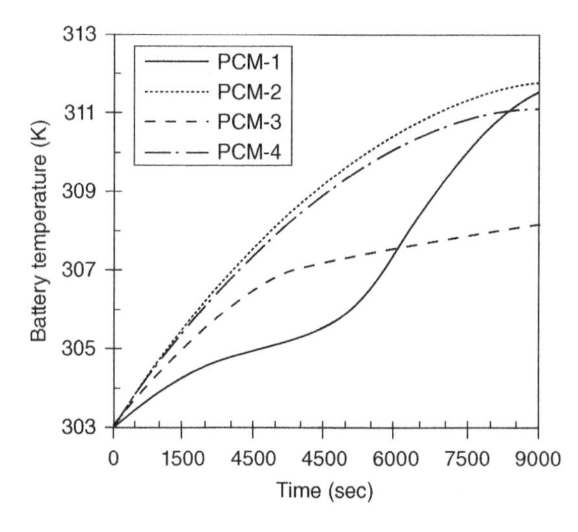

Figure 7.43 Battery temperature change for single PCM shell (non-insulated walls) (adapted from Ramandi *et al.*, 2011).

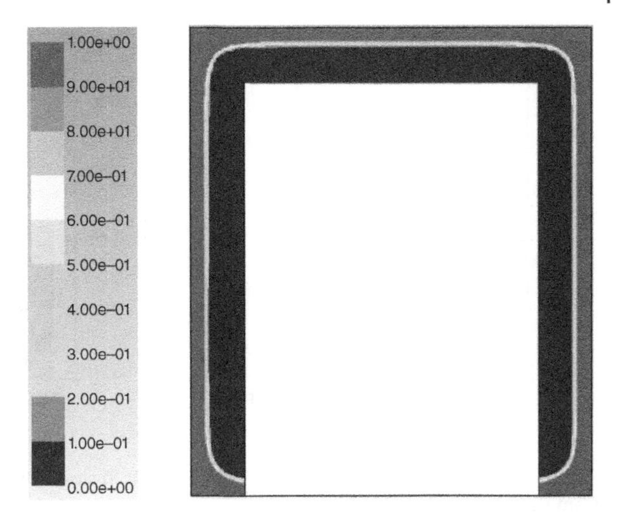

Figure 7.44 Liquid Fraction of PCM-1 after 30 minutes for single PCM shell (non-insulated) (Ramandi *et al.*, 2011).

(Figure 7.37), at any time and for any trend, there is a significant difference between an insulated and non-insulated system. At any time, the liquid fraction of the non-insulated system is higher than the insulated system. In the non-insulated system, since the environment temperature (318 K) is higher than the PCM temperature at any time, heat is transferred by convection to the shell. This undesired heat increases the temperature and hence liquid fraction of thePCM.

Figure 7.44 shows the contours of liquid fraction for PCM-1 after 30 minutes of electric vehicle operation. It can be seen that the melting process has started from the region close to the surroundings, not from the interface between the battery and PCM. This means that the PCM has melted because of the heat transfer from the environment, not from the heat generated in the battery. From Figure 7.43, the system is not beneficial (except PCM-3), because the trends indicate that the final temperature of the battery is above 311 K. The heat transfer from the environment to the PCM shell has decreased the heat absorption potential of the PCM. So it cannot absorb all battery heat generated and as a result, is not able to control the battery temperature during electric vehicle operation. PCM-3 is an exceptional case because the final battery temperature in this case is still acceptable. It is a result of higher latent heat, which can absorb more heat generated in the battery. So PCM-3 is the only PCM which can be used in the thermal management system with non-insulated walls.

7.5.4.2 Part II: Exergy Analysis

A number of exergy results were also obtained. Two important aspects in investigating the performance of the system are exergy efficiency and exergy destruction. The ambient temperature is anticipated to have a large effect on the overall performance of the system. From the previous studies the system with non-insulated walls could not be an alternative for the thermal management systems, so it is not considered in the exergy analysis. The two cases to be investigated in this section are: a) single PCM shell system with insulated walls; and b) double PCM shells system with insulated walls.

7.5.4.2.1 a) Single PCM Shell System with Insulated Walls For the single PCM shell system with insulated walls, Figures 7.45a and 7.45b illustrate the differences in exergy

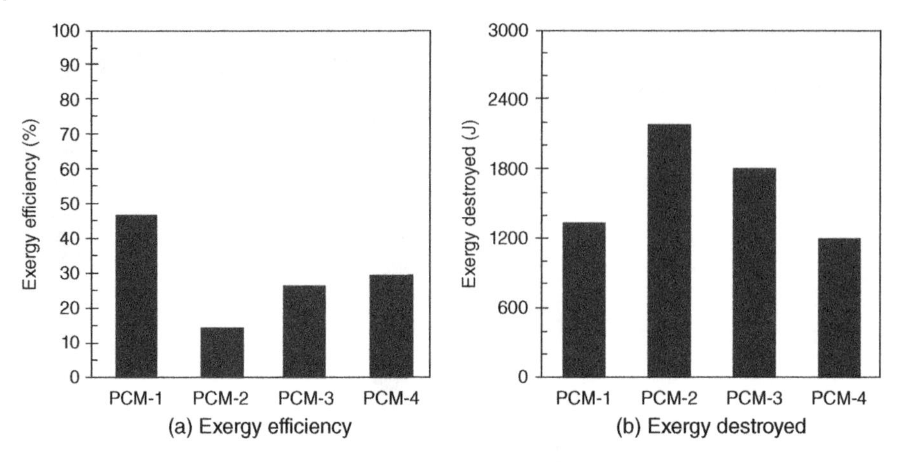

Figure 7.45 Exergy analysis result for different PCMs in single PCM shell (insulated walls) (adapted from Ramandi *et al.*, 2011).

efficiencies and exergy destruction for different PCMs with the conditions stated previously for the CFD analysis. As shown in these figures, in all cases the exergy efficiency is below 50%. The system including PCM-1 has the largest efficiency. As a result, the thermal management systems, involving PCM with the lower melting temperature, have higher exergy efficiency. The exergy destroyed is a result of both the ambient temperature effect and the internal entropy generation. Since the ambient temperature is identical in all cases, the exergy destroyed is also the same. Figure 7.45b shows the exergy destroyed for different PCM systems. The system including PCM-2 has the largest amount of exergy destroyed. Since the exergy transfer due to heat transfer from the battery is similar for different PCM systems, the largest exergy destruction results in the smaller exergy efficiency.

7.5.4.2.2 b) Double PCM Shells System with Insulated Walls For double PCM shells with insulated walls, Figures 7.46a and 7.46b demonstrate the differences in exergy

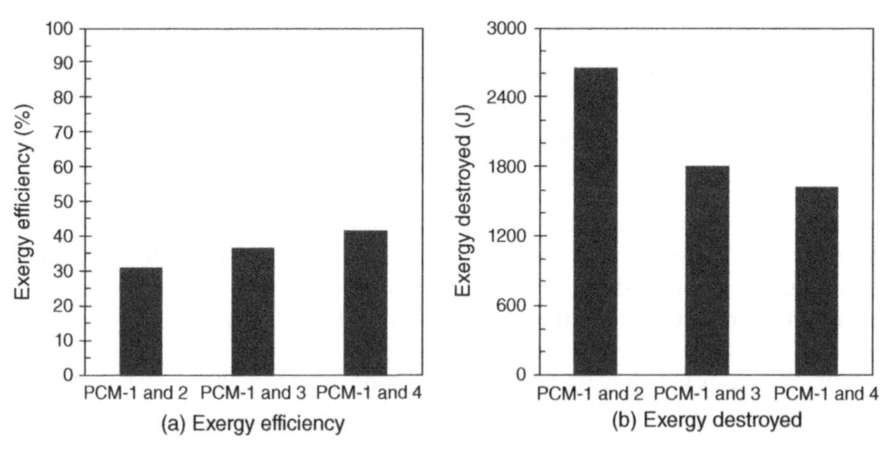

Figure 7.46 Exergy analysis result for different PCMs in double PCM shell (insulated walls) (adapted from Ramandi *et al.*, 2011).

efficiencies and exergy destruction for different PCMs with the conditions stated previously. The double PCM shell systems involve two different PCMs in two different shells. The first shell in all cases contains a similar PCM (PCM-1). Because the exergy analysis of a single PCM shells system showed that the system involving PCM-1 as the phase change material, has the highest exergy efficiency. Since the goal was to find the best combination of PCM shells in order to maximize the efficiency of the double PCM shells system, PCM-1 was selected as the first shell and its combination with PCM-2, PCM-3 and PCM-4 was the second shell analyzed. The results show that the exergy efficiency in all cases is in the range of 30–40% and the system involving the combination of PCM-1 and 4 has the highest efficiency. The second shell content does not have significant influence on the exergy efficiency. Comparing to Figure 7.45a, the double PCM shell system has higher exergy efficiency than a single PCM shell system (except PCM-1).

7.5.4.2.3 c) Effects of Dead State Temperature Figures 7.47a and 7.47b demonstrate the effect of the ambient environment on exergy efficiencies and exergy destruction for different PCMs (single PCM shell system). The reason for this investigation is to determine the effect of ambient temperature on the efficiency. Since the exergy destroyed (and consequently the exergy efficiency) is dependent on the reference temperature, they are anticipated to have a significant effect on the overall system performance. In order to determine the effects of the reference temperature on exergy efficiency and exergy destroyed, the ambient temperature was varied from 290 K to 302 K, in increments of 4K. The results are summarized in Figures 7.47a and 7.47b. Results in these figures indicate that the exergy efficiencies decrease as the ambient temperatures increase, while the destroyed exergy increases gradually. On the other hand, the exergy destroyed contents, unlike exergy efficiencies, are not affected significantly. Moreover, the results showed that the exergy efficiencies for different ambient temperatures vary from 15% to 85%.

Figures 7.48a and 7.48b demonstrate the effects of reference environment on exergy efficiencies and exergy destruction for different PCMs and the double PCM shell system. From Figure 7.48a, the exergy efficiencies are largely affected by varying the ambient

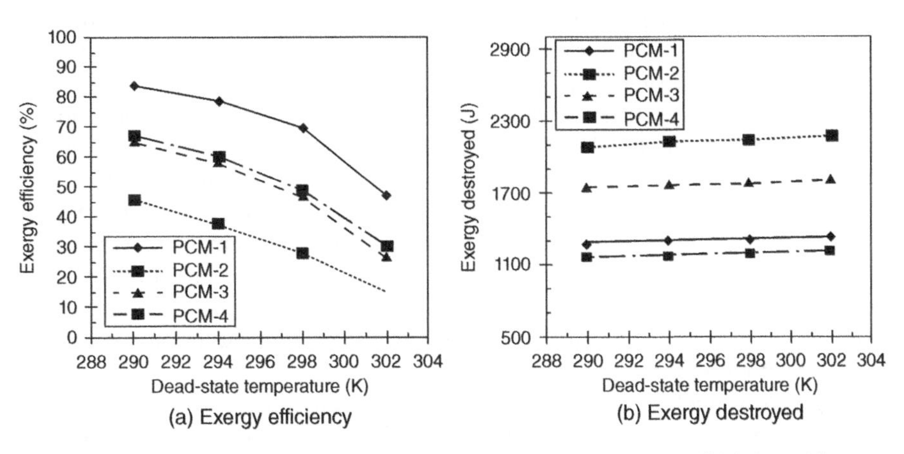

Figure 7.47 Effects of ambient temperature (single PCM shell; insulated walls) (adapted from Ramandi *et al.*, 2011).

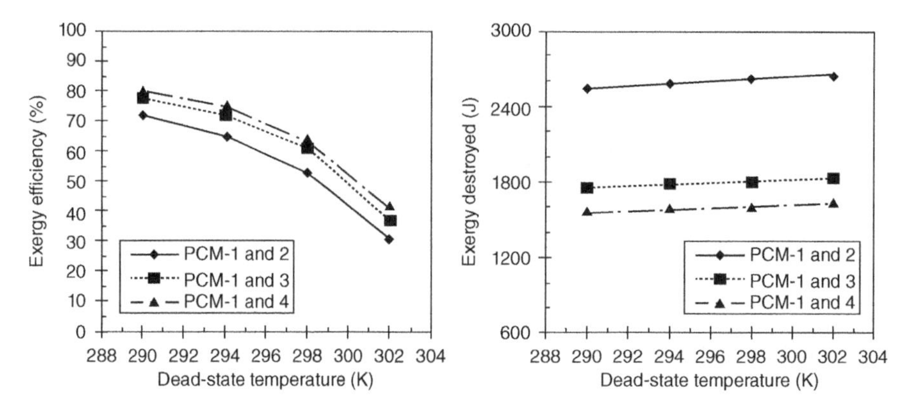

Figure 7.48 Effect of ambient temperature (double PCM shell; insulated walls) (adapted from Ramandi *et al.*, 2011).

temperatures. As the temperature increases, the exergy efficiencies decrease. Moreover, for all ambient temperatures, the efficiencies are very similar. Unlike the exergy efficiencies, the exergy destruction is largely unaffected when the ambient temperature rises. Comparing to the single PCM shell system, it appears that the double shell system is more efficient. Except for the ambient temperature of 302 K, the exergy efficiencies for different combinations are well over 50%. Generally for different ambient temperatures, the exergy efficiencies vary from 30% to 80%, respectively.

7.5.5 Closing Remarks

A passive thermal management system for electric vehicle batteries consisting of encapsulated phase change materials was studies in this case study. The configuration is based on double series PCM shells. A combination of computational fluid dynamics (CFD) and second law analysis was used to evaluate and compare the new system versus single PCM shells. A model validation was performed to ensure the model is a viable simulation of actual processes. The modeling predictions were compared with a previous similar study and the data showed good agreement. Three cases were investigated (single PCM shell system with insulated walls; double PCM shells system with insulated walls; and single PCM shell system with non-insulated walls).

The following concluding remarks are obtained based on the newly developed model in this study. For the single PCM shell system with insulated walls, the four selected PCMs can control the battery temperature effectively. It was found that the width of the shell can be reduced to lower the costs. For the double PCM shells systems, there was a significant difference between the insulated and non-insulated system. Unlike the insulated system, the melting process started from the region close to the surroundings, not from the interface between the battery and PCM. The heat transfer from the environment to the PCM shell in the non-insulated case was disadvantageous because the final battery temperature was about 313 K.

In the single PCM shell system with insulated walls, the system including PCM-1 was the more efficient system. The system including PCM-2 had the largest quantity of exergy destruction. For the double PCM shell systems with insulated walls, the system involving the combination of PCM-1 and 4 was the most efficient system. In addition, it was found that the second shell content did not have a significant influence on the exergy efficiencies. In all cases, the exergy efficiencies decrease when the ambient temperatures increase, while the exergy destruction increases gradually. Overall, the double PCM shell system was more efficient than single PCM shell systems in terms of the exergy efficiency.

7.6 Case Study 5: Experimental and Theoretical Investigation of Novel Phase Change Materials For Thermal Applications

7.6.1 Introduction

The demand for electric vehicle battery thermal management is increasing significantly as the battery chemistries with higher specific energy/power are used under severe driving and ambient conditions. Thus, the conventional air and/or liquid cooling systems are starting to become considerably large, complex or expensive to keep the batteries operating within their ideal temperature ranges. In this regards, PCMs become highly useful for these applications, both as primary and auxiliary cooling system, as they can have heat storage capacities more than an order of magnitude (per unit volume) higher than most sensible storage materials in response to slight changes in temperature. Thus, they can be compact, safe and inexpensive solutions to many thermal related issues in the battery pack.

In this regard, R134a clathrates can be a viable option for such applications. Clathrates of several refrigerants already exist and are used as PCMs. The existing refrigerant based PCMs take a very long time to charge. They can also have limited capacity and have undesired operating temperatures. The motivation behind the case study is to change the effective thermal properties of the PCMs in order to improve the charging and discharging times. It is believed that the latent heat can also be improved to enhance the capacity. The improvement in the desired parameters can be achieved by adding metal particles, salt particles and liquid additive.

Thus, in this case study, an experimental investigation is conducted to test the thermal properties, behaviour and characteristics of R134a clathrates with additives, as phase change materials (PCMs). PCMs' charging characteristics are analysed and evaluated for electric vehicle battery cooling applications. The formation of refrigerant clathrates is investigated due to their potential use in active as well as in passive cooling applications. PCMs are formed using R134a clathrate and distilled water with different refrigerant fractions and additives. The main objective of using additives is to study their potential for enhancing the clathrate formation and their thermal properties under direct contact heat transfer. PCMs are formed in glass tubes to determine their freezing onset time, transformation time and thermal properties. The thermal properties determined are the liquid phase thermal conductivity, mushy phase thermal conductivity, and specific latent heat of the PCMs. Different refrigerant R134a fractions are used to form clathrate and several additives (such as ethanol, sodium chloride, magnesium nitrate hexahydrate, copper and aluminum) with various mass fractions are used. Discharge tests are also conducted for which the PCMs are used to cool the hot air as well as to cool down an electric vehicle battery. Energy and exergy analyses are performed to assess PCMs' performance (Zafar, 2015).

7.6.2 System Description

This section describes the experimental set up to determine the latent heat, phase change temperature, thermal conductivity and heat capacity of the PCM. It also provides the system description for the equipment and apparatus to be used for the experiments.

For the experiments, a cold constant temperature bath is used as a constant temperature source. The refrigerant clathrate with additive, named PCM, are formed in glass tube. The tubes are submerged in the constant temperature water bath for which the temperature is set at 276 K and 278 K. The constant temperature bath works by providing cold energy and heat simultaneously to the distilled water in the bath to maintain its temperature at a set value. The graphic illustration of the experimental system is shown in Figure 7.49.

A refrigeration system with cooling coils around the water bath pumps out the heat. A controller constantly monitors the water temperature in the bath while continues to provide the desired heat to maintain the desired temperature. The bath is converted into constant energy bath for thermal properties experiments. A constant cold and hot energy is provided to the water in the bath to maintain the amount of energy. A stirrer is also used which circulated the water in the bath. Without the stirrer, the water near the hot or cold source would change its temperature while the water away from the source would see its effect later.

The PCM is formed in the glass tubes. First the glass tube is filled with distilled water and the desired additive. The exact mass of the tube with its constituents is measured using a high accuracy digital weighing scale. The tube is sealed and then vacuumed to get rid of excess air. The last step is to fill the desired refrigerant using a needle valve that allows one way flow. The glass tube is then submerged into the cold temperature water bath for charging. The tubes are visually observed after regular interval to observe the onset and end set of freezing. The freezing times, PCM temperatures and pressures are recorded for each test. Onset of freezing is usually east to detect as the top layer starts

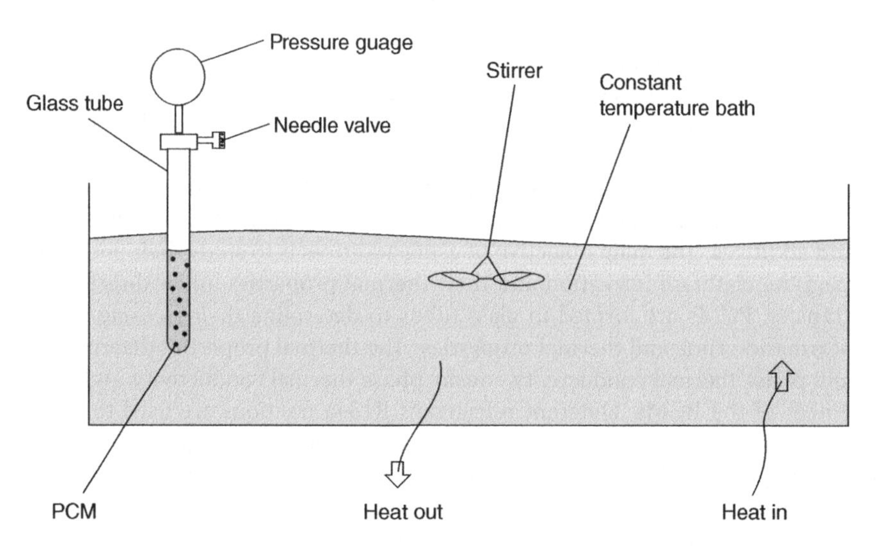

Figure 7.49 A schematic diagram of the proposed PCM testing system (Zafar, 2015).

freezing. The end set is challenging to pin point so it is important to continue observing the PCM until after the last observed changes in the PCM structure. PCM usually rises as it freezes so height is observed for the end set.

The K-type thermocouples are attached to a reader to read the temperatures. For initial charging test, only one temperature reading is taken. For thermal property tests, temperatures are taken at two different locations. The tube is comprehensively tested for leaks and provisions are made to make sure there are no leaks. It is important to use a glass tube since the onset of phase change needs to be observed visually. The illustrative figure of the glass tube, its connections and used systems are shown in Figure 7.50.

The PCM discharge using hot air is also part of the experimental investigation as it helped cross check the values of thermal properties. Hot air at 315 K is used to melt the PCM in the glass tube. The process of melting the PCM from solid to liquid is the discharge of energy hence called PCM discharge. The temperature of the PCM is recorded

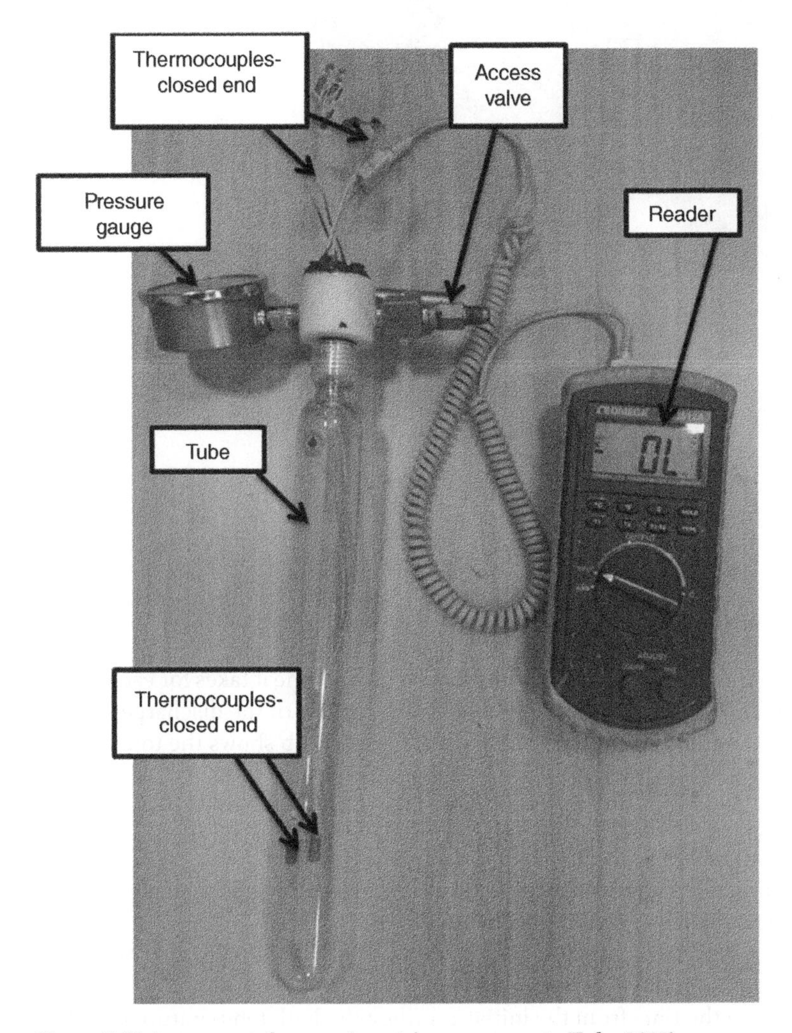

Figure 7.50 Instruments for experimental measurements (Zafar, 2015).

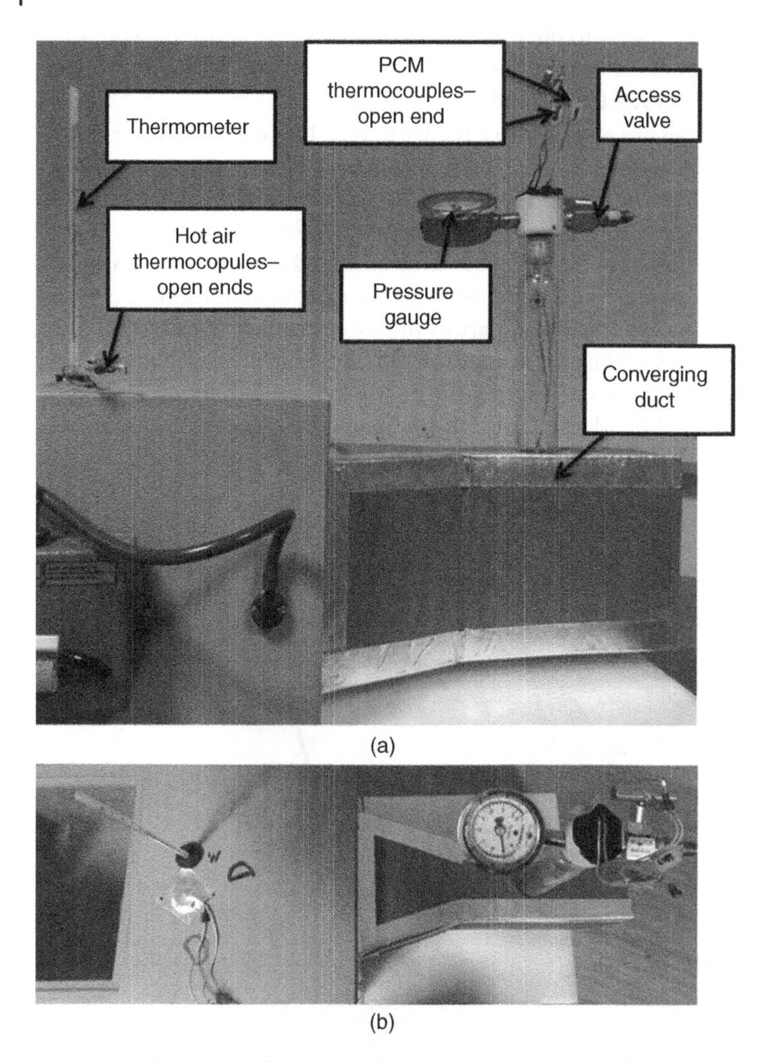

Figure 7.51 Illustration of the system for discharge (a) side view, (b) top view (Zafar, 2015).

at regular interval and the phase of the PCM is observed. The time it takes for each PCM to discharge is noted. Figure 7.51 shows the illustrative picture of the discharge set up. Figure 7.51a shows the side view of the set up while Figure 7.51b shows the top view of the same set up.

The procedure to determine the thermal properties of the PCM are described as follows:

- Used 16.24 kg of distilled water with the density of 998 kg/m^3 and specific heat of 4200 J/kg K in the constant temperature bath.
- Operated the bath with constant supply of chiller and heater initially at room temperature, that is, ambient temperature.
- Started noting down the time from the initiation when the bath temperature is ambient (T_{amb}) until the time bath temperature does not change (T_{ss}).

- The amount of energy is calculated by the equation $Q = m_{\text{bath}} C_{\text{Pwater}} (T_{\text{amb}} - T_{\text{ss}})$.
- The value of Q is divided by time, in seconds, to determine the heat rate $\dot{Q}$ in J/s or Watts.
- The PCM tubes are submerged in the bath.
- Temperature (T_i), time (t_i) and phases (solid, mushy or liquid) of the PCM is recorded after regular interval.
- Two probes are inserted for temperature readings 14 mm away from each other. One is at the center location r_1 and the other one 14 mm away at r_2.
- The center location reads temperature T_{core} of the tube cross-section while the other is 14 mm away, T_{away}, from it.
- The PCM is discharged using hot air.
- The temperatures T_{core} and T_{away} are recorded during discharge after regular interval.
- Specific heat, Cp, is determined using the governing equation $Q = m_{\text{PCM}} C_P (\Delta T_i)$.
- Latent heat is determined by adding the total amount of heat Q from the time of onset until the end set.
- Thermal conductivity, k, is determined using $k = \frac{\dot{Q} \ln\left(\frac{r_2}{r_1}\right)}{2\pi l \, \Delta T \, \Delta t}$

A 6s LiPo 5000 mAh 60C battery is used to conduct the battery cooling tests. An aluminum jacket is made to house the battery and filled by the desired PCM. Since no observations are needed for these tests, the jacket can be of non-transparent material. Aluminum is used since it is light and can easily be welded. Welding is preferred over binding with glue since sealing the jacket with glue binder is very difficult. The jacket has the pressure gauge and it is sealed using a gasket from the threaded regions. Figure 7.52 shows the dry fit close up of the jacket and battery.

The battery, while inside the PCM jacket, is connected to a motor with load which is operated at maximum power. A voltmeter is attached to the battery to read its voltage while a thermocouple is placed between the battery cells to read the battery temperature. Battery is charged to its maximum voltage of 4.18 Volts and discharged until it reaches cutoff voltage of 3.6 Volts. It is not safe to operate the battery below the cutoff. Figure 7.53 shows the set up of the battery cooling experiment with labels of components used.

7.6.2.1 Experimental Layouts

This sections presents the layout of the experiments conducted for this case study. Three different sets of experiments are conducted for this case study. First set of experiments are constant temperature bath charging where PCMs are charged using the cold bath. Tubes are submerged in the cold bath and the PCM temperatures are recorded after regular interval until it is completely charged. The second set of tests are conducted to determine the thermal properties of the PCMs. Tubes are submerged in the cold constant energy bath with two temperature probes. Once again the temperature readings are recorded after regular interval for both the probes. Third set of experiments are discharge tests. The PCMs are discharged using hot air and electrical battery. For hot air discharge, temperatures inside the tube are recorded after regular internal until the PCM is completely discharged. For battery cooling tests, a special jacket is made to house the selected battery. Battery is placed in the jacket and operated on a load. The cutoff temperature, cutoff time and cooling time until the battery reached ambient temperature is recorded.

Figure 7.52 dry fit set up of the battery and the PCM jacket (Zafar, 2015).

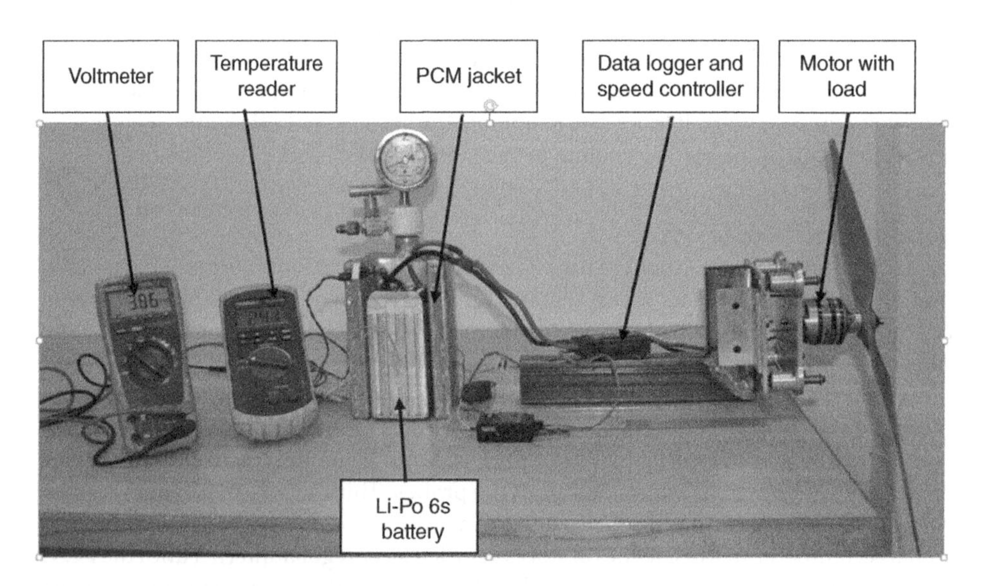

Figure 7.53 Illustration of the battery cooling test set up (Zafar, 2015).

First set of experiments are conducted to achieve three primary goals. First goal is to find out which refrigerant, among R134a and R141b, is better at forming clathrates. The time it takes to form the clathrate and the formation pressure established determined which refrigerant is better. The second goal is to determine the better bath temperature for charging the clathrate. Bath temperature of 276 K and 278 K are used. Although it is clear that it takes more energy to cool the bath down to 3 °C (276 K), as compared to 278 K. But it is to be established if the charging time for the clathrate is the same or different. Once the charging times are determined, the energy and exergy evaluations established which bath temperature requires the least amount of energy. The third goal is to find out the most appropriate refrigerant mass fraction for clathrate formation. Charging times and clathrate structure are some of the parameters used to establish the most appropriate mass fraction. Refrigerant mass fractions of 0.15, 0.2, 0.25, 0.3, 0.35 and 0.4 are used to form the clathrate. For the sake of simplicity, clathrate with percent refrigerant is used to describe the clathrate. Energy and exergy values are also evaluated to determine the amounts required to form clathrate with each refrigerant mass fraction. Figure 7.54 shows the experimental layout of the first set of experiments.

For the second set of experiments, PCMs are formed by adding five different additives in 0.35 mass fraction R134a clathrate. The five additives used are copper, aluminum, ethanol, magnesium nitrate hexahydrate and sodium chloride. The refrigerant clathrate without any additive is considered to be the base PCM. These experiments are conducted to achieve four primary goals. The first goal is to find out the charging times of each PCM. The charging times included the onset and end set of PCM freezing. Using the charging onset and end set times, energy and exergy values of charging are evaluated.

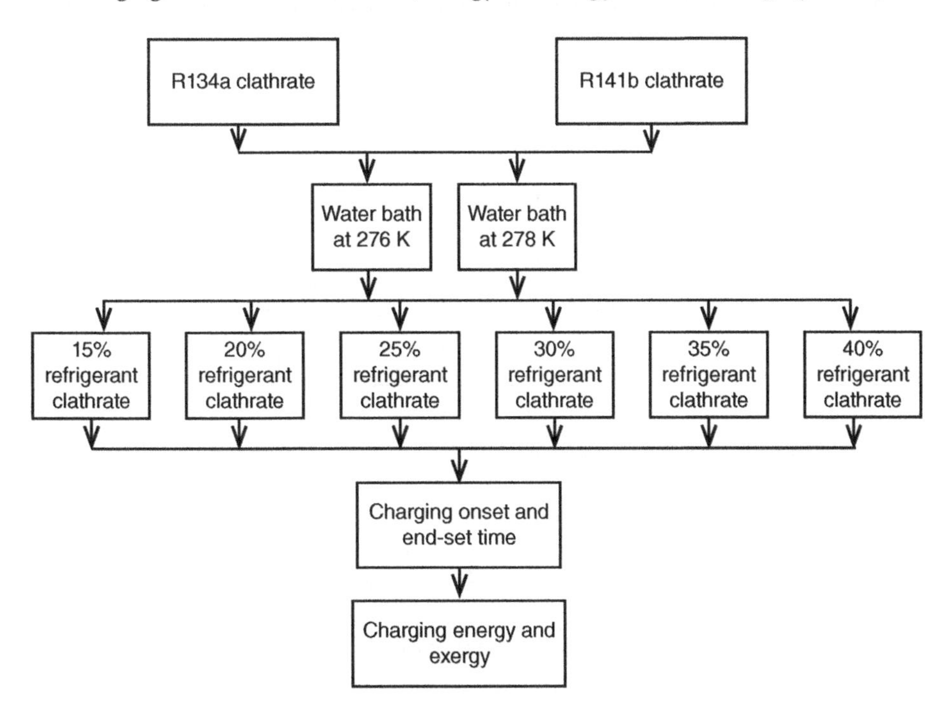

Figure 7.54 Experimental layout of the first set of experiments (modified from Zafar, 2015).

The second goal is to find out the charging times of each PCM with additive mass fractions varied from 0.01 to 0.05 with an interval of 0.01. It is desired to find out the effects of additive fractions on the charging times, charging energy values and charging exergy values. The charging times included the onset and end set of PCM freezing. The third goal is to find out the discharging times of each PCM using hot air. The discharging times only included the end set of PCM freezing since onset almost starts immediately. Using the discharging times, energy and exergy values of discharging are evaluated. The fourth goal is to find out the discharging times of each PCM using battery heat. A special jacket is made to house the selected battery and PCMs are formed inside the jacket. Parameters recorded for this set of tests for each PCM are battery run times, battery temperature at cutoff voltage and time for the battery to cool to a minimum possible temperature to recharge the battery safely. Using the cooling times, energy and exergy values of discharging are also evaluated. Figure 7.55 shows the experimental layout of the second set of experiments.

The third set of experiments is conducted to determine the thermal properties of each PCM and establish which additive is better. Five different additives are used namely copper, aluminum, ethanol, magnesium nitrate hexahydrate and sodium chloride. Tubes with two temperature probes are used for these experiments. Charging times for each PCM is recorded along with its temperatures for each temperature probe. Using the temperature data, thermal properties of each PCM is determined. Thermal properties determined included specific heat, latent heat, liquid phase thermal conductivity and solid phase thermal conductivity. Comparing the charging times and structure of each PCM, performance of each PCM is predicted. Soft fluffy structure takes short time to charge but it cannot provide cool energy for long. Alternatively, a hard solid structure

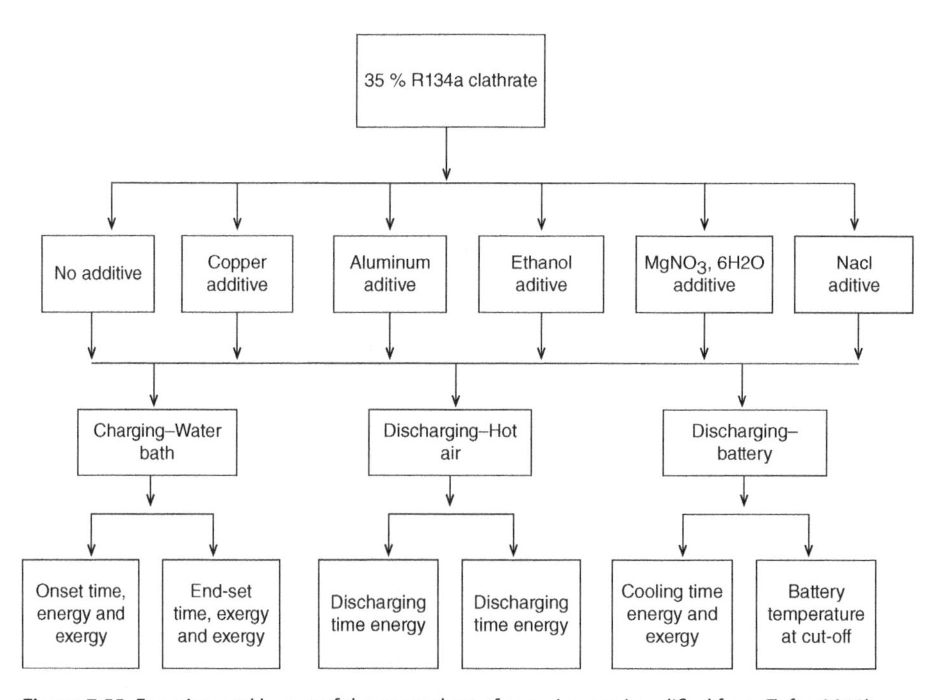

Figure 7.55 Experimental layout of the second set of experiments (modified from Zafar, 2015).

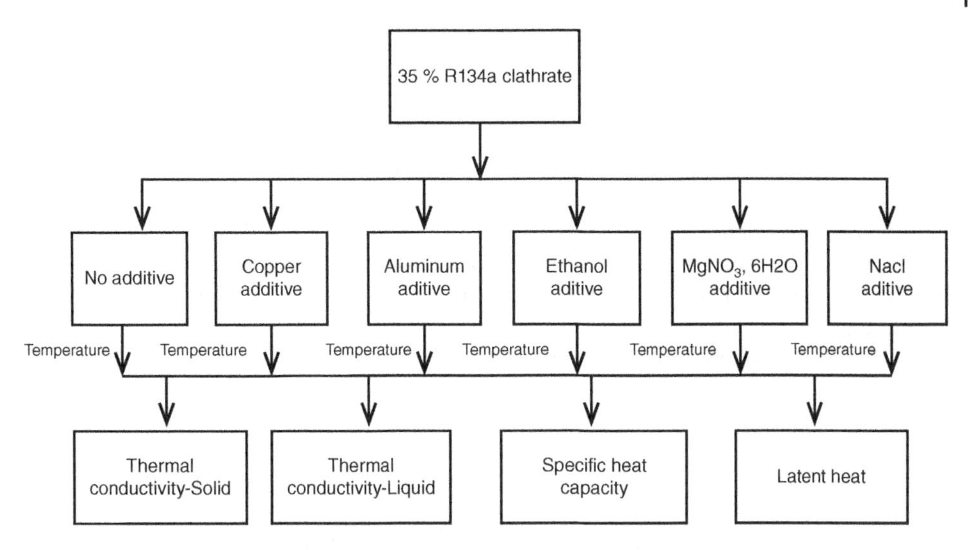

Figure 7.56 Experimental layout of the third set of experiments (modified from Zafar, 2015).

may take longer to charge but it can provide cool energy longer. Charging times are used to evaluate the energy and exergy values of each PCM. Figure 7.56 shows the experimental layout of the third set of experiments.

7.6.2.2 Challenges

Refrigerant is stored under high pressure since at atmospheric pressure, it vaporizes. In order to use the refrigerant as a liquid, it has to be maintained at high pressure. For R134a, the pressure at 298 K is 455 kPa which is the pressure inside the container. Storing the refrigerant in the tank is not a concern but keeping the glass tube pressurized is a big challenge. It is to be noted that a transparent testing container is required to visually see the onset and end set of freezing. If the visual part is not required, welded metal tubes would have been a better option. There are limited numbers of tubes and the test procedure requires the tube to be vacuumed out before every test run. This requirement meant there has to be a needle valve to fill the refrigerant and a main threaded stopper. The two openings had to be tightly sealed or else the refrigerant leaked. Clathrate formation does not take place whenever there is even a minor refrigerant leak. The challenge magnified for the thermal properties' experiments. Since two more openings are required for two sets of thermo couples, the chances of refrigerant leaks magnified. It is particularly tricky to seal the thermocouple wire since it has to be a single wire that needs to perfectly seal with the cork material. The sealing usually did not last more than two runs and for discharging, when the pressure reached 590 kPa, it leaked for every second run.

7.6.3 Analysis

This section presents the overall and part analyses of the PCMs under investigation. It also presents the energy and exergy analyses of the entire system. A single-component pure substance changes its phase at a precise temperature. The material temperature does not change until the material has completely changed its phase. Such materials are considered as one-region materials as the temperature change occurs only when the material is in a particular phase region. In a multi-component PCM, as the one under investigation, phase change occurs over a range of temperature, instead of being at a particular temperature. This phase between the solid and liquid region is described as a mushy region where the temperature as well as the phase, both, changes. Since such PCMs would have a solid, liquid and mushy region, the problem associated with such materials is described as multi-region or moving boundary problems. Predicting the behavior of phase-change systems is difficult due to its inherent non-linear nature at moving interfaces, for which displacement rate is controlled by the latent heat lost or absorbed at the boundary (Dutil *et al.*, 2011)

The governing equation for heat can be written as follows:

$$Q = (mC_p\Delta T)_{solid} + m\lambda + (mC_p\Delta T)_{liquid} \tag{7.47}$$

where m is the mass of the PCM, C_P is the specific heat, T is the temperature and λ is the specific latent heat of PCM.

The amount of heat rejected from the PCM tube to the bath is given as follows:

$$Q = m_{bath}C_{p,water}(T_{amb} - T_{ss}) = m_{bath}(h_{amb} - h_{ss}) \tag{7.48}$$

where T_{amb} is the ambient bath temperature and T_{ss} is the steady state temperature of the water bath. Similarly h_{amb} is the ambient bath specific enthalpy and h_{ss} is the steady state specific enthalpy. Specific heat, C_P, is determined using the energy balance equation as follows:

$$Q = m_{PCM}C_P\Delta T_i \tag{7.49}$$

where ΔTi is the temperature difference of the same location in the PCM.

Thermal conductivity, k, is determined using

$$k = \frac{\dot{Q}\ \ln\left(\frac{r_2}{r_1}\right)}{2\pi l\,\Delta T} \tag{7.50}$$

where, r_1 and r_2 is the inner and outer radius of the PCM inside the tube, 'l' is the height of the PCM inside the tube, ΔT is the difference in temperature between location r_1 and r_2.

In order to theoretically estimate the thermal properties, empirical equations are used to validate the trend. Specific heat can be calculated by adding the product of the mass fraction and specific heat of the pure species present in the PCM.

$$C_{P,PCM} = w_1C_{P,1} + w_2C_{P,2} + w_3C_{P,3} \tag{7.51}$$

where $C_{P,i}$ is the heat capacity of species i and *w* its mass fraction.

The other important parameter is the heat of fusion which can be described for two-dimensional problem, as follows:

$$\frac{\partial h}{\partial t} = \frac{\partial}{\partial x}\left(\beta\frac{\partial h_{latent}}{\partial x}\right) - \rho_L\lambda\frac{\partial f}{\partial t} \tag{7.52a}$$

where 'h' is the enthalpy, β is the thermal diffusivity, λ is the spacific latent heat of fusion, 't' is time, ρ is density, and 'f' is liquid fraction describes as follows:

$$f = \begin{cases} 0 & T < T_m \text{ solid,} \\ [0,1] & T = T_m \text{ mushy,} \\ 1 & T > T_m \text{ liquid.} \end{cases} \tag{7.52b}$$

with the thermal diffusivity β defined as:

$$\beta = \frac{k}{C_P \rho} \tag{7.53}$$

where 'k' is the thermal conductivity coefficient and ρ is the density.

Substituting Equation (7.53) into Equation 7.52b yields the enthalpy of fusion with respect to thermal conductivity and density as follows:

$$\frac{\partial h}{\partial t} = \frac{\partial}{\partial x}\left[\left(\frac{k}{C_P \rho}\right)\frac{\partial h}{\partial x}\right] - \rho_L \lambda \frac{\partial f}{\partial t} \tag{7.54}$$

Note that PCM in solid phase has higher density as compared to liquid phase PCM while it is in the middle during the 'mushy' region. In order to further simplify the Equation (7.54) to obtain a preliminary result, it could be assumed that the heat of fusion does not change with respect to the time hence treating it as steady state condition. With this assumption, the equation becomes as follows:

$$\rho_L \lambda \frac{\partial f}{\partial t}\left(\frac{k}{C_P \rho}\right)\frac{\partial^2 h}{\partial x^2} \tag{7.55}$$

Assuming the change is linear with respect to time for 'f' and distance for 'h', equation (9) can be simplified to:

$$\rho_L \lambda \frac{\Delta f}{\Delta t}\left(\frac{k}{C_P \rho}\right)\frac{\partial^2 h}{\partial x^2} \tag{7.56}$$

Since it is difficult to predict the exact density during phase change, it is important to have the density eliminated with something that tends to stay constant during phase change. Unlike density, mass of the material tends to stay the same so an expression can be arrived upon without density of the PCM.

Since 'h' is the enthalpy, substituting mass per unit volume in place of density would yield the enthalpy value. The equation becomes:

$$\rho_L \lambda \frac{\Delta f}{\Delta t}\left(\frac{k}{C_p m}\right)\frac{\partial^2 H}{\partial x^2} \tag{7.57}$$

where 'm' is the mass and 'H' is the volumetric enthalpy of fusion.

The thermal conductivity 'k' for any material, in Cartesian coordinates, is defined as:

$$\dot{Q} = \frac{kA\Delta T}{\Delta l} \tag{7.58}$$

where $\dot{Q}$ is the heat rate, A is the cross-sectional area and l is the distance between two temperature readings.

The thermal conductivity 'k' for the refrigerant clathrate can be described as follows (ACE Glass Incorporated, 2014):

$$k = 0.4 \times \exp\left(\frac{T}{T_{cm}} - 1\right)\left(\frac{D_1}{D_2}\right)^2 \times M^{0.5} w_1 w_2 \left(\frac{k_2}{M_2^{0.5}} - \frac{k_1}{M_1^{0.5}}\right) \tag{7.59}$$

where D is the dipole moment, w is the mass fraction and M is the molar mass of the species.

Refrigerants are mixed with water to form the clathrate in liquid phase. Adding solid additives, to improve thermal conductivity, improves the thermal transport properties of the PCM. It requires a different set of equations to predict the thermodynamic properties of the PCM with solid additives, be it salts or nanoparticles.

Several models are presented to predict the thermal conductivity of the fluids containing small solid particles The thermal conductivity of the PCM having solid particles is not easy to calculate as it yields complex parameters upon which the values are based. One of the models is the extension of Effective Medium Theory (EMT) described as (Mattea *et al.*, 1986):

$$\sum_{i=1}^{n} V_i \frac{k_m - k_{ith}}{k_i + \left(\frac{Z}{2} - 1\right) k_m} = 0 \tag{7.60}$$

where V is the volume fraction, Z is the coordination number, k_{ith} is the thermal conductivity of the ith element and k_m is the thermal conductivity of the mixture.

Another model is a two component three dimensional model for isotropic material thermal conductivity as follows (Morley and Miles, 1997):

$$k = k_b \left[\frac{1 - J}{1 - J(1 - V_d)}\right] \tag{7.61}$$

where

$$J = V_d^2 \left(1 - \frac{k_d}{k_b}\right) \tag{7.62}$$

where k is the overall thermal conductivity of the system, k_b and k_d are the thermal conductivities of the continuous and discontinuous component(s), respectively, and V_d is the volume fraction of the discontinuous phase.

The third model that can be used to predict the thermal conductivity of the PCM is based on potential theory as follows (Heldman and Singh, 1981):

$$k = k_b \left[\frac{1 - (1 - a(k_d/k_b))b}{1 + (a - 1)b}\right] \tag{7.63}$$

where

$$a = \frac{3k_b}{2k_b + k_d} \tag{7.64}$$

and

$$b = \frac{V_d^3}{V_d^3 + V_c^3} \tag{7.65}$$

In Equations 7.63–7.65, k_c and k_d are the thermal conductivities of the continuous and discontinuous component(s), respectively. V_d is the volume fraction of the discontinuous phase while V_c is the volume fraction of the continuous phase.

Some models propose prediction of thermal conductivity based on the effect of interfacial layer formed around the nanoparticle, which makes calculations more difficult (Rizvi *et al.*, 2013). An effective model to present the improvement in thermal conductivity of fluid with nanoscale particles is through Single Phase Brownian Model (SPBM) as follows (Prasher *et al.*, 2006):

$$k = k_b(1 + A\ \mathrm{Re}^m \mathrm{Pr}^{0.333} V_d)\frac{[k_d(1+2\alpha) + 2k_b] + 2V_d[k_d(1-\alpha) - k_b]}{[k_d(1+2\alpha) + 2k_b] - V_d[k_d(1-\alpha) - k_b]} \tag{7.66}$$

where V is the volume fraction while A and m are constants. A is independent of fluid type while m depends on the fluid and particle type. Biot number α is:

$$\alpha = 2R_b k_c / d_p \tag{7.67}$$

The equation yields the thermal conductivity of the fluid with nanoparticles that incorporates the conduction contribution of the particles, particle-fluid thermal boundary resistance and the convection contribution. Overall, this Equation (7.67) takes care of the localized convection due to Brownian motion as well; something that previous models failed to address. When the additives are used that are not nanoscale particles, the equation simplifies to only the static thermal conductivity as follows:

$$k = k_b\frac{[k_d(1+2\alpha) + 2k_b] + 2V_d[k_d(1-\alpha) - k_b]}{[k_d(1+2\alpha) + 2k_b] - V_d[k_d(1-\alpha) - k_b]} \tag{7.68}$$

For an uncertainty analysis, the standard deviation is described as follows:

$$\sigma = \sqrt{\frac{\Sigma(x_j - \bar{x})^2}{n-1}} \tag{7.69}$$

where $\bar{x}$ is the mean value and n is the number of trials.

The errors in calculated values, ΔR, is determined using the following equation:

$$\Delta R = \frac{\partial R}{\partial x_1} + \frac{\partial R}{\partial x_2} + \ldots + \frac{\partial R}{\partial x_n} \tag{7.70}$$

where x is the individual parameter making up the calculated result R.

The thermoeconomic analysis on the PCMs is also conducted with the following equation (Abusoglu and Kanoglu, 2009):

$$f_{TE} = \frac{Z_k}{Z_k + \xi Ex_{dst}} \tag{7.71}$$

where Z_k is the total cost of the items used in the PCM in dollars, ξ is the energy cost in \$/J and f_{TE} is the thermoeconomic factor.

Three primary systems are used to conduct the tests and attain the test results. Cold water bath is used for charging the PCMs, hot air for discharge the PCM and battery cooling system to see the effects of PCM cooling. Their analyses are presented in this section.

7.6.3.1 Analysis of Constant Temperature Bath

Cold bath water absorbs the heat from the PCM during the charging process. The bath has a built in refrigeration system that works on basic vapor compression refrigeration cycle. Since it is a close loop system, the mass balance equation for every single component of the constant temperature bath is as follows:

$$\dot{m}_i = \dot{m}_e \tag{7.72}$$

Compressor pressurises the refrigerant while increasing the refrigerant's temperature, enthalpy and entropy. Compressor takes the electrical energy in to produce the work. The work done by the compressor is increasing the pressure and temperature of the refrigerant. The energy balance equation simplifies as follows:

$$\dot{m}_i h_i + \dot{W}_{\mathrm{ele}} = \dot{m}_e h_e + \dot{Q} \tag{7.73}$$

For all the practical systems, exergy is destroyed due to irreversibilities present in it. The exergy balance equation can be written as follows:

$$\dot{m}_i ex_i + \dot{W}_{ele} = \dot{m}_e \mathrm{ex}_e + \dot{E}x_{\mathrm{dst}} + \dot{E}x^Q \tag{7.74}$$

The refrigerant system enters the condenser. The condensore rejects the heat to the surrounding, ideally only changing the refrigerant phase. However, practicaly, the temperature of the refrigerant also drops while the pressure remains the same. The energy balance equation for the condenser refrigerant is given as follows:

$$\dot{m}_i h_i = \dot{m}_e h_e + \dot{Q} \quad \text{(Phase change involved)} \tag{7.75}$$

Similarly, the exergy balance equation is as follows:

$$\dot{m}_i ex_i = \dot{m}_e ex_e + \dot{E}x_{dst} + \dot{E}x^Q \tag{7.76}$$

After condensor, the refrigerant enters the expansion valve. Expansion valve decompresses the refrigerant droping its pressure, subsequently its pressure as well. The enthaly across the expansion valve is designed to remain the same. The energy balance equation is described as follows:

$$\dot{m}_i h_i + \dot{Q} = \dot{m}_e h_e \tag{7.77}$$

The exergy balance equation is described as follows:

$$\dot{m}_i ex_i + \dot{E}x^Q = \dot{m}_e ex_e + \dot{E}x_{dst} \tag{7.78}$$

After the expansion valve, the refrigerant system enters the evalorator. Avaporator coils are wraped around the water bath to cool the bath. As the refrigerant goes through the evaporator, it absorbs the heat from the water bath. This addition of heat ideally only changing the refrigerant phase. However, practicaly, the temperature of the refrigerant also rises while the pressure remains the same. The energy balance equation for the condenser refrigerant is given as follows:

$$\dot{m}_i h_i + \dot{Q} = \dot{m}_e h_e \text{(Phase change involved)} \tag{7.79}$$

Similarly, the exergy balance equation is as follows:

$$\dot{m}_i ex_i + \dot{E}x^Q = \dot{m}_e ex_e + \dot{E}x_{dst} \tag{7.80}$$

7.6.3.2 Analysis of Hot Air Duct

Hot air gives away the heat to the PCM during the discharging process. The hot air duct has a built in electric heater that heats up the air flowing across it. Since it a simple duct with one inlet and one exit (ignoring leaks), the mass balance equation can be written as follows:

$$\dot{m}_i = \dot{m}_e \tag{7.81}$$

The air goes across the eletric heater where its temperature is raised while its pressure remains the same. Heater takes the electrical energy to produce heat. The heat is provided to the flowing air. The energy balance equation simplifies as follows:

$$\dot{m}_i h_i + \dot{Q} = \dot{m}_e h_e \tag{7.82}$$

For all the practical systems, exergy is destroyed due to irreversibilities present in it. The exergy balance equation can be written as follows:

$$\dot{m}_i \text{ex}_i + \dot{E}x^Q = \dot{m}_e \text{ex}_e + \dot{E}x_{\text{dst}} \tag{7.83}$$

7.6.3.3 Analysis of Battery Cooling

When electrical energy in lithium polymer battery is discharged as it runs an electrical motor, it tends to heat up. PCM is used to absorb the heat from the battery and cool the battery down. The battery has a resistance, so when the current is drawn from that battery, the current goes through a certain resistance heating up the battery. For the battery cooling test, no mass transfer is associated with it as only the heat transfer is experienced. The heat from the battery goes through the PCM jacket to the PCM. The PCM heats up and changes its phase by absorbing the energy. The heat released by the battery is associated with the battery's internal energy change which can be described as follows:

$$m\Delta u_{\text{ch}} = \dot{Q}\Delta t \tag{7.84}$$

The exergy balance equation can be written with respect to the chemical exergy as follows:

$$m\Delta \text{ex}_{\text{ch}} = +\dot{E}x^Q \Delta t \ + \dot{E}x_{\text{dst}} \tag{7.85}$$

7.6.3.4 Energy and Exergy Analyses

The general energy and exergy balance equations for the PCM charging and discharging are presented in this section. Charging is described as the process of solidification while discharge is described as the process of melting. PCM gives away the heat during the process of charging while it takes in the heat during the process of discharging. The schematic diagram of the charging, storage and discharging process for the PCM is shown in Figure 7.57.

For charging the PCM, mass balance equation for the charging fluid can be described as follows:

$$(\dot{m}_{\text{in}})\ \Delta t = (\dot{m}_{\text{out}})\ \Delta t = (\dot{m}_c)\ \Delta t \tag{7.86}$$

where $\dot{m}$ is the mass flow rate, subscript 'c' is refers to the charging fluid while 'in' and 'out' refers to the incoming and exiting fluid, respectively.

The energy balance equation between the charging fluid and PCM for the flow can be described as follows:

$$\begin{aligned}&[(\dot{m}_c h_c)\Delta t]_{\text{out}} - [(\dot{m}_c h_c)\Delta t]_{\text{in}} \\ &\quad = [(\dot{m}_{PCM} h_{PCM})\Delta t]_{\text{in}} - [(\dot{m}_{PCM} h_{PCM})\Delta t]_{\text{out}} + \dot{Q}_{\text{gain}}\Delta t\end{aligned} \tag{7.87}$$

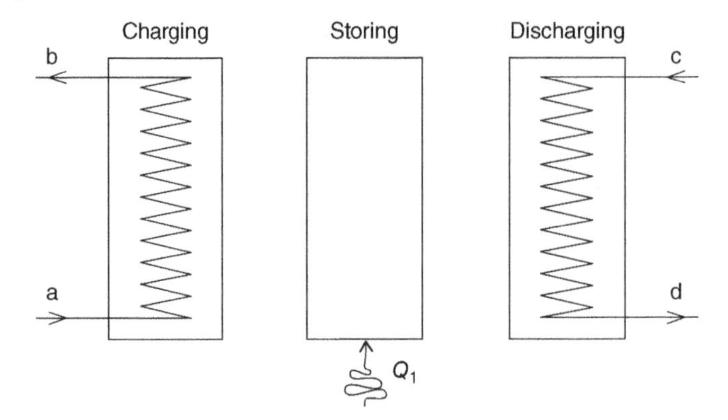

Figure 7.57 Schematic diagram of the charging, storage and discharging process for the PCM (adapted from Dincer and Rosen, 2013).

where 'h' is the specific enthalpy of the substance while subscript 'gain' referrers to the heat gained from the surrounding over the period of Δt.

When the charging is done using stationary fluids to a stationary PCM over a time period, the equation becomes as follows:

$$(m_c h_c)_f - (m_c h_c)_i = (m_{PCM} h_{PCM})_i - (m_{PCM} h_{PCM})_f + \dot{Q}_{\text{gain}} \Delta t \tag{7.88}$$

The entropy balance equation between the charging fluid and PCM for the flow can be described as follows:

$$(\Delta \dot{m}_c \Delta s_c \Delta t) = [(\dot{m}_{PCM} s_{PCM})\Delta t]_{\text{in}} - [(\dot{m}_{PCM} s_{PCM})\Delta t]_{\text{out}} + \frac{\dot{Q}_{\text{gain}}}{T_0}\Delta t + \dot{S}_{gen}\,\Delta t \tag{7.89a}$$

where 's' is the specific entropy of the substance, T_0 is the ambient temperature and $\dot{S}_{gen}$ is the entropy generation rate. When the charging is done using stationary fluids to a stationary PCM over a time period, the equation becomes as follows:

$$(m_c s_c)_f - (m_c s_c)_i = (m_{PCM} s_{PCM})_i - (m_{PCM} s_{PCM})_f + \frac{\dot{Q}_{\text{gain}}}{T_0}\Delta t + \dot{S}_{gen}\,\Delta t \tag{7.89b}$$

The exergy balance equation between the charging fluid and PCM for the flow can be described as follows:

$$(\Delta \dot{m}_c \Delta \text{ex}_c \Delta t) = [(\dot{m}_{PCM} ex_{PCM})\Delta t]_{\text{in}} - [(\dot{m}_{PCM} ex_{PCM})\Delta t]_{\text{out}} + \dot{E}x_{\text{ dst}}\Delta t + \dot{E}x^{Q}\Delta t \tag{7.90}$$

where 'ex' is the specific exergy, $\dot{E}x_{\text{Dst}}$ is the exergy destruction and $\dot{E}x^{Q}$ is the thermal exergy loss of the substance.

The thermal exergy loss can be described as:

$$\dot{E}x^{Q} = \left(1 - \frac{T_0}{T}\right)\dot{Q} \tag{7.91}$$

When the charging is done using stationary fluids to a stationary PCM over a time period, the equation becomes as follows:

$$(m_c\text{ex}_c)_f - (m_c\text{ex}_c)_i = (m_{PCM}ex_{PCM})_i - (m_{PCM}ex_{PCM})_f + \dot{E}x_{\text{dst}}\Delta t + \dot{E}x^{\text{Q}}\Delta t \quad (7.92)$$

For discharging the PCM, mass balance equation for the charging fluid can be described as follows:

$$(\dot{m}_{\text{in}})\,\Delta t = (\dot{m}_{\text{out}})\,\Delta t = (\dot{m}_c)\,\Delta t \quad (7.93)$$

The energy balance equation between the discharging fluid and PCM for the flow can be described as follows:

$$[(\dot{m}_{PCM}h_{PCM}\Delta t]_{\text{out}} - [(\dot{m}_{PCM}h_{PCM})\Delta t]_{\text{in}} = [(\dot{m}_c h_c)\Delta t]_{\text{in}} - [(\dot{m}_c h_c)\Delta t]_{\text{out}} + \dot{Q}_{\text{gain}}\Delta t \quad (7.94)$$

When the discharging is done using stationary fluids to a stationary PCM over a time period, the equation becomes as follows:

$$(m_{PCM}h_{PCM})_f - (m_{PCM}h_{PCM})_i = (m_c h_c)_i - (m_c h_c)_f + \dot{Q}_{\text{gain}}\Delta t \quad (7.95)$$

When the heat is being absorbed from a stationary solid, the energy balance equation becomes as follows:

$$(m_{PCM}h_{PCM})_f - (m_{PCM}h_{PCM})_i = \dot{Q}\Delta t \quad (7.96)$$

The entropy balance equation between the discharging fluid and PCM for the flow can be described as follows:

$$(\Delta\dot{m}_{PCM}\Delta s_{PCM}\Delta t) = [(\dot{m}_c s_c)\Delta t]_{\text{in}} - [(\dot{m}_c s_c)\Delta t]_{\text{out}} + \frac{\dot{Q}_{\text{gain}}}{T_0}\Delta t + \dot{S}_{gen}\Delta t \quad (7.97)$$

When the discharging is done using stationary fluids to a stationary PCM over a time period, the equation becomes as follows:

$$(m_{PCM}s_{PCM})_f - (m_{PCM}s_{PCM})_i = (m_c s_c)_i - (m_c s_c)_f + \frac{\dot{Q}_{\text{gain}}}{T_0}\Delta t + \dot{S}_{gen}\Delta t \quad (7.98)$$

When the heat is being absorbed from a stationary solid, the entropy balance equation becomes as follows:

$$(m_{PCM}s_{PCM})_f - (m_{PCM}s_{PCM})_i = -\frac{\dot{Q}_{\text{supply}}}{T_0}\Delta t + \dot{S}_{gen}\Delta t \quad (7.99)$$

The exergy balance equation between the discharging fluid and PCM for the flow can be described as follows:

$$(\Delta\dot{m}_{PCM}\Delta ex_{PCM}\Delta t) = [(\dot{m}_c ex_c)\Delta t]_{\text{in}} - [(\dot{m}_c ex_c)\Delta t]_{\text{out}} + \dot{E}x_{\text{dst}}\Delta t + \dot{E}x^{\text{Q}}\Delta t \quad (7.100)$$

When the discharging is done using stationary fluids to a stationary PCM over a time period, the equation becomes as follows:

$$(m_{PCM}ex_{PCM})_f - (m_{PCM}ex_{PCM})_i = (m_c ex_c)_i - (m_c ex_c)_f + \dot{E}x_{\text{dst}}\Delta t + \dot{E}x^{\text{Q}}\Delta t \quad (7.101)$$

When the heat is being absorbed from a stationary solid, the equation becomes as follows:

$$(m_{PCM}ex_{PCM})_f - (m_{PCM}ex_{PCM})_i = \dot{E}x_{dst}\Delta t - \dot{E}x^Q{}_{supply}\Delta t \tag{7.102}$$

where $\dot{E}x^Q{}_{supply}$ is the thermal exergy.

In order to determine the efficiencies, it is first important to describe the useful input and required output of the system.

For the charging process, the heat absorbed by the charging fluid $Q_{in,c}$ is described as:

$$[(\dot{m}_c h_c)\Delta t]_{out} - [(\dot{m}_c h_c)\Delta t]_{in} = Q_{in,c} \tag{7.103}$$

While heat given out by the PCM $Q_{out,PCM}$ is

$$(m_{PCM}h_{PCM})_i - (m_{PCM}h_{PCM})_f = Q_{out,PCM} \tag{7.104}$$

The thermal exergy absorbed by the charging fluid $Ex^Q_{in,c}$ is described as:

$$[(\dot{m}_c ex_c)\Delta t]_{out} - [(\dot{m}_c ex_c)\Delta t]_{in} = Ex^Q_{in,c} \tag{7.105}$$

While thermal exergy given out by the PCM $Ex^Q_{out,PCM}$ is defined as:

$$(m_{PCM}ex_{PCM})_i - (m_{PCM}ex_{PCM})_f = Ex^Q_{(out,PCM)} \tag{7.106}$$

For the discharging process, the heat absorbed by the PCM $Q_{in,PCM}$ is described as:

$$(m_{PCM}h_{PCM})_f - (m_{PCM}h_{PCM})_i = Q_{in,PCM} \tag{7.107}$$

The heat released by the discharging fluid $Q_{out,c}$ or the heat emitted by the stationary solid is described as:

$$(m_c h_c)_i - (m_c h_c)_f = Q_{out,c} = \dot{Q}\Delta t \tag{7.108}$$

The thermal exergy absorbed by the PCM $Ex^Q_{in,PCM}$ is described as:

$$(m_{PCM}ex_{PCM})_f - (m_{PCM}ex_{PCM})_i = Ex^Q_{in,PCM} \tag{7.109}$$

The thermal exergy released by the discharging fluid $Ex^Q_{out,c}$ or the thermal exergy released by the stationary solid is described as:

$$(m_c ex_c)_i - (m_c ex_c)_f = Ex^Q_{out,c} = \dot{E}x^Q_{supply}\Delta t \tag{7.110}$$

The overall system efficiencies can now be described since useful output and required inputs have been established. The required input is the energy/exergy released by the charging material to change the phase or charge the PCM. The useful output is the energy/exergy absorbed by the discharging material which in turn is absorbed by the PCM. The overall system's energy efficiency can be described as:

$$\eta_{oa} = \frac{Q_{in}}{Q_{out}} \tag{7.111}$$

The overall system's exergy efficiency can be described as:

$$\Psi_{oa} = \frac{Ex_{in}}{Ex_{out}} \tag{7.112}$$

7.6.4 Results and Discussion

This section provides the results and discussion of the study conducted on novel PCMs. Experimental results of charging, discharging, battery cooling test results, thermoeconomic, uncertainty analysis, analytical study, optimization and validation are presented here. Energy and exergy evaluation of the charging and discharging process is also given in this section.

After the literature review, the two refrigerant candidates were determined to be R141b and R134a. Initially, the tests are conducted to find out if the selected refrigerant formed clathrate at temperature above 0 °C (273 K). Furthermore, the most appropriate refrigerant percent composition is determined. Later, several different additives are added with the refrigerant clathrates to test the improvement in the thermal properties. Sodium chloride (NaCl), magnesium nitrate hexahydrate ($Mg(NO_3)_2.6H_2O$), aluminum particles, copper particles and ethanol are selected as additives. The additives are added to study the improvement in the thermal properties of the clathrate based on the before mentioned refrigerants. The thermal properties are determined for a variable fraction of additives since their solubility changes with the change in temperature (Hernandez, 2011; DuPont, 2004).

The safety, handling and storing characteristics of the used materials are discussed as follows:

- R134a is a class common refrigerant used in a variety of domestic and commercial applications. It is classified as A1 in the American Society of Heating, Refrigeration, and Air Conditioning Engineers safety group. It means that R134a has low toxicity and does not propagate flame. R134a operates at about 340 kPa at 298 K so containment is required to prevent leaks and needs to be stored under 323 K. R134a however has high global warming potential and more environmentally friendly alternatives will soon be available in the market.
- R141b is nonflammable and has a very low operating pressure. It does not require any specialized pressurized system to contain it but it must be stored below 323 K.
- Ethanol is considered flammable so it must be kept away from ignition, spark or extreme high temperatures. Ethanol is toxic if ingested and it causes irritation to skin and eyes.
- Magnesium nitrate hexahydrate may be flammable at high temperatures. It can cause irritation to skin and eyes in case of contact. It is considered hazardous to ingest or to inhale it.
- Aluminum is slightly hazardous in case of skin contact as it causes irritation. It does not cause any irritation to the eyes and is considered non-hazardous in case of ingestion. Aluminum is considered nonflammable.
- Copper is considered very hazardous, if ingested. It is hazardous in case of inhalation, skin contact and eye contact as it causes irritation. It can be flammable at high temperatures.
- Sodium chloride is slightly hazardous in case of skin contact as it causes irritation. It can also cause irritation to the eyes and is considered non-hazardous in case of ingestion. Sodium chloride is considered nonflammable.

7.6.4.1 Test Results of Base PCM

Experiments are conducted to determine the onset and end set times of R134a and R141b clathrates without additives. Refrigerant mass fraction is varied from 0.15 to 0.4 with 0.05 intervals. Figure 7.58 shows the graphical illustration of water and refrigerant mass for each fraction.

Different fractions of refrigerant and water ratios are tested to find out the most appropriate combination. Table 7.27 shows the values of water and refrigerant mass used for each fraction. The total mass of the tested clathrate is maintained at the value of 80 grams.

Refrigerant clathrate is tested at two different bath temperatures for charging. Refrigerant clathrates are tested at 3 °C (276 K) and 5 °C (278 K). Water at 5 °C (278 K)

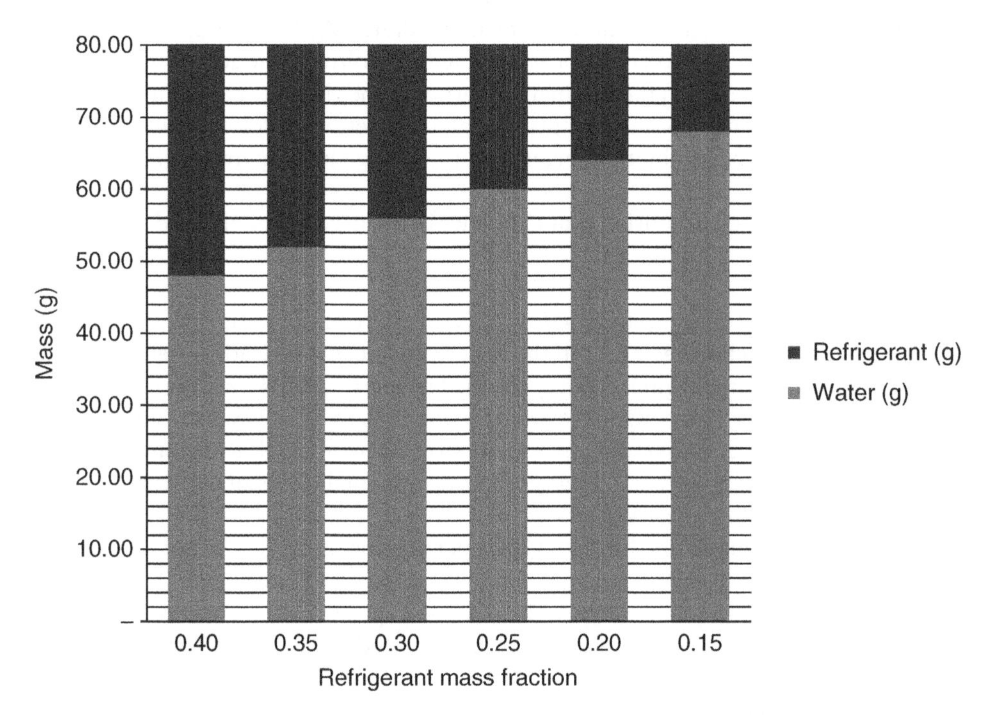

Figure 7.58 Graphical illustration of water and refrigerant masses for each fraction (Zafar, 2015).

Table 7.27 Values of water and refrigerant mass for each fraction.

Mass (g)	Refrigerant mass fraction	Water (g)	Refrigerant (g)
80	0.40	48	32
80	0.35	52	28
80	0.30	56	24
80	0.25	60	20
80	0.20	64	16
80	0.15	68	12

Source: Zafar, 2015.

temperature requires lower energy while 3 °C (276 K) water temperature requires greater energy to bring the bath to their desired temperatures. Charging times are important to determine as they yield the total energy used to form the PCMs. Less charging time means low energy while more time means large amount of energy required to form the PCMs. R141b did not form clathrate at 3 °C (276 K) for all the tested refrigerant fractions. No more tests are conducted on R141b since it failed to form the clathrate at lower temperature.

Figure 7.59 shows the formed R134a clathrate in the glass tubes at 300 kPa and 276 K. Figure 7.59a represents the clathrate with 0.15 refrigerant mass fraction, (b) is 0.2, (c) is 0.25, (d) is 0.3, (e) is 0.35 and (f) is 0.4. In Figure 7.59a,b, water can be seen that has not contributed towards clathrate formation. This is due to lack of refrigerant in the tube which leaves some water to remain liquid and not form clathrate. From Figure 7.59c-f, an almost complete utilisation of water can be observed. The illustrations shows that refrigerant mass fractions of 0.15 and 0.2 do not form complete clathrate hence should not be considered for any further analysis.

Figure 7.60 shows the R134a clathrate onset and end set average times for clathrate formation at different refrigerant mass fractions. Onset time is the time clathrate takes to start freezing while end set is when the process of freezing is complete. It is to be noted that complete freezing does not necessarily mean everything in the tube is frozen. For some fractions, either the water or the refrigerant remains liquid and does not freeze at the water bath temperature of 276 K. Refrigerant mass fractions of 0.15 to 0.40 are

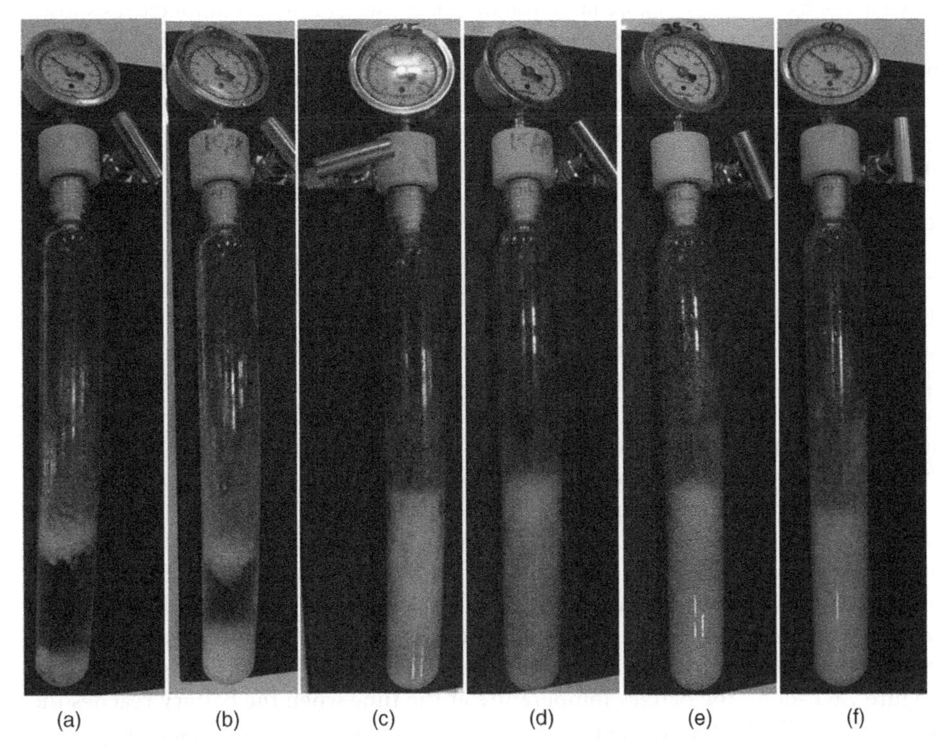

Figure 7.59 R134a clathrates in tubes (a) 0.15, (b) 0.2, (c) 0.25, (d) 0.3, (e) 0.35 and (f) 0.4 refrigerant mass fraction at 276 K (Zafar, 2015).

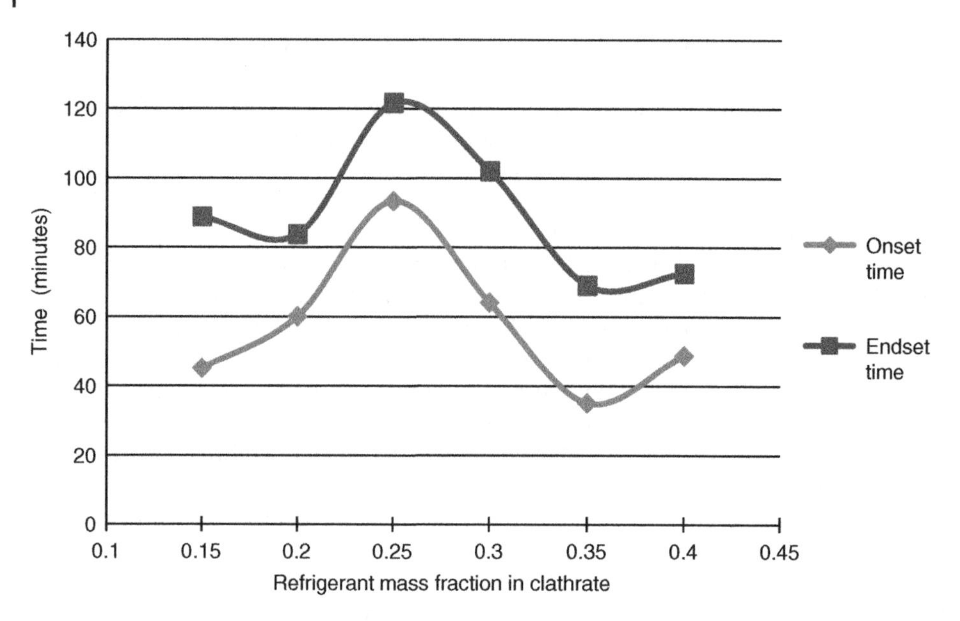

Figure 7.60 R134a clathrate times for onset and end set at different refrigerant mass fractions at 278 K (modified from Zafar, 2015).

shown in the figure. Below 0.25, a large fraction of the water remains unmixed, as shown in Figure 7.60. Above 0.4, the refrigerant does not have enough water to mix with hence does not get utilized. The graph shows that the charging time reduces until 0.35 refrigerant mass fraction and then it starts to increase. From the tests, it is concluded that 0.35 is the most optimal mass fraction for refrigerant since it takes the least amount of time.

7.6.4.2 Results of Battery Cooling Tests

The experiments are conducted to determine discharge capability of the PCMs. The PCMs are discharged using hot air to find out their discharge capacity. Furthermore, a battery jacket is made to cool down the battery using the PCMs. Cooling down the battery using the PCMs shows the potential application.

Figure 7.61 shows the total time battery takes to cool down using each PCM type. The pattern bars in the above graph shows the averaged value of the time readings for each case. The battery is considered to have cooled down and ready to be charged again when the battery temperature reaches 26 °C. When the battery jacket is empty, it tends to prevent battery cooling compared to the case when the battery is exposed to ambient air. The results show that PCM with ethanol additive cools down the battery fastest with copper, aluminum and no additive PCM have similar cooling time. Salt additives have higher cooling time compared to other additives because of their mildly soft structure and low cool energy storage capacity. It is interesting to note that no additive base clathrate PCM performs very well, compared to other PCMs, while cooling the battery.

Figure 7.62 shows the battery temperature at the time when the battery reaches the cutoff voltage. The battery temperature at cutoff is recorded for each case. The pattern bars in the above graph shows the averaged value of the temperature readings for each case. It is important to find this temperature since it effects the total cooling time of

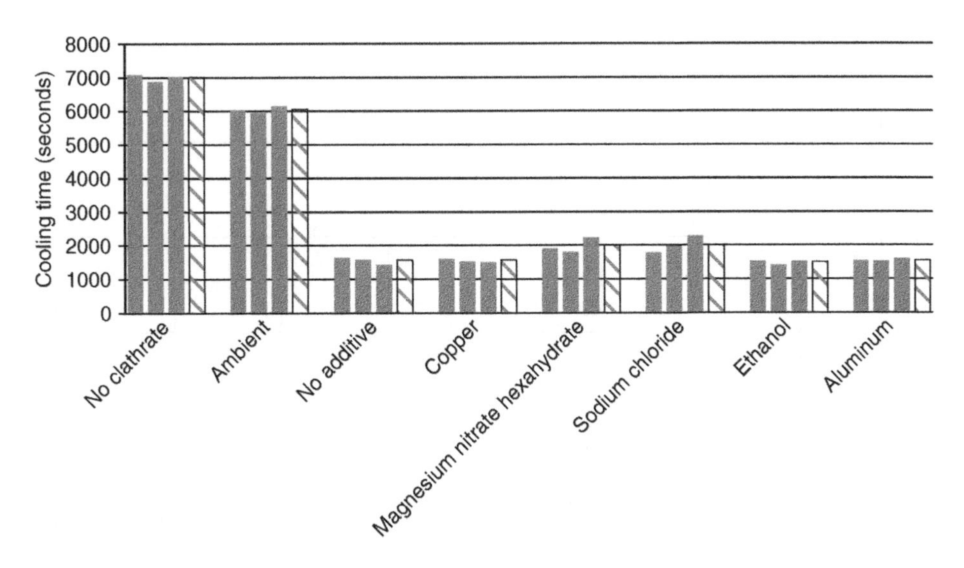

Figure 7.61 Total battery cooling down time using each PCM (adapted from Zafar, 2015).

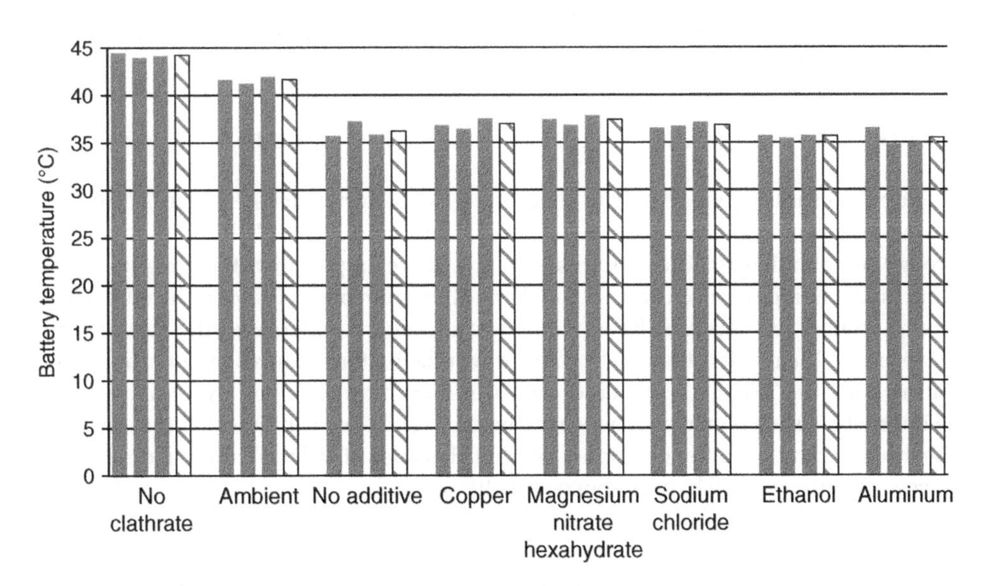

Figure 7.62 Battery temperature at cutoff voltage (adapted from Zafar, 2015).

the battery. It is also of significance since a cooler battery can be operated longer or more power burst can be extracted from it. The results show that empty jacket with no clathrate in it tends to heat up the battery when compared to the case when the battery is exposed to ambient air. This heating of battery can be associated due to the heat being trapped in the jacket with little dissipation to the surroundings. PCM with ethanol and aluminum additive helps battery maintain the lowest temperature. PCMs with salt additives keep the battery temperature higher than the other PCMs. Overall; the PCMs reduce the battery temperature by approximately 6 °C.

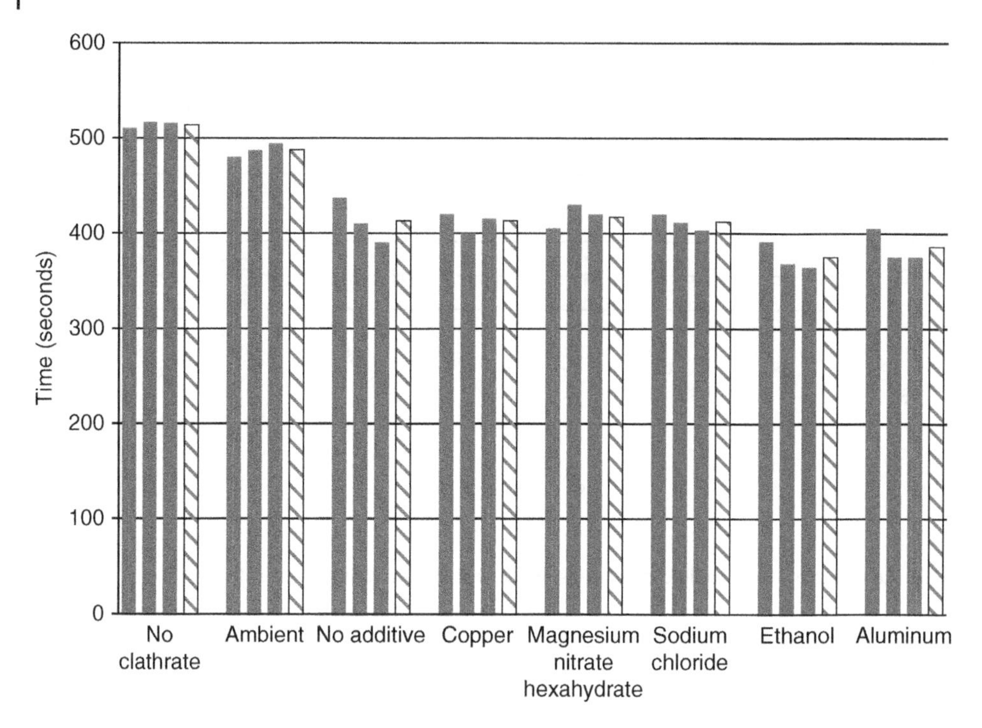

Figure 7.63 Battery run times until the cutoff voltage (adapted from Zafar, 2015).

Figure 7.63 shows the battery run times until the battery reaches the cutoff voltage. For the "Ambient" case, no jacket is used as it is desired for the battery to be cooled through surrounding air. The pattern bars in the above graph shows the averaged value of the time readings for each case. Ethanol and aluminum based PCMs reduced the run time to around 380 seconds. Jacket without clathrate case increased the run time to 510 seconds. The results show that cooling down the battery reduces its run time while heating it up, the "No clathrate" case, improves its run time. Under low temperatures, the internal resistance increases hence the battery yields low run time. Whereas, under higher temperature, it gives out higher run time because of reduced internal resistance. However, heating up the battery shortens its life span.

7.6.4.3 Results of Energy and Exergy Analyses on Base Clathrate

Energy and exergy analyses are conducted on the base PCM (R134a clathrate without any additive). These analyses determined the amount of energy and exergy required to charge the PCMs. The analyses are also conducted to compare the two different bath temperatures of 276 K and 278 K for charging. Table 7.28 shows the energy and exergy values calculated for the onset and end set of charging the base clathrate. The energy and exergy values calculated are for water bath temperature of 278 K.

Figure 7.64 shows the energy and exergy values of R134a clathrate at different refrigerant mass fractions for 278 K bath temperature. The energy and exergy values are for the charging of the R134a clathrate without any additive. The energy and exergy analyses are done based on the end set time of R134a clathrate. Refrigerant fractions of 0.15 to 0.4 are shown in the figure. For 0.15, 0.2 and 0.25 mass fractions, the excess water

Table 7.28 Energy and exergy values for onset and end set times of charging process of base clathrate at 278 K bath temperature.

End Set at 278 K				Onset at 278 K			
$\dot{Q}$ (W)	Q (kJ)	$\dot{Ex}$ (W)	Ex (kJ)	$\dot{Q}$ (W)	Q (kJ)	$\dot{Ex}$ (W)	Ex (kJ)
95	639	68	457	95	398	68	284
95	531	68	379	95	355	68	254
95	590	68	421	95	398	68	284
95	583	68	416	95	398	68	284
95	533	68	380	95	355	68	254
95	546	68	390	95	384	68	274

Source: Zafar, 2015.

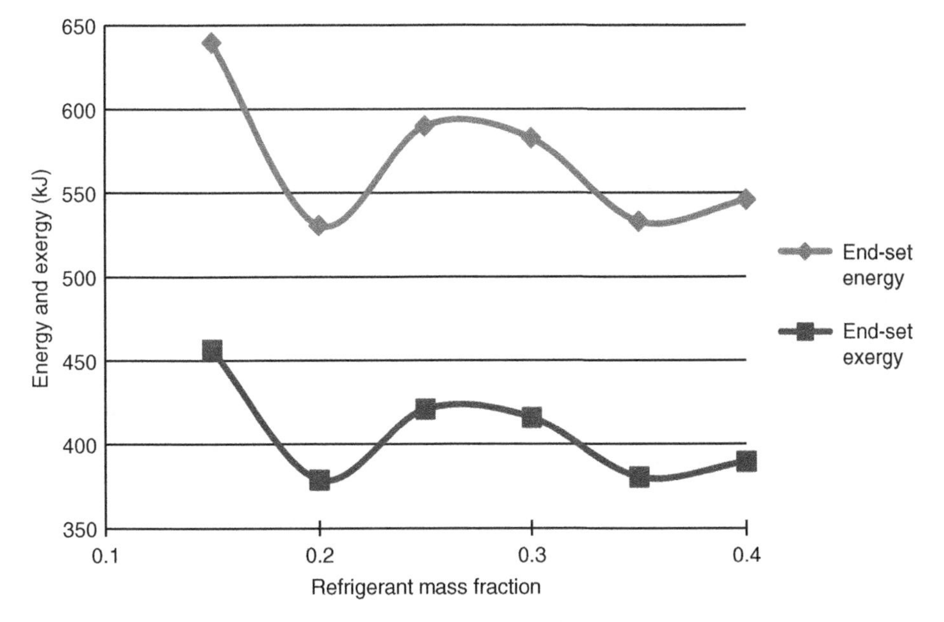

Figure 7.64 End set energy and exergy values at different refrigerant mass fractions for 278 K bath temperature (adapted from Zafar, 2015).

or the refrigerant remains liquid and does not freeze at the water bath temperature of 278 K. Because these three fractions do not form clathrate comprehensively, they are not considered for any further analysis. The charging time reduces until 0.35 refrigerant mass fraction and then it starts to increase, as is the case with the charging time. The analysis shows that 0.35 refrigerant mass fraction is the most optimal mass fraction for clathrate charging since it takes the least amount of energy. The energy required at 0.35 refrigerant fraction is 532 kJ and the exergy is at 380 kJ. Table 7.29 shows the energy and exergy values calculated for the end set of charging process of the base clathrate. The energy and exergy values calculated are for water bath temperature of 276 K.

Table 7.29 Energy and exergy values for onset and end set times of charging process of base clathrate at 276 K bath temperature.

End Set at 276 K				Onset at 276 K			
$\dot{Q}$ (W)	Q (kJ)	$\dot{Ex}$ (W)	Ex (kJ)	$\dot{Q}$ (W)	Q (kJ)	$\dot{Ex}$ (W)	Ex (kJ)
107	572	68	360	107	290	68	183
107	540	68	340	107	387	68	243
107	784	68	494	107	601	68	379
107	657	68	414	107	412	68	260
107	444	68	280	107	225	68	142
107	467	68	294	107	314	68	198

Source: Zafar, 2015.

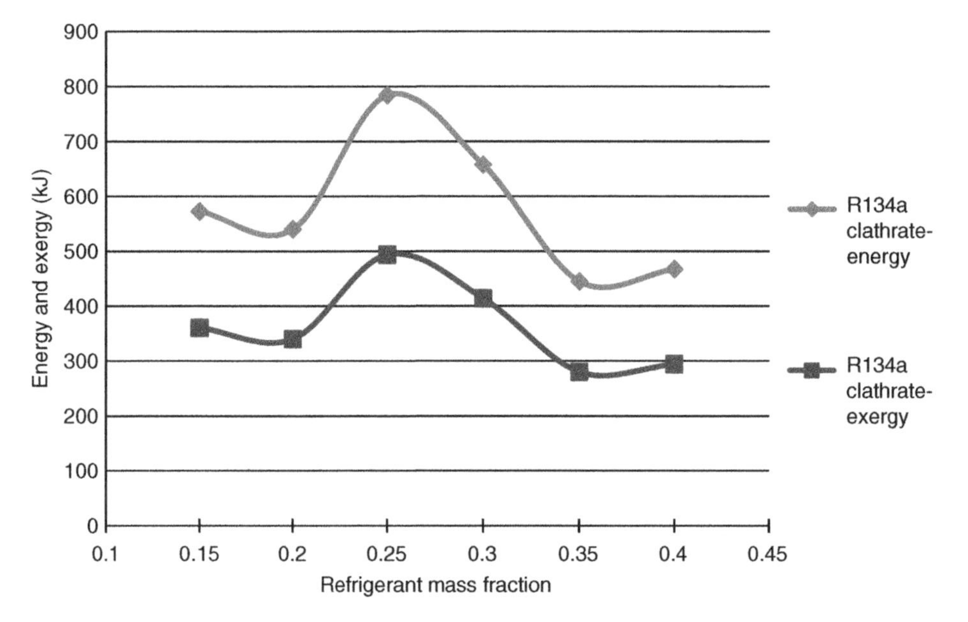

Figure 7.65 End set energy and exergy values at different refrigerant mass fractions for 276 K bath temperature (adapted from Zafar, 2015).

Figure 7.65 shows the R134a clathrates' energy and exergy values at different refrigerant mass fractions for 276 K bath temperature. The energy and exergy values are for the charging of the R134a clathrate without any additive. The energy and exergy analyses are done based on the end set time of R134a clathrate. Refrigerant mass fractions of 0.15 to 0.4 are shown in the figure. For 0.15 and 0.2 mass fractions, the excess water or the refrigerant remains liquid and does not freeze at the water bath temperature of 276 K. Because these two fractions do not form clathrate comprehensively, they are not considered for any further analysis. The graph shows that the charging time reduces until 0.35 refrigerant mass fraction and then it starts to increase, as is the case with the charging time. The analysis shows that 0.35 refrigerant mass fraction is the most optimal

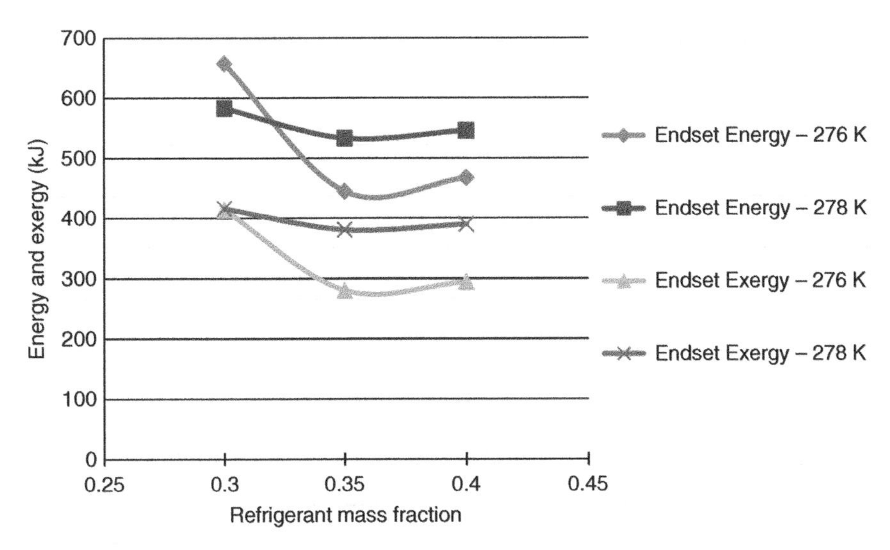

Figure 7.66 Base PCM end set energy and exergy values for charging process at 276 K and 278 K (adapted from Zafar, 2015).

mass fraction for clathrate charging since it takes the least amount of energy. The energy required at 0.35 refrigerant mass fraction is 445 kJ and the exergy is at 280 kJ.

Figure 7.66 shows end set energy and exergy amounts required to charge the base PCM with water bath temperatures of 276 K and 278 K. The graph shows the energy and exergy for refrigerant mass fraction of 0.3, 0.35 and 0.4. Both energy and exergy amounts required to charge the base PCM is lower when the bath temperature is 276 K compared to when the bath temperature is 278 K. Although it takes greater energy to cool the bath to 276 K, the PCM takes shorter time to get charged at lower temperature. Since it takes shorter to charge the PCM at 276 K, the total energy and exergy amounts required to charge the PCM is lower at this bath temperature.

7.6.4.4 Results of Thermoeconomic Analysis

Using the equations described in the previous section, thermoeconomic analyses are conducted on the PCMs used in the experiments. Thermoeconomic analyses include the evaluation of thermoeconomic factor and cost-benefit analyses for the PCMs. Figure 7.67 shows the variation of thermoeconomic variable, f_{TE} as it changes with respect to each PCM. Thermoeconomic variable, f_{TE}, is studied for each PCM including its energy, containment and PCM components costs. For thermoeconomic factor, higher the value, more feasible it is. The results show that magnesium nitrate hexahydrate based PCM has the highest thermoeconomic factor while sodium chloride based PCM has the lowest thermoeconomic factor. The low thermoeconomic factor for magnesium nitrate hexahydrate based PCM is due to its low exergy destruction. Similarly, high thermoeconomic factor for sodium chloride based PCM is due to its high exergy destruction.

Figure 7.68 shows the energy cost of producing the PCM and amount saved using the PCM. The energy cost is calculated using the electricity unit rate of \$0.32. Ethanol, having the highest efficiency, gives the highest return in terms of discharge energy. Sodium chloride has the greatest difference between charging and discharging price due to its

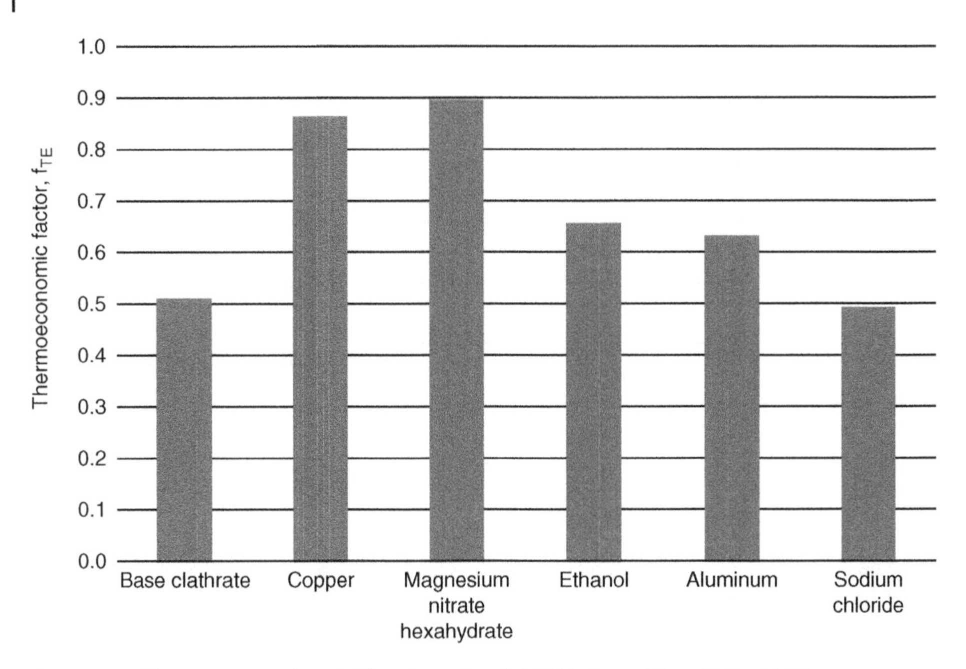

Figure 7.67 Thermoeconomic variable values of each PCM (adapted from Zafar, 2015).

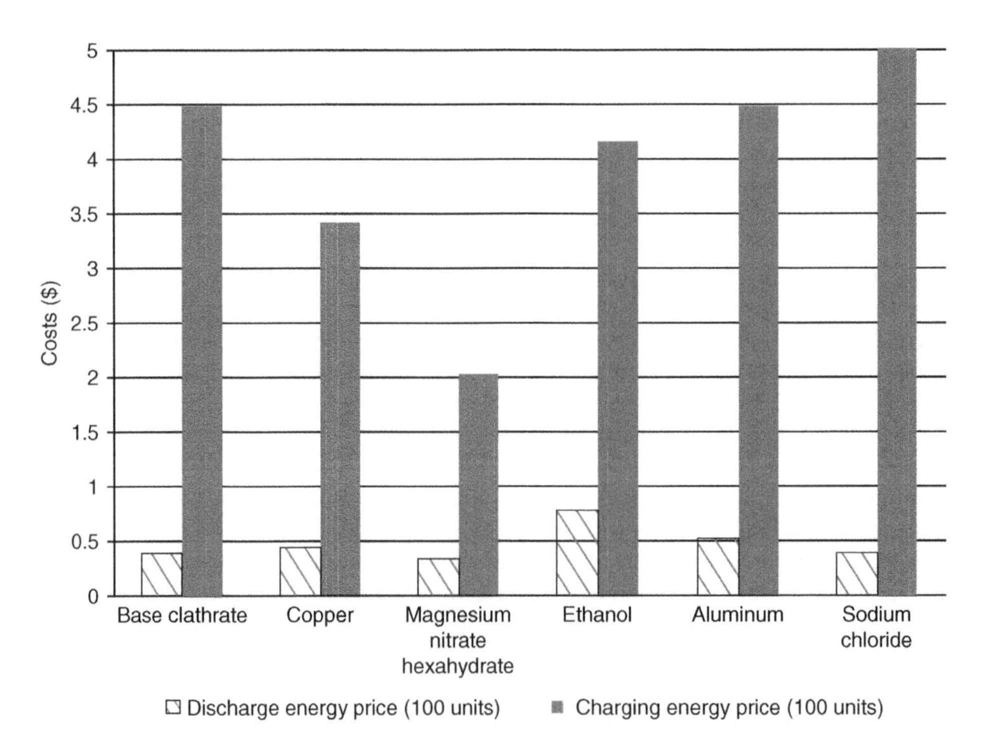

Figure 7.68 Energy costs of producing and using PCM (adapted from Zafar, 2015).

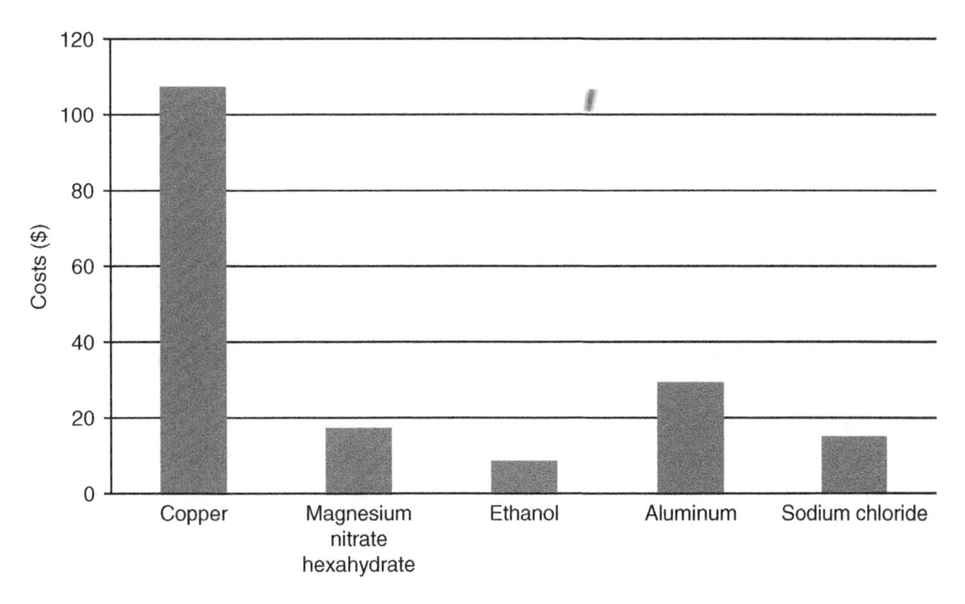

Figure 7.69 Costs of producing 100 units of each PCMs (adapted from Zafar, 2015).

low efficiency. As it can be seen from the graph that the energy cost of charging 100 units is not very high and it remains below $5. Magnesium nitrate hexahydrate shows to have the lowest energy cost but it also has the lowest return.

Figure 7.69 shows the costs of producing 100 units of each PCM with additives compared to the base PCM without any additive. This figure includes the price for 80 gram additive and the costs of energy to charge it. The energy cost is calculated using the electricity unit rate of $0.32. Copper additive proves to be the most expensive primarily due to the price of copper particles. Magnesium nitrate hexahydrate has low cost since it does not take long to get charged. It should be noted that magnesium nitrate hexahydrate does not produce a lot either so its low price is somewhat deceiving. Ethanol has the second lowest cost due to its low price. Ethanol additive makes the most economical PCM since it costs relatively low to produce it yet it gives the highest efficiency.

7.6.5 Closing Remarks

Experimental studies are conducted on refrigerant clathrates for use in cooling applications. Refrigerant R134a and R141b are used to form the clathrate. Sodium chloride, magnesium nitrate hexahydrate, aluminum particles, copper particles and ethanol is used as an additive to determine their impact on the refrigerant clathrate. Charging time, discharging time, thermal properties, energy, exergy and battery cooling characteristics are evaluated for the PCM. The major findings in this case study are as follows:

- R141b does not form clathrate at tested temperatures of 276 K.
- Refrigerant R134a mass fraction of 0.35 requires the lowest time to form the complete clathrate with bath temperature of 276 K and 278 K.

- Additives copper, aluminum, magnesium nitrate hexahydrate and ethanol decrease the onset time of the PCM. Whereas, sodium chloride increases the onset time.
- The magnesium nitrate hexahydrate forms the clathrate fastest, followed by copper, ethanol, aluminum and then sodium chloride. The copper and aluminum additives show an improvement of 71%, magnesium nitrate hexahydrate 57% and ethanol by 28%.
- The magnesium nitrate hexahydrate and copper accelerate the clathrate formation while aluminum and ethanol does not affect the charging time much.
- The sodium chloride delays the clathrate formation time while at 0.04 and 0.05 sodium chloride mass fraction, it does not allow clathrate formation.
- Increasing the additive fraction does not help speed up the process while in some cases it slows the charging.
- Energy and exergy values follow the same trend as the charging time.
- Greater charging time does not necessarily mean that the PCM is ineffective for cooling applications.
- The liquid phase specific heat capacity of the base PCM is found to be 3140 J/kg K. Additives reduce the liquid phase specific heat capacity of the base PCM with sodium chloride having the most adverse effect.
- The liquid phase thermal conductivity of the base PCM is found to be 0.09 W/m K. All the additives increased the liquid phase thermal conductivity of the base PCM. Copper additive improved the thermal conductivity by 350%, followed by aluminum at 230%, magnesium nitrate hexahydrate at 134%, ethanol at 60% and sodium chloride at 50%.
- The solid phase thermal conductivity of the base PCM is found to be 0.33 W/m K. Except for sodium chloride; all other additives increase the solid phase thermal conductivity of the base PCM. Copper additive improved the thermal conductivity by 74%, followed by aluminum at 52%, magnesium nitrate hexahydrate at 40% and ethanol at 9%. Sodium chloride reduced the solid phase thermal conductivity by 12%.
- Magnesium nitrate hexahydrate additive lowers the specific latent heat of the PCM by 68%. The specific latent heat of the PCM with ethanol is 600% more than the base PCM.
- Compared to analytical results, all the experimental values for the liquid phase specific heat are found within 14% difference. For liquid phase thermal conductivity, the experimental values are found within 26% difference.
- Ethanol additive lasts the longest during discharge for 435 seconds while magnesium nitrate hexahydrate lasts the shortest at 180 seconds.
- PCM with ethanol additive cools down the battery fastest with copper, aluminum and no additive PCM have similar cooling time. Salt additives have higher cooling time compared to other additives.
- PCM with ethanol and aluminum additive helps battery maintain the lowest temperature. PCMs with salt additives keep the battery temperature higher than the other PCMs. Overall; the PCMs reduce the battery temperature by approximately 5 K.

Further improvements can be investigated in future studies which can include experimental results to validate the predictions of analytical studies. Experiments can be conducted to compare the results and possibly suggest further improvements in the equations predicting the thermal conductivities. Due to the highly empirical nature of thermal conductivity equations, the thermal conductivity prediction highly depends on the substances being studied, its phase and its surrounding environment. The analyses, results and discussions presented in this case study can be used to compare the effectives of the refrigerant based PCMs to other PCMs under research. Some results and developments can be directly used for practical passive cooling applications. However, there are several recommendations for the future work.

- Refrigerant clathrate of other refrigerants can be made and studied for their performance on passive cooling applications. R134a refrigerant used in this study has a high global warming potential and it will soon be phased out. New refrigerants such as R1234yf or R32 can be used for a similar study as they have low global warming potential and will soon replace existing refrigerants.
- The studied PCMs are primarily studied for passive cooling applications for electronics. However, they can be used in a cold thermal energy storage system to provide cool energy to cooling fluids. These cooling fluids can then be used for space cooling applications. Colt thermal energy storage system's performance can be evaluated with each additive. The additives may help speed up the charging process, which for a large system, could yield significant improvement.
- Obtaining the thermal properties is a huge challenge. The existing methods to measuring the thermal properties are difficult to apply for this set up as it had pressurized refrigerant. An improved method of obtaining the thermal properties, possibly based on wheat stone bridge circuit can be developed. Such a mechanism would greatly reduce the number of tests results required to obtain thermal properties' values.
- Repeatability and cycling stability tests can also be conducted by future researchers to see if the performance of the PCMs changes. The PCMs can be frozen and melted for several cycles to see if the amount of heat release or absorbed bas changed. The impact of cyclic phase change can also be studied on the freezing temperature and latent heat.
- Visually inspecting the onset and end set phase is something that can be improved. Instead of observing the onset and end set phase, a more improved method is to charge all the PCMs for a specific amount of time, irrespective of its additive.
- Studying the crystal growth can be look at as each additive produces a unique crystal shape which can be studied for its shape and size.
- Discharge the PCM at different temperatures and recording the discharge time is another aspect that can be studied by future researchers. This study would determine the capacity of the PCM for each environmental condition.

Nomenclature

A	area (m^2)
c_p	specific heat capacity (J/kg°C)
D	Dipole moment, Debye
d_p	diameter of the particle, m
dE/dT	temperature coefficient (V/°C)
dT/dx	temperature gradient (°C /m)
E	open-circuit potential (V)

Ex	exergy (J)
ex	specific exergy (J/kg)
$\dot{ex}$	specific exergy rate (W/kg)
f_{TE}	thermoeconomic cost factor
h	specific enthalpy (J/kg)
h	heat transfer coefficient (W/m^2°C)
I	current (A)
LHV	lower heating value (MJ/kg)
k	thermal conductivity coefficient (W/m K)
L	latent heat (kJ/kg)
l	length, height (m)
M	molar mass (g/mol)
m	mass (kg)
$\dot{m}$	mass flow rate (kg/s)
N	number
p	pressure (kg/m s^2)
Pr	Prandtl number
Q	heat (KJ)
$\dot{Q}$	heat generation rate (W)
q	heat flux (W/m^2)
R	resistance (Ω)
Ra	Rayleigh number
R^2	coefficient of determination
R_b	thermal boundary resistance (m^2 K/W)
Re	Reynolds Number
S	source term (J/m^3 s)
$\dot{S}$	entropy rate (W/K)
s	specific entropy (J/kg K)
T	temperature (°C or K)
$\overline{\overline{T}}$	average temperature (K)
t	time (s)
u	specific internal energy (J/kg)
U	heat transfer coefficient (W/m^2 K)
V	volume fraction
V	cell voltage or cell potential (V)
$\dot{W}$	work rate, power (W)
w	mass fraction
x	distance (m)
Z	coordination number
Z_k	cost of items ($)

Greek Symbols

α	thermal diffusivity (m^2/s)
β	thermal expansion coefficient, fraction of a given type of vehicle in a fleet, liquid fraction

Δ	change in variable
ρ	Density (kg/m^3)
η	Energy efficiency
λ	Specific latent heat of fusion (J/kg)
ξ	Energy cost ($/J)
ε	emissivity
i	layer index
μ	dynamic viscosity (kg/ms)
ν	kinematic viscosity (m^2/s)
$\vec{v}$	velocity vector (m/s)
∇	differential operator
ψ	exergy efficiency
I	stress tensor
∇^2	Laplacian operator

Subscripts

ac	actual
Amb	ambient
As	average surface
b,bat	battery
bf	base fluid
bs	battery surface
c	charge, charging, cell
car	car
cm	critical
$conv$	convection
d	discharge, destroyed, discontinuous
dst	destruction
e	Electrical, exit
f	fluid, final
fc	fuel cell
gen	generated, generation
h	heat
i,j	indexes
in	internal, inlet, initial
l	liquid phase
m	mass, melting, mixture
max	maximum
n	negative electrode
oc	open circuit
out	outlet
P	solid
p	positive electrode, constant pressure
PCM	phase change material
rad	radiation
ref	reference

s	solid phase
sys	system
T	temperature
th	thermal
w	water, wall
x,y,z	Cartesian coordinate directions
∞	ambient

Acronyms and Abbreviations

Act	Actual
APS	Accelerator pedal position
ARC	Accelerated rate calorimeter
BC	Boundary conditions
BPS	Brake pedal position
BEV	Battery electric vehicle
BTMS	Battery thermal management system
C	Discharge rate, Capacity
CAD	Computer aided design
CFD	Computational fluid dynamics
Dischg	Discharge
DOD	Depth of discharge
EV	Electric Vehicle
FEM	Finite element model
HEV	Hybrid electric vehicle
LFP	Lithium phosphate
$LiCoO_2$/LCO	Lithium-cobalt-dioxide
$LiFePO_4$	Lithium-phosphate
LiBOB	Lithium bis (oxalate) borate
LiPo	Lithium polymer
LPM	Lumped parameter model
LPV	Linear parameter varying
$Mg(NO_3)_2$	Magnesium nitrate
NaCl	Sodium chloride
PCM	Phase change material
PDE	Partial differential equation
PE	polyethylene
PSAT	Power train system analysis toolkit
RESS	Rechargeable electricity storage system
Sim	Simulation
SLE	Special limits of error
SOC	State of charge
TCE	Thermal control element
TES	Thermal energy storage

References

Abousleiman R, Al-Refai A, Rawashdeh O. (2013). Charge capacity versus charge time in CC-CV and pulse charging of Li-ion batteries. SAE Technical Paper 2013-01-1546, doi:10.4271/2013-01-1546.

Abusoglu A, Kanoglu M. (2009). Exergetic and thermoeconomic analyses of diesel engine powered cogeneration: Part 1 – Formulations. *Applied Thermal Energy* **29**:234–241.

ACE Glass Incorporated. (2014). Pressure Tube 185 mL. *Item Number* 8648–33. Available at: http://www.aceglass.com/page.php?page=8648 [Accessed August 2016].

Al-Hallaj S, Kizile R, Lateef A, Sabbah R, Farid M, Selman JR. (2006). *Passive Thermal Management Using Phase Change Material (PCM) for EV and HEV Li-ion Batteries*, Illinois Institute of Technology, Chicago, USA.

Al-Hallaj S, Selman JR. (2000). A novel thermal management system for electric vehicle batteries using phase-change material. *Journal of Electrochemical Society* **147**:3231–3236.

Alawadhi EM. (2001). *Thermal Analysis of a PCM Thermal Control Unit for Portable Electronic Devices: Experimental and Numerical Studies*, PhD dissertation, Pittsburg, USA.

Basrur S, Pengelly D, Campbell M, Macfarlane R, Li-Muller A. (2001). *Toronto air quality index: health links analysis*. Report. Toronto Public Health.

Damodaran V, Murugan S, Shigarkanthi V, Nagtilak S, Sampath K. (2011). *Thermal management of lead acid battery (Pb-A) in electric vehicle, in SAE International*, Detroit, USA.

Dhingra R, Overly J, Davis G. (1999). *Life-cycle environmental evaluation of aluminum and composite intensive vehicles*. Report. University of Tennessee. Center for Clean Products and Technologies.

Dincer I, Rosen M. (2013). *EXERGY Energy, Environment and Sustainable Development*. London. Elsevier.

DuPont™ Suva® refrigerants. (2004). *Mutual Solubility of Select HCFCs and HFCs and Water*. DuPont USA.

Dutil Y, Rousse DR, Salah NB, Lassue S, Zalewski L. (2011). A review on phase-change materials: Mathematical modeling and simulations. *Renewable and Sustainable Energy Reviews* **15**:112–130.

FLUENT Software user guide. (2006). *Fluent Inc*. Available at: http://users.ugent.be/~mvbelleg/flug-12-0.pdf [Accessed August 2015].

Granovskii M, Dincer I, Rosen MA. (2006). Economic and environmental comparison of conventional hybrid, electric and hydrogen fuel cell vehicles. Journal of Power Sources 159:1186:1193.

Gu WB, Wang CY. (2000). Thermal-electrochemical modeling of battery systems. *Journal of The Electrochemical Society* **147**:2910–2922.

Heldman DR, Singh RP. (1981). *Food Process Engineering*. Westport: AVI Publishing Co., Inc., 87–157.

Hernandez O. (2001). *SIDS Initial Assessment Report for 12th SIAM*. UNEP Publications. Paris, France.

Houghton JT, Meira Filho LG, Callander BA, Harris N, Kattenberg A, Maskell K. (1996). *Climate Change 1995*. Cambridge University Press, New York, USA.

Incropera FP, Dewitt DP. (2007). *Fundamentals of Mass and Heat Transfer*. Chicago.
Lo J. (2013). Effect of Temperature on Lithium-Ion Phosphate Battery Performance and Plug-in Hybrid Electric Vehicle range" MA.Sc thesis, University of Waterloo, UK.
Mattea M, Urbicain MJ, Rotstein E. (1986). Prediction of Thermal Conductivity of Vegetable Foods by the Effective Medium Theory. *Journal of Food Science* **51**:113–116.
Morley MJ, Miles CA. (1997). Modelling the thermal conductivity of starch-water gels. *Journal of Food Engineering* **33**:1–14.
Panchal S, Dincer I, Agelin-Chaab M, Fraser R., Fowler M. (2016). Experimental and Theoretical Investigation of Temperature Distributions in a Prismatic Lithium-ion Battery. *International Journal of Thermal Sciences* **99**:204–212.
Pehnt M. (2001). Life cycle assessment of fuel cell stacks. *Int. J. Hydrogen* **26**:91–101.
Ramadass P, Haran B, White R, Popov BN. (2011). Capacity fade of Sony 18650 cells cycled at elevated temperatures: Part I. *Cycling performance. Journal of Power Sources* **112**:606–613.
Pesaran AA, Bruch S, Keyser M. (1999). *An Approach for Designing Thermal Management Systems for Electric and Hybrid Vehicle Battery Packs, Proceedings of the 4th Vehicle Thermal Management Systems*, London, UK.
Prasher R, Bhattacharya P, Phelan PE. (2006). Brownian-motion-based convective-conductive model for the effective thermal conductivity of nanofluids. *ASME Journal of Heat Transfer*, **128**:588–595.
Ramadass P, Haran B, White R, Popov BN. (2008). Capacity fade of Sony 18650 cells cycled at elevated temperatures: Part I. *Cycling performance. Journal of Power Sources* **112**:606–613.
Ramandi MY, Dincer I, Naterer GF. (2011). Heat transfer and thermal management of electric vehicles with phase change materials. *Heat and Mass Transfer* **47**:777–788.
Rizvi IH, JainA, Ghosh SK, Mukherjee PS. (2013). Mathematical modelling of thermal conductivity for nanofluid considering interfacial nano-layer. *Heat and Mass Transfer* **49**:595–600.
Rosen MA, Dincer I, Pedinelli N. (2000). Thermodynamic Performance of Ice Thermal Energy Storage Systems. *Journal of Energy Resources Technology* **122**:205–211
Rantik M. (1999). *Life cycle assessment of five batteries for electric vehicles under different charging regimes*. Report. KFB-Stockholm.
Satyam P. (2014). *Impact of Vehicle Charge and Discharge Cycles on the Thermal Characteristics of Lithium-ion Batteries*, MA.Sc thesis,University of Waterloo, UK.
Zafar S. (2015). Experimental and Theroretical Investigation of Novel Phase Change Materials for Thermal Applications, PhD thesis, University of Ontairo Institute of Technology.
Smith K, Wang CY. (2006). Power and thermal characterization of a lithium-ion battery pack for hybrid-electric vehicles. *Journal of Power Sources* **160**:662–673.
Australian Greenhouse Office. (2005). Weighting methodologies for emission from transport fuels. Available at: www.greenhouse.gov.au/transport/comparison/pubs/3ch1.pdf [Accessed March 2005].

8

Alternative Dimensions and Future Expectations

8.1 Introduction

In this book, electric vehicles and architectures and their thermal management systems, along with battery chemistries, are initially introduced to the readers to provide the necessary background information followed by a thorough examination of various conventional and state-of-the-art electric vehicle thermal management systems that are currently used or potentially proposed to be used in the industry. Through these latter chapters, the readers are provided with the tools, methodology and procedures to select the right thermal management designs, configurations and parameters for their battery applications under various operating conditions, and are guided to set up, instrument and operate their TMSs in the most efficient and cost effective manners. In this final chapter, a further step is taken over the current technical issues and limitations, and a wider perspective is adopted by examining other (and more subtle) factors that will ultimately determine the success and wide adoption of these technologies and elaborating what can be expected to see in the near future in terms of EV technologies and trends as well as the compatible thermal management systems.

8.2 Outstanding Challenges

Despite the potential advantages and capabilities of EV technologies, these vehicles still represent a small, but steady increasing market share of the currently produced vehicles due to the significant challenges that impede the widespread adoption of these technologies. Even though electric (and hybrid electric) vehicles, and their technologies (with the focus on BTMSs) are analyzed and compared against conventional vehicles and systems on purely technical, economic and environmental grounds in the previous chapters; it is also important to provide information on key actors that lay outside of these domains but still have significant effect on the acceptance and market penetration of these technologies to provide the readers with a more complete understanding of the challenges EVs are facing today.

8.2.1 Consumer Perceptions

It is crucial to take consumers' views on H&EVs into account to better identify the wider barriers of success for these vehicles since, regardless of the technological advantages of

Thermal Management of Electric Vehicle Battery Systems, First Edition.
İbrahim Dinçer, Halil S. Hamut and Nader Javani.

EVs, customers tend to resist new technologies that are not widely visible in the market, considered unfamiliar and not fully proved (Egbue and Long, 2012). Thus, customer acceptance is critical to the success of this sustainable transportation alternative. As owning a car has been traditionally and historically connected with the feeling of independence and universal access; purchasing a vehicle has been "more than mere rational choice" (Ozaki and Sevastyanova, 2011). This is linked to various aspects such as emotional attitudes, social esteem and branding and is influenced by knowledge/experience, expression of identity and membership as well as the perceived image by the consumers. Various studies reveal that the perceived innovation characteristics play an important role in explaining the rate of adoption and have been tried to be explained under the following main categories: relative advantage (perceived superiority in terms of financial, status, etc.), compatibility (perceived consistency with previous values, norms, practices, etc.), complexity (perceived difficulty to use and understand), trialability (degree of experimentability), observability (degree of adoption visibility) and perceived risks (expected probability of economic and social loss) (Labay and Kinnear, 1981).

Among these, many surveys and studies determine that the main factors that prevent consumers from considering EV technology are the lack of knowledge of potential adopters as well as the relatively low tolerance for risk and uncertainty (Diamond, 2009). Moreover, studies have repeatedly confirmed that a part of the population is always uncomfortable with technological change and uncertainty, and make choices with "notions of tradition and familiarity" rather than embracing new technologies regardless of their capabilities (Sovacool and Hirsh, 2009). Thus, it is imperative to understand the significance of the existing traditions and the symbolic dimensions of newly introduced vehicles as well as the fact that it takes time for the new products to relay its meaning to customers (Heffner *et al.*, 2007). Currently, researchers in this area mostly agree that the general stereotypes of EVs and their drivers tend to be mainly unfavorable in the past. Among these, the main ones are tied to the perceived range anxiety and the corresponding issues with insufficient charging infrastructure. However, recent car trial programs highlight that consumers are tend to be overcautious when planning their journeys and the range of these vehicles are within the actual traveled distances for the most part. This discrepancy can be mainly tied to general public's lack of familiarity with the technology and its capabilities and the associated feeling that leads to consumers to think the vehicle is under performing. In addition, a general perception of having lower maximum speeds and less powerful acceleration also exists, mainly based on the initial versions of these vehicles that were developed decades ago as EVs can now achieve equal or better performance levels compared to conventional vehicles today. This shows that the current technologies and capabilities of H&EVs have not yet been relayed to the consumers sufficiently. Furthermore, considerable misconceptions also exist on the complexity, safety, financial implications and benefits of the technology (Burgess *et al.*, 2013). The technology is also perceived by many consumers as relatively difficult to understand/operate as well as less safety than the alternatives (ICEVs) even though most studies show they are equally straightforward to operate and just as safe as conventional vehicles in many cases. Furthermore, some consumers tend to misjudge the long-term monetary benefits of these vehicles as many of them tend to be more focused on the purchase price rather than the overall cost of ownership. Even though the initial cost of these vehicles are tend to be higher than their conventional counterparts (especially without government incentives), this may be offset by lower

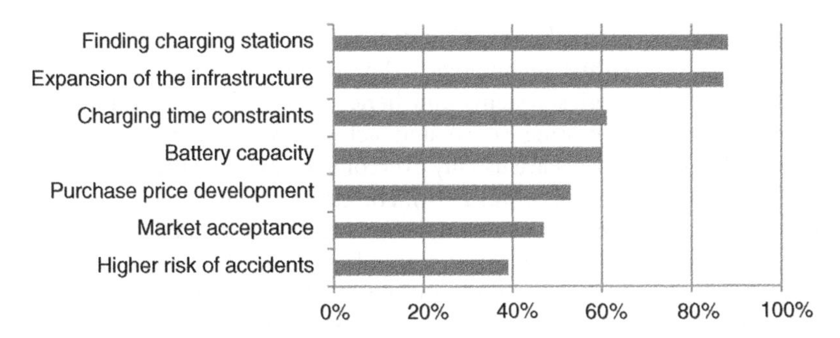

Figure 8.1 Risk barriers perceived by consumers regarding electric vehicle technology (adapted from Bessenbach and Wallrapp, 2013).

operating and maintenance costs and additional incentives which are not often taken into account. Based on several surveys, the main aspects of the H&EV technology that are negatively perceived by the consumers are summarized in Figure 8.1.

Regardless of the aforementioned negative consumer perceptions on these technologies, there are also many positive associations; especially for the technology being efficient and environmental benign, as it is generally accepted by the public. Therefore, the potential for H&EVs to establish social benefits by reducing fossil fuel consumption and GHG emissions are appealing to customers with high environmental awareness. It is also associated with conducting socially desirable behavior, sharing common values as well as being forward-thinking and open to new ideas (Graham-Rowe *et al.*, 2012). It is imperative to understand that the future market penetration success of these technologies is highly dependent on rectifying the aforementioned negative misconceptions and reinforcing the positive attributes in the mind of the consumers. The most effective method for this is through educating the public on H&EVs and their capabilities and getting them in these vehicles to enable them to see and drive this technology in order to replace their pre-existing perceptions with first-hand experience.

8.2.2 Socio-Technical Factors

Even though perceptions of individuals play an imperative role in the success of H&EVs, it should be noted that these perceptions are usually significantly affected by the cultural and social dynamics the individual takes part in. Thus, a wider frame of reference needs to better understand the factors that have a more fundamental impact on these technologies. In this regard, the socio-technical factors, that take account of the technology and the society it is embedded within, including cultural, social and political aspects, are also needed to be evaluated as they play an important role in reinforcing (or impeding) the successful transition from conventional vehicles to H&EVs. These technologies are not only supported by the wider technological system it is incorporated by, but also by the social rules and conventions that reinforces it. Thus, the alignment of preexisting technologies, regulations, user patterns and cultural discourses result in socio-technical regimes that develops the main criteria behind the decision-making processes (Smith and Stirling, 2010).

In this regard, the prevalent technological resolutions are largely determined by the society, and progress among predominant existing technologies are constrained by the collective and continuous change. This is done as a result of socio-cognitive frameworks that allow incremental improvements but prohibit more discontinuous and divergent frameworks (Bakker *et al.*, 2012) Thus, technological development therefore reinforces and reproduces existing technologies/systems and hence resists the emerging ones through both passive (through existing rules) and active (by incumbent players) mechanisms. These difficulties barriers are usually overcome by external pressures that destabilize the existing regime (Bakker *et al.*, 2014). For electric vehicle technologies, these pressures are mainly driven by environmental concerns the cost and dependency of fossil fuels. Thus, the protection of the new technology from the hostile market environments (such as competition with CVs) through public and private funding and/or policies become crucial until the technology becomes developed enough to be widely commercialized to the public.

In order to better understand socio-technical factors, it is important to examine the interrelation of the interests, expectations and the strategies between the stakeholders as well as political institutions. The stakeholders can vary from the national/local governments, car manufacturers/importers, energy producers to charging infrastructure developers. For a successful transition to EV technologies, the emerging interest of the new entrants must outweigh the vested interests of the incumbent actors and the configurations of the emerging system must align with interest of the stakeholders. In the past EVs have encountered strong opposition from car manufacturers, oil companies and repair businesses that have invested considerable financial resources in the supply and production of infrastructure for conventional vehicles.

As an example, the large market penetration of EVs can provide significant benefits to electricity grid operators if the EV technology is reinforced with additional charging systems (such as smart grid). Once the transition starts to take place, the stakeholders tend to develop a strategic response to gear the transition towards influencing the configuration of these emerging technologies (such as the installation of the charging infrastructure) to match their interest and expectations, which can be the funding, location, type of the charging infrastructure as well as the charging behavior. Thus, these dynamics can be best understood with the forces that have significant impact on the EV development and market penetration, and can be done through observing the key parameters in this regard;such as the increase rivalry among car manufacturers, the technological dispersion among firms (such as tier suppliers, research institutes or providers of infrastructure) and the presence of new entrants (such as various new EV and subsystem manufacturers) (Wessling *et al.*, 2014).

In addition to the stakeholders, the political institutions are also one of the key players in this socio-technical environment as their decisions can significantly impact this technological transition. The impact of political institutions are generally categorized under 4 items, namely *collective actions, high density of institutions,* enhancing *asymmetries of power* and *complexity and opacity of politics* (Foxon, 2007). The collective actions principle is based on the dependency of the actions of these institutions to each other such as the H&EV rules and regulations associated in country and state levels. Moreover, since the public policies are generally extensive, legally binding constraints on behavior, they are cumbersome to alter after being implemented, which may act as a barrier for H&EVs since most of the preexisting policies are established for conventional vehicles. In addition, when the authorities are in position to generate changes in the rules, they are likely to do so to enable outcomes that are favorable to themselves. Finally, links between actions and outcomes can make politics inherently ambiguous, where it becomes hard to rectify the determined problems. These influences can have significant impact on the speed and the outcome of the decisions, rules and regulations that affect the acceptance and market penetration of H&EV technologies.

8.2.3 Self-Reinforcing Processes

As explained in the previous sections, the characteristics of early markets, the institutional and regulatory factors governing its introduction and the expectation of consumers usually favor incumbent technologies against newcomers both through consumer's perceptions and socio-technical factors (Foxon, 2007). In order for the EV technology to prove itself to the public and gain a wide market penetration, it also has to fight against the self-reinforcing processes that further favor the existing technologies even though potential "superior" alternatives may exist. These processes are developed mainly as a result of increasing number of sales and can be classified under *economies of scale, learning effects, adaptive expectations* and *network externalities* (also called co-ordination effects) and *complementary effects* (Arthur, 1989; Sydow and Screyogg, 2013). It is important to understand even though they will initially be impeding the market penetration of electric vehicle technologies, as more H&EVs are being sold, the same processes will start acting in the new technology's favor when used wisely.

As more vehicles are being sold, the most obvious impact is on the reduction of the cost of each system/component (and hence the vehicle) by firms taking advantage of the economies of scale (units costs declining through spreading over increasing production volume), which makes the new technology with limited market penetration harder to financially compete with the existing technology. Moreover, as the production and market experience is increased through larger sales, the specialized skills/knowledge accumulates (e.g., faster production with higher productivity and lower scrap rates) which in turn also further reduces the associated cost for the subsequent iterations of the product. Moreover, on the production side, as the sales increase, both the producer and the user acquire more confidence on the quality, performance and/or longevity of the existing technology, which makes it harder to find alternatives. In addition, on the marketing side, the firms acquire more in depth knowledge and understanding of the heterogeneity of the demand, where the preferences are no longer fixed individually, but vary along with the expectations or activities of others. Consumers gain practice with the technology, learn to employ it in an efficient way and incorporate them in their existing structure; making it possible to offer products that can simulate better sales. Furthermore, a social construction of meaning develops among the society as more products are being sold, making the product more desirable based on the established (positive) social meaning, especially if the consumers believe that that other agents will also chose the same technology. In addition, as the usage of these vehicles increase, so do the direct benefits and the value due to the complementary goods (as it would in the increased network of charging stations) as the technology becomes more valuable when it is more widely used. The direct network effects occur when "the number of network participants directly influences a product's utility", whereas the indirect network effects occur when the utility of a technology depend on the complementary goods and services provided by independent firms (Meyer, 2012). The self-reinforcing nature of direct and indirect network effects for H&EV technologies is summarized in Figure 8.2.

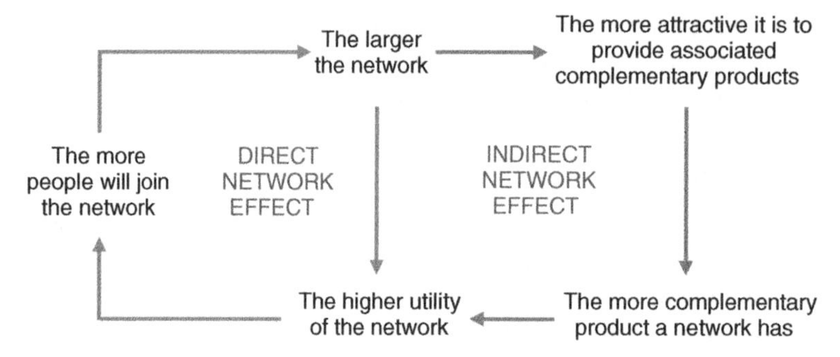

Figure 8.2 Direct and indirect networks effects (adapted from Meyer, 2012).

Moreover, as the vehicle volumes increase, the concerns over the resale value of the vehicles and the charging infrastructures are being reduced, thus further increasing the rate of sales. Finally, as a result of the synergy resulting from the interactions of multiple separate actors and factors, the impact of these self-reinforcing processes bring additional surpluses than the total sum, as various unintended side consequences and side effects also occur.

It is imperative to understand the impact of these factors as they can amplify the initial variations in the market share and result in one technology achieving complete market dominance at the expense of the other, commonly referred as the "lock-in" effect, which can determine the difference between the take-up and failure of newer technologies in the market. This phenomenon can be understood with the well-known example of the QWERTY keyboard layout, that was initially developed to reduce the typing speeds in order to avoid jamming of early developed mechanical typewriters, but now dominated the computer technology even though "superior" alternative designs can be developed.

In the sections above, the perceived drawbacks, outstanding social-technical challenges and current opportunities for a wider adopted EV technology is elaborated. Based on the aforementioned past evolution and the future expectation of this technology, the typology of the innovation modes/phases of the market evolution is provided in Figure 8.3.

In Figure 8.3, the bottom left side (Quadrant A) portrays the gradual change in the existing products within the established regime without any significant innovation, such as the early improvements on the ICE technology with increases the fuel efficiency and reduces the emissions in the conventional vehicles. The top left side (Quadrant B) represents the disruptive innovations with low level of change where the change enables alternative producers' capabilities that can co-exist with the dominant design in the market, such as the early versions EVs produced. Bottom right side (Quadrant C) shows high levels of innovation but the incumbent retain a strong market position. These are mainly mild hybrids that transformed the way the conventional vehicles are being built, but have not played a role in replacing them. In this region, the social context of the market, user practices and business models become an important player in establishing the rate of this market transformation. Finally, the top right (Quadrant D) reveals the disruptive innovations with high levels of change that can replace the existing technology with an alternative one (Dijk, 2014). Thus, through both the improvements in its technology (with high capacity, longer lasting batteries, etc.), infrastructure (both through charging and communication infrastructure as well as associated standards,

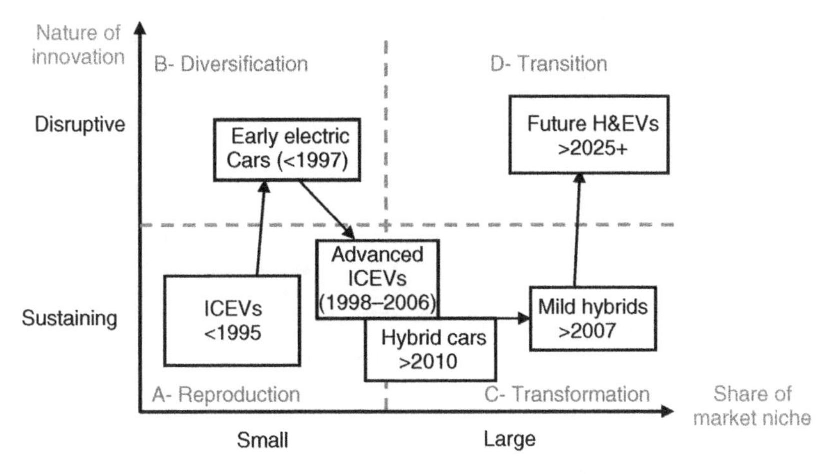

Figure 8.3 Past evolution and future expectation of the H&EV technologies (adapted from Dijk, 2014).

rules, regulations, operating practices and training)as well as aforementioned individual perceptions and socio-technical challenges (with respect to individual, cultural, social and political aspects), electric vehicle technology is expected to start placing itself in this quadrant in the next decade. The EV technologies and trends that will play a key role in this transition are provided in the next section.

8.3 Emerging EV Technologies and Trends

In order to overcome the aforementioned challenges, various technologies and trends are currently being developed that is expected to reduce the perceived disadvantages of electric and hybrid electric vehicle technologies and widen their market penetration. Brief information on these emerging technologies is provided in this section.

8.3.1 Active Roads

As mentioned in previous sections, since the energy densities of the electric batteries are significantly lower than that of the gasoline, wireless power transfer is crucial in the widespread commercialization of EVs in the market as it provides a safe and convenient way of charging the vehicle. Although this convenient way of charging in the parking lots is important, the main strength of this technology is with the implementation of wirelessly charging the vehicle in motion. Thus, wireless in-motion charging of EVs have been a research interest for a long time and was proposed in 1950s to supply power to mining cars and later rail cars to eliminate the pantograph systems (Miller *et al.*, 2014).

In these systems, as the vehicle is traveling, it remotely picks up electricity through the magnetic field from power transmitters buried underground. Each power transmitter contains an inverter and inductive cables that are placed under the roadway. As the vehicle moves, they "connect" to the grid and receive brief but high-level charge that can be as high as the power output of DC fast charging (current commercial wireless charging power levels are equivalent to Level 2 charging). Unlike stationary wireless charging where the primary and secondary coils are brought into perfect alignment;

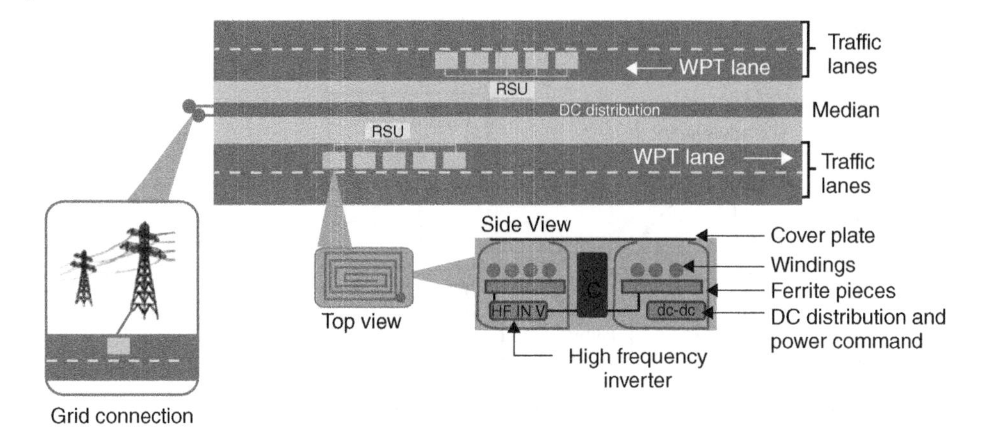

Figure 8.4 The concept of dynamic wireless charging system on special highway lanes (adapted from Miller *et al.*, 2014).

quasi-dynamic and dynamic charging needs driver assistant systems in order to adjust the vehicle speed and horizontal and vertical alignments to optimize the energy transfer efficiency. In semi-dynamic charging, energy is transferred from coils on the road-side (primary)to the ones on the slowly moving vehicle (secondary), whereas in dynamic charging it is done via a special driving lane that incorporates primary coils with higher power levels and can charge vehicles that are travelling at medium to high speeds (Taiber, 2014). Dynamic charging can virtually remove need for conventional power sources or even home charging. The concept of dynamic wireless charging being installed on special highway lanes is shown in Figure 8.4. On the other hand, even with the semi-dynamic charging can also provide significant advantages to passenger vehicles in congested areas or electric buses in large cities where they can be quickly charged in motion or while pausing to pick up passengers.

Currently, the active road technology is still at its infancy and the main focus is still on the unidirectional stationary charging where units with 85% system efficiency are already established. However, with the technical expertise obtained from stationary charging, active roads can be envisioned to be on used on highways in the near future which can have considerably impact on increasing the all-electric vehicle range and mitigating the corresponding emissions. Even though the infrastructure associated with this technology would have significantly high, they could be amortized via telematics-based road charging systems as well as with savings associated with reducing the size of the battery. On the other hand, additional considerations such as cooling the charging system that corresponds to additional cost, weight and space should also be taken into account.

8.3.2 V2X and Smart Grid

In the future, if PHEVs become widespread and take over a large share of the vehicle market, without any measures taken on the gird, it will have significant impact on the demand peaks, reserve margins and electricity prices and lead to reduction of power quality and mismatch between supply and demand. Thus, as mentioned in Section 1.7.4, one of the ways of overcoming these issues is utilizing vehicle-to-grid (V2G) and even

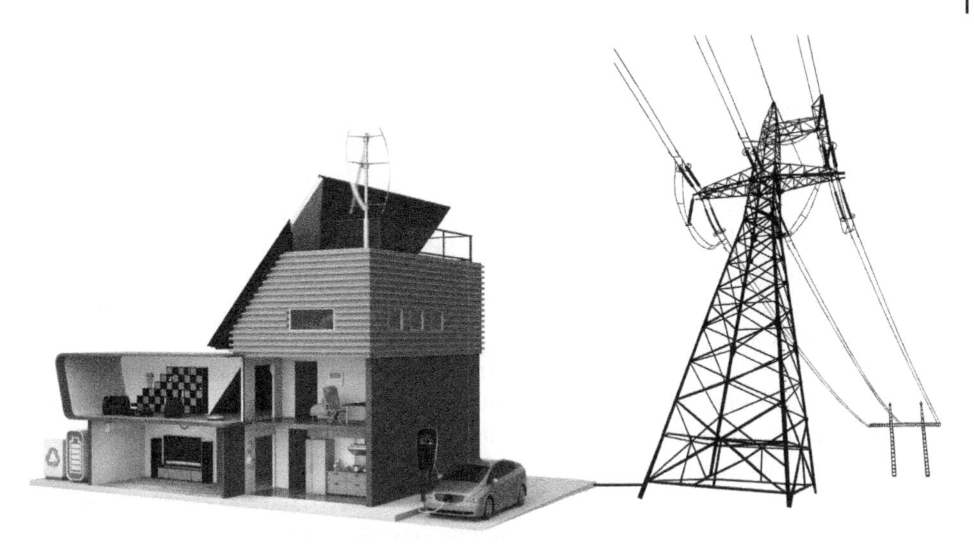

Figure 8.5 Futuristic concept of smart grid and V2X systems.

vehicle-to-everything (V2X), and smart grid technology to regulate the charging times and use the vehicles as storage/back up devices to level the electricity utilization during the day and provide peak power when the demand is high and therefore enhance the sustainability and resilience of the electric power infrastructures (Peng, 2012). Thus, PHEVs would not only cause an overload on the grid, but also help shaving the peak demand when the vehicle is not in use, which is estimated to be around 95% of the time. For example, it is estimated that, for 1 million grid connected electric vehicles (with an average of 16 kWh capacity) discharged at a rate of 2 kWh can supply the grid with 2 GW of power within 2 hours (Electrification Coalition, 2009). A futuristic concept of a smart grid and V2X systems are shown in Figure 8.5.

Moreover, since the price of electricity is higher during peak times than it is at night, the owner of these vehicles can make profit from the difference. In addition, the vehicles can also absorb the excess renewable energy during off peak hours to transmit it into the grid (again for a higher price). Since most of the renewable sources have intermittent properties (e.g., more than half of the total power from wind energy power plant does not have sufficient stability to connect to the grid) it makes the integration of these renewable sources, especially wind and solar, easier. Even though the amount of this profit is estimated to vary significantly with respect to various studies conducted on the literature, it can be said that integration of V2X and smart grid will increase the benefits and ease the transition of PHEVs to the market.

8.3.3 Battery Swapping

Even though EV technologies have progresses a long way since their first invention, there are still inherent to their current characteristics that limits the full use of these technologies such as range, long charging times and initial costs of the electric batteries. The high cost of the electric battery is still one of the largest impediments to the proliferation of EVs into the vehicle market. For PHEVs, it is estimated that increasing the battery energy capacity from 30 to 60 km (20–40 miles) of all electric range would provide an extra 15%

reduction in fuel consumption but also almost doubles the cost (Markel and Simpson, 2006). On the other hand, it is estimated that the battery costs should be reduced by half in order to equate the cost of ownership of PHEVs to conventional vehicles. Various strategies are being conducted to improve the capacity of the batteries, while reducing their weight and cost, such as advanced battery chemistries, optimized cell and battery designs and manufacturing techniques that are mentioned in previous sections, these strategies are not likely to produce short term impact on the cost of EVs. Therefore, various other methods of mitigating these undesired characteristics of the technology are also being investigated in parallel. Among these alternative charging mechanisms and methods also play an important role.

As mentioned in Chapter 1, currently electric vehicle charging is mainly conducted by directly charging from a power system through the charger. Significant research is currently being taken on to improve the charging behavior of AC charging, DC charging, AC and DC hybrid charging and wireless charging. Among these, AC charging has slow charging speeds with 6–8 hours. DC charging is much faster, but has expensive equipment and low availability on the streets, and may have negative impact on the capacity and cycle life of the batteries. AC/DC hybrids tend to have issues associated with communications between charging couplers, EVs and charging devices. And finally, stationary wireless charging technology has improved significantly in the past decade; however, it does not provide much help with the aforementioned issues except for providing extra safety and convenience for the users. Furthermore, the actual solution of semi-dynamic or dynamic wireless charging is still at its infancy for it to be widely used at the moment (as shown in Section 8.3.1). Thus, these battery charging technologies do not currently answer the need for powering EVs for long distances and in short enough duration.

On the other hand, battery swapping (also called battery exchange) is a relatively new but promising technology that can replace the depleted battery of the vehicle with a fully charged one without the hassle of waiting for the battery to be charged. This technology can overcome the aforementioned issues associated with limited battery capacity and long charging times and can provide both a standalone solution and/or can work in parallel with the alternative solutions. In battery exchange concept, the time it takes to exchange the battery is considerably less than the previous methods and it is convenient for battery management and maintenance. Moreover, it can prolong the cycle life and reduce its costs and even the environmental impact associated with battery use. An illustration of a battery swapping station is shown in Figure 8.6.

World's first battery exchange capable electric bus has started in Seoul city in 2010 and currently this technology exists to a certain degree mainly in China. Beijing Olympic, Shangai World Expo and Guangzhou University City and Hangzhou Gucui Road are some of the EV battery exchange stations in this country. Beijing Institute of Technology and Xuji Corporation Harbin Institute of Technology also conducted significant work on battery exchange technologies for electric passenger vehicles and/or buses. In 2012, battery exchange stations for electric scooters and motorcycles were opened in Taiwan, where the battery swap process is done manually by the driver. Moreover, Renault partnered with Better Place, the creator of one of the first network of stations to replace the batteries in passenger electric cars, which got bankrupt mainly due to certain issues with the technology, low number of electric vehicles sales at the time and the lack of public knowledge of the product.

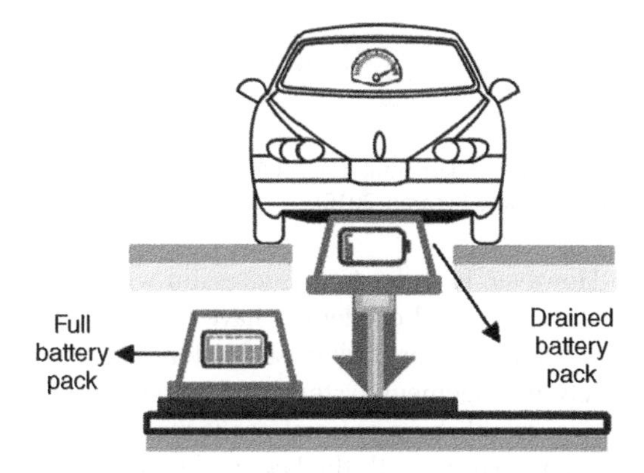

Figure 8.6 Illustration of a battery swapping station (adapted from Davies, 2013).

Even though battery swapping can provide significant advantages, especially in densely populated areas with limited home and/or public charging availability, it requires service/swap stations, specialty technical staff and even more importantly standardization to ensure consistency across vehicles types and jurisdictions. Moreover, this technology also requires significant recalibration of the controller built inside the battery in order for this process to be "plug-and-play" and provide an optimal performance to the vehicle. Furthermore, an optimal control strategy for the coordinated charging of the battery service/swap stations would also be needed in order to minimize the power losses or maximizing the grid load factor.

8.3.4 Battery Second Use

Even though significant research is being conducted to increase the performance, capacity and cycle life of EV batteries, their vehicle lifetime will eventually end as their capacity is reduced through aging of the battery. As the vehicle battery market is predicted to grow tens of thousands of megawatt hours and exceed tens of billions USD annually in the next decade there will be significant amount of retired EV batteries with still 80% of their initial capacity remaining. Even though this remaining capacity may not be enough to provide the desired all electric range for the vehicle, it is still sufficient to support mobile vehicles (utility and recreational vehicles, public transportation) low-energy buildings and utility reliability as well as various grid ancillary services (both grid- based and off-grid stationary) for mitigating the intermittency of various renewable energy sources and provide balancing, spinning reserve and load following. Thus, significant studies and business models are currently being created to reuse of EV batteries in second-use (also called second life) applications by attaining supplementary services and income from the battery after it completes its mission in the vehicle. With increasing use of renewable energy in the electricity generation mix of many countries, the need for such grid stabilization/reliability and peak load reduction is expected to increase. Thus, retired EV batteries would provide significant uses in these sectors by being removed from the vehicle at the end of its vehicle lifetime and get aggregated with other batteries to megawatt hour sized system (unlike V2G applications where the battery is utilized while still being in the vehicle). These applications can considerably increase the total lifetime value of the battery, and therefore lessen its associated cost to the automotive user. In order to achieve this, the simplified steps shown in Figure 8.7 need to be taken.

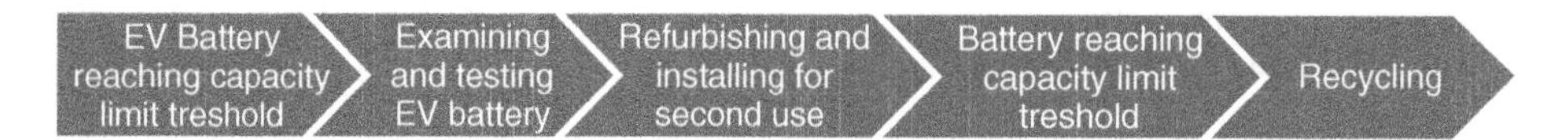

Figure 8.7 Fundamental steps of EV battery between primary use and recycling (adapted from Neubauer and Pesaran, 2010).

However, in order for a successful wide-range second use application of retired EV batteries, several challenges have to be overcome in this process, including; drawing new standards for pack classification and grading, designing new energy, thermal and safety management systems, accurately predicting the decaying conditions used batteries, establishing optimal profit making models/smart cost management structures and conforming with the laws and regulations (Lih *et al.*, 2012). Even after these steps are essentially taken, several barriers exist for second use applications as listed below (Neubauer and Pesaran, 2010):

- Sensitivity to uncertain degradation rates in second use
- Determining the operating requirements of selected second use application
- High cost of battery refurbishment and integration
- Low cost of alternative energy storage solutions
- Accurately identifying financial data for revenues and operating costs
- Lack of market mechanisms and presence of regulation
- Perceptions of used batteries

As these issues are resolved, the EV batteries can be repurposed to fit the needs of the second use. In this process, first the modules (with minimal BMS) and integral structures and interfaces of the battery are removed from the vehicle. Then the batteries are visually examined, then diagnosed and tested with and without its BMS. After the cells are graded and the faulty ones are replaced; the battery is repacked and relabeled with respect to its second use application. The associated repurposing procedure is shown in Figure 8.8.

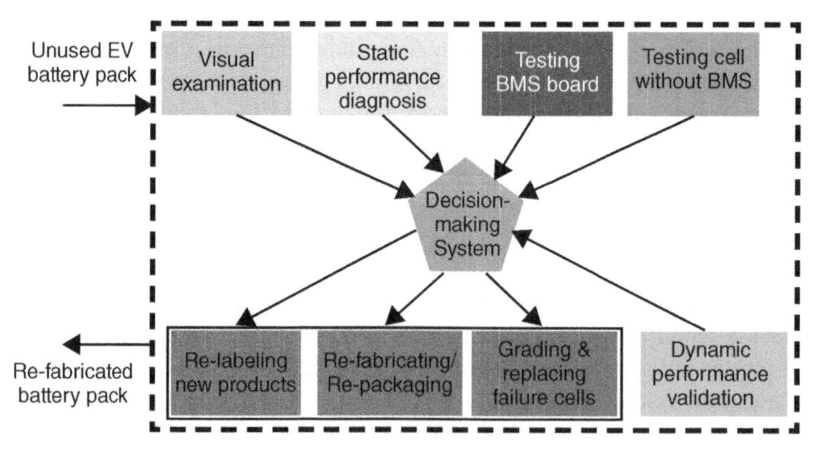

Figure 8.8 Schematics of the repurposing processes for a used EV Li-ion battery pack (adapted from Lih *et al.*, 2012).

Based on the current estimations, the immediate impact of second-use batteries on the current EV costs would be negligible today, but can potentially reach over 10% in the next couple of years. However, further analyses should be conducted on long term effect of battery degradation and second use application to provide more accurate results.

8.4 Future BTM Technologies

In the previous chapters, various state-of-the-art battery thermal management systems that are commercially available or will be in the near future are examined, and are thoroughly evaluated based on various criteria in the latter chapters. In this final section, a brief introduction of various TMSs that are developed to be used in a wide range of applications, ranging from microprocessor and nuclear reactors to space shuttles, will be introduced to the readers in order to tie the loose ends in the battery thermal management technologies and provide the big picture in these systems. These technologies may not be currently practical to be used in BTMSs due to issues ranging from the size and the geometry of the targeted surface area, the discrepancy in the effective operating temperature of the systems, cost, safety as well as the interactions with the battery chemistry or electronics in the pack. However, with the expected improvements in these technologies, they would be viable options to complement or replace the current state-of-the-art systems in the future.

8.4.1 Thermoelectric Materials

In the past, various efforts have been made to use thermoelectric materials (since 1820s when it was first discovered) to convert the waste heat into electricity in various applications as well as the reverse process of cooling/heating by applying electricity. However these materials had very limited commercial use due to the utilized materials having high cost and low efficiencies of conversion (2% to 5%) between heat and electricity (known as thermoelectric figure of merit or ZT), which is governed by the Carnot efficiency based on the second law of thermodynamics. In the recent past, through effective method of nanostructuring for thermal conductivity reduction, has led to ZT values up to 1.8, but haven't managed to overcome the generally desired threshold of 2 (Quick, 2012).

Recent advancements in material science have led to the development of new materials, based on the environmentally stable common semiconductor telluride that is expected to convert 15% to 20% of waste heat into electricity. By considering sources of scattering heat-carrying photons on all relevant length scales, from atomic and nanoscale to mesoscale boundaries, maximum lattice thermal conductivity reduction and large thermoelectric performance improvements have been achieved. Through these methods, enhancements beyond nanostructing are made possible and thermoelectric figure of merits over 2 became practically possible (Biswas, 2012). Currently it is ideal for any application that uses large combustion engines, such as heavy manufacturing industries, refineries and coal or gas fired power plants, large ships or tankers, where the waste heat temperatures can range from 400°C to 600°C (Fellman, 2012). However, it is already started to be tested to recover the waste heat from the vehicle tailpipes (by various OEMs including BMW) to feed it back into the vehicle to supply the auxiliary units, including thermal management in hybrid electric vehicles.

Even though the current effective temperature ranges are too high to be practically used for capturing battery heat dissipation, it is expected in the near future that significant improvement will be made to achieve efficient conversions at lower temperatures ranges. Thus, a wide range of applications are foreseen to become possible for incorporating thermoelectric materials directly into the BTMSs mentioned in the previous chapters to increase the efficiency and reduce the required cooling power.

8.4.2 Magnetic Cooling

Magnocolaric effect (MCE) increases the temperature of certain materials, with magnocaloric properties, when being exposed to magnetic field and decreases when it is no longer applied. The temperature where this effect is the strongest is material specific and the effect is highly reversible and virtually instantaneous. Other properties such as mass and shape of the material also plays important roles in the generated power. MCE has been discovered in 1881 and magnocaloric systems have been built as early as 1930s, by controlling the magnetic field through series of magnetization/demagnetization cycles that are being exerted on the used alloys that has MCE properties. As temperature difference is being established in the material, stabilized temperature differences are achieved after various repetitions, where the produced cooling energy can be captured via a heat transfer fluid, such as water. The MCE systems have various advantages from reduced energy use to elimination of hazardous fluids. However, until recently, these systems have failed to be used in many practical cooling applications as the utilized superconducting magnets themselves need to be cooled to low temperatures, which reduce the efficiency and cost effectiveness of the systems.

Recently, researchers have developed new types of thermal management technology using magnets that are more environmentally friendly and are 20% to 30% more efficient than the state-of-the-art technologies being used today with the help of a new type of nickel-manganese alloys for magnets that can function at room temperatures (GE Reports, 2014). By arranging these magnets in a series of various cooling stages, the temperatures of a water-based fluid flowing in between is reduced by up to 80°C. Even though the system is currently "the size of a cart", it is expected to be more compact with even larger cooling capacities at lower costs to be used in practical applications at the end of the decade. Thus, they would able to be incorporated with a coolant fluid (such as water-glycol mix) for the heat transfer between the hot and cold sources in the battery to be used in BTMS applications.

In addition, significant advancements are also being made in the method of achieving the magnetocaloric effect, by simply rotating the crystal of $HoMn_2O_5$ within a constant magnetic field. This method can significantly reduce the energy absorption of the cooling system and enable building simplified, efficient and compact BTMSs in the near future (Balli *et al.*, 2014). An illustrative example of a magnetic thermal management system diagram is provided in Figure 8.9 that can be utilized for electric vehicle batteries in the near future.

8.4.3 Piezoelectric Fans/Dual Cooling Jets

Piezoelectric effect (PE), "the linear electromechanical interaction between mechanical and electrical state in crystalline materials", has been observed since 1880s, but have not practically used until the First World War, where it was utilized in sonar devices

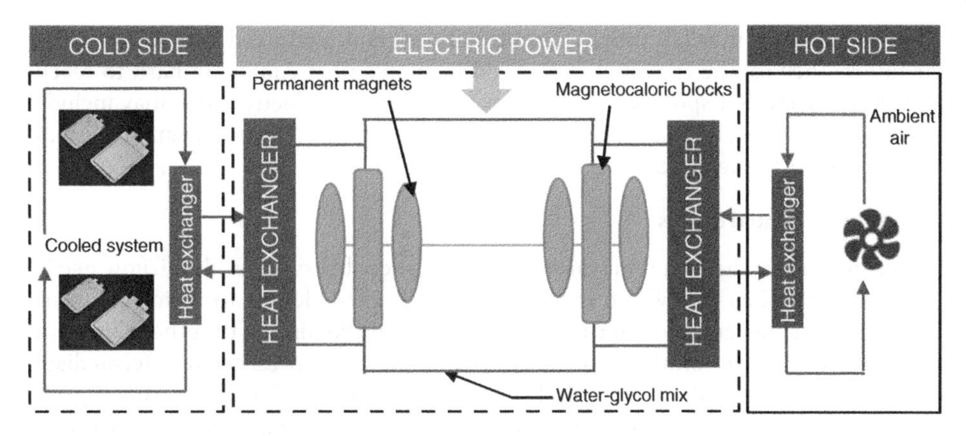

Figure 8.9 Illustrative example of a magnetic thermal management system diagram integrated into EV battery pack (adapted from Cooltech, 2015).

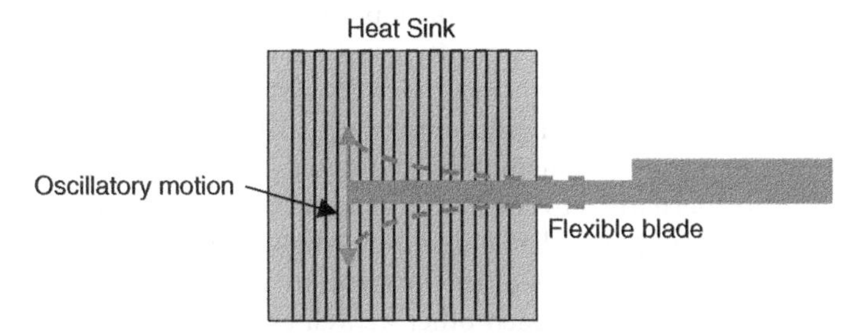

Figure 8.10 Representation of piezoelectic fan having horizontal vibration over a heat sink.

(Barecke *et al.*, 2011). In the next decades, man-made materials have been discovered that have piezoelectric constants many times higher than the natural materials, and in the past decade they have been used to harvest kinetic energy from pedestrians for various applications.

Currently, piezoelectric fans are used in various cooling applications by inducing flow with vibrating flexible blades near its base end and making oscillatory motion at the free end of the blade when an input signal is applied. Moreover, piezoelectric fins are being used with heat sinks to enhance the convection heat transfer coefficient from the fins by creating enhanced air flows, as demonstrated in Figure 8.10. These systems stand out among most other thermal management methods due to its simplicity, small size, geometric flexibility and low power consumption. Moreover, they have no moving parts, which enable them to be quieter and have better reliability and lifetime compared to traditional rotary fans. With the recent advancements in the piezoelectric materials, it has been alternative device for cooling microelectronics, cell phones, tablet PCs and various mobile devices. Moreover, recent breakthroughs in piezoelectric technology have led to the increased use of dual piezoelectric cooling jets (DCJ), that act like micro-fluidic bellows that provide high-velocity jets of air that can increase the heat transfer rate over an order of magnitude over natural convection, enabling even higher reliability and lower maintenance costs. They have been very effective for cooling hot-spots such as precise locations for industrial spot cooling and are starting to be used in various chips, radio frequency components and even batteries.

Even though they may not fully replace traditional fans for forced air cooling in the near future, they can be expected to be used to supplement cooling in hard to reach stagnant areas and hot spots where rotational fans are less effective. This may include the hot spots in the cells near the terminals as well as the battery management system microprocessors in the pack.

8.4.4 Other Potential BTMSs

In order to increase the heat removal capability of air cooled systems, the airflow needs be increased considerably, which may not always be possible due to various reasons including power concerns, cost and acoustic noise. Liquid cooling on the other hand, can provide more effective solutions, however are limited by the presence of intermediate materials and interfaces that may limit the heat transfer capabilities. Thus various developments have been made to overcome these drawbacks and improve the heat transfer capabilities of the thermal management systems. For these reasons, various other potential thermal management systems are briefly introduced below that can provide solutions on these aforementioned issues in future BTMS applications.

Thermosyphons, which are heat pipes assisted by gravity, are the simplest type of heat pipes and have been used in various applications in heat recovery systems. Unlike most traditional water cooling systems, they have a self-sustaining cycle that does not rely on a pump as it uses the convection for the movement of heated water. Two-phase closed thermosyphon systems are cheap alternatives to conventional liquid cooling systems as no pump is required to be installed and are reliable due to not having moving parts. However, they do have limitations with respect to the orientation and placement of the evaporator and condenser with respect to each other as well as the requirement of the system being relatively motionless. They can also be used in connection with phase change materials as well as jet impingement structures mentioned in the previous section that directs the coolant into the surface that needs to be cooled. The thermosyphon modes and jet impingement modes can be altered depending on the surface cooling requirements to provide more effective TMSs.

Another way of avoiding to use an external pump is to utilize the electric potential to manipulate the movement of liquid droplets on a dielectric surface to precisely deliver them to the hot spots, known as electrowetting. By continuous, fast repetition of this process, hot surfaces can be cooled considerably over very short periods of time. In many studies, it has been found that the heat dissipation of these systems can be comparable to microchannel cooling of systems and are being used in microprocessor applications.

Moreover, low-cost versions of the liquid metal cooling (LMC), that has been widely used for nuclear reactors, is also being developed for various electronics and devices. LMC can potentially have order of magnitude higher effective thermal conductivity than copper and are cost efficient and reliable over large periods of time. They are effective to conduct the waste heat away to the heat sinks using thin pipes filled with various liquid metal coolants (such as compounds of gallium, indium and tin ectectic). The technology is comparable to heat pipes and vapor chambers, but is gravity independent, mechanically flexible and can achieve higher performances with similar costs.

Finally, other technologies that are currently being developed for space applications that provide cutting-edge thermal management and insulation for various associated systems. Among these mechanical devices called heat straps are used to allow improved conduction while maintaining mechanical flexibility and use materials with effective mass and thermal conductivity such as ultra-high conductivity carbon fibers, artificial diamond films, carbon or boron nitride nanotubes. Furthermore, mechanical or gas based devices called heat switched are also used to allow or prevent thermally-conductive links in space shuttles. They can be either passive (open/close at specified preselected temperatures) or by active command (NASA, 2015).

8.5 Concluding Remarks

In this last chapter, a wide perspective is adopted in electric vehicle technologies and thermal management systems with the focus on their challenges, emerging technologies and potential thermal management systems. Since purely technical, economic and environmental aspects are discussed in the previous chapters, the remaining outstanding challenges and trending technologies that might provide the necessary solutions are elaborated. Finally, various TMS technologies that are currently under development for an extensive range of applications are introduced to provide an indication of what the BTMS might incorporate in the future.

Nomenclature

Acronyms

BMS battery management system
BTMS battery thermal management system
CV conventional vehicle
DCJ dual cooling jets
EV electric vehicle
GHG greenhouse gas
H&EV hybrid and electric vehicle
ICE internal combustion engine
LMC liquid metal cooling
MCE Magnocolaric effect
PE Piezoelectric effect
PHEV plug-in hybrid electric vehicle
TMS thermal management system
V2G vehicle-to-grid
V2X vehicle-to-everything

Study Questions/Problems

8.1 Elaborate on the reasons why purchasing a vehicle may be more than a mere rational choice? What are the widely accepted negative and positive images associated with electric vehicle technologies in your country, state or community?

8.2 What do you see as the socio-cognitive frameworks that prohibit discontinuous and divergent transitions of the electric vehicle technologies? Identify a "self-reinforcing process" that impedes the wider market penetration of this technology and elaborate on the necessary steps to overcome it.

8.3 What do you think is the most crucial emerging technology associated with EVs? What can be done to accelerate the acceptance of this technology or trend?

8.4 Select a thermal management technology that is not currently being commercially utilized for BTMSs and identify the advantages and drawbacks of this technology to replace the ones used in EV batteries. What can be done to overcome these drawbacks?

8.5 Which thermal management systems provided in section 8.4 is the (a) easiest, (b) most effective and (c) most technologically ready to be used with existing BTMSs? Elaborate on the reasons.

References

Arthur WB. (1989). Competing technologies, increasing returns, and lock-ins by historical events. *The Economic Journal* **99**:116–31.

Barecke F, Wahab MA, Kasper R. (2011). Integrated Piezoceramics as a Base of Intelligent Actuators, Advances in Ceramics - Electric and Magnetic Ceramics Bioceramics Ceramics and Environment.

Bakker S, Lente H, Engels R. (2012). Competition in a technological niche: the cars of the future. *Technology Analysis & Strategic Management* **5**:421–434.

Bakker S, Maat K, Wee B. (2014). Stakeholders interests, expectations, and strategies regarding the development and implementation of electric vehicles: The case of Netherlands. *Transportation Research Part A* **66**:52–64.

Balli M, Jandl S, Fournier P, Gospodinov MM. (2014). Anisotropy-enhanced giant reversible rotating magnetocaloric effect in $HoMn_2O_5$ single crystals. *Applied Physics Letters* **104**: 232402

Bessenbach N, Wallrapp S. (2013). Why do Consumers resist buying Electric Vehicles? Copenhagen Business School, Thesis.

Biswas K, He J, Blum ID, Wu C, Hogan TP, Seidman DN, Dravid VP, Kanatzidis MG. (2012). High-performance bulk thermoelectrics with all-scale hierarchical architectures. *Nature* **489**: 414–418.

Burgess M, King N, Harris M, Lewis E. (2013). Electric vehicle drivers' reported interactions with the public: Driving stereotype change. *Transportation Research Part F* **17**:33–44.

Cooltech.(2015). The Magnetic Refrigeration System Diagram. Available at: http://www.cooltech-applications.com/magnetic-refrigeration-system.html [Accessed June2015].

Davies C. (n.d.). Tesla battery-swap demos this week confirms Elon Musk Article. Available at: http://www.slashgear.com [Accessed July 2015].

Diamond, D. (2009). The impact of government incentives for hybrid-electric vehicles: Evidence from U.S. states. *Energy Policy* **37**:972–983.

Dijk M. (2014). A socio-technical perspective on the electrification of the automobile: niche and regime interaction. *International Journal of Automotive Technology and Management* **14**:158–171.

Egbue O, Long S. (2012). Barriers to widespread adoption of electric vehicles: An analysis of consumer attitudes and perceptions. *Energy Policy* **48**:717–729.

Electrification Coalition. (2009). Revolutionizing Transportation and Achieving Energy Security, Electrification Roadmap.

Fellman M. (2012). Thermoelectric material is the best at converting heat waste to electricity article. Available at: http://www.northwestern.edu [Accessed August 2015].

Foxon TJ. (2007). Technological lock-in and the role of innovation. In: *Handbook of Sustainable Development*, G Atkinson, S Dietz, E Neumayer (eds.). Edward Elgar Publishing Limited UK.

GE Reports.(2014). Not Your Average Fridge Magnet. Available at: http://www.gereports.com/post/75911607449/not-your-average-fridge-magnet [Accessed June 2015].

Graham-Rowe E, Gardner B, Abraham C, Skippon S, Dittmar H, Hutchins A. (2012). Mainstream consumers driving plug-in battery-electric and plug-in hybrid electric cars: A qualitative analysis of responses and evaluations. *Transportation Research Part A* **46**:140–153.

Heffner RR, Kurani KS, Turrentine TS. (2007). Symbolism in California's early market for hybrid electric vehicles. *Transportation Research Part D* **12**:396–413.

Labay DG, Kinnear TC. (1981). Exploring the Customer Decision Process in the Adoption of Solar Energy Systems. *Journal of Consumer Research* **8**:271–278.

Lih W, Yen J, Shieh F, Liao Y, (2012). Second Use of Retired Lithium-ion Battery Packs from Electric Vehicles: Technological Challenges, Cost Analysis and Optimal Business Model. International Journal of Advancements in Computing Technology (IJACT), Vol. **4**, Issue 22.

Markel T, Simpson A. (2006). *Cost-Benefit Analysis of Plug-In Hybrid Electric Vehicle Technology*. National Renewable Energy Laboratory.

Meyer TG. (2012). *Path Dependence in Two-Sided Markets*. PhD Dissertation, Freie Universitat, Berlin.

Miller JM, Onar OC, White C, Campbell S, Coomer C, Seiber L, Sepe, R., Steyerl A. (2014). Demonstrating Dynamic Wireless Charging of an Electric Vehicle. IEEE Power Electronics Magazine, March.

NASA Technology Roadmaps, TA 14. (2015). Thermal Management Systems, May 2015 Draft. Available at: http://www.nasa.gov/sites/default/files/atoms/files/2015_nasa_technology_roadmaps_ta_14_thermal_management.pdf [Accessed June 2015].

Neubauer J, Pesaran A. (2010). PHEV/EV Li-ion Battery Second-Use Project. NREL Report.

Ozaki R, Sevastyanova K. (2011). Going Hybrid: An Analysis of consumer purchase motivations. *Energy Policy* **39**:2217–2227.

Peng, M. (2012). A review on the economic dispatch and risk management of the large-scale plug-in electric vehicles (PHEVs)-penetrated power systems. *Renewable and Sustainable Energy Reviews* **16**:1508–1515.

Quick D. (n.d.). World's most efficient thermoelectric material developed. *Gizmag*. Available at: http://www.gizmag.com/most-efficient-thermoelectric-material/24210 [Accessed June 2015].

Smith A, Stirling A. (2010). The politics of social—ecological resilience and sustainable socio--technical transitions. Ecology and Society 15:1–11.

Sovacool BK, Hirsh RF. (2009). Beyond batteries: an examination of the benefits and barriers to plug-in hybrid electric vehicles (PHEVs) and a vehicle-to-grid (V2G) transition. Energy Policy 37(3):1095–1103.

Sydow J, Screyogg G. (2013). Self-Reinforcing Processes In and Among Organizations, Self-Reinforcing Processes. In: *Organizations, Networks, and Fields – An Introduction*, J Sydow, G Screyogg (eds.). Palgrave MacMillan, UK.

Taiber JG. (n.d.). Overview about Wireless Charging of Electrified Vehicles – basic principles and challenges, Transportation Electrification Community Report. Available at: http://tec.ieee.org [Accessed July 2015].

Wessling JH, Faber J, Hekkert MP. (2014). How competitive forces sustain electric vehicle development. *Technological Forecasting & Social Change* **81**:154–164.

Index

Note: page numbers in italics refer to figures; page numbers in bold refer to tables.

Printed and bound by CPI Group (UK) Ltd, Croydon, CR0 4YY

16/04/2025

14658474-0004